Photokinetics

Mounir Maafi

Photokinetics

A New Perspective

Springer

Mounir Maafi
Leicester School of Pharmacy
De Montfort University
Leicester, UK

ISBN 978-3-031-98984-1 ISBN 978-3-031-98985-8 (eBook)
https://doi.org/10.1007/978-3-031-98985-8

This Springer imprint is published by the registered company Springer Nature Switzerland AG
The registered company address is: Gewerbestrasse 11, 6330 Cham, Switzerland

To my parents,
to my family,
to Kenzy.

Preface

Why a Book on Photokinetics?

The subject of photokinetics is an underdeveloped area of physical chemistry for reasons that are not completely clear. Thus far, only one specialised book was completely dedicated to the subject. Related textbooks seldom give an account or include independent chapters on this particularly important topic. A situation that does not objectively correlate with the fact that photochemistry, being over 150 years old, is one of the oldest subjects of modern chemistry. This does corroborate neither the ever-increasing number of scientific papers in the field nor the extended application of photokinetics to technology and modern challenges.

As a tool, photokinetics was needed since the earliest stages of photochemistry. With time, it has now settled, in a way or in another, virtually in all areas of chemical sciences and a large section of biology and engineering (at the least in relation to photostability of the respective materials). Photokinetics impact on important aspects of a number of new technologies, promotes it as a science for the future.

Nonetheless, at the present time, basic questions such as What is the order of a photoreaction? or What is its integrated rate law? do not find simple answers as usually found handy for the sister kinetics of thermal reactions.

This book endeavors to shed light on many new aspects in photokinetics, attempts to answer some of the still hanging questions, and brings a fresh evaluation of a set of ideas that are used in photokinetics and which might have been considered, rightly or wrongly, as established standard concepts.

The intention of this book is to make the subject of photokinetics accessible and enjoyable for the largest number of under- and post-graduate students, as well as for confirmed researchers. It could also be useful to many non-specialists, particularly those investigators involved in many other fields of science that require photokinetics.

The chapters of this book will take the reader on a journey from the most basic principles of photokinetics to the latest state-of-the-art concepts in the field. Whenever relevant, appropriate sections provide a comprehensive review of the literature on the themes treated in order to expand the reader's view on the subject. Additionally, it often describes the difficulties encountered in dealing with photokinetics.

Throughout the contents of this book, three aspects are sought. The first relates to targeting specific issues for which, as much as possible, a detailed description is given. The second is to make it easy for the curious mind to learn, discuss, and/or evaluate the issues selectively. Finally, it allows reviewing and implementing a new set on concepts. One of the major aims here is to induce an amount of targeted and genuine discussions in the photokinetics community that would eventually incite and hopefully lead to an agreed set of concepts, principles, and procedures.

What Has Been Achieved and What Remains to Build?

The conventional wisdom in the field of photokinetics has settled on a number of issues that don't seem to have been properly scrutinised to become accepted concepts. For instance, it does not assign kinetic orders to characterise photoreactions and does not seem to consider the matter relevant, considering the very scarce discussions of this issue in the literature, e.g., photochemistry textbooks lack exclusively dedicated chapters to shed light on these matters. Research and academic literature in the area is remarkably unaffected by a quasi-total absence of integrated rate laws to map out the experimental kinetic traces obtained for photoreactions, in a clear contradiction with the general practice adopted for thermal kinetic analysis. A single formula is officially recommended for the experimental determination of the photochemical quantum yield of a photoreaction, namely the integral quantum yield equation (there is also the differential quantum yield equation but this one is barely used in practice). The integral quantum yield equation is supposed to calculate this quantity for any photoreaction despite that this equation does not have a known origin, fitting with fundamentals of physical chemistry (its derivation from or relation to the rate law is still obscure). Also, despite it is known that the integral quantum yield equation does not apply in some cases, notably for the photoreversible reaction, no equivalent alternatives were proposed and no discussion has set out the contours of such integral quantum yield equation's field of applicability. The conventional wisdom also falls short to providing complete answers to many other questions, such as Is Kasha's rule universal in photochemistry? Are there kinetic differences between using mono- or polychromatic light for irradiation? How the reactant initial concentration influences the kinetics? and What determining factors lead to variation in kinetic behaviour?

The purpose of this book is to attempt addressing such issues through detailed analyses and ventures to propose rationalised and concise answers to each. The approach adopted here aspires to shed light on some aspects that do not benefit from much attention in the literature and endeavours to construct a rationale to describe photokinetics with standard tools. This book intends to provide a valuable reference for students, researchers, and scientists in academia and industry.

Layout: Chapters and Parts

In organising this book, I opted for an organisation in several parts, each including a number of chapters. There are seven parts focused on particular themes, detailed over 23 chapters. *Part I* overviews some basic aspects related to photokinetics and when adequate, comparisons with thermal kinetics. In *Part II*, four particular themes were discussed to some extent, because they, either or both, are seemingly considered as concepts or they do not receive enough coverage in the field. *Part III* describes the mathematical bases for photokinetics under isosbestic monochromatic irradiation, where photokinetic elucidation methods are laid out with a particular attention to solving the identifiability problem. *Part IV* aims at showing the mathematical difficulties encountered in solving photoreactions' systems of non-linear differential equations when the reactive medium is subjected to monochromatic and non-isosbestic light. Several approaches to overcome this hurdle are described in *Part V*, notably by a semi-empirical method. The theoretical description of the latter method is applied to experimental data in *Part VI*. The treated subjects in *Part VII* put emphasis on a breakthrough, represented by a general integrated rate law for photothermal reactions. It applies to both monochromatic and polychromatic light irradiation, irrespective of the reaction mechanism. Methods for quantifying the effects of altering reaction conditions, elucidating kinetics, and performing actinometry are also detailed in this part.

Future Expectation

The journey within this book reveals gaps in the photokinetics knowledge, some of which have been filled. It proves as well that a rationalisation of photokinetics along the conventional rules of physical chemistry and mathematics is possible. The shift in paradigm introduced here embodies a serious alternative perspective on the subject of photokinetics and opens a litany of avenues for future research. This book, however, only sets the first milestones on a long path towards a solid consolidation of photokinetics.

Leicester, UK

Mounir Maafi

Competing Interests The author has no competing interests to declare that are relevant to the content of this manuscript.

Contents

Part VII A General Photokinetic Model Equation for Photo-thermal Reactions Under Monochromatic and Polychromatic Lights

Part I
Basics of Photokinetics

Chapter 1
Kinetics: A Preview

1.1 Introduction

The concepts operating the evolution of reactions and the quantification of the changes occurring with reaction time make up the subject of kinetics. This general idea is valid for all reactions, including thermal and photochemical processes. By a detailed kinetic investigation, many specificities are revealed that help differentiate between the types of reactions. This overview seeks to shed light on some specificities of kinetics.

1.2 The Object of Chemical Kinetics

Chemical reactions proceeding in the dark are driven by heat and so are dubbed *thermal reactions*. These thermal reactions speed up with an increase in the amount of heat energy received by the reactive medium. In practice, such an acceleration of the reaction occurs when the temperature of the reactive medium increases (it is much easier to measure the variation of temperature as an indication of such heat exchange). It is therefore essential to study these reactions at constant temperature. Only at a temperature of absolute zero Kelvin, there is no reaction progress, but for any temperature over that value, there should always be some advancement of reaction for a given thermally active system.

One important aspect of the investigation of chemical reactions focuses on the quantification of the evolution of the species concentrations with reaction time, i.e. the basic information for a kinetic study. This includes the patterns and the relative speeds of both reactants and products during the reaction progress. Generally speaking, a kinetic study essentially relates to the above aspects and often includes answering questions in relation to the mechanism undergone by the reaction and/or

M. Maafi, *Photokinetics*, https://doi.org/10.1007/978-3-031-98985-8_1

determining the numerical values of the quantities and properties of the species involved in that reaction mechanism.

Kinetic investigations cannot be undertaken without employing one type or another of analytical chemistry techniques. These are the means that will allow following the time evolution of the species' amounts within the reactive system, evidencing their respective appearance or disappearance in the medium. Routine analytical chemistry techniques could and are generally used for kinetic analyses, but a special emphasis must be directed towards the choice of the technique and its performance, as well as its limitations, in order to reach an adequate description of the kinetic problem at hand. Analytical chemistry techniques could also be used qualitatively to separate and identify the products of the reaction with the aim of unravelling, partially or fully, the reaction mechanism.

The variations of the analytically measured quantity of the species (e.g. concentrations) with reaction time provide a number of information on the mechanism as well as on individual species, and hence on the reaction.

The second important tool, necessary for the quantification of reaction behaviours, is the mathematical modelling of the reaction kinetics. The basic procedure starts with the identification of the system of differential equations or the rate-laws describing the reaction. This step means that the mechanism of the reaction is known since the system of differential equations is mechanism-specific.

In summary, kinetics is the study of the changes in rate of a chemical reaction in response to an external stimulus (heat, light, etc.). A combination of the information collected on the reaction evolution with time (using chemical analysis techniques) and the mathematical model equations represented by the solved reaction rate-law is the core strategy that is adopted to conduct a kinetic investigation. The expected results are mainly the quantification (by numerical values) of the attributes of the reaction (reaction order, reaction rate-constant, spectroscopic and/or chemical properties, etc).

The gained knowledge from the kinetic analysis could be further used to obtain other aspects of the reaction, such as the activation energy (using the Arrhenius empirical relationship), or allow predictions of the reaction behaviour whenever some conditions are changed (e.g. the percentage transformation of a species within a period of time).

If the mechanism undergone by the reaction is not known with certainty, then the kinetic analysis is employed in a retrospective way to gain more insight into the operating mechanism. The classical approach globally consists of assuming a mechanism, solving its rate-law and using the equations obtained to fit the experimental data recorded on the reaction evolution. In this approach, when the fitting is acceptable, the preconceived mechanism at hand is adopted for the reaction as being plausible (this classical method is not necessarily always effective, as we shall see later).

Kinetic investigation is an essential part of understanding, quantifying and controlling the reactivity of species. It is unavoidable when quantitative data are required. Today, this includes almost all areas of chemistry and biology, at least to assess the extent of the initial materials' degradation, but can become pivotal to control/direct the reaction towards specific aims.

1.3 Photokinetics

Photochemical reactions are driven by light (i.e. they do not operate in the dark). The most famous of these is the naturally occurring photosynthesis, which completely vanishes at night, leaving the stage to the permanent plant respiration (occurring over a 24-hour period).

Although photochemical reactions are necessarily governed by light, they can also be sensitive to heat, either because the reactant or its photoproducts undergo thermal reactions in addition to their photochemical processes, or because their excited-state transformations involve energy barriers. The latter means that even a pure photochemical reaction (that does not involve any apparent thermal steps) can depend on temperature whenever there is an energy barrier to be crossed in the excited state, for instance, to form an intermediate. This clearly suggests that photoreactions should, in principle, be studied at constant temperature.

Photoreaction speeds up when subjected to a stronger, higher-intensity radiation. Basically, the radiation intensity is an account of the number of photons reaching the reactive system. In this context, the photon is considered a reactant in a photoreaction. Theoretically, moderate temperature (or heat) plays no role in the emergence of molecular photoexcitation but could impact the formed excited-intermediate transformation or the reactivity involving bimolecular species. Hence, it is necessary to envisage quantitative investigations of photochemical reactions under a constant flow of photons (a constant incident radiation intensity) and at a constant temperature. Thermal reactions, if part of the underlying reaction mechanism in the dark, should coexist with photoreaction when light is provided to the reactive system.

Photokinetics is the study of the kinetics of photoreactions. The basic means required to undertake photokinetic studies are similar to those considered for the kinetic investigation of thermal reactions (Chaps. 2 and 3). However, the mathematics of photokinetics must incorporate parameters relative to the radiation intensity. This includes both the amount of light impinging on the system from the light source and the amount of photons absorbed by the species present in the reactive medium. Such considerations significantly modify the mathematical formulation and treatment of photokinetic equations compared to those of the thermal reactions. Ample details of these aspects are presented in the upcoming *parts*.

The purpose of photokinetics is to determine the photoreactions' parameters, mechanism and reaction evolution characteristics during irradiation time. The behaviour of photoreactions is determined by both *intrinsic* and *extrinsic* features. The intrinsic ones are specificities of the reaction and its reactants and products, of which the quantum yield and the absorption coefficients are essential. The extrinsic parameters influencing the reactions are mainly found among the experimental conditions, which are under the control of the experimentalist, including the initial concentration of the reactant, the intensity and type of the incident radiation, the volume of the reactive medium and the optical path length of the excitation light in the sample.

Since the extrinsic parameters of the reaction are controlled by the experimentalist, quantitative photokinetics is often considered as basically a means to determine the numerical values of the intrinsic parameters (reaction rate-constants, quantum yields, reaction order, absorption coefficients, etc.). However, as we shall see further in this book, photokinetics might turn out to be an excellent tool to also assess, detect and quantify a number of effects of photochemical processes. Despite the relevance of photokinetics, it has, unfortunately, remained an underdeveloped subject of photochemistry from the viewpoint of formal description. A situation that incites further and adequate attention and improvement of this subject. This book is a contribution attempting to fill part of this gap.

1.4 Laws of Photochemistry

The three fundamental laws of photochemistry are the Grotthuss–Draper, Stark–Einstein, and Bunsen–Roscoe laws [1].

The Grotthuss–Draper law states that only light absorbed by a compound can lead to its possible structural photo-transformation. In substance, this principle can be summarised by '*no light absorption = no photoreaction*'. Nonetheless, light absorption does not necessarily mean that a reaction will occur. The absorption of light is a gain of energy that is translated, at the molecular level, by an excitation of the molecule from its electronic ground state to one of its excited states. Photochemistry is not the only or the most important pathway available to all molecules. Many pathways dissipate the excess energy gained by the molecule, from light excitation, for instance, by photophysical routes. For many molecules, no measurable photoreaction happens after absorption of a photon of light (in a given period of time).

The Grotthuss–Draper law clearly underlines an important fact, that is, the wavelength span of the irradiation light provided to the sample must overlap the region of absorption of the investigated molecule if eventual photoreaction processes are to be observed and quantified. It somehow implicitly stresses the importance of correctly quantifying the wavelength spans of both the light supplied to the reactive system and the electronic absorption spectra of each individual species of the reactive system. A good understanding of the possible photoreactivity of each species will in part depend on such overlap.

The Stark–Einstein law raises a fundamental property of the light interaction with matter; that is, one molecule can only absorb one photon, leading to its excitation. The transition of a molecule between the ground and its excited state, during excitation, is due to the absorption of a unique photon, even though the light source might provide a continuous flow of photons. If the molecule returns to its original form after de-excitation, it could absorb another photon. The Stark–Einstein law applies to laboratory and industrial experiments that are carried out with conventional lamps. It also justifies why the number of photons at each irradiation wavelength must be known with accuracy. Conversely, this law can be breached for very intense light sources, such as lasers, which can induce multiple light absorptions by one molecule.

The final Bunsen–Roscoe law stresses that the impact of light on a photochemical system is directly proportional to the total number of photons received by the system (energy dose), irrespective of the time interval taken to deliver that number of photons. Simply put, this law can also be interpreted as the velocity of the reaction being directly dependent on the number of photons provided to/absorbed by a given reaction. It also indicates that successive and staggered steady-state doses produce the same effects as continuous irradiation if the total number of photons reaching the considered photoreaction in both cases is the same. This applies to pure photoreactions. The situation can be very different, however, when concomitant thermal reactions compete with the photochemical process in photothermal reactions. The Bunsen–Roscoe law is a strong plea for the unreliability of comparison between experiments if the number and/or the type (wavelength span and pattern) of the photons driving such processes are not exactly the same (since the velocities of photoreactions depend on both).

These general laws guide the design and quantification of photoreactions. Therefore, photokinetics must respect each of them at all times in order to conform to the spirit of photochemistry and to produce results and interpretations that are reliable and useful.

1.5 Differences in Kinetics of Thermal and Photoreactions

The compulsory presence of light for photoreactions to occur is an obvious difference with thermal reactions. This operational difference has a significant impact on the mathematical description of the kinetic equations. It imposes that the radiation intensity is explicitly expressed within the differential equation of the species' rate-laws since the number of photons provided to the reactive system directly affects the behaviour of the overall reaction. Notice that the equivalent equations relative to thermal reaction kinetics do not explicitly involve a term for heat (or temperature). The Arrhenius methodology is used in a second step to relate already determined reaction rate-constants to the respective temperatures of the medium (i.e. it is not possible to determine the activation energy directly from the kinetic equations).

The light intensity is not the only factor that constitutes a difference between the kinetics of these reactions. For example, unlike the general approach describing thermal kinetics, kinetic equations of photoreactions must involve the dimensions of the reactor. Most usually, the volume of the irradiated solution makes part of the mathematical formulation of the differential equations, hence, this feature (the volume) influences the performance of the reaction. The volume of the reactive system, especially in solution, has virtually no effect on the progress of a thermal reaction.

Of similar importance is the distance crossed by the excitation beam through the sample. The so-called *irradiation optical path length* does not have an equivalent for thermal reactions.

The energy supplied to a thermal reactive system is delivered in a single form, heat. In the case of light irradiation, the energy is delivered in the form of light, but there are several specifications of the beam that variably change the behaviour of the reaction (such as monochromaticity, or wavelength range). Hence, the properties (other than the intensity) of the light used are important factors in photokinetics.

In addition, as we are going to see later in this book, the kinetic behaviour of photoreactive species is affected by the other species present in the reactive medium. For instance, the products formed during the photo-process affect the kinetics of the reactant, and proof will be given that even spectator molecules, present in the reactive medium, but both thermally and photochemically stable, might have a substantial and quantified effect on the kinetics of the photoreactive species. No evidence of such a phenomenon exists for thermal reactions (i.e. the spectator molecules do not, at all, contribute to the reaction).

The initial concentration of the reactant does not usually change the velocity of a monomolecular thermal reaction obeying 0th- or 1st-order kinetics. This is different for photoreactions, where this experimental factor generally influences the velocity of photoreactions, except for the simplest photoreaction $X \rightarrow Y$, as we shall see in Chap. 11.

Thermal reactions' rate-equations do not explicitly involve the yield of the reaction, whereas those characterising photoreactions must include the photochemical yield of the considered species. It is called the *efficiency* or *quantum yield*, which is a specific property of the molecule at a given wavelength. The quantum yield of a photoreaction-step must be defined for strictly monochromatic light (Chap. 5).

Finally, the rate-laws of thermal reactions are usually linear differential equations, contrasting with the usually non-linear equations of photokinetics.

Overall, we shall see that photoreactions depend on the conditions set out for the experiment as well as on the intrinsic properties of the species (reactive or unreactive) present in the medium, including the properties of the irradiation light (Chaps. 8, 11, and 19).

We can add two more important criteria that distinguish between the two kinetics. On the one hand, in photokinetics, it is sometimes possible to target one particular reactant amongst the components of the (reactive) medium by a careful selection of the wavelength of the irradiation beam. Such selectivity cannot be performed in thermal reactions because the heat provided to the reactive medium engages all the components of this medium without a possibility of distinction between species. On the other hand, if the thermal reactions are most often characterised by a reaction order, photoreactions, according to the literature, do not have standard or specifically known reaction orders.

The above features make up compelling evidence that the kinetic descriptions are not interchangeable between thermal and photochemical systems. This conclusion will be supported and confirmed, in further chapters, by theoretical analyses and experimental data.

1.6 Relating Rate-Laws and Integrated Rate-Laws

The rate of change (rate or $r_{Y_j}(t)$) of a specific reactant ($X = Y_0$) or product (Y_j, *with* $j \neq 0$) is expressed by the equation of variation of its concentration ($C_{Y_j}(t)$) with time in the form of a differential, $r_{Y_j}(t) = dC_{Y_j}(t)/dt$. The rate also has a formulation translating its dependence on the properties of the reaction, such as $r_{Y_j}(t) = k\ C_{Y_j}(t)$, where k is a constant, as for 1st-order thermal reaction. The obtained differential equation defines the *rate-law* of the considered species' reaction (for our example, species Y_j, and $r_{Y_j}(t) = dC_{Y_j}(t)/dt = k\ C_{Y_j}(t)$).

The rate-law captures the essence of the kinetic behaviour of the considered reactive species (reactant or product). It faithfully describes the evolution of the concentration of the targeted species during reaction time (i.e. from start to end), as it must encompass, explicitly and/or implicitly, all the parameters affecting the reaction kinetics of that species.

The concentration is explicitly mentioned in the concentration rate-equation. Whereas the reaction rate-constant (k), as a blind constant or a parameter of the equation, cannot be analytically expressed for thermal reactions by an explicit and precisely defined equation. Therefore, the actual relationship between or the weighing of the different factors influencing the reaction cannot be mathematically identified within k for thermal reactions. Although its numerical value, obtained experimentally, expresses implicitly the overall sum of contributions of the various reaction parameters.

The rate-law, in its differential formulation, expresses the variation of the gradient (or the tangent) to the curve $C_{Y_j}(t)$. This configuration is most often not easily manipulated to extract reaction parameters (such as k) from the experimental data, which are generally presented as concentration values, measured at given reaction times ($C_{Y_j}(t)$).

The solutions of differential equations, such as the rate-laws of a reaction, are equations (conversely to the solutions of algebraic equations, which are obtained as numerical values). Conventionally, the solution of a differential equation is an equation that links the dependent variable (present in the numerator, y) to the independent variable (x) shown in the denominator of the differential (dy/dx). In the case of reaction kinetics, the solution of the rate-law is an algebraic equation relating the variables concentration and time (its formula is free from differentiation, e.g. $dC_{Y_j}(t)/dt$). The mathematical process of solving differential equations requires an integration procedure. The solution, hence obtained as an equation, represents the *integrated rate-law* of the considered reactive species. The integrated rate-law fits the experimental data ($C_{Y_j}(t)$) of the species Y_j. Because the integrated rate-law offers a much simpler handling of the experimental data compared to the rate-law, it is preferably used by kineticists.

It is important to underline here that solving the differential equations might sometimes be a tedious task and might even be impossible for many cases. A rigorous mathematical methodology should always prevail in solving individual or system of rate-laws since an erroneous equation will be detrimental to a valid understanding, quantification and/or prediction of the kinetics.

In this context and whenever possible, it is essential to bring tangible proof of the impossibility of achieving a solution to the differential equations at hand. This allows identifying, with certainty, the unsolvable cases for which alternative methods might be useful (Chap. 12 and Part V).

1.7 Meaning of the Order of a Reaction

The order of a reaction is a concept allowing the reaction behaviour to be identified. Practically, the order of a reaction is defined by a mathematical equation that fits and predicts the evolution of the concentration of a reactant or a product with reaction time.

Classically, in thermal kinetics practice, the reaction order of simple thermal conversions can be determined by testing the experimental data of a reactant (or a product) against the different mathematical models at hand. The order can also be determined from the differential equation of the process if its integrated rate-law fits the experimental data. However, the order of a reaction cannot be determined theoretically in the sense that it is not possible to predict by examining the reaction mechanism. The order of a reaction is a property of the reactive system that is, primarily, an experimentally confirmed concept.

Hence, the difference in the reactions' behaviours indicates a difference in the reactions' orders. This is established by a difference in the mathematical formulae of the integrated rate-laws describing the reactions. The mathematical difference resides in the fact that these reactions' different integrated rate-law equations are not linearly dependent and cannot be worked out from one another. They are distinct laws, completely independent mathematical entities.

For simple thermal reactions, the rate-laws are proportional to some coefficients of the reactants' concentrations. The differential equations of these rate-laws are written as a product between the reaction rate-constant (k) and the concentrations (e.g. C_X and $C_{X'}$) of the reactants (X and X'), where each of these concentration terms is raised by a numerical exponent factor (e.g. 0, 1, or 2), as illustrated in Sect. 3.5.

In summary, the order of a reaction defines the particular kinetic behaviour of that reaction, is embodied by the integrated rate-law, cannot be predicted, in general, from solely considering the mechanism of the reaction, and is rather an experimental feature that must be confirmed by an excellent correspondence between the reaction's experimental data and the integrated rate-law.

1.8 Tentative Explanation of the Rate-Constant of a Reaction

For each type of reaction (different by their kinetic orders), the behaviour of the reaction over time must be characterised by a time-independent (constant) factor. This constant can be seen as globally indicating the amount of variation of the concentration in a given time interval that is specific to the actual reaction order. The amount of variation is expressed by a function of concentration, $f(C_X(t))$. The ratio $f(C_X(t))/t$ is always constant because $f(C_X(t))$ is unequivocally proportional to the time of reaction. This link is univocal because it is defined by the order of the reaction and its integrated rate-law. This means that the mathematical expression for the rate-constant of the reaction in the form $f(C_X(t))/t$, will differ with the order obeyed by the reaction. Let us compare the rate-constants of the three basic thermal reactions, obeying different orders (derived from the equations of Sect. 3.5).

For the 0th-order reactions, the rate-constant is simply the gradient of the linear kinetic trace over time.

$$k\left(0^{th}\right) = \frac{C_X(0) - C_X(t)}{t} \tag{1.1}$$

whereas, $f(C_X(t))$ has a logarithmic form for 1st-order reactions.

$$k(1^{st}) = \frac{-\ ln\ \frac{C_X(t)}{C_X(0)}}{t} \tag{1.2}$$

The function $f(C_X(t))$ involves a fractional term of the concentration for 2nd-order reactions.

$$k(2^{nd}) = \frac{\frac{1}{C_X(t)} - \frac{1}{C_X(0)}}{t} \tag{1.3}$$

The time and concentration invariance of the reaction rate-constant, k, turns it into an objective parameter of the reaction. In fact, k gives a quantitative measure of how fast the reaction progresses. Meaning, the reaction is faster when k is higher and *vice versa*. But this statement is only true if two conditions are met. Such a comparison between k values is only allowed for reactions obeying the same kinetic order (according to Sect. 1.7 as well as to Eqs. (1.1)–(1.3), it is meaningless to compare the numerical values, $k(0^{th})$, $k(1^{st})$ and/or $k(2^{nd})$). Also, in order to comply with the definition of the rate-law (e.g. $r_{Y_j}(t) = dC_{Y_j}(t)/dt = k\ C_{Y_j}(t)$), the compared reactions must have the same starting concentration $C_X(0)$. We shall see that such a principle applies, as well, in photokinetics.

The rate-constant is then a discrete feature of the reaction, which, strictly speaking, does not inform about the actual changes in concentration or the absolute value of the reaction-rate [2].

1.9 Definitions of Quantum Yield and Quantum Efficiency

The International Union of Pure and Applied Chemistry, *IUPAC* [3, 4] defines the quantum yield of a photoreaction where species Y_j is transformed into species $Y_{j'}$ ($\Phi_{Y_j \to Y_j{}'}$, with $j \neq j'$) as the '*Number of defined events, occurring per photon absorbed by the system at a specified wavelength*'. It also stipulates that $\Phi_{Y_j \to Y_j{}'}$ can be evaluated by two equations, set out for the so-called integral and differential photochemical quantum yields.

The integral quantum yield is presented as being '*the number of events divided by the number of photons absorbed at a specified wavelength*'. It equates, for a photochemical reaction, to '*the amount of reactant consumed or product formed,* $n_{tot.\ rect.\ mol.}$ (# of reacted molecules), *divided by the amount of photons absorbed,* $n_{tot.\ abs.\ h\nu}$'. It is given by the following formula:

$$\Phi_{Y_j \to Y_j{}'}(\lambda) = \frac{n_{tot.rect.mol.}}{n_{tot.abs.h\nu}} \tag{1.4}$$

The differential quantum yield, for a homogeneous system, expressed in terms of rates, is given by the following equation:

$$\Phi_{Y_j \to Y_j{}'}(\lambda) = \frac{\frac{dx}{dt}}{q^0_{n,p,\lambda}\left(1 - 10^{-A(\lambda)}\right)} \tag{1.5}$$

where dx/dt is the rate of change of a measurable quantity x (spectral or any other property that measures the progress of a reaction), and $q^0_{n,p,\lambda}$ is the spectral amount of photons (mol or its equivalent, as the non-SI-unit, *einstein*) incident (prior to absorption) per time interval (defining the spectral photon flux, amount basis). The superscript '0' (zero) in the symbol of the spectral amount of photons (q^0), emphasises the fact that the incident number of photons is measured *prior to absorption*. In Eq. (1.5), $A(\lambda)$ is the decadic absorbance at the excitation wavelength, λ.

The *IUPAC* recommendations [3, 4] also give three clarification notes in relation to the differential quantum yield, for which it is required that (1) when x in Eq. (1.5) is either a number concentration ($C = n/V$), or an amount concentration, it is convenient to use in the denominator the rate of absorbed spectral photons per volume (absorbed photon flux density) number basis, or chemical amount basis, respectively; that (2) Φ can be used for photophysical processes or photochemical reactions; and that (3) strictly, the term quantum yield applies only for monochromatic excitation.

In homogeneous photochemistry, and when the quantum yield is *independent of excitation wavelength* in the spectral region used, quantum efficiency is identical to quantum yield, which is always referred to a monochromatic excitation.

The quantum yield is considered equal to the quantum efficiency of a photoreaction in a homogeneous medium if and only if the excitation is monochromatic. The term is also accepted for polychromatic light whenever the quantum yield is independent of excitation wavelength in the spectral region used.

1.10 Some Commonly Used Standards in Photokinetics

It is interesting to review some of the concepts that are held as standards in photokinetics. Perhaps the most common one considers the quantum yield to be independent of the irradiation wavelength. That is, irrespective of the wavelength of the monochromatic irradiation light impinging on the sample, the quantum yield of a reaction step, such as $X \rightarrow Y$, has a unique value. This concept follows, in fact, from the assessment that Kasha's rule (or Kasha-Vavilov's rule) applies to photochemistry and photochemical processes as much as it applies to photophysical processes. We shall discuss the validity of such an affirmation in Chap. 4.

The second widely held concept in the practice of experimental photochemistry states that the quantum yield value for the reaction-step $X \rightarrow Y$, is determined experimentally by a unique formula, for example, Eq. (1.4). Accordingly, Eq. (1.4) would apply irrespective of the reaction mechanism of the photoreaction at hand or the reaction conditions, as long as the irradiation light is monochromatic. In practice, the application of Eq. (1.4) might have been extended to polychromatic light since the quantum yield is considered wavelength-independent. Such a generalisation, for both or either case, has not been supported by a detailed derivation of Eq. (1.4). This aspect will be discussed in Chap. 5.

One can add another convention in the field by which photoreactions do not seem to obey a particular known kinetic order, and they have no established rate-constant (except for the many studies which employ 0th-, 1st-, and 2nd-order kinetics for photokinetic data). These considerations will be explored further in Chap. 7.

Lastly, when a proposed mathematical equation or a numerical integration calculation fits data obtained experimentally, then usually the model is deemed valid (e.g. fitting the data to 0th-, 1st-, and 2nd-order kinetic equations). Chapters 6, 9, and 10 attempt to clarify such issues and the consistency of this approach.

The few points above do not represent an exhaustive list of all conventions in the field. We shall explore some of their aspects in the following chapters.

References

1. Bonfield E, Knauber T, Lévesque F, Moschetta EG, Susanne F, Edwards LJ (2020) Photons as a 21st century reagent. Nat Commun 11:804. https://doi.org/10.1038/s41467-019-13988-4
2. Lente G (2015) Deterministic kinetics in chemistry and systems biology: the dynamics of complex reaction networks. Springer, Cham. https://doi.org/10.1007/978-3-319-15482-4

3. Braslavsky SE (2007) Glossary of terms used in photochemistry. Pure Appl Chem 79:293–465. https://doi.org/10.1351/pac200779030293
4. Braslavsky SE, Braun AM, Cassano AE, Emeline AV, Litter MI, Palmisano L, Parmon VN, Serpone N (2011) Glossary of terms used in photocatalysis and radiation catalysis (IUPAC Recommendations 2011). Pure Appl Chem 83:931–1014. https://doi.org/10.1351/PAC-REC-09-09-36

Chapter 2
Some Technical Aspects of Photokinetics

2.1 Introduction

Conducting photokinetic measurements and analyses requires the general background knowledge needed for kinetics of thermal reactions, but not only that. Some specific conditions and features proper to the kinetics of photoreactions must be part of the kinetic toolkit. We will review here a few relevant aspects needed for photokinetic investigation.

2.2 The Medium

Photokinetic studies can be performed on all types of samples, physically in gas, solid or liquid states. Solution samples are experimentally much easier to manipulate as they do not require complicated technical equipment or preparation. Also, routine laboratory analytical techniques, such as spectrophotometry, fluorimetry, HPLC, NMR and volumetric analysis, are suitable for the analysis of solutions. Photokinetic studies are often performed in homogeneous solutions, where the reactive medium must be well and continuously stirred. Throughout this book, we will only deal with photokinetics in homogeneous solutions.

2.3 The Mechanisms of Photoreactions

Photomechanisms can be divided into two types. In one type, the reaction starts from a single reactant, and in the other, it is initiated by a combination of two species. In all cases, the reaction must be initiated by light and so must involve at least one photochemical step.

M. Maafi, *Photokinetics*, https://doi.org/10.1007/978-3-031-98985-8_2

Scheme 2.1 A Φ-shaped mechanism for photo- and photothermal reactions

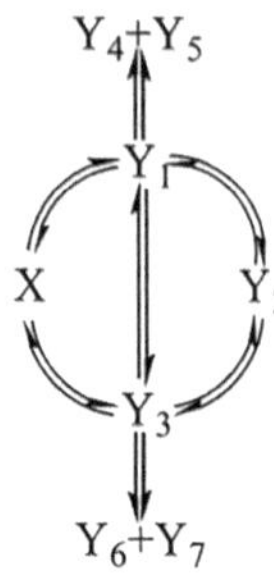

For the general first type of reactions, a reactive photomechanism involves ($n_{\rm sp}$ + 1) reactive species, Y_j (with $0 \le j \le n_{\rm sp}$, and $Y_0 = X$ for the reactant), interlinked by a set of n_Φ photochemical and n_Δ thermal reaction-steps. The Φ-shaped reaction mechanism shown in Scheme 2.1 includes more than 100 submechanisms representative of many reactions encountered in experimental photochemistry. Each arrow in Scheme 2.1 corresponds to either or both a photochemical and a thermal reaction-step (it shows 28 reaction-steps in total). The reactive species starting the reaction-step are labelled Y_j and those ending the reaction-steps, $Y_{j'}$. The index j (and j') can take values between 0 and $n_{\rm sp}$. We adopt here the convention, by which the reactants are labelled X for the first type of mechanisms, and X and X' for the second type. The photoproducts are always labelled by Y_j with $j \neq 0$. Each photoreaction-step between any two species ($j \neq j'$) of the reaction mechanism Y_j and $Y_{j'}$ (or *vice versa*) is labelled by the corresponding photochemical quantum yield for a forward reaction-step $\Phi^{\lambda_{irr}}_{Y_j \to Y_{j'}}$, or by $\Phi^{\lambda_{irr}}_{Y_{j'} \to Y_j}$ for a reverse reaction-step, measured for the irradiation wavelength, λ_{irr}.

Similarly, the thermal reaction steps are labelled by the thermal reaction rate-constant $k^{\Delta}_{Y_j \to Y_{j'}}$ (or $k^{\Delta}_{Y_{j'} \to Y_j}$). This labelling conventions avoid confusion with other parameters used in photokinetics, facilitate the mathematical presentation of the equations and the mathematical operations.

Photoreactions involving exclusively photochemical reaction-steps are labelled $XY_{n_{sp}}(n_\Phi \Phi)$, thermal reactions are marked by $XY_{n_{sp}}(n_\Delta k)$, and those involving both photochemical and thermal processes, i.e. photothermal reactions, are identified as $XY_{n_{sp}}(n_\Phi \Phi, n_\Delta k)$. Usually, such a distinction is seldom used in photokinetics, even though it adds clarity to the description (because thermal and photoreactions have different kinetic behaviors, Chap. 19). The distinction shall be adopted here.

Each species Y_j is at the start or the end of subsets $n_{\Phi j}$ and $n_{\Delta j}$ of the total sets n_Φ and n_Δ of the global mechanism reaction-steps. For instance, the possible reaction-steps for species Y_1 ($j = 1$) in the $XY_3(2\Phi, 1k)$ consecutive photothermal reaction mechanism, $X \to Y_1 \leftrightarrows Y_2$, involving two photochemical steps and one thermal reaction ($k_{Y_2 \to Y_1}$), has $n_{\Phi 1} = 0, 1$ (i.e. $X \to Y_1$ and $Y_1 \to Y_2$), and $n_{\Delta 1} = 1$ (i.e. $Y_2 \to Y_1$), where the numbers in the subsets correspond to the j values for species (Y_j) at the start or the end of the considered reaction-step (j' would be the other species in that reaction-step). The photochemical quantum yields (Φ) and thermal rate-constants (k) are two parameters with different physical meanings but are ubiquitous in photokinetics because they are considered to be the best quantifiers of the respective reaction-steps.

2.4 Intrinsic and Extrinsic Reaction Parameters

The kinetics of a photoreaction is fully defined by two types of parameters, namely, extrinsic and intrinsic coefficients (Sect. 1.3). These are, in fact, the parameters making up the rate-laws describing the species reactions.

The extrinsic parameters are controlled by the experimentalist and represent the experimental conditions imposed on the reaction. They are generally both accessible and varied, almost at will. Their list includes any parameter whose value can be independently set by the experimentalist, such as the initial concentration of the reactant, $C_X(0)$ expressed in mol/L or M, the optical path lengths of both the excitation (or irradiation) light [1], l_{irr} in cm, and the monitoring (or observation) l_{obs} (in cm), inside the sample (Fig. 2.1), the amount of photons of the irradiation light at wavelength λ_{irr} that is supplied to the sample, $P_0^{\lambda_{irr}}$ in einstein s^{-1} dm^{-3} (the photon flux per irradiated area and volume of the sample, see Sect. 2.17), the area from which the irradiation light reaches the sample, S_{irr} in cm^2, the volume of the sample subjected to irradiation, V_{irr} in dm^3, the wavelength of the monochromatic light that serves as the irradiation light of the sample, λ_{irr} in nm, the wavelength at which the absorption of the reaction is monitored (or observed), either λ_{irr} or λ_{obs} in nm, the temperature of the medium (T in °C), and the rate of stirring of the solution (that is intended to be vigorous and continuous).

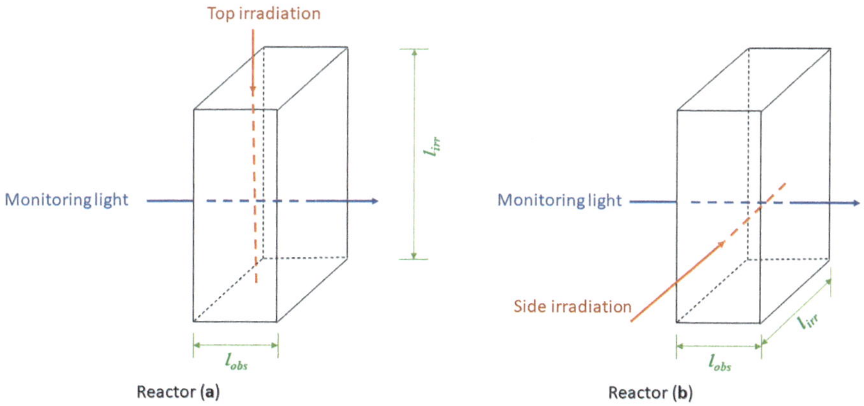

Fig. 2.1 Two examples of possible system configurations, where the photochemical reactor is subjected to a collimated irradiating light, impinging perpendicularly to the probing (observation) light. (**a**) A top irradiation configuration where l_{irr} is larger than l_{obs}, (**b**) a lateral irradiation ($l_{irr} > l_{obs}$)

The intrinsic parameters are specific features of the different species involved in the photoreaction. They arise from the physical nature of each species. There are mainly three properties, namely, the species absorption coefficient (or absorptivity) at a given wavelength, $\varepsilon_{Y_j}^{\lambda_{irr}}$ in M^{-1} cm^{-1}, the dimensionless photochemical quantum yield of a given reaction step (e.g. $\Phi_{Y_j \to Y_{j'}}^{\lambda_{irr}}$), and the thermal rate-constant of a given thermal reaction-step at a given temperature, $k_{Y_j \to Y_{j'}}^{\lambda_{irr}}$ or $k_{Y_{j'} \to Y_j}^{\lambda_{irr}}$ in s^{-1}. In a photothermal reaction mechanism, there are as many quantum yields and rate-constants as there are photochemical (i.e. n_{Φ}) and thermal (i.e. n_{Δ}) reaction-steps, respectively. The absorptivity can be estimated by computational chemistry methods, but its values are easily recorded experimentally by spectrophotometry and the Beer-Lambert law (using the corresponding absorbances). The absolute values of quantum yields and the rate-constants of the thermal processes are still exclusively determined experimentally (the theoretical estimation of their values requires tedious tasks and only provides approximations). The values of the three parameters cannot be precisely accessed by examining the chemical structure/reaction mechanism of the reactive species.

2.5 Characteristics of the Light Beam Used for Irradiation

The light energy supplied to a reaction, inducing phototransformation of species in the reaction medium, requires a degree of specification in order to correctly account for the photokinetics in play. In other words, the term ‘light beam’ or ‘excitation radiation’, as such, is not informative enough, as some features must clearly be identified.

Obviously, the radiation intensity (of the light beam) is important and must be known with precision in photokinetics. The intensity of radiation is the number of photons it delivers to the reactive system or the reactor. As the number of photons increases with exposure time, the latter becomes an important factor that defines the total energy procured to the system. Also, because the frequency of short-wavelength photons is higher than that of long-wavelength ones (the former are more energetic than the latter), an uneven distribution of the number of photons is generally recorded over the wavelength span of a lamp (i.e. for a similar spectral irradiance value for the lights of two monochromatic wavelengths $\lambda_1 < \lambda_2$, the light beam of wavelength λ_1 carries a larger number of photons). The spectral irradiance is the power of the light per square area per wavelength (as defined in Sects. 2.17 and 2.18). A count of the photons at each wavelength is mandatory for consistent photokinetic studies. For polychromatic lights, the total number of photons (as the sum of photons’ numbers at each specific frequency) is not sufficiently informative to carry out extensive and/or reliable photokinetic investigation (see Part VII). However, despite its importance, the number of photons reaching the reactor is not the only property of the light beam required for such photokinetic studies.

The light driving the photoprocess can be of two types. The first type corresponds to a light which has "one" frequency (f) or wavelength (λ), called a *monochromatic light*. Even though the name hints at a single (mono) colour (chrome), the monochromaticity generally applies to light of one wavelength in either ultraviolet or visible spectral regions (UV-Vis light being most relevant to photochemistry). Technically, a monochromatic light can only be obtained by the use of a monochromator (see Sect. 2.7). Lasers also provide monochromatic light. However, the reduced availability of the latter (both monochromators and lasers) in common photochemistry laboratories has limited their usage in photokinetics studies. Recently, LED lights (light-emitting diodes) have become more and more available for photochemists. However, LED lights are seldom monochromatic (see examples in Fig. 2.3) and make part of the second type of light, the *polychromatic light* (for poly = many and chrom = colour). Similarly, the light produced from either unfiltered or filtered lamps, including commonly used fluorescent lamps for indoors lighting, is not possibly monochromatic. This is also obviously the case for natural sunlight. In photokinetics, polychromatic light is the most common type of irradiation used since the dawn of photochemistry (both from natural and artificial origins). Albeit, polychromatic light is a non-specific radiation. Conversely to monochromatic beams, a polychromatic light activates, statistically and at the same time, all the molecular sites susceptible to lead to photochemistry.

Polychromatic light beams cover a spectral domain, $\Delta\lambda$ (e.g. ranging between λ_a and λ_b, that necessarily includes more than a single wavelength, $\Delta\lambda = \lambda_b - \lambda_a \neq 1$), with either a continuous or a discontinuous distribution over the $\Delta\lambda$ domain. Continuous-light patterns over a large section of the visible domain are exhibited by sunlight or indoor fluorescent lamps. A discontinuous polychromatic light presents a series of different bands in the *UV* and/or visible regions, such as emissions of arc lamps (e.g. xenon and mercury lamps). Lorentzian or Gaussian-like curves characterise the light emitted by LEDs with relatively narrow half-bandwidth (ranging from a few to a few tens of nanometres, Fig. 2.3). Light sources most likely differ by their profiles even if their emission domains are identical.

The nature of the light used in photokinetics can then be defined by two fundamental properties, namely, the wavelength domain it covers and the number of photons it supplies at each single frequency of that domain. The light provided to the sample must overlap a region of the absorption spectrum of the studied species in order to expect photochemistry (Grotthuss-Drapper law, Sect. 1.4). Also, from an experimental as well as from a mathematical description viewpoints, it is important to consider the way light impinges on the sample. This relates to whether the light beam is collimated or not (see Sect. 2.6).

The relevance of the above factors in photokinetics comes from their individual or cumulative effects on the kinetic behaviour of reactions. They strongly indicate that comparison between photoreactivities cannot be objectively undertaken without a precise knowledge of these parameters.

In summary, to be fully conducted, a photokinetic investigation requires the knowledge of five characteristics of the light beam used for the irradiation of the sample: (1) whether it is mono- or polychromatic, (2) whether it is collimated, (3) its wavelength span, (4) the incident number of photons, at each wavelength, reaching the sample, and (5) the time of exposure.

2.6 Collimated Light

A *collimated light* has all its rays, falling onto the reactor, practically parallel (conversely to a diffuse light whose rays diverge, Fig. 2.2 [4]). The distance separating the light source from the receiving system (e.g. a reactor) is determinant for the degree of collimation. The sun, as a source, provides a collimated light for a receiving system on Earth (due to the great distance). A laboratory lamp, irradiating a relatively close-by reactor, delivers a non-collimated, divergent light. Collimation of a beam is best achieved by lenses or mirrors that align the light rays in parallel fashion. For instance, the diffuse light entering the monochromator in Fig. 2.4 is collimated by the optical system (the collimating mirror (C)) before impinging on the grating.

In photokinetics, the direction of the collimated beam is perpendicular to the surface of the reactor containing the investigated sample. The importance of light collimation in photokinetics stems from three aspects: (a) it provides a uniform distribution of the light reaching the sample through the irradiation area, (b) the rays of the light beam cross the same distance inside the sample, and (c) it confers mathematical simplicity to the rate-equations. It is most convenient to express the latter in one direction (e.g. the x-axis corresponding to the direction of irradiation). A diffuse light delivers rays in different directions which travel various distances inside the sample, and hence, the rate-law describing the system must take into account such parameters (generally by incorporating a greater number of integrals, hence, becoming more complicated to solve). In the subsequent chapters, we only consider irradiation with a collimated light.

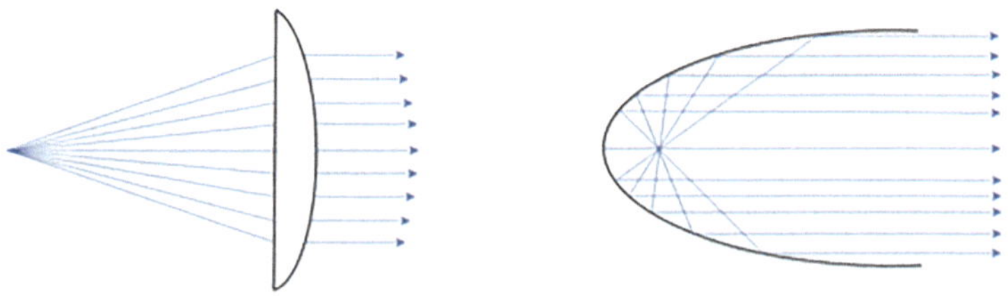

Fig. 2.2 Collimation of a diffuse light by a lens (left), or a parabolic reflector (right)

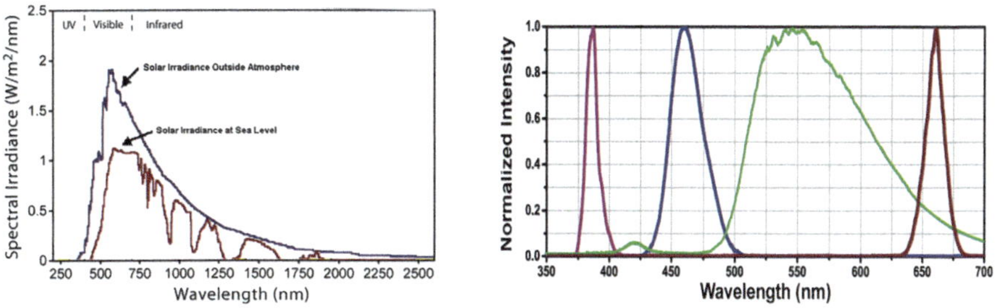

Fig. 2.3 (*Left*) Profiles of sunlight outside the atmosphere and at sea level [2]. (*Right*) Profiles of various LEDs emitting at different visible regions [3]

2.7 Measuring Lamp Profiles

As discussed above, in addition to the excitation radiation wavelength span, it is essential to know the relative distribution of the beam intensity over the wavelength range of the polychromatic light, particularly useful over the region where it overlaps the absorption spectrum of the reactive system under investigation.

The profile of a light beam is obtained by a special instrumentation, radiometers or spectroradiometers. They measure the radiant energy or the radiant power of a light source, even though their individual methods of measurement and/or wavelength span might differ. Most useful instruments measure the *irradiance* of the source, i.e. the area density of the radiant flux, the radiant flux per unit area at a specified point on a specified surface that is incident on, passing through, or emerging from that point in the surface [6]. The irradiance is expressed in Watt m^{-2} or J s^{-1} m^{-2}, i.e. power per unit area or energy per unit time per unit area. The spectral irradiance refers to the irradiance per nm. The light source profile is generally presented as a plot of spectral irradiance *vs.* wavelength (Fig. 2.3).

2.8 The Monochromator

The techniques of spectroscopic analysis are usually performed by submitting the analytical sample to a screening over a given range of light energy (e.g. light wavelengths, wavenumbers). The sample is hence successively subjected to light beams of, for instance, increasing wavelength that covers a selected interval between λ_1 and λ_2. In order to technically achieve such a wavelength screening, the light needs to be split into components of the narrowest possible wavelength range (since it is quasi-impossible to produce a purely monochromatic light). The instrument facilitating such a consecutive selection of one wavelength after the other is called a *monochromator*. Its basic principle is to disperse a polychromatic light provided by a light source (e.g. a lamp) into its constituent wavelengths [7]. For example, a

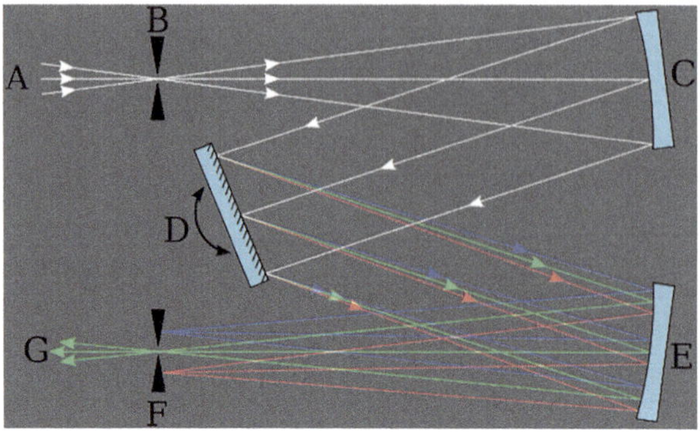

Fig. 2.4 An illustration diagram of Czerny–Turner monochromator [5]. Incident white light (A), entrance (B) and exit (F) slits, concave mirrors (C and E), diffraction gratings (D), and exit green light (G)

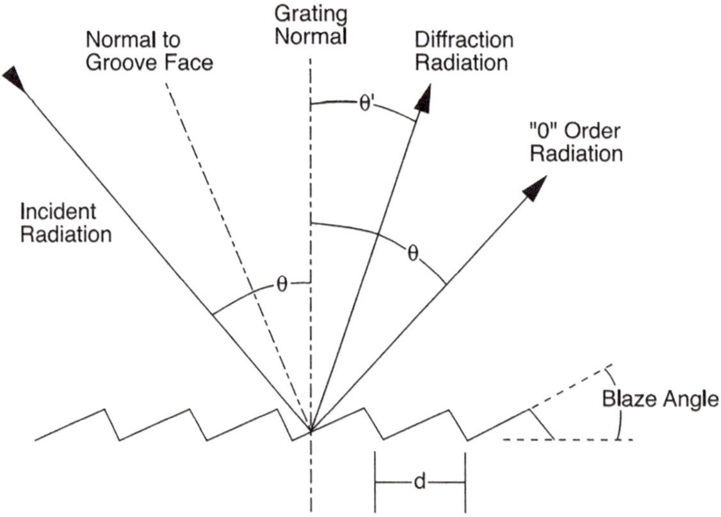

Fig. 2.5 Description of a ruled grating [8]

monochromator selects a single, coloured light (e.g. green) from the original white light.

The light dispersion can be achieved by a prism, but nowadays gratings are preferably used in modern instrumentation. The monochromator is a delicate disposition of a combination of components including gratings, mirrors, entrance and exit slits (Fig. 2.4).

The grating is the element of the monochromator responsible for the diffraction of the light. It consists of a series of closely spaced, parallel grooves of a sawtooth profile (Fig. 2.5) on a reflective surface (engraved by a diamond tool mounted on a ruling engine, coated with aluminium and an oxidation protective thin layer of SiO_2, and where the grooves are separated by a distance d that must be larger than the wavelength λ to be diffracted). Each groove behaves as an independent source of light, diffracting a different wavelength of light at a different angle. The diffuse polychromatic light entering the monochromator (through slit (**B**) in Fig. 2.4) is collimated by the concave mirror (**C**). The position or orientation of the grating, guides, sequentially, different narrow bands of wavelengths towards the exit slit (**F**). Decreasing the width of the exit slit, decreases the bandwidth of the emerging beam and the energy reaching the receiving system (e.g. the reactor).

A maximum constructive interference occurs for the angle θ' that satisfies the grating equation (Eq. (2.1)), which includes a proportionality of the product of the diffraction order (n_d) and wavelength (λ), with the product of the distance between the grooves and the incidence (θ) and diffraction (θ') angles.

$$n_d\,\lambda = d\,(\sin\,\theta - \sin\,\theta') \tag{2.1}$$

The resolution of the grating defines its ability to separate between two consecutive peaks at a distance $\Delta\lambda_d$ (the smaller $\Delta\lambda_d$, the higher the resolution). The resolution (Eq. (2.2)) is an inverse proportionality between the number (N_{gr}) of grooves of the grating, and $\Delta\lambda_d$. Accordingly, a resolution of 10^4 for a first-order ($n = 1$) diffraction, requires 10^4 grooves in the grating (e.g. for a ruled 10 cm length of the grating, 1000 grooves/cm are needed).

$$\frac{\lambda}{\Delta\lambda_d} = n_d\,N_{gr} \tag{2.2}$$

2.9 Isosbestic and Non-isosbestic Irradiations

These notions are exclusive to irradiation with monochromatic light. They relate to a feature observed on the successive spectra of the reactive medium through its progress during reaction time. For quantitative reactions, one or more points of the spectra, at specific wavelengths, do not have their absorbances change with time.

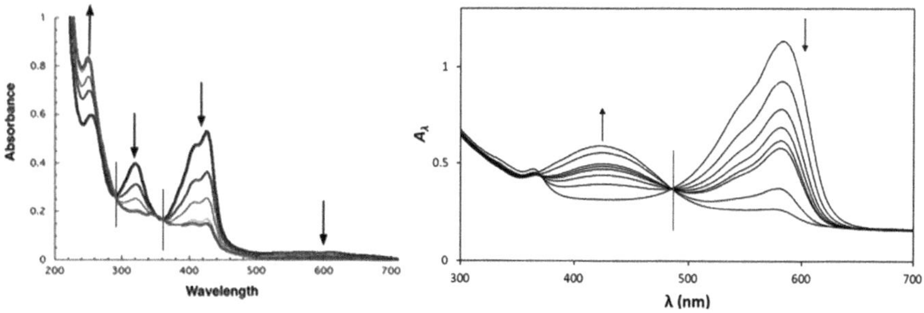

Fig. 2.6 (Left) Transformation of the UV-visible spectra of a dihydropyrene under irradiation with light >455 nm in cyclohexane [9]. (Right) Absorption spectra recorded for a phenol red-based CO_2 sensitive plastic film as a function of increasing gas-phase levels of $\%CO_2$ [10]. The arrows and the lines indicate, respectively, the direction of evolution of the absorbance, and the isosbestic points

These points, characterised by a time invariant absorbance value, are called *isosbestic points* (Fig. 2.6), and the corresponding wavelengths are dubbed *isosbestic wavelengths* (λ_{isos}). Irradiation at a particular isosbestic wavelength is called an isosbestic irradiation (Chap. 8). Conversely, the remaining shape of the spectra evolves with time, where the absorbance at each wavelength is variable with time. If the monochromatic irradiation light has a wavelength different from those characterising isosbestic points, it is dubbed a *non-isosbestic irradiation* ($\lambda_{n\,-\,\text{isos}}$).

A particular feature emerges at isosbestic points. The invariance of the absorbance means that the absorption coefficient of any of the $n_{\text{sp}}+1$ species absorbing at the isosbestic wavelength are equal ($\varepsilon_{Y_j}^{\lambda_{\text{isos}}}=\varepsilon_{Y_0}^{\lambda_{\text{isos}}}$) but their respective concentrations vary with time. In this case, the absorbance at the isosbestic point only depends on the total concentration, i.e. the initial reactant concentration ($C_X(0)=C_{Y_0}(0)$), as

$$A_X^{\lambda_{\text{isos}}}=\sum_{j=0}^{n}A_{Y_j}^{\lambda_{\text{isos}}}=\varepsilon_{Y_0}^{\lambda_{\text{isos}}}\,l_{irr}\,C_{Y_0}(t)+\varepsilon_{Y_1}^{\lambda_{\text{isos}}}\,l_{irr}\,C_{Y_1}(t)+\ldots+\varepsilon_{Y_n}^{\lambda_{\text{isos}}}\,l_{irr}\,C_{Y_n}(t) \tag{2.3}$$

or,

$$A^{\lambda_{\text{isos}}}=\varepsilon_{Y_0}^{\lambda_{\text{isos}}}\,l_{irr}\,(C_{Y_0}(t)+C_{Y_1}(t)+\ldots+C_{Y_n}(t))=\varepsilon_{Y_0}^{\lambda_{\text{isos}}}\,l_{irr}\,C_{Y_0}(0)=\varepsilon_X^{\lambda_{\text{isos}}}\,l_{irr}\,C_X(0) \tag{2.4}$$

2.10 Quantifying Light Absorption: The Beer-Lambert Law

The *Beer-Lambert law* is a relationship that expresses the variation of the sample absorbance as a function of the concentration of the absorbent species. The absorbance of the sample is directly proportional to the concentration if (1) the sample contains one absorbing species or several species with a constant concentration ratio, (2) the incident light is strictly monochromatic at λ, (3) the light beam is collimated, and (4) the concentration of the species is relatively low.

In these conditions, the incident light travelling through an infinitesimally thin layer (dx) of the solution will decrease in intensity (or power, P) by a factor dP in proportion to both the absorbing species concentration (C) and the layer thickness dx, as

$$dP=\,-\alpha\,P\,C\,dx \tag{2.5}$$

The minus sign in Eq. (2.5) indicates an absorption of the light (i.e. a decrease in dP as dx increases). The decrease in dP is also proportional to P, meaning that the higher the incident intensity (P_0), the higher the absorption. The α factor is a proportionality constant.

The differential equation can be rearranged and integrated to solve for P, assuming the boundary conditions, at $x=0$, $P=P_0$, and the power is P at $x=l$ [7, 11].

$$-\int_{P^0}^{P}\frac{dP}{P}=\alpha\, C\int_0^l dx \tag{2.6}$$

which finally gives,

$$\log\left(\frac{P_0}{P}\right)=\left(\frac{\alpha}{\ln 10}\right)C\,l \tag{2.7}$$

Or, set in its usual form (with ε being the decadic absorption coefficient)

$$A=\log\left(\frac{P_0}{P}\right)=\varepsilon\, l\, C \tag{2.8}$$

The power of the incident light is decreased as the light passing through each succeeding thin layer of the sample is reduced because of the fraction of the light absorbed in the previous layer. Equation (2.8) represents a gem for spectrophotometry as applied in analytical chemistry and holds the name of 'the Beer-Lambert Law'. The absorbance (A) is dimensionless, the concentration is given in moles per litre (M = mol/L or mol/dm^{-3}), and the optical path length (l) is expressed in cm. The dimension analysis of the equation assigns the non-SI unit $M^{-1}cm^{-1}$ for the quantity ε, called the *absorption coefficient* (or molar absorptivity) of the absorbing species at the considered wavelength (ε, relates to the light absorption capacity of a given species at a given wavelength).

Since both absorbance and absorption coefficient depend on the selected wavelength, and the optical path length corresponds to the section of the sample crossed by the irradiation light, Eq. (2.8) must always refer to these features, as:

$$A_\lambda=\varepsilon_\lambda\, l_{irr}\, C \tag{2.9}$$

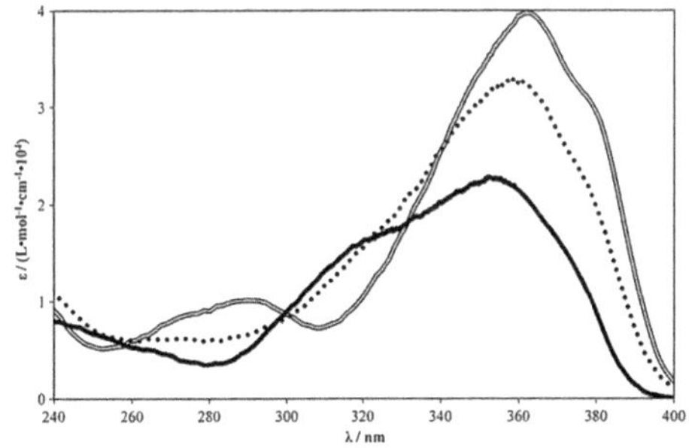

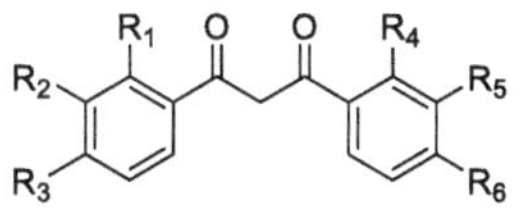

	R_1	R_2	R_3	R_4	R_5	R_6
—	H	OMe	H	H	OMe	H
•••	H	OMe	H	H	H	OMe
=	H	H	OMe	H	H	OMe

Fig. 2.7 Illustration of absorption spectra (ε units) of three, *UV* and *Vis* absorbing, methoxy and dimethoxy 1,3-diphenylpropane-1,3-diones (at given concentrations (C) of the species, temperature (T) of the medium, and optical irradiation path length (l_{irr}) in ethanol) [12]

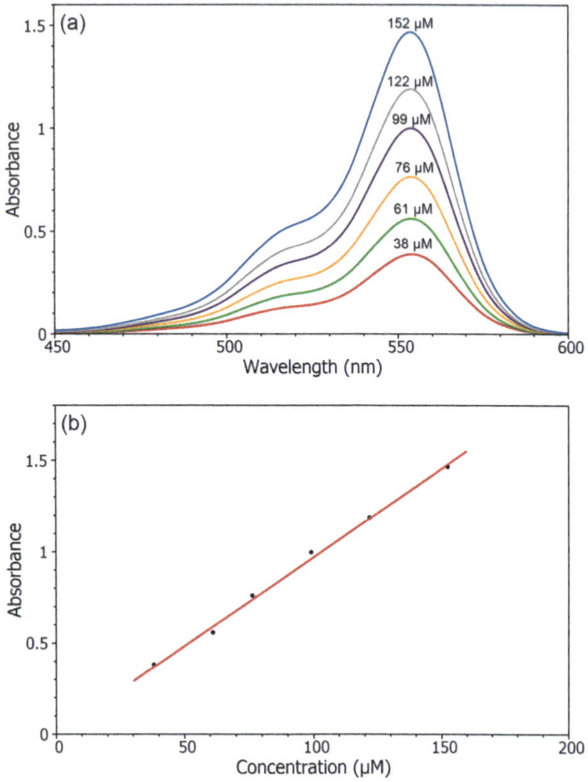

Fig. 2.8 (**a**) Absorption spectra of a series of differently concentrated samples of the same species. Spectra obtained at given constant temperature of the medium (T), optical irradiation path length (l_{irr}), solvent, and experimental conditions. (**b**) Plot of the absorbance (measured at a given wavelength λ) against the sample concentration [13]

2.11 The Beer-Lambert Law and Calibration Graph

Practically, the absorption spectrum, obtained for a solution containing a species at a concentration C, is a plot showing the variation of the absorbance (A_λ) or the absorption coefficient (ε_λ) with wavelength (λ), Fig. 2.7.

Overlapping successive absorption spectra for analytical samples of the same species but of different concentrations will show an increasing spectrum pattern with the increasing concentration (Fig. 2.8).

Plotting the experimentally measured absorbances (Fig. 2.8a), at the maximum wavelength (or, in principle, at any individual wavelength of the spectrum), for the differently concentrated samples, will result in a linear pattern (Fig. 2.8b), as predicted by the Beer-Lambert law. The slope of this graph is equal to the constant ($\varepsilon_\lambda\, l_{irr}$).

In photokinetics, the Beer-Lambert law (Eq. (2.9)) must be valid in order to derive a valid rate-law for the photoconversion induced by light irradiation.

2.12 The Beer-Lambert Law as a Tool for Analytical Chemistry

The latter graph representing the variation of the absorbance versus concentration (Fig. 2.8b) is called *the calibration graph*. Note that the calibration graph is valid for the conditions stated above (conditions (1)–(4), Sect. 2.10), i.e. the incident light is strictly monochromatic and collimated, the concentration of the species is not high, the temperature of the medium and the optical irradiation path length are constant, and the experimental conditions overall (except the concentration but including the wavelength at which the absorbances are measured) are the same for all samples.

The calibration graph is the tool used for the determination of the unknown concentration ($C_{\text{unk.}}$) of a given sample of the same species. The concentration of the unknown sample ($C_{\text{unk.}}$, Eq. (2.10)) is worked out from both its experimental absorbance ($A_{\lambda,\ \text{unk.}}$, measured on the spectrum of the unknown at the selected wavelength λ) and the Beer-Lambert law previously established (Eq. (2.9)).

$$C_{\text{unk.}} = \frac{A_{\lambda,\text{unk.}}}{\varepsilon_\lambda\ l_{irr}} \tag{2.10}$$

Hence, the Beer-Lambert law is represented here by the linear equation of the calibration graph. Notice that the theoretical Beer-Lambert equation (Eq. (2.9)) corresponds to a line without an intercept, i.e. the calibration graph line must cross the zero-zero point of the referential. In analytical chemistry, the (0,0) point must not be included in the table of the experimental data points (C, A_λ) of the calibration series. Yet, the obtained line must go through zero, as a first validation of the experiment. In practice, there is always a small intercept that has a negligible value (but which should be taken into account when performing calculations, for accuracy).

It is obvious that the data used for the calibration graph must be corrected for the amount of the blank experiment. The latter is, generally, made up of the solvent. However, for complex samples, it is represented by the matrix of the sample, i.e. the sample without the analyte (the species considered for analysis). It is recommended to ascertain that the matrix has no effect on the measurements. Indeed, interferences might occur, leading to wrong conclusions. Most of the interferences cannot be deduced from the spectra, but they can be evidenced by a separate standard addition experiment. The latter should be compulsory for any sample involving more than a solvent in the matrix if reliable results are sought.

The calibration graph can be characterised by a few calibration parameters, including the linearity range, the correlation coefficient of the line, and the limits of detection and quantification of the species. In the case where more than one species absorbs the incident light (where the ratios of these species concentrations are constant), the Beer-Lambert law can be derived from a modified differential equation to Eq. (2.5) that includes the presence of multiple "n" compounds (Y_0, Y_1, …Y_n, with $n \geq i \geq 0$).

$$dP_{Y_i}^{\lambda} = -\sum_{i=0}^{n} \alpha_{Y_i}^{\lambda}\, P^{\lambda}\, C_{Y_i}\, di\, dx = -P^{\lambda} \sum_{i=0}^{n} \alpha_{Y_i}^{\lambda}\, C_{Y_i}\, di\, dx \tag{2.11}$$

That can be rearranged, for integration, to

$$dP_{Y_i}^{\lambda} = -\int_{0}^{l_{irr}} P^{\lambda} \sum_{i=0}^{n} \alpha_{Y_i}^{\lambda}\, C_{Y_i}\, di\, dx = -P^{\lambda} \sum_{i=0}^{n} \alpha_{Y_i}^{\lambda}\, C_{Y_i}\, di \int_{0}^{l_{irr}} dx \tag{2.12}$$

We can proceed by the integration over dx since none of the terms of the latter equation depends on the length crossed by the light inside the sample.

$$\int_{P_0^{\lambda}}^{P^{\lambda}} \frac{dP^{\lambda}}{P^{\lambda}} = -\sum_{i=0}^{n} \alpha_{Y_i}^{\lambda}\, C_{Y_i}\, di \int_{0}^{l_{irr}} dx \tag{2.13}$$

The remaining integration is performed similarly to that of Eq. (2.5), except that now n compounds are present in the medium.

$$ln\, \frac{P_0^{\lambda}}{P^{\lambda}} = l_{irr} \left(C_{Y_0}\, \alpha_{Y_0}^{\lambda} + C_{Y_1}\, \alpha_{Y_1}^{\lambda} + \ldots + C_{Y_n}\, \alpha_{Y_n}^{\lambda} \right) \tag{2.14}$$

Transforming ln into log gives

$$\log \frac{P_0^{\lambda}}{P^{\lambda}} = l_{irr} \left(C_{Y_0} \frac{\alpha_{Y_0}^{\lambda}}{\ln 10} + C_{Y_1} \frac{\alpha_{Y_1}^{\lambda}}{\ln 10} + \ldots + C_{Y_n} \frac{\alpha_{Y_n}^{\lambda}}{\ln 10} \right) = l_{irr} \sum_{i=0}^{n} C_{Y_i}\, \varepsilon_{Y_i}^{\lambda} \tag{2.15}$$

The final form of the total absorbance at the given wavelength λ, of the monochromatic incident light, corresponds to the sum of the individual absorbances of the Y_n species present in the medium.

$$A_{\mathrm{tot}}^{\lambda}(t) = \varepsilon_{Y_0}^{\lambda}\, l_{irr}\, C_{Y_0}(t) + \varepsilon_{Y_1}^{\lambda}\, l_{irr}\, C_{Y_1}(t) + \ldots + \varepsilon_{Y_n}^{\lambda}\, l_{irr}\, C_{Y_n}(t) = \sum_{i=0}^{n} A_{Y_i}^{\lambda}(t) \tag{2.16}$$

This defines the additivity equation of the Beer-Lambert law at a given wavelength. It is always valid when the concentrations of the different species fall within the linearity ranges of their individual calibration graphs. A calibration graph of the mixture can be built if and only if the ratios between the species concentrations are preserved (because the linearity of the individual equations is preserved). If such ratios differ between samples, the absorption data at a given wavelength cannot possibly be used to construct a calibration graph (because now the absorbance data have a non-linear trend).

Conversely to the case of monochromatic/collimated light, there is no known standard Beer-Lambert law equation that applies when (collimated or uncollimated) polychromatic light is used.

2.13 When the Beer-Lambert Law Fails to Apply

The complete concept of a calibration graph, as observed experimentally, is that the latter is linear only over a relatively small range of concentrations. For all compounds, beyond a certain concentration (not relatively high), the calibration graph deviates from linearity (Fig. 2.9). The point of divergence is situated at the higher limit of the straight line (C_{LOL}), that corresponds to the highest value on the linearity range. It is also an experimental fact that the calibration graph is not linear at very small concentrations. The lower limit of linearity is given by the concentration corresponding to the limit of quantification (C_{LOQ}). It is then only understandable that analytical chemistry requires, as a good practice, that the linearity range ($[C_{LOQ} - C_{LOL}]$) is given as part of the analytical parameters.

In the non-linear sections of the calibration graph, the variation of the absorbance is not any more linearly proportional to the monitoring radiation intensity. The sample absorbs either more or less than it should have if the proportionality was preserved (green line in Fig. 2.9). This translates as a small variation in dP can no longer be considered as proportional to C ($dP \neq -\alpha P C dx$) as previously stated in the fundamental Eq. (2.5) (hence, the subsequent equations (Eqs. (2.6)–(2.16)) are no longer valid).

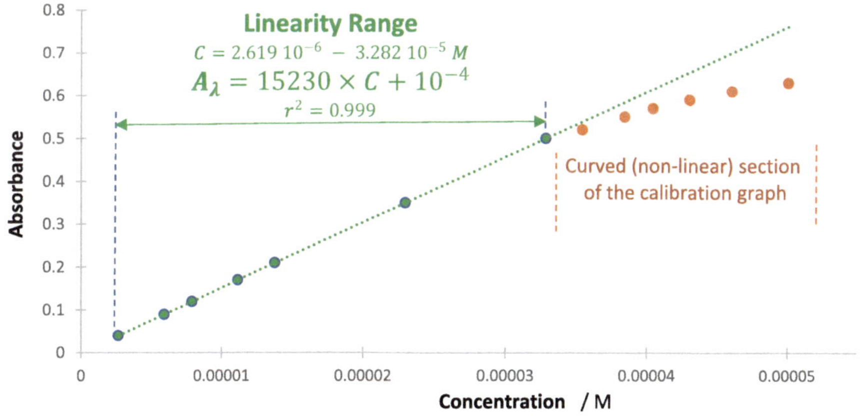

Fig. 2.9 A typical full calibration curve showing the linearity range limits and the curved section. The dots (green or orange) are the actual measured absorbance values for the corresponding medium concentrations. The extrapolation of the line over the linearity range is shown here only for reference to the non-linearity of the measurements beyond that range (but cannot be used for quantification)

Also, the physical phenomenon of absorption, when the light is polychromatic, should be described by another differential equation which must be substantially different to Eq. (2.5). Hence, we are led to conclude that the Beer-Lambert law is only useful for quantification in the span of the linearity range when the absorbed light is monochromatic (as rightly adopted in analytical chemistry). It also suggests that this law, as it stands in Eq. (2.9), is not a general law (since it is unable to render a full account of the absorption phenomenon as a function of any concentration due to the curvature of the calibration graph). Such a deviation is a proof of the limited applicability of the Beer-Lambert law in its classical form (Eq. (2.9)) and the likelihood of occurrence of phenomena other than just a quantitative absorption of the monitoring light by the molecules under study.

A full explanation for the observed curvature of the calibration graph and the reasons why the ratio absorbance/absorbent concentration varies with concentration are not fully understood. However, several tentative interpretations have been proposed.

In 1969 [14], such a behaviour was suggested to have either or both chemical, physical, and/or instrumental origin. The study introduced a method to distinguish between these three possible effects but without proposing an alternative to the Beer-Lambert law. Instead, it was suggested that such deviations from linearity may be attributed to three main causes: the light source being far from monochromatic, the medium being highly scattering, and/or the concentrations of analytes being very high.

Another more practical attempt focused on mapping out the curvature of the non-linear calibration plot by a Gaussian regression analysis [15]. It led to establishing absorbance/absorbent concentration relationships, but with no explanation of the possible causes of the curved shape. In the same perspective, more sophisticated mathematical methods have also been applied for such practical purposes, such as support vector regression, random forests, neural networks, PCR and PLS [16–19].

More recently, a finer analysis using Maxwell's wave equation derived Beer's law from dispersion theory [20]. It showed the law to be a limiting case. Indeed, absorption, molar attenuation coefficients, and absorption cross sections were all proven not to be properties of the material, so much so not to be additive (while all these quantities are supposed by the Beer-Lambert law to be both inherent properties of molecules and additive). In this model, the deviation from linearity scales up with oscillator strength, i.e. with both local electric fields and nearfield effects. A special finding of the study showed that the parameters emerging from the derivation to linearity do not exactly match with those found in the formulation of the Beer-Lambert law.

A further overview of the deviation of the Beer-Lambert law has advanced an understanding of the spectra and their quantitative interpretation on the basis of oscillators' positions, their strengths and damping constants [21]. By using the electromagnetic theory, this study found that it was possible to reach a better quantification of both band shift and intensity variation that occur on the absorption spectrum. It, hence, concluded that the classical formula of the Beer-Lambert (conveying a proportionality of the absorbance with the absorbent concentration, Eq. (2.9)) is rather an ideal absorption law that is only valid for relatively low concentration domains, indicated by the linear section of the calibration curve.

Other authors added to the variation of the absorption coefficients the possible light absorption by some emerging species in the solution caused by the high concentration of the medium, e.g. complexes [22], or quenching [23].

In summary, the Beer-Lambert law only applies when the absorbed light is monochromatic (this law is not appropriate when polychromatic light is absorbed by the sample). Likewise, the Beer-Lambert law does not apply when the concentration value falls outside the linearity range of the calibration graph. At high concentration, the absorption coefficient values characterising a molecule will be different from those recorded when the species concentration belongs to the linearity range. Unfortunately, such new values of ε, expected for highly concentrated media, are not possibly accessible experimentally. Also, the possible occurrence of other absorbers, other than the reaction species, in a highly concentrated reactive medium adds an extra disadvantageous complication for employing the Beer-Lambert law in its classical form.

Furthermore, even if the literature does not, in general, raise the point relative to a possible variation of both the quantum yield values and pattern with wavelength, it is conceivable to consider that such a change might occur when the concentration is high. Also, by analogy with the alteration of ε (discussed above when the concentration is high), it might be postulated that the quantum yield value might also differ between (low and high) concentration conditions.

2.14 Variation of the Absorbance During the Photoreaction

The monitoring of a photoreaction progress can be performed by a variety of analytical chemistry techniques. Nonetheless, it is always preferable, and even highly recommended, to acquire the electronic absorption spectra of the reaction at timely intervals using UV-Vis spectrophotometry. The recommendation of measuring the absorption spectra of the reaction stems from a number of reasons. Amongst those, and in principle, the absorption spectrum delimitates the spectral wavelength range useful for irradiation. Also, UV-Vis measurements allow the selection of the initial concentration of the reactant so that the concentrations of reactant and products, evolving during the photoreaction progress, fall within the respective linearity ranges of their spectrophotometric calibration graphs. From a descriptive viewpoint, the total absorbance ($A_{tot}^{\lambda_{irr}/\lambda_{irr},T}$) makes part of the differential equation of the photoreaction, and the Beer-Lambert law for spectrophotometry is the basis for the determination of the amount of absorbed photons by the reactive medium (Sect. 2.18).

The set of data collected in kinetics is, in general, represented by a monitored timely change of a property or properties of the reaction (i.e. concentration or any

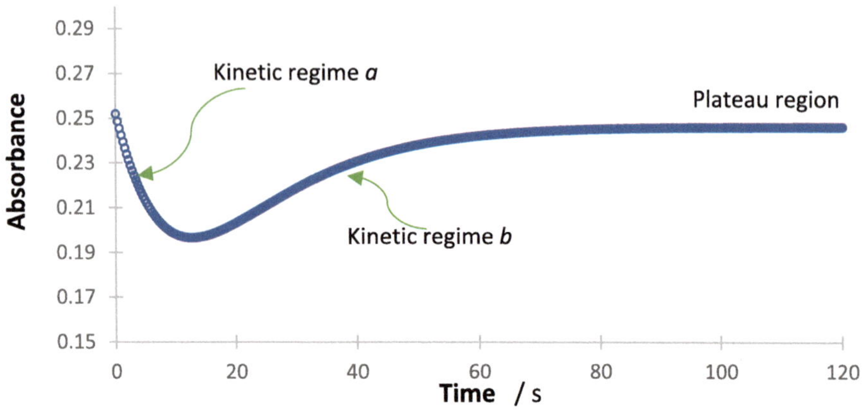

Fig. 2.10 Example of a total absorbance kinetic trace

quantity related to the concentration, as the analytically determined properties are, such as the absorbance). The time variation of any quantity related to the amount of a species in the reactive medium constitutes a kinetic trace (plots as $A_{\mathrm{tot}}^{\lambda_{irr}/\lambda_{\mathrm{obs}},T}(t)$ or $C_{Y_j}^{\lambda_{irr},T}(t)$ *vs*. *t*, Fig. 2.10). A trace usually has a curved shape of some type that might or might not be monotonically increasing or decreasing but always reaches a plateau after a long reaction time. Non-monotonical kinetic traces present one or more changes in the direction of their curvatures (inflexion points). The analysis of such features can be informative.

The temporal change of the total absorbance (e.g. $A_{\mathrm{tot}}^{\lambda_{irr}/\lambda_{\mathrm{obs}},T}(t)$ *vs*. *t*) at a given non-isosbestic observation wavelength (λ_{obs}) is a useful kinetic trace of the reaction. Any variation of the absorbance value during the reaction, shown by an increase or a decrease of the trace, is indicative of a change in the sample composition. It could be due to either a change of concentration of the constituents of the sample or to a change of their absorptivity if the species concentrations, at a given time, are constant, or any combination of both aspects. The phototransformation of a reactant into a product, $X \rightarrow Y$, means a change of both concentration and absorptivity so that the total absorbance measured at a time t_1 is different from that recorded at a subsequent time t_2 (e.g. for $t_1 = 0s$ and $t_2 = 50s$, $A_{\mathrm{tot}}^{\lambda_{irr}/\lambda_{\mathrm{obs}},T}(0) = \varepsilon_X^{\lambda_{\mathrm{obs}}}\, l_{irr}\, C_X(0) \neq A_{\mathrm{tot}}^{\lambda_{irr}/\lambda_{\mathrm{obs}},T}(50) = \varepsilon_X^{\lambda_{\mathrm{obs}}}\, l_{irr}\, C_X(50) + \varepsilon_Y^{\lambda_{\mathrm{obs}}}\, l_{irr}\, (C_X(0) - C_X(50))$). A variation of the absorbance is, at the least, interpreted as a qualitative confirmation of the occurrence of a reaction (i.e. and confirmation of at least one reaction-step).

The direction in which the changes take place, as an increase or a decrease of the magnitude of the absorbance at λ_{obs}, is indicative of a change in the sum of ($\varepsilon\, C(t)$) values of the species present in the medium (l_{irr} being the same). The increase of a total absorbance compared to its initial value necessarily means the increase of that sum, and *vice versa* if a decrease occurs. Overall, inflexions of the trace are qualitative proofs of the occurrence of a variation of the electronic molecular structures of the measured species (i.e. different from that of the species measured before the inflexion point).

Since each inflexion is attributable to the emergence (or the disappearance) of a new product(s), the kinetic traces of the studied reaction acquired at various observation wavelengths might have different profiles. This might allow the analyst to obtain more information on the number of products formed.

In addition to the latter analysis, the experimentalist might propose some extra experiments in order to test hypotheses, with the perspective to acquire better knowledge on the reaction and/or the reaction mechanism.

When the absorbance, measured at the non-isosbestic wavelength (λ_{obs}), does not change at all or in negligible proportion after a long period of irradiation, it can be interpreted either as the absorbance of the end-product(s) if the reaction has reached completion, or as due to an equilibrium "photostationary" state, *pss* (that might involve more than two species, and be established between thermal and photochemical reaction-steps).

Overall, a careful analysis of the absorption spectra and the traces will not only attest to the occurrence of photoreactivity but also, and perhaps more importantly, it helps identify with certainty the involvement of a minimum number of products. This contributes to developing a better understanding of the mechanism of the photoreaction. However, identifying the various inflexion points on the traces does not by itself allow one to reach a definite conclusion about the full mechanism in action. The variation of the traces' profiles, notably, depends on the rates of the various reaction-steps involved as well as on the magnitude of the differences between the absorptivities of the absorbing species at the observation wavelength. Hence, such an analysis only confirms the minimum reaction-steps occurring in the mechanism.

In Fig. 2.10, the absorbance decreases up to *ca.* $t \approx 12\ s$, then increases up to *ca.* $t \approx 80\ s$, and reaches a plateau region at long reaction time. The information of this trace allows to deduce that the reactant present at $t = 0$, transforms (through at least one reaction-step) into at least one product. The latter is (are) characterised by a lower absorptivity compared to that of the reactant. This process constitutes the first kinetic regime (*a*) observed on the trace. The first product(s) go on to generate a second product(s) through a different (or different) reaction-steps from the one involving the reactant species (the kinetic regime *b*). The second product(s) has a higher absorptivity than that of the first product, as indicated by the increase of the absorbance from $t \approx 12\ s$. The trace stabilised at an apparently constant absorbance value at the plateau region (i.e. after a long reaction time). The plateau indicates that the reaction has either reached completion or an equilibrium (or *pss*) state. Therefore, the minimum qualitative information obtained from the trace of Fig. 2.10 consists of the involvement of at least three species (a reactant and two products) and two reaction steps in the reaction mechanism. One of the reaction-steps starts at the reactant, and the second involves at least one of the products generated from the reactant. The true mechanism of the reaction at hand might be more extensive than this picture (but certainly includes the features described above).

It is also interesting to point out that, irrespective of the type of light used for irradiation (collimated, uncollimated, mono- or polychromatic light), a smooth pattern is always observed for the temporal variation of the absorbance measured at a given observation wavelength. This particular feature reveals that it is not possible to identify the light type to which the sample was exposed by analysing the resulting observed trace, $A_{tot}(t)$. This raises a caution on the usage of the $A_{tot}(t)$ data in fundamental photokinetic equations if the acquisition of the latter does not correspond to the requirements and conditions stipulated for the derivation of those equations (e.g. using $A_{tot}(t)$ obtained with polychromatic light in equations written for exclusively monochromatic light).

2.15 Solving the Kinetics

One of the purposes of a kinetic analysis is the determination of the mechanism of the reaction investigated (it is often the case that the true mechanism of the studied photoreaction is unknown). For that purpose, the analysis of kinetic data will, in principle, encompass unravelling both the reaction mechanism and reaction parameters (numerical values), as well as a model allowing a more predictive ability of its evolution and control. A full kinetic solution, from the perspective of a pure kinetic analysis, includes, if at all possible, the uniqueness of the three criteria, namely, a defined reaction order, a defined reaction mechanism, and a defined set of values for the reaction parameters. This "uniqueness" principle is paramount in kinetic studies.

The tools available for a kinetic analysis include the spectroscopic spectra and analytical chemistry data corresponding to the timely progression of the reaction, a sufficient number of traces, the experimental conditions, and any information from the literature.

We have seen in the previous section that the analysis of the absorbance traces can only provide the minimum number of reactive species in addition to a minimum number of reaction-steps (e.g. the reactant conversion by light). In the general case, the full mechanism (species and specific reaction-steps) cannot be unravelled with certainty by using any number of such traces (except for very simple reactions).

The best way to identify the reactive species involved in a reaction is to use a combination of separation, spectrometry, and/or spectroscopic analyses. For instance, the species might be separated by chromatography (e.g. HPLC), their individual molecular masses may be obtained by spectrometry (MS, LC – MS... etc.), and their molecular structures can be identified by spectroscopic techniques such as NMR and FTIR (in combination with the fragmentation observed by MS). This classical procedure works generally well, but might be limited or much more tedious for some photothermal reactions than others (e.g. when involving thermally unstable species).

Unravelling the number and nature of the reaction steps occurring between the reaction species is a much more complicated task. Except for the simplest photo- or photothermal reactions (e.g. $X \rightarrow Y$, $X \leftrightharpoons Y$), it is seldom easy to identify the complete true set of the reaction steps operating in a reaction, either or both by using kinetic and analytical data. From a kinetic point of view, the observed kinetic data and traces can be well attributed to more than one mechanism. This is known as the distinguishability problem (see Chaps. 6 and 9). Also, qualitative and quantitative analytical chemical analyses rarely offer extended possibility to know with certainty the reaction-steps in action. As a result, this section of photokinetics can be considered as a stand-alone research area that needs to be developed by proposing standard procedures. It also explains the many complex reaction mechanisms that are still not fully understood (e.g. a number of photochromes make part of this list).

Despite the facts laid out above, in practice, a procedure seems to have been adopted in the literature. This *ideal* or conventional approach can be summarised by the following steps: (a) a mechanism for the reaction under study is proposed based on the kinetic analysis of the data at hand, (b) once a mechanism is proposed, the corresponding rate-law equations are written down, (c) a solution is approximated from the equations (analytical integrated rate-law are not known), (d) the numerical values of the set for the parameters values are determined, (e) the obtained solution (mechanism and parameters' values) is used to generate a *theoretical* trace from the proposed integrated rate-law and the set of values of the parameters, (f) if a good fitting is obtained between experimental and theoretical values (or traces), then the common wisdom will consider the proposed mechanism and the values of the reaction parameters as valid, representing the true features of the reaction, but (g) if there is no agreement between experimental data and theoretical model, then a new mechanism is proposed, and the procedure (*Points* (a)–(g)) is reconducted.

However, this approach is often hindered by a number of problems. We have already mentioned above that the kinetic analysis can only provide the minimum number of reaction-steps (which is far from providing a full mechanism, and weakening the proposal stated in *Point* (a)). The possible occurrence of a distinguishability problem also casts doubt on the conclusion that might be reached in *Point* (f). Because the integrated rate-law (*point* (c)) cannot, most often, be derived analytically (see Chap. 12), in which case, the order of the reaction cannot be known. Finally, the values of the reaction parameters (*point* (c)) might be uncertain due to an identifiability problem (Chaps. 6 and 9). In summary, this procedure is an approximation that may not always lead to consistent results.

Partial information can be delivered by other approaches, such as using initial velocity or steady-state methods or power expansion series (Chap. 13). Alternatively, numerical integration methods (NIMs) can provide a robust mathematical background for the problem posed by the impossibility of analytical integration of the rate-law without relying on approximations. Despite its higher performance, the NIM-based strategy has to be applied with great caution in order to avoid misinterpretations and incorrect results. We will discuss some NIMs aspects in Chaps. 14 and 19.

Another and very popular way to reach the full kinetic solution is an indirect approach. That is, instead of attempting to find the integrated rate-law, some known integrated rate-laws are applied to the kinetic data at hand so as to retain the one that fits well with the experimental data. The principle applied here is that the kinetic data of a given reaction cannot obey more than one kinetic model. This principle is considered to act as a selection rule for the identification of the reaction order. The conventional procedure adopted here also considers that once the supposedly true order is known, the set of reaction parameters can be worked out in a unique way. Traditionally, in thermal reaction kinetics, the actual experimentally obtained reaction data are tentatively and successively fitted with the equations corresponding to the three conventional (0th, 1st, and 2nd) orders of the thermal reactions. The agreement of the experimental data with one of these equations (generally, the best agreement is defined on the basis of the highest value of the correlation coefficient) defines the order of the reaction as stated by the aforementioned principle. This practice, which is ubiquitous in chemical kinetics, has surprisingly also found a relatively large application in photokinetics. Such an approach leads, in fact, to a dramatic situation, as it is redundant in photokinetics because photoreaction orders are unknown (no known integrated rate-laws). Therefore, the supposed principle applied in this approach is not fully proven and should be reconsidered.

The above succinct review proves that photokinetics requires, perhaps, a serious discussion and coherent research efforts to improve it towards being an investigative tool.

2.16 Spectators

A spectator molecule (one of n_{SPM} molecules, SPM_r, with $n_{\mathrm{SPM}} \geq r \geq 1$) is a species present in the reactive medium but does not chemically interact with the reactants and products of the photothermal reaction, has a constant concentration, is photo- and thermally inert, but absorbs at the irradiation wavelength or wavelength domain. In thermal kinetics, a spectator of this type would have no effect on the reactivity of the species involved in the reaction mechanism. However, the situation is different in photokinetics. In the cases where a non-zero absorbance of the spectator molecule (s) at the irradiation wavelength λ_{irr}, $A_{\mathrm{SPM}_r}^{\lambda_{irr}} \neq 0$ (Eq. (2.17)), is considered, it must necessarily make part of the total absorbance of the medium (Eq. (2.16)), since the fraction of the light absorbed by the spectators is lost for the reactive species.

$$A_{\mathrm{tot}}^{\lambda_{irr}}(t) = \sum_{r=1}^{n_{\mathrm{SPM}}} A_{\mathrm{SPM}_r}^{\lambda_{irr}} + \sum_{j=0}^{n_{\mathrm{sp}}} A_{Y_j}^{\lambda_{irr}}(t) \tag{2.17}$$

For this reason, an increase in the absorbance of the spectators is expected to result in a decrease of all species' reactivities (i.e. rate-constants and initial rates), as depicted in Chaps. 16 and 23.

Notice that no conditions on $A_{\mathrm{SPM}_r}^{\lambda_{irr}}$ to belong to the linearity ranges of SPM_r calibration graph are necessary, since $A_{\mathrm{SPM}_r}^{\lambda_{irr}}$ does not contribute to the mathematical formalism of the photokinetics other than by its actual constant value. However, very high concentrations of *SPMs* might cause unsuitable interactions (e.g. formation of ground or excited-state complexes).

2.17 Usefulness of the Beer-Lambert Law for Photons' Count

In addition to the important role played by the Beer-Lambert law in quantitative analytical chemistry, it has a primordial role in photokinetics. In particular, as part of the equation, allowing the count of the absorbed photons by the photoreactive species.

For a monochromatic irradiation light of wavelength λ_{irr}, the Beer-Lambert law can be amended to include the fraction of that amount of light absorbed ($P_{a,X}^{\lambda_{irr}}(t)$) by the species (e.g. X) at time t. This fraction of absorbed light is related to the incident ($P_0^{\lambda_{irr}}$) and transmitted ($P_{tr,X}^{\lambda_{irr}}(t)$) radiations by

$$P_{tr,X}^{\lambda_{irr}}(t) = P_0^{\lambda_{irr}} - P_{a,X}^{\lambda_{irr}}(t) \tag{2.18}$$

and therefore, Eq. (2.8) becomes

$$A_X(t) = -\mathrm{Log}\left(\frac{P_{tr,X}^{\lambda_{irr}}(t)}{P_0^{\lambda_{irr}}}\right) = -\mathrm{Log}\left(\frac{P_0^{\lambda_{irr}} - P_{a,X}^{\lambda_{irr}}(t)}{P_0^{\lambda_{irr}}}\right) \tag{2.19}$$

Rearranging the latter equation, we obtain

$$P_{a,X}^{\lambda_{irr}}(t) = P_0^{\lambda_{irr}}\left(1 - 10^{-A_X^{\lambda_{irr}}(t)}\right) \tag{2.20}$$

Hence, Eq. (2.20) is an expression of the absorbed light of wavelength λ_{irr} by species X. It can be used to quantify theoretically as well as practically the fraction of light absorbed by the studied species X if $P_0^{\lambda_{irr}}$ and $A_X^{\lambda_{irr}}(t)$ are known.

However, it is important to stress out that $P_{a,X}^{\lambda_{irr}}(t)$ is determined by Eq. (2.20) only and only if Eq. (2.19) (or Eq. (2.8)) stands. Since the Beer-Lambert law (Eq. (2.19)) is the equation of a line (Eq. (2.8)), then Eq. (2.20) is valid only on the linearity range of the spectrophotometric calibration graph of the species, plotted for the absorbance at λ_{irr}. This is important because there will be a strong inconsistency to calculate $P_{a,X}^{\lambda_{irr}}$ for concentrations of X outside the linearity range, as, in such a case, Eq. (2.19) does not apply (Sect. 2.13).

2.18 Calculating the Number of Photons Entering the Reactor

Determining the number, $P_0^{\lambda_{irr}}$, of photons of an irradiation/monochromatic light of wavelength λ_{irr}, reaching the reactive medium is an essential parameter of photokinetics, as photons are an inherent part of the photoreaction kinetics.

As the incident radiation intensity, $P_0^{\lambda_{irr}}$, is related to the energy supplied by the light source, the latter is a good starting point to define this quantity, $P_0^{\lambda_{irr}}$ [24].

The incident radiation of the monochromatic light beam, which can be measured by a physical actinometer, by a photometer or a spectroradiometer, as a spectral irradiance of the light source, E_{sp-irr}, expressed in W cm^{-2} nm^{-1} or J s^{-1} cm^{-2} nm^{-1} (i.e. energy (*J*) per unit time, *t* (*s*), per unit area (1 cm^{-2}), and per wavelength, λ_{irr} (nm^{-1}), and where the unit *J*/*s* corresponds to an energy flux). For a monochromatic radiation, the unit, per wavelength (nm^{-1}) can be overlooked in the dimension of the spectral irradiation.

In experimental photochemistry, the number of photons is preferred over the energy of the beam. The former allows an adequate quantitative metric because the reaction occurs between this number of photons and the number of photoactive species in the medium. The number of photons is determined as follows.

The energy ($E_{h\nu}^{\lambda_{irr}}$ in *J*) carried by one photon of the radiation is a function of the wavelength λ_{irr}.

$$E_{h\nu}^{\lambda_{irr}} = \frac{h\,c}{\lambda_{irr}} \tag{2.21}$$

where *h* is Planck's constant expressed in J s ($h = 6.62608\ 10^{-34}$ J s), *c* the velocity of light given in m/s (299792458 m s^{-1}), and λ_{irr} in m.

An einstein (a mole) of photons of a given wavelength, carry a total energy $E_{\text{mol}\ h\nu}^{\lambda_{irr}}$. And, there is an Avogadro's number of photons in a mole, $N_a = 6.02214\ 10^{23}$, expressed in einstein^{-1}. The latter equation becomes

$$E_{\text{mol}\ h\nu}^{\lambda_{irr}} = N_a\,E_{h\nu}^{\lambda_{irr}} = N_a\frac{h\,c}{\lambda_{irr}} \tag{2.22}$$

Applying this formula to the physically measured energy of the radiation, provided by the spectral irradiance, E_{sp-irr}, gives the number of photons (number of einstein or moles of photons) carried by the considered radiation per unit of time and area. This quantity represents the incident flux of the monochromatic light source through a unit area, $flx_{h\nu}^{\lambda_{irr}}$.

$$flx_{h\nu}^{\lambda_{irr}} = \frac{E_{sp-irr}}{E_{mol\ h\nu}^{\lambda_{irr}}} = \frac{E_{sp-irr}\,\lambda_{irr}}{N_a h\,c} \tag{2.23}$$

The dimension analysis of Eq. (2.23), $\left[flx_{h\nu}^{\lambda_{irr}}\right] = \left[E_{sp-irr}\,\lambda_{irr}/N_a h\,c\right] = (\mathrm{J\,s^{-1}\,cm^{-2}\,m})/(\mathrm{einstein^{-1}\,J\,s\,m\,s^{-1}}) = \mathrm{einstein\,s^{-1}\,cm^{-2}}$, confirms a flux dimension (einstein s^{-1}) per unit area.

Therefore, in order to express the number of photons entering the reactor per *second* for a given experimental setup, with a specific irradiated area of the sample, it is necessary to multiply the incident flux per unit surface $\left(flx_{h\nu}^{\lambda_{irr}}\right)$, by the actual irradiated area of the sample, S_{irr} (in cm^2).

$$flx_{h\nu,S_{irr}}^{\lambda_{irr}} = flx_{h\nu}^{\lambda_{irr}}\,S_{irr} \tag{2.24}$$

Finally, $flx_{h\nu,S_{irr}}^{\lambda_{irr}}$ is to be reported to the actual volume (V_{irr} in dm^3) of the sample subjected to irradiation. The derived quantity, representing the $P_0^{\lambda_{irr}}$, is expressed as

$$P_0^{\lambda_{irr}} = \frac{flx_{h\nu}^{\lambda_{irr}}\,S_{irr}}{V_{irr}} \tag{2.25}$$

whose dimension analysis gives $\left[P_0^{\lambda_{irr}}\right] = (\mathrm{einstein\,s^{-1}\,cm^{-2}\,cm^2})/\left(\mathrm{dm^3}\right) = \mathrm{einstein\,s^{-1}\,dm^{-3}}$.

$P_0^{\lambda_{irr}}$ is, therefore, the number of photons entering the reactor per second, per irradiated area and volume of the investigated sample.

The dimension of $P_0^{\lambda_{irr}}$ is adequate to be used in a rate-law describing a photochemical reaction (*dC*/*dt*, e.g. *vide infra* Eq. (3.37)). The dimension characterising the left-hand side term of the rate-law, corresponding to a differentiation of the concentration with time, as mol/(L *s*) or mol $dm^{-3}\,s^{-1}$, and the right-hand term of that equation (Eq. (3.37)) has the unit of $P_0^{\lambda_{irr}}$ (both quantum yields, $\Phi^{\lambda_{irr}}$, and factors multiplying $P_0^{\lambda_{irr}}$ (e.g. Eq. (3.38)) are dimensionless). Therefore, the same unit is found for either side of the differential equation (the similarity claim is consistent, since for photons count, mol is equivalent to einstein).

The IUPAC [25] defines a quantity $q_{n,p,\lambda}^0$ (Eq. (1.5)) as the radiation spectral photon flux on amount basis, with a dimension of einstein $s^{-1}\,nm^{-1}$ or einstein s^{-1} for a monochromatic light. In the expression given therein for the rate-law, the spectral photon flux is divided by the volume irradiated ($q_{n,p,\lambda}^0/V$, in einstein $s^{-1}\,dm^{-3}$). The latter quantity has the same dimension as $P_0^{\lambda_{irr}}$ (Eq. (2.25)). However, notice that Eq. (2.25) takes into account the irradiated area of the sample (the two equations will be equal if $S_{irr} = 1\ cm^2$).

The value of the incident radiation intensity, $P_0^{\lambda_{irr}}$, can change due to a variation of either the incident flux of the light source or the light source itself ($flx_{h\nu}^{\lambda_{irr}}$), the irradiated area ($S_{irr}$), or the volume ($V_{irr}$) of the sample exposed to the light. In this context, it is obviously important to ensure a vigorous stirring of the reactive medium, throughout irradiation, for consistency with the value of V_{irr} in Eq. (2.25). Notice also that the ratio (V_{irr}/S_{irr}), in Eq. (2.25), is not necessarily equal to the optical path length crossed by the light beam inside the sample l_{irr}

(it would only be the case if the whole volume is exposed to the light through one of its full sides, but in general, $V_{irr}/S_{irr} \geq l_{irr}$).

This then commends that the three properties are reported in the investigation in order to complete the adequate photokinetic analysis. Unfortunately, most often, these quantities are missing from published studies. This might explain, at least in part, the difficulties in comparing photokinetic results when obtained in different laboratories.

Practically, the changes on $flx_{h\nu}^{\lambda_{irr}}$, S_{irr} and V_{irr} can be controlled by the experimentalist by respectively varying the electric potential of the lamp, the blinded surface, or the volume irradiated of the sample (these are part of the extrinsic parameters).

As an illustration, let us calculate the $P_0^{\lambda_{irr}}$ for a monochromatic beam of wavelength 264 nm, impinging on a sample of a 2 mL volume, the light reaches the sample through an irradiated area of 1.5 cm^2. The recorded power of the light (0.44 mW cm^{-2}) is measured on a radiant power meter. Using first Eq. (2.23), we obtain

$$\begin{aligned} flx_{h\nu}^{264} &= \frac{E_{sp-irr}\,\lambda_{irr}}{N_a h\, c} \\ &= \frac{0.44\ 10^{-3}\ J\,s^{-1}\ cm^{-2} \times 264\ 10^{-7} m}{6.02214\ 10^{23}\ einstein^{-1} \times 6.62608\ 10^{-34}\ J\,s \times 299792458\ m\,s^{-1}} \end{aligned} \tag{2.26}$$

or,

$$flx_{h\nu}^{264} = 0.97102\ 10^{-7}\ \text{einstein cm}^{-2}\ \text{s}^{-1} \tag{2.27}$$

and then, Eq. (2.25) gives

$$P_0^{264} = \frac{flx_{h\nu}^{264}\ S_{irr}}{V_{irr}} = \frac{0.97102\ 10^{-7}\ \text{einstein cm}^{-2}\ \text{s}^{-1} \times 1.5\ \text{cm}^2}{2 \times 10^{-3}\ \text{dm}^3} \tag{2.28}$$

or,

$$P_0^{264} = 7.2826 \times 10^{-5}\ \text{einstein dm}^{-3}\ \text{s}^{-1} \tag{2.29}$$

2.19 Solving Photoreactions' Rate-Laws

Textbooks on thermal kinetics usually include a presentation on the ways to analytically integrate the differential equations of simple systems such as the unimolecular $X \rightarrow Y$, the reversible $X \leftrightharpoons Y$, and the consecutive ($X \rightarrow Y_1 \rightarrow Y_2$) reactions. Beyond these trivial examples, the solutions corresponding to more complicated mechanisms and their systems of differential equations become very scarce in the literature and rather a matter of specialisation. The reason for that is mainly technical, as the solutions are harder to achieve, often requiring complicated mathematical approaches. A complete kinetic solution of a cyclic reaction $XY_2(6\Phi, 6k)$ involving three species (X, Y_1 and Y_2) might represent an example of such difficulty (see Chap. 8).

Conversely, the systems of differential equations that describe photochemical systems have not received considerable attention in the past. It is striking to observe that not a single of these differential equation systems, fully solved through a rigorous analytical derivation, can be found in photochemistry books. In general, only approximate solutions or numerical integration initiatives are available.

The reasons for the absence of analytical solutions in photokinetics can be tracked down to mainly technical difficulties. Indeed, the unusual mathematical presentation of the differential equations compared to that encountered in thermal kinetics, might explain such a situation. As a matter of fact, most of photokinetics' differential equations are unsolvable using the standard integration methods (Chap. 12).

Beyond the fact that this contrasts with the normal practice in thermal kinetics, the absence of integrated rate-laws meant that photoreactions were not possibly characterised by reaction orders. Of course, way too often, photoreactions are assigned the same reaction orders that characterise thermal reactions, but this approach is not consistent except for a monochromatic/isosbestic irradiations (as we shall see in Chap. 8).

Nonetheless, even for the unsolvable differential equations, it is conceptually a requirement that photoreactions obeyed proper photochemical reaction orders that are not necessarily the same as those established for thermal reactions (see Chaps. 8 and 11).

Nonetheless, it would be helpful to know the integrated rate-laws of reactions whose differential equations are analytically solvable and specify the reactions for which it is not the case (Part IV). As we shall see, approximated integrated rate-laws under strict conditions could be developed (Part V). It is always more convenient to work with reliable (even if not analytically derived) explicit equations.

2.20 Metrics in Photokinetics

In general, the most common metric in thermal kinetics is the reaction rate-constants (k). However, this typical approach might be compromised by unsolvable identifiability issues (as we shall see in forthcoming Chaps. 6 and 9). In order to circumvent such a hurdle, the use of the initial reaction-rate of the reactant and/or those of the species emanating from the reactant would represent a more appropriate means for the quantification of photoreactivity. This metric would prove to be useful in photokinetics, as described in subsequent chapters.

2.21 Actinometry and Kinactinometry

Actinometry is the measure of the incident light flux per unit area and volume ($P_0^{\lambda_{irr}}$) at λ_{irr}, that enters a given reactor (a slab-shaped reactor for our study). A chemical actinometer is a standardised reactive system able to deliver $P_0^{\lambda_{irr}}$. The standardisation can be achieved kinetically by quantifying the linearly correlated variation of $P_0^{\lambda_{irr}}$ with the rate-constant of a given reaction-step ($k_{ij}^{\lambda_{irr}}$) of the global photoreaction, and/or the initial reactant-rate ($r_{0X}^{\lambda_{irr}}$ or $r_{0A}^{\lambda_{irr}}$, as a metric). Such a linear correlation is an advantage provided by a description of the photokinetic traces with equations of the type described in Chaps. 8, 11, 14 and 19. The correlation between the number of photons entering the reactor, $P_0^{\lambda_{irr}}$, and $r_{0Y_j}^{\lambda_{irr}}$ follows from the relationship between $P_0^{\lambda_{irr}}$ and the number of photons absorbed by the considered species, $P_{aY_j}^{\lambda_{irr}}(0)$.

The qualitative effect of increasing $P_0^{\lambda_{irr}}$ is an increase in the photoreaction reactivity. The quantitative assessment of such an effect is provided by proportional changes in the values of $k_{ij}^{\lambda_{irr}}$. Also, an increase of $P_0^{\lambda_{irr}}$ leads to a reduction of the overall time necessary for the reaction to reach completion, and hence a reduction of $t_{1/2}$. Also, a variation of $P_0^{\lambda_{irr}}$ causes a proportional change in the value of $r_0^{\lambda_{irr}}$.

The determination of the values of $P_0^{\lambda_{irr}}$ is essential for quantitative photochemistry and fundamental in converting a reactive system into an actinometer. The value of $P_0^{\lambda_{irr}}$ is obtained from a rearranged initial rate-equation (an example is given by Eq. (2.30)).

$$P_0^{\lambda_{irr}} = \frac{-r_{0X}^{\lambda_{irr},T}}{\left(\Phi_{X\to Y_1}^{\lambda_{irr}} + \Phi_{X\to Y_3}^{\lambda_{irr}}\right)\left(1 - 10^{-A_X^{\lambda_{irr}}(0)}\right)} \tag{2.30}$$

Hence, photokinetic data are used to perform actinometry, and this methodology is dubbed *kinactinometry*.

We will see detailed usage of reaction rate-constant and the initial reaction rate, together with examples of application to real samples in actinometry, in subsequent chapters.

2.22 Fitting the Kinetic Traces

Photokinetic investigation often requires data treatments and fitting of kinetic plots. The fitting of linear data series is performed by routine software such as Excel. Non-linear series are also fitted by software, but these are more elaborate.

The fitting of non-linear curves, such as kinetic traces, whose results are presented in this book, is obtained by the application of a Levenberg-Marquardt algorithm (LMA), provided by Matlab software. The goodness of fit is evaluated by three features: the correlation coefficient, r, the sums of squared errors, SSE, and the root-mean-square errors, RMSE.

In all cases, an excellent fit of the traces by the model equations must be evidenced by high (>0.995) correlation coefficient values, and relatively low values for both $SSE \leq 10^{-9}$ and $RMSE \leq 10^{-5}$.

2.23 An Experimental Setup for Monochromatic Irradiation

The experimental setup used in our group for monochromatic irradiation studies is detailed here as an example of the setups employed in the field.

The light source is an Ushio 1000 W Xenon arc-lamp held in a housing shell model *A*6000. The lamp is powered by an *LPS*-1200 power supply. Tap water circulating through a pipe system cools the lamp setting. The use of a high-power polychromatic lamp is necessary to ensure that the output monochromatic light has enough power to induce photoreactions. In general, the output monochromatic lights have a power ranging from a fraction of, to a few mW (that correspond, in such a setup, to typical $P_0^\lambda = 10^{-8}$-10^{-5} einstein dm^{-3} s^{-1}).

The wavelength of the monochromatic light is selected by a nub on the monochromator box. The latter, is a model 101, consisting of a special *f*/2.5 monochromator with a 1200 groove/mm at 300 nm blaze grating.

The monochromator is connected to the sample by an optical fibre, guiding the monochromatic light towards the sample. The light impinges on the sample from the top (i.e. the area irradiated is equal to $S_{irr} = 1\ cm^2$). The distance travelled by the light beam inside the sample (l_{irr}) is equal to the length of the liquid solution in the reactor (l_{irr} is hence determined by the actual volume of the solution). The reactor is a 1-cm cuvette, the kind of cells typically used for spectrophotometry analysis. In general, the experiment is conducted with $l_{irr} = 1 - 2$ cm and $V_{irr} = 1 - 2$ cm^3. The monitoring light of the spectrophotometer crosses the cuvette on its side (fixed length of $l_{\rm obs} = 1$ cm), so that analysis and excitation beams have perpendicular directions (Fig. 2.1), and their respective optical path lengths, $l_{\rm obs}$ and l_{irr}, are most often different. The photoreaction progress is advantaged when $l_{irr} > l_{\rm obs}$ as the rate of the reaction increases with higher values of l_{irr}.

It is important to mention here that the monitoring light might contribute to the overall progress of the photoreaction. Such an effect becomes significant for reaction endowed with fast kinetics. A careful consideration of these effects is recommended. The input of the monitoring light can be quantified by subjecting the sample to the monitoring light but without excitation light in the same conditions of analysis and measuring the resulting advancement of the reaction.

2.24 Polychromatic Irradiation Setup

Polychromatic light irradiation comes with many different features depending on the conditions employed for a given experiment. In general, the light power of the lamps used for polychromatic light (~a few to several *W*) is much higher than that usually delivered in monochromatic light experiments (Sect. 2.23). The original polychromatic light, incident from a light source, might be filtered either in terms of spectral range (e.g. using bandpass filters) or in terms of intensity (e.g. using light intensity attenuating filters).

In our team experimental setups, spectrally unfiltered 6 W UV-lamps are individually housed in a designed handle placed on the top of an irradiation cabinet (UPV *C*10-*E*6 mini UV viewing contrast control). A reactor with an $S_{irr} = 1$ cm^2, containing a vigorously stirred solution sample, is positioned inside the cabinet. The interior of the cabinet is completely shielded from environmental light, so that only the light provided by the *UV*-lamp reaches the photosensitive sample.

The reactor/lamp distance is *ca.* 15 cm. It was supposed that the light beam interacting with the sample is collimated and not undergoing scattering within the sample. The reactor sides were covered with aluminium foil (except for the top 1-cm^2 area), typically, 2-mL irradiated sample volumes ($l_{irr} = 2$ cm) were used.

In order to tune the intensity of the lamp light to suitable values, one or more copper grid-mesh tiles are placed on the top of the 1-*cm* reactor. A procedure required for actinometric measurements.

2.25 Blank Experiment

One essential piece of the photolysis investigation is the blank experiment. It is part of the chemical analyses that are conducted by analytical techniques. The blank experiment consists of a measurement of a solution that includes all the constituents of the sample but the analyte, i.e. the photoreactive reactant or system. In general, this corresponds to the solvent in which the reactant is dissolved. The signal or the spectrum resulting from the blank experiment must later be deduced from that of the analytical sample (e.g. constituted of the solvent and the photoactive reactant) to obtain the signal or the spectrum of the analyte only. This procedure is mandatory, usually employed to construct the calibration graph in analytical chemistry or in kinetic studies where the analytical sample is exposed to light. When significant, the effect of the monitoring light on the photoreaction progress must be included in the blank experiment.

For the samples containing one or several absorption competitors (spectators) in addition to the solvent and the reactive system, the blank experiment will be run on a solution that contains both the absorption competitors and the solvent. These techniques have been applied in our team and are part of the selected experimental results presented in this book.

2.26 Conclusion

Some of the important features of photokinetics have been reviewed in this chapter. The consideration of this type of features is essential for the coherence of the photokinetic investigation. Another essential part of kinetic analysis is the mathematical framework describing the physical system. This will be treated in the next chapter.

References

1. Maafi M (2008) Useful spectrokinetic methods for the investigation of photochromic and thermo-photochromic spiropyrans. Molecules 13:2260–2302. https://doi.org/10.3390/molecules13092260
2. Bozzetti M, de Candia G, Gallo M, Losito O, Mescia L, Prudenzano F (2010) Analysis and design of a solar rectenna, vol 2004. IEEE, Piscataway. https://doi.org/10.1109/ISIE.2010.5637486
3. Thorlabs (2024) Available at https://www.thorlabs.com/newgrouppage9.cfm?objectgroup_id=3836
4. Ryer A (2024) Light measurement handbook. Available at https://homepages.inf.ed.ac.uk/rbf/CVonline/LOCAL_COPIES/RYER/index.html
5. Horiba (2024) https://www.horiba.com/gbr/scientific/technologies/spectrometers-and-monochromators/spectrometers-monochromators-and-spectrographs/
6. McCluney R (2003) Radiometry and photometry. In: Meyers RA (ed) Encyclopedia of physical science and technology. Academic, New York, pp 731–758. https://doi.org/10.1016/B0-12-227410-5/00648-7
7. Harris DC (2003) Quantitative chemical analysis.6th edn. Freeman and Co., New York
8. Thorlabs (2024). Available at https://www.thorlabs.com/catalogpages/802.pdf
9. Sawada T, Kubo S, Nanamura K (2019) Chirality and circular polarized properties of photochromic polyaromatic molecules. In: Akitsu T (ed) Chirality from molecular electronic states. IntechOpen, London
10. Yusufu D, Mills A (2018) Spectrophotometric and digital colour colourimetric (DCC) analysis of colour-based indicators. Sensors Actuators B Chem 273:1187–1194. https://doi.org/10.1016/j.snb.2018.06.131
11. Parnis JM, Oldham KB (2013) Beyond the Beer–Lambert law: the dependence of absorbance on time in photochemistry. J Photochem Photobiol A Chem 267:6–10. https://doi.org/10.1016/j.jphotochem.2013.06.006
12. Zawadiak J, Mrzyczek M (2010) UV absorption and keto–enol tautomerism equilibrium of methoxy and dimethoxy 1,3-diphenylpropane-1,3-diones. Spectrochim Acta A 75:925–929. https://doi.org/10.1016/j.saa.2009.12.040
13. Edenborough Instruments (2024). Available at www.edinst.com/blog/the-beer-lambert-law/
14. Buijs K, Maurice MJ (1969) Some considerations on apparent deviations from Lambert-Beer's law. Anal Chim Acta 47:469–474. https://doi.org/10.1016/S0003-2670(01)95647-8
15. Young JC (2015) A Gaussian accommodation of Beer's law deviations. Chem Educ 20:206–213. https://doi.org/10.1333/s00897152648a
16. Wu W, Walczak B, Massart DL, Heuerding S, Erni F, Last IR, Prebble KA (1996) Artificial neural networks in classification of NIR spectral data: design of the training set. Chemom Intell Lab Syst 33:35–46. https://doi.org/10.1016/0169-7439(95)00077-1

17. Santana FB, Neto WB, Poppi RJ (2019) Random forest as one-class classifier and infrared spectroscopy for food adulteration detection. Food Chem 293:323–332. https://doi.org/10.1016/j.foodchem.2019.04.073
18. Mekonnen BK, Yang W, Hsieh TH, Liaw SK, Yang FL (2020) Accurate prediction of glucose concentration and identification of major contributing features from hardly distinguishable near-infrared spectroscopy. Biomed Signal Process Control 59:101923. https://doi.org/10.1016/j.bspc.2020.101923
19. Mamouei M, Budidha K, Baishya N, Qassem M, Kyriacou PA (2021) An empirical investigation of deviations from the Beer–Lambert law in optical estimation of lactate. Sci Rep 11:13734. https://doi.org/10.1038/s41598-021-92850-4
20. Mayerhofer TG, Hofer S, Popp J (2019) Deviations from Beer's law on the microscale – nonadditivity of absorption cross sections. Phys Chem Chem Phys 21:9793–9801. https://doi.org/10.1039/C9CP01987A
21. Mayerhofer TG, Pahlow S, Popp J (2020) The Bouguer-Beer-Lambert law: shining light on the obscure. ChemPhysChem 21:9793–9801. https://doi.org/10.1002/cphc.202000464
22. Tonnesen HH (ed) (2004) Photostability of drugs and drug formulations. CRC Press, Boca Raton
23. Menzel JP, Noble BB, Blinco JP, Barner-Kowollik C (2021) Predicting wavelength-dependent photochemical reactivity and selectivity. Nat Commun 12:1691. https://doi.org/10.1038/s41467-021-21797-x
24. Maafi M (2023) On photokinetics under monochromatic light. Front Chem 11:1233151. https://doi.org/10.3389/fchem.2023.1233151
25. Braslavsky SE (2007) Glossary of terms used in photochemistry. Pure Appl Chem 79:293–465. https://doi.org/10.1351/pac200779030293

Chapter 3
A Mathematical Tool-Kit

3.1 Introduction

As kinetics is part of physical chemistry, it naturally involves mathematical treatments. The kinetic traces are modelled by equations and formulae that link the chemical phenomenon taking place in the reaction to observed outputs. A great deal of kinetics is based on analytical chemistry analyses, which equally requires the use of mathematical tools. A few examples of this principle have been discussed in Chap. 2. The purpose of the present chapter is to give a succinct overview of some methods, procedures and tools that might be considered as the minimum mathematical background necessary for a consistent kinetic investigation.

3.2 Ordinary Differential Equations and Kinetics

The kinetic-DNA of a reaction is confined within the differential equations describing that reaction. Each equation is called a rate-law or a rate-equation. A rate-law expresses the variation of a reaction quantitative parameter with time (e.g. dC/dt) as a function of that parameter, its derivatives, and or time (e.g. Eq. (3.1)). Hence, it becomes obvious that kinetics cannot be properly and fully studied without involving differential equations. In this context, it is perhaps useful to overview the main types of ordinary differential equations that find use in kinetics, in general, and photokinetics in particular.

An ordinary differential equation (ODE) is a differential equation that involves one or more functions of an independent variable and may also contain the derivatives of those functions [1]. The term ordinary is employed here, in contrast with another class of differential equations, called the partial differential equations, that involve more than one independent variable and partial derivatives. We will restrict the present description to the ordinary differential equations. The independent

M. Maafi, *Photokinetics*, https://doi.org/10.1007/978-3-031-98985-8_3

variable in kinetics is usually t, "the reaction time", and the dependent variable is C, "the concentration of a species". The dependent variable is differentiated relative to the independent variable, also known as the derivative, that can be represented by various symbols (e.g. dC/dt). It is also possible to proceed to successive derivations of the independent variable, as second, third, and up to the nth derivative, which is reflected in the notation d^2C/dt^2, d^3C/dt^3,..., d^nC/dt^n, respectively. The symbols used to represent the derivatives have historically varied between authors and or the task at hand. The most common ones are named after their respective mathematician authors, including Leibniz's notation, dC/dt, d^2C/dt^2,..., d^nC/dt^n, Lagrange's notation, C', C'', C''', ...,$C^{n'}$, and Newton's notation, $\dot{y}, \ddot{y}, \dddot{y}, \ldots$ The usual notation in kinetics is Leibniz's.

In general, the ODE including a function F of t, C and derivatives of C, is written as

$$d^nC/dt^n = F\left(t, C, dC/dt, d^2C/dt^2, \ldots, d^{n-1}C/dt^{n-1}\right) \tag{3.1}$$

It is called an ordinary explicit differential equation of order n. If the left-hand side term of this equation is equal to 0, it is then called an implicit differential equation, taking the form

$$F\left(t, dC/dt, d^2C/dt^2, \ldots, d^{n-1}C/dt^{n-1}, d^nC/dt^n\right) = 0 \tag{3.2}$$

In photokinetics, explicit differential equation are mostly used.

A differential equation is autonomous if it does not explicitly involve the independent variable (e.g. t) in the function F. This kind of equation is mostly encountered in both thermal and photokinetics.

A linear ordinary differential equation is written as a linear combination of the derivatives of the dependent variable (e.g. C).

$$d^nC/dt^n = \sum_{i=0}^{n-1} \theta_i(t)\, d^iC/dt^i + \rho(t) \tag{3.3}$$

with $\theta_i(t)$ are coefficients of the linear combination, or continuous functions of the independent variable, t. $\rho(t)$ is also a continuous function of t, called the source term. Depending on its value, further types of *ODE* can be distinguished.

The homogeneous ordinary differential equations have no source term ($\rho(t) = 0$). The solution of a linear homogeneous equation is called a complementary function ($\overline{A}$). Conversely, the non-homogeneous equation has a non-zero source term ($\rho(t) \neq 0$). Its solution is called the particular solution (A^*). So, the complete or general solution of a linear differential equation is, $C = \overline{A} + A^*$. Finally, when the differential equation cannot be written as a linear combination of terms, as given by Eq. (3.3), the differential equation is a non-linear differential equation.

In general, thermal reaction kinetics are described by linear ordinary differential equations, whereas photokinetics mostly involves non-linear ordinary differential equations (except for isosbestic irradiation, see Chap. 8, where they are linear ODEs).

Defining the correct rate-law of the reaction plays a pivotal role in kinetics since the solutions of the differential equations are themselves algebraic equations expressing the dependent variable as a function of the independent variable (e.g. $C(t)$), offering a precise temporal evolution of each species during the reaction time. The solution equations $C(t)$, called the integrated rate-laws, are essential in determining the fate and order of the reaction. The former allows to estimate the amount of reactant (or product) consumed (or produced) at a given time after the start of the reaction. The latter, the order of the reaction, is a property of the reaction that consists of specific mathematical formulae that distinguish between different kinetic behaviours of reactants and reactions (Sect. 1.7).

3.3 System of Differential Equations

Irrespective of the studied reactive system, an individual rate must be written for each of the n_{sp} + 1 species involved in the mechanism of the considered photothermal reaction, $XY_{n_{sp}}(n_{\Phi}\Phi, n_{\Delta}k)$. Therefore, every reaction is described by a system of differential equations. For instance, the simplest reaction $XY(1\Phi)$, $X \rightarrow Y$, is fully described by two differential equations, i.e. one relative to the concentration of X ($r_X^{\lambda_{irr}}(t) = dC_X^{\lambda_{irr}}(t)/dt$), and the other for the concentration of Y ($r_Y^{\lambda_{irr}}(t) = dC_Y^{\lambda_{irr}}(t)/dt$, which is, for the simplest reaction, equal to $r_Y^{\lambda_{irr}}(t) = -r_X^{\lambda_{irr}}(t)$).

It is also sometimes possible to combine the n_{sp} + 1 species' individual differential equations into a single one. This is the case of the rate of the total absorbance considered at a given irradiation (λ_{irr}) and observation (λ_{obs}) wavelengths. This differential equation is a linear combination of the species' concentration differential equations, as

$$r_A^{\lambda_{irr}/\lambda_{obs},T}(t) = \frac{dA_{tot}^{\lambda_{irr}/\lambda_{obs},T}(t)}{dt} = l_{obs} \sum_{j=0}^{n_{sp}} \varepsilon_{Y_j}^{\lambda_{obs}} r_{Y_j}^{\lambda_{irr}}(t) \tag{3.4}$$

Solving a system of differential equations is a mathematical procedure leading to extracting the different algebraic equations or the solutions, i.e. the integrated rate-laws, expressing the concentration of each of the n_{sp} + 1 species as an explicit function of time ($C_{Y_j}^{\lambda_{irr}}(t) = f(t)$). When such a solving procedure is conducted free from any assumptions, the solution is dubbed an *analytical solution*, or the solution is obtained by a *closed-form integration*. Standard methods and procedures are available in textbooks for solving a number of systems of differential equations. Examples of straightforward procedures are described in Sect. 3.5. Some reaction kinetics need more complicated methods, others require the development of original approaches, and many are simply impossible to solve. As a general rule, analytical solutions of differential equations and closed-form integrations are possible if and only if the integrands have known antiderivatives (see examples in the next section).

Photokinetics is described by systems of non-linear differential equations for which there are no known standard mathematical solving procedures and usually leads to integrands that do not have known antiderivatives. A situation that incites the use of alternative methods. Two of those are discussed below (Sects. 3.6 and 3.7).

3.4 Integrals and Antiderivatives

Irrespective of the procedure selected for the purpose of solving the reaction rate-laws, represented by a system of differential equations, there will be a stage where an integration needs to be performed.

Conceptually, an integral corresponds to the area under the curve drawn by a function on a graph, where the area is delimited on the y-axis by that curve and by two limit values, x_1 and x_2, on the x-axis, which is the integration variable x (or dx under the integral). The integral is called, in this case, a *definite integral*. The function ($f(x)$) to be integrated is called the *integrand*. The result of integration is a function ($g(x)$), that, if differentiated, gives back $f(x)$, i.e. the derivative of the former is equal to the latter, as $g'(x) = f(x)$. The function $g(x)$ is called the *primitive* or the *antiderivative* of $f(x)$. The final numerical value of the integral is obtained by subtraction of the values of $g(x)$ calculated for the two limit values of x (Eq. (3.5)).

$$\int_{x_1}^{x_2} \underbrace{f(x)}_{\text{integrand}}\ dx = \underbrace{g(x)}_{\text{antiderivative}}\Bigg|_{x_1}^{x_2} = g(x_2) - g(x_1) \tag{3.5}$$

When the integration limits on the x-axis are not defined (by numerical values), the integral is dubbed an *indefinite integral* (Eq. (3.6)). The *constant* of integration is unknown (Eq. (3.6)), but can, eventually, be defined when the two limit values of x are introduced in the equation, as was shown in Eq. (3.5).

$$\int f(x)\, dx = g(x) + \text{constant} \tag{3.6}$$

According to what was stated above, the integral of a derivative is the differentiated function.

Table 3.1 Examples of integrand functions and their antiderivatives

#	$f(x)$[a] integrand	$g(x)$ antiderivative
1	k	$k\,x$
2	x^n	$\frac{x^{n+1}}{n+1}$
3	e^x	e^x
4	$\ln(x)$	$x\ \ln(x) - x$
5	$\frac{f'(x)}{f(x)}$	$\ln \lvert f(x) \rvert$
6	$\frac{1}{x}$	$\ln(x)$
7	$\log_a(x)$	$\frac{\ln(x)}{\ln(a)}$
8	$\frac{1}{x^2+k^2}$	$\frac{1}{k}\ \arctan\frac{x}{k}$
9	$\frac{1}{k^2-x^2}$	$\frac{1}{2\,k}\ \mathrm{Ln}\left\lvert\frac{k+x}{k-x}\right\rvert$
10	$\frac{2\,a\,x+b}{a\,x^2+b\,x+c}$	$\ln\lvert a\,x^2 + b\,x + c\rvert$
11	$\frac{1}{a+b\,e^{c\,x}}$	$\frac{x}{a} - \frac{x}{a\,c}\ \ln(a + b\,e^{c\,x})$

[a] a, b, c, and k are numerical constants

$$\int f'(x)\,dx = f(x) + \text{constant} \tag{3.7}$$

Also, the integral of a sum is the sum of integrals, as

$$\int (f(x) + g(x))\,dx = \int f(x)\,dx + \int g(x)\,dx \tag{3.8}$$

Many books and websites provide tables of antiderivatives, which are relatively easy to access (examples are given in Table 3.1).

However, despite the fact that a large number of antiderivatives are known, not all integrands have each a (known) antiderivative.

One of the handy methods of integration consists of separating the variables of the differential equation by rearranging the equation in such a way that the right-hand side of the new equation involves one variable only (e.g. y, dy), while the left-hand side of the equation is expressed by the other variable only (e.g. x, dx). When this is possible, the two terms of the equation are then integrated separately. For example, the following differential equation

$$dy/dx = h(x)\,y \tag{3.9}$$

is rearranged into

$$dy/y = h(x)\,dx \tag{3.10}$$

and integrated as

$$\int dy/y = \int h(x)\,dx \tag{3.11}$$

Many complicated integrands can be rearranged into one of the common forms, such as those given in Table 3.1. Sometimes there is a need to employ additional mathematical techniques to transform the integrand into a useful form.

The procedure of completing the squares can be used in the case of quadratic equations, as follows:

$$\begin{aligned} a\,x^2 + b\,x + c &= a\left(x + \frac{b}{2a}\right)^2 + \left(c - \frac{b^2}{4a}\right) = a\left[\left(x + \frac{b}{2a}\right)^2 + \left(\frac{c}{a} - \frac{b^2}{4\,a^2}\right)\right] \\ &= a\left[\left(x + \frac{b}{2a}\right)^2 \pm k^2\right] \end{aligned} \tag{3.12}$$

The transformation undertaken in Eq. (3.12) is very useful for fractional integrands that consist of a constant (m) dividing a quadratic equation, as

$$\int \frac{m\,dx}{a\,x^2 + b\,x + c} = \frac{m}{a} \int \frac{dx}{\left(x + \frac{b}{2a}\right)^2 \pm k^2} \tag{3.13}$$

Proceeding with the integral of Eq. (3.13), a change of variable is helpful, if

$$Z = x + \frac{b}{2a} \qquad \text{with} \qquad dZ = dx \tag{3.14}$$

It allows the integral in Eq. (3.13) to be written in one of the canonical forms (#8 or #9) given in Table 3.1.

$$\int \frac{m\,dx}{a\,x^2 + b\,x + c} = \frac{m}{a} \int \frac{dZ}{Z^2 \pm k^2} \tag{3.15}$$

The change of variable can turn out to be a very efficient means for the integration of complicated functions. For example, the antiderivative of case #10 in Table 3.1 is also readily found if the following change of variables is undertaken.

$$Z(x) = a\,x^2 + b\,x + c \tag{3.15}$$

with

$$dZ(x) = (2a\,x + b)\,dx \tag{3.17}$$

which transforms the integrand #10 (Table 3.1) into the same format as #5, hence, the antiderivative is straightforward (Eq. (3.18)).

$$\int \frac{2\,a\,x+b}{a\,x^2+b\,x+c}dx = \int \frac{dZ}{Z} = \ln\left|Z^2\right| = \ln\left|\left(a\,x^2+b\,x+c\right)^2\right| + \text{constant} \quad (3.18)$$

Partial fractions can facilitate the integration by breaking down a ratio into simple fractions (when the degree of the function in the numerator is lower than that of the function in the denominator). Let us consider an integral whose integrand has no known antiderivative. The integral of the function on the left-hand side of Eq. (3.19) cannot be directly resolved, except if, for instance, this improper fraction is decomposed into partial fractions. The factors a to j are known constants. The coefficients A_1 to A_4 are unknown constants that have to be identified by solving Eq. (3.19).

$$\frac{a\,x^2+b}{(c\,x+d)\,(e\,x+f)\,(g\,x^2+h\,x+j)} = \frac{A_1}{c\,x+d} + \frac{A_2}{e\,x+f} + \frac{A_3\,x+A_4}{g\,x^2+h\,x+j} \quad (3.19)$$

Furthermore, the last fraction of the left-hand side of Eq. (3.19) can be written as

$$\frac{A_3\,x+A_4}{g\,x^2+h\,x+j} = \frac{A_3\,x+h}{g\,x^2+h\,x+j} + \frac{A_4-h}{g\,x^2+h\,x+j} \quad (3.20)$$

And supposing that the quadratic equation (divided by g) has real solutions (α_1 and α_2), we can implement the partial fraction method again and define the values of A_5 and A_6 in Eq. (3.21).

$$\frac{A_4-h}{g\,x^2+h\,x+j} = \frac{1}{g}\frac{A_4-h}{x^2+\frac{h}{g}\,x+\frac{j}{g}} = \frac{A_4-h}{g}\frac{A_5}{x+\alpha_1} + \frac{A_4-h}{g}\frac{A_6}{x+\alpha_2} \quad (3.21)$$

The integral of the fraction of Eq. (3.19) is now much accessible, taking the form of a sum of simple integrals.

$$\int \frac{a\,x^2+b}{(c\,x+d)\,(e\,x+f)\,(g\,x^2+h\,x+j)}dx = A_1\int\frac{dx}{c\,x+d} + A_2\int\frac{dx}{e\,x+f} +$$
$$\frac{A_3}{2\,g}\int\frac{2\,g\,x+h}{g\,x^2+h\,x+j}dx + A_5\,\frac{A_4-h}{g}\int\frac{dx}{x+\alpha_1} + A_6\,\frac{A_4-h}{g}\int\frac{dx}{x+\alpha_2} + \text{constant} \quad (3.22)$$

Integration by parts is another efficient procedure to successfully reach a form of integrands that have known antiderivatives. This technique is applied when the integrand can be written as a product of two functions, with one of them being a derivative of a known function, as $f(x) = g(x)\, h'(x)$. Subsequently, using the differentiation rule of a product, we have

$$(g(x)\, h(x))' = g(x)\, h'(x) + g'(x)\, h(x) = f(x) + g'(x)\, h(x) \tag{3.23}$$

we then work out the integral of $f(x)$ as

$$\int f(x)\, dx = \int (g(x)\, h(x))'\, dx - \int g'(x)\, h(x)\, dx \tag{3.24}$$

or

$$\int f(x)\, dx = -\int g'(x)\ h(x)\, dx + g(x)\, h(x) \tag{3.25}$$

If the integrand of the integral on the right-hand side has a known antiderivative, then the integral of $f(x)$ will be solved.

3.5 Notorious Kinetic Orders of Thermal Reactions

Let us consider the following reaction example.

$$X + X' \rightarrow Y_1 + Y_2 + \cdots + Y_j$$

The rate-law of the reactant X, is

$$\text{rate of } (X) = r_X(t) = \frac{dC_X(t)}{dt} = k\ C_X^{\alpha}(t)\ C_{X'}^{\beta}(t) \tag{3.26}$$

The values of the exponent factors α, β, …, etc. define the individual orders of the respective reactants. Whereas, the sum of the individual reaction orders $(\alpha + \beta + \cdots)$ defines the global order of the reaction. All these numbers can be integers or positive real numbers but not negative. In general, the probability of three species reacting together to give products is very low but for complex or multiple reaction paths, one given species might well be involved in more than one reaction.

The rate-laws of the three most famous reaction kinetic behaviours (kinetic orders) encountered in thermal kinetics, for simple uni- or bimolecular reactions (of the types $X \rightarrow Y$ and $2\,X \rightarrow Y$, respectively), are mapped out by the following general rate-law (with $Y_0 = X$ and $Y_1 = Y$).

$$r_{Y_j}^T(t) = \pm \frac{dC_{Y_j}^T(t)}{dt} = k_{Y_j}^T \left(C_X^T(t)\right)^\alpha \tag{3.27}$$

where the reaction rate ($r_{Y_j}^T$, expressed in mol L^{-1} s^{-1}), concentration ($C_X^T(t)$, in mol L^{-1}) and reaction rate-constant ($k_{Y_j}^T$) are specific to both the constant medium temperature (T) at which the reaction is performed and to species, Y_j, targeted in the reaction mechanism. The signs plus and minus indicate, respectively, formation or depletion of that species during the reaction. The exponent α, is the order of the reaction of Y_j. It may take three integer values: 0, 1 or 2, which define the zeroth-, first- and second reaction kinetic order, respectively.

For a 0th-*order reaction*, $\alpha = 0$, and the rate of the reaction is constant throughout the reaction time, as in the right-hand side term of Eq. (3.27), where $\left(C_X^T(t)\right)^\alpha = \left(C_X^T(t)\right)^0 = 1$. The rate-law, hence, takes the simpler form given by Eq. (3.28). The reaction rate-constant has the same dimension as the reaction rate, according to the dimension analysis: $\left[r_{Y_j}^T\right] = \left[dC_{Y_j}^T(t)/dt\right] = \text{mol}\, L^{-1}\, s^{-1} = \left[k_{Y_j}^T\right] = M\, s^{-1}$ (for time given in s).

$$r_{Y_j}^T(t) = \pm \frac{dC_{Y_j}^T(t)}{dt} = k_{Y_j}^T \tag{3.28}$$

By rearranging the above equation, the variables of the rate-law can be separated as

$$dC_{Y_j}^T(t) = \pm k_{Y_j}^T\, dt \tag{3.29}$$

The integration is achieved on both terms of the equation

$$\int_{C_{Y_j}^T(0)}^{C_{Y_j}^T(t)} dC_{Y_j}^T(t) = \pm k_{Y_j}^T \int_0^t dt \tag{3.30}$$

between the limits $t = 0$ and t. The solution of the differential equation is the typical linear equation characteristic of the 0th-order integrated rate-law. A minus sign is considered when $Y_j = Y_0 = X$, and a plus sign for $Y_j = Y_1 = Y$ as depicted in the mechanisms of Scheme 3.1 without X).

$$X + X' \rightarrow Y_1 + Y_2 + \cdots + Y_j$$

Scheme 3.1 A general mechanism for a thermal bimolecular reaction

$$C_{Y_j}^T(t) = C_{Y_j}^T(0) \pm k_{Y_j}^T\, t \tag{3.31}$$

In the rate-law of 1st-*order reactions*, a value of 1 is assigned for α. A dimension analysis of the equation reveals that, $\left[r_{Y_j}^T\right] = \left[k_{Y_j}^T \times C_{Y_j}^T(t)\right] = \text{mol}\, L^{-1}\, s^{-1}$, hence, the reaction rate-constant must be expressed in $\left[k_{Y_j}^T\right] = s^{-1}$.

$$r_{Y_j}^T(t) = \pm \frac{dC_{Y_j}^T(t)}{dt} = k_{Y_j}^T\, C_{Y_j}^T(t) \tag{3.32}$$

The integration proceeds by variable separation, as

$$\int_{C_{Y_j}^T(0)}^{C_{Y_j}^T(t)} \frac{dC_{Y_j}^T(t)}{C_{Y_j}^T(t)} = \pm k_{Y_j}^T \int_0^t dt \tag{3.33}$$

whose solution is the characteristic mono-exponential integrated rate-law.

$$C_{Y_j}^T(t) = C_{Y_j}^T(\infty) + \left(C_{Y_j}^T(0) - C_{Y_j}^T(\infty)\right)\, e^{-k_{Y_j}^T\, t} \tag{3.34}$$

The case of the 2nd-*order kinetics* ($\alpha = 2$) which is considered here as a bimolecular reaction involving two molecules of the same reactant ($X + X \rightarrow Y$). The individual or partial orders of the species are considered to be equal to unity, and the total order is equal to 2. The reaction rate-constant in Eq. (3.35) has a dimension of $\left[k_{Y_j}^T\right] = M^{-1}\, s^{-1}$, since $\left[r_{Y_j}^T\right] = M\, s^{-1} = \left[k_{Y_j}^T \times \left(C_A^T(t)\right)^2\right] = \left[k_{Y_j}^T\right] \times M^2$.

$$r_{Y_j}^T(t) = \pm \frac{dC_{Y_j}^T(t)}{dt} = k_{Y_j}^T\, \left(C_X^T(t)\right)^1\, \left(C_X^T(t)\right)^1 = k_{Y_j}^T\, \left(C_X^T(t)\right)^2 \tag{3.35}$$

Here as well, the integration is performed by the variable separation method and leads to the typical reciprocal integrated rate-law of 2nd-order reactions.

$$C_{Y_j}^T(t) = \frac{C_{Y_j}^T(0)}{1 + C_{Y_j}^T(0)\, k_{Y_j}^T\, t} \tag{3.36}$$

The integrated rate-laws (Eqs. (3.31), (3.34) and (3.36)) are exclusive, i.e. obeyed by the respective reaction kinetic type they describe (either 0th-, 1st- or 2nd-order reactions).

The exclusivity of the three above equations comes mainly from the impossibility of deriving any one of them from any combination of the other two. They are mathematically distinct or independent from each other. In fact, this was expected from the evident differences between the differential equations (Eqs. (3.28), (3.32) and (3.35)) they were derived from. It is a general rule that *mathematically different solutions (integrated rate-laws) should necessarily be expected for different differential equations (or rate-laws).*

Extending this mathematical principle to photokinetic differential equations (*vide infra* Eq. (3.45)), one can conclude that because of the significant differences between photo- and thermal rate-laws, it is possible to predict with certainty that the integrated rate-laws characterising these reactions must necessarily have different algebraic forms (see Chap. 11).

As will be mentioned below, the occurrence of the absorbed light in the differential equations of photoreactions causes *de facto* their non-linearity, whereas most of the thermal reactions' differential equations are linear. The unexpected consequence of this non-linearity, as detailed in Chap. 12, is an impossibility to solve such rate-laws analytically (with an exception of the simplest reaction). Whereas the linear equations describing thermal reactions are generally solvable (even if the methods used for that purpose might sometimes be tedious to carry out).

The non-solvability of photokinetics' differential equations meant a lack of integrated rate-laws for photokinetics, the very equations that define reaction orders. As a matter of fact, photoreactions have not been assigned specific kinetic orders in the related photochemistry literature. This contrasts with the commonly known and ubiquitously used kinetic orders characterising thermal reactions.

3.6 Editing Differential Equations for Photoreactions

The standard differential equation (DE) describing the photoreaction of species (X), involved in a single reaction-step, $X \rightarrow Y$ at a medium temperature T, subjected to a monochromatic light at λ_{irr}, and whose photoproduct (Y) is transparent ($\varepsilon_Y^{\lambda_{irr}} = 0$) to the incident radiation, is simply written as a rate ($r_X^{\lambda_{irr},T}(t)$) equal to the product of the photochemical quantum yield of the photoreaction-step, $\Phi_{X \rightarrow Y}^{\lambda_{irr}}$ (which is a constant), and the fraction of the incident radiation intensity that is absorbed by X, $P_{a,X}^{\lambda_{irr}}(t)$ (which is variable with time). Notice that in Eq. (3.37), the concentration is labelled by the irradiation wavelength in addition to the medium temperature. Such a consideration is imposed by the fact that the quantities in both terms of the equation are wavelength dependent, and hence, the variation of the concentration must also be dependent on the irradiation wavelength.

$$r_X^{\lambda_{irr},T}(t) = \frac{dC_X^{\lambda_{irr},T}(t)}{dt} = -\Phi_{X \rightarrow Y}^{\lambda_{irr}} P_{a,X}^{\lambda_{irr}}(t) \tag{3.37}$$

The quantum yield of such a photoreaction ($\Phi_{X \rightarrow Y}^{\lambda_{irr}}$) is an intrinsic property, specific to the reactive species, X. It is independent of the concentration of the species, the temperature of the reaction medium, the intensity of the provided monochromatic light used for irradiation and/or shape or volume of the reactor. It expresses the efficiency of the reaction. According to the IUPAC definition given earlier (Sect. 1.9), $\Phi_{X\to Y}^{\lambda_{irr}}$ is a constant at λ_{irr}, that is obtained from the ratio of the number of molecules of X that have photoreacted, divided by the number of photons absorbed (Eq. (1.4)). An insight into the limits of application of that "general" equation (Eq. (1.4)) for the calculation of the quantum yield is available in Chap. 5.

The fraction of the incident radiation provided by the light source, $P_0^{\lambda_{irr}}$, that is effectively absorbed by the reactant ($P_{a,X}^{\lambda_{irr}}$), is directly worked out from the Beer-Lambert law (on the linear part of the calibration graph of X) as previously detailed in Sect. 2.17 (Eq. (2.20)).

Therefore, the rate-law of the reaction becomes,

$$r_X^{\lambda_{irr},T}(t) = \frac{dC_X^{\lambda_{irr},T}(t)}{dt} = -\Phi_{X\to Y}^{\lambda_{irr}} P_0^{\lambda_{irr}} \left(1 - 10^{-\varepsilon_X^{\lambda_{irr}} l_{irr} C_X^{\lambda_{irr},T}(t)}\right) \tag{3.38}$$

with l_{irr} being the optical path length of the irradiation light inside the sample (Fig. 2.1).

In its formulation, Eq. (3.38) indicates that there is a single light-absorbing species in the medium since the absorbance corresponds to only one species $\left(A_X^{\lambda_{irr},T}(\mathrm{t}) = \varepsilon_X^{\lambda_{irr}} l_{irr} C_X^{\lambda_{irr},T}(t)\right)$.

In a more general approach, let us first consider that there is a series of several absorbing compounds in the medium (since the reaction may, for instance, involve at a given time, a reactant and a series of absorbing photoproducts all present in the medium). The latter are characterised by individual absorbances, $A_{Y_0}^{\lambda_{irr}}$, $A_{Y_1}^{\lambda_{irr}}$, $A_{Y_2}^{\lambda_{irr}}$, …, $A_{Y_j}^{\lambda_{irr}}$,…, $A_{Y_n}^{\lambda_{irr}}$. In this case, the total absorbance $\left(A_{tot}^{\lambda_{irr}}\right)$ corresponds to the sum of the individual contributions of all the species present in the reaction medium (as given by Eq. (2.16)).

Let us next consider the total amount of light that is effectively absorbed by the reactive medium, $P_{a,tot}^{\lambda_{irr}}(t)$. It must correspond to the cumulative absorbance of the species.

$$P_{a,tot}^{\lambda_{irr}}(t) = P_0^{\lambda_{irr}} \left(1 - 10^{-A_{tot}^{\lambda_{irr},T}(t)}\right) \tag{3.39}$$

It is interesting to notice here two important aspects of Eq. (3.39). First, $P_0^{\lambda_{irr}}$ is the constant "total" incident radiation intensity of the monochromatic light entering the reactor (Sect. 2.18), and second, that the value of $P_{a,tot}^{\lambda_{irr}}$ is time-dependent since $A_{tot}^{\lambda_{irr}}$ is variable during the reaction progress. $P_{a,tot}^{\lambda_{irr}}$ value approaches zero only when $A_{tot}^{\lambda_{irr}}$ is very small (i.e. no or very little light absorption when the medium is cleared of all absorbing species, and $10^{-A_{tot}^{\lambda_{irr},T}(t)} \cong 1$, so that, $P_{a,tot}^{\lambda_{irr}} \cong 0$). However, its value tends to that of $P_0^{\lambda_{irr}}$ if $A_{tot}^{\lambda_{irr}}$ is exceedingly high (where $10^{-A_{tot}^{\lambda_{irr},T}(t)} \cong 0$ with $A_{tot}^{\lambda_{irr},T}(t) \gg 1$). But for the latter case, the Beer-Lambert law will most likely cease to apply (see Sect. 2.13).

This might simply mean that the total absorbed fraction of the incident radiation is higher at the beginning than at the end of the reaction but this is true only in the case where $A_{tot}^{\lambda_{irr}}$ is supposed to decrease during the progress of the reaction until total depletion of the species (i.e. the absorptivities of the photoproducts are lower than the reactant's). However, there are other cases where this is not true. The latter statement would not stand, for instance, if all or some of the end-products of the reaction absorb the available light, or the reaction reaches a steady photostationary state (so, in both these cases, $A_{tot}^{\lambda_{irr}}$ is never equal to zero). $P_{a,tot}^{\lambda_{irr}}$ has a higher value at the end of the reaction than at its start, if the value of $A_{tot}^{\lambda_{irr}}$ increases with time.

Therefore, it is not possible to predict a general/standard evolution trend for the total absorbed light of a medium ($P_{a,tot}^{\lambda_{irr}}$). It might adopt variable trends during reaction time, including adopting increasing or decreasing exponential-like patterns depending on the relative absorptivities and concentrations of the initial molecule and photoproducts at various reaction times.

In any case and from a photokinetic point of view, the total absorbed light by the medium is of little interest, since for a given reaction, it is necessary to write a rate-law for each individual species of the reaction. Therefore, it is necessary to specify exactly the amount of light absorbed by the particular photoreactive compound of the considered reaction-9step that has a specific photochemical quantum yield (otherwise, a general Φ would not have a physical meaning).

The specification of a particular species absorption is established by the absorption fraction (or absorbance fraction, f_{Y_j}, Eq. (3.40)) that corresponds to the absorbance of the species of interest, Y_j, relative to the total absorbance of the medium (i.e. only a fraction of $P_{a,tot}^{\lambda_{irr}}$ corresponds to the absorption by Y_j). Obviously, $f_{Y_j} = 1$ if in the reactive medium, only the species Y_j absorbs the incident light, but in general, the sum of all the absorption fractions of the species present in the reaction medium should be equal to unity ($\sum f_{Y_j} = 1$).

$$f_{Y_j}(t) = \frac{A_{Y_j}^{\lambda_{irr},T}(t)}{A_{tot}^{\lambda_{irr},T}(t)} \tag{3.40}$$

Accordingly, when more than one absorbing species is involved ($A_{Y_0}^{\lambda_{irr}}$, $A_{Y_1}^{\lambda_{irr}}$, $A_{Y_2}^{\lambda_{irr}}, A_{Y_j}^{\lambda_{irr}}, \ldots A_{Y_n}^{\lambda_{irr}}$) in the mechanism, the absorbed light by any species Y_j, $P_{a,Y_j}^{\lambda_{irr}}$, must only account for the fraction of the light absorbed by that species, as

$$P_{a,Y_j}^{\lambda_{irr}} = P_0^{\lambda_{irr}} \left(1 - 10^{-A_{tot}^{\lambda_{irr},T}(t)}\right) \frac{A_{Y_j}^{\lambda_{irr},T}(t)}{A_{tot}^{\lambda_{irr},T}(t)} \tag{3.41}$$

and the rate-law for species Y_j involved in the photoreaction-step $Y_j \rightarrow Y_{j'}$ (where $j \neq j'$), is given by

$$r_{Y_j}^{\lambda_{irr},T}(t) = \frac{dC_{Y_j}^{\lambda_{irr},T}(t)}{dt} = -\Phi_{Y_j \to Y_{j'}}^{\lambda_{irr}} P_0^{\lambda_{irr}} \left(1 - 10^{-A_{tot}^{\lambda_{irr},T}(t)}\right) \frac{A_{Y_j}^{\lambda_{irr},T}(t)}{A_{tot}^{\lambda_{irr},T}(t)} \tag{3.42}$$

Equation (3.42) can be rearranged into

$$r_{Y_j}^{\lambda_{irr},T}(t) = -\Phi_{Y_j \to Y_{j'}}^{\lambda_{irr}} P_0^{\lambda_{irr}} A_{Y_j}^{\lambda_{irr},T}(t) \frac{1 - 10^{-A_{tot}^{\lambda_{irr},T}(t)}}{A_{tot}^{\lambda_{irr},T}(t)} \tag{3.43}$$

In this formulation, we notice that the fraction in the last term of Eq. (3.43) will be the same irrespective of the species considered (i.e. the same for all species' rate-laws of a given reaction). It is usual, in the photokinetic community, that this fraction is coined the *photokinetic factor*. It is labelled here by *PKF*. Hence, Eq. (3.43) can be expressed as

$$\begin{aligned} r_{Y_j}^{\lambda_{irr},T}(t) &= -\Phi_{Y_j \to Y_{j'}}^{\lambda_{irr}} P_0^{\lambda_{irr}} A_{Y_j}^{\lambda_{irr},T}(t) \frac{1 - 10^{-A_{tot}^{\lambda_{irr},T}(t)}}{A_{tot}^{\lambda_{irr},T}(t)} \\ &= -\Phi_{Y_j \to Y_{j'}}^{\lambda_{irr}} P_0^{\lambda_{irr}} A_{Y_j}^{\lambda_{irr},T}(t)\, PKF^{\lambda_{irr},T}(t) \end{aligned} \tag{3.44}$$

The photokinetic factor is dimensionless and time-dependent (evolves during the reaction progress, since $A_{tot}^{\lambda_{irr},T}(t)$ varies with reaction time). It is responsible for the non-linearity characterising these differential equations. Because $PKF^{\lambda_{irr},T}(t)$ is common to all the rate-laws of the photoreactive species belonging to the considered reaction; it plays the role of a coupling factor between these differential equations. Incidentally, the presence of the photokinetic factor in the system of the rate-laws describing a given photoreaction constitutes an additional difference between thermally and light-driven reactions. The photokinetic factor is also relevant because it accounts for all absorbing species present in the reaction medium. We shall see that this includes both reactive and non-reactive species (see Sect. 2.16 and Chap. 16). A feature that confers to *PKF*, an additional important property of pondering the species reactivities.

Overall, a careful consideration of Eq. (3.44) indicates that it describes only the photoreaction's species that deplete during the reaction time, as clearly indicated by the presence of the minus sign. We can now write the general rate-law that describes any species Y_j of the considered reaction mechanism, irrespective of the number of reaction-steps the species is involved in. This encompasses depleting and generating reaction-steps that, respectively, start and end at Y_j (the terms corresponding to the former carry a minus sign, whereas those of the latter, a plus sign). In Eq. (3.45), the indices j must always differ from j'. $n_{sp} + 1$, is the total number of species in the reaction mechanism. Notice that when the incrementation of j' leads to a reaction-step that does not make part of the set of reaction-steps linked to Y_j in the reaction mechanism, the corresponding term vanishes from Eq. (3.45), since its quantum yield is equal to zero.

$$r_{Y_j}^{\lambda_{irr},T}(t)=\frac{dC_{Y_j}^{\lambda_{irr},T}(t)}{dt}=\sum_{j'=0}^{n_{sp}}\left(\Phi_{Y_{j'}\to Y_j}^{\lambda_{irr}}P_0^{\lambda_{irr}}A_{Y_{j'}}^{\lambda_{irr},T}(t)-\Phi_{Y_j\to Y_{j'}}^{\lambda_{irr}}P_0^{\lambda_{irr}}A_{Y_j}^{\lambda_{irr},T}(t)\right)\mathrm{PKF}^{\lambda_{irr},T}(t) \tag{3.45}$$

The dimension analysis of Eq. (3.45) reveals that the left-hand-side term (the rate) has a dimension of M s^{-1}, and the right-hand-side term has the dimension of $P_0^{\lambda_{irr}}$ since the other coefficients are dimensionless, $\left[\Phi_{Y_{j'}\to Y_j}^{\lambda_{irr}}\right]=\left[\Phi_{Y_j\to Y_{j'}}^{\lambda_{irr}}\right]=\left[A_{Y_{j'}}^{\lambda_{irr},T}(t)\right]=\left[A_{Y_j}^{\lambda_{irr},T}(t)\right]=\left[PKF^{\lambda_{irr},T}(t)\right]=1$. We have seen that the unit of the photon flux per irradiated area and volume of the sample that enters the reactor, $\left[P_0^{\lambda_{irr}}\right]=\text{einstein s}^{-1}\,\text{dm}^{-3}$ (Eq. (2.25)), is equivalent to that of the rate because the *einstein* unit equates the Avogadro number, or a *mole* (see Sect. 2.18).

$$\left[r_{Y_j}(t)\right]=M\,s^{-1}=\left[P_0^{\lambda_{irr}}\right]=\text{einstein dm}^{-3}\,\text{s}^{-1} \tag{3.46}$$

Even though Eq. (3.45) is a non-linear differential equation, it, nonetheless, turns into a linear differential equation for a single, particular situation. That is, when the non-linear photokinetic factor becomes a constant. The only situation where this is possible is when the irradiation is performed with a monochromatic light whose wavelength coincides with that of an isosbestic point ($\lambda_{irr}=\lambda_{isos}$). Here, the rate-laws will be similar to those encountered in thermal chemical kinetics. We shall discuss the related features in Part III.

In practice, sometimes instead of measuring the concentration of the species, the photoreaction progress is monitored by the variation of the total absorbance of the medium, $A_{tot}^{\lambda_{irr}/\lambda_{obs},T}$, that is now measured at a given observation wavelength (λ_{obs}), that belongs to the electronic absorption spectrum of the reaction medium. $A_{tot}^{\lambda_{irr},T/\lambda_{obs}}$ is measured through the observation optical path length of the monitoring light inside the sample (l_{obs}). Here, both λ_{obs} and l_{obs} may or may not be equal to their respective irradiation counterparts (i.e. λ_{irr} and l_{irr}). The species concentrations, however, are always referring to irradiation features. Hence, a labelling with a double wavelength, $\lambda_{irr}/\lambda_{obs}$, is necessary to specify which total absorbance is measured.

$$A_{tot}^{\lambda_{irr},T/\lambda_{obs}}(t)=\sum_{j=0}^{n_{sp}}\varepsilon_{Y_j}^{\lambda_{obs}}\,l_{obs}\,C_{Y_j}^{\lambda_{irr},T}(t) \tag{3.47}$$

The corresponding rate-law, written in terms of absorbance, is modified accordingly, taking the following form.

$$r_A^{\lambda_{irr},T/\lambda_{obs}}(t) = \frac{dA_{tot}^{\lambda_{irr},T/\lambda_{obs}}(t)}{dt}$$

$$= \sum_{j'=0}^{n_{sp}} (\varepsilon_{Y_{j'}}^{\lambda_{obs}} \Phi_{Y_{j'} \to Y_j}^{\lambda_{irr}} P_0^{\lambda_{irr}} A_{Y_{j'}}^{\lambda_{irr},T}(t) - \varepsilon_{Y_j}^{\lambda_{obs}} \Phi_{Y_j \to Y_{j'}}^{\lambda_{irr}} P_0^{\lambda_{irr}} A_{Y_j}^{\lambda_{irr},T}(t)) \, l_{obs} \, PKF^{\lambda_{irr},T}(t)$$

(3.48)

In the latter formulation (Eq. (3.45)), it is implicit that the solvent is totally transparent to the incident light. This is a usual assumption in solution studies, even if in a pure sense, the effects of the medium or more generally of the matrix or the sample holder should be considered (Sect. 2.25). For instance, reflectance of the incident light (e.g. by the solution and/or by the sample holder) will cause a reduction of $P_0^{\lambda_{irr}}$ by a factor $\rho_a \leq 1$. If the phenomenon occurs within the sample, then an attenuation of the incident light inside the homogeneous sample may be formulated by an exponential factor $\rho_b \geq 0$. Accordingly, the absorbed light by a species (Eq. (3.41)) may be amended as follows (where the absorbance label indicates that irradiation and observation wavelengths correspond to λ_{irr}).

$$P_{a,Y_j}^{\lambda_{irr}} = P_0^{\lambda_{irr}} \left(1 - 10^{-A_{tot}^{\lambda_{irr},T/\lambda_{irr}}(t)}\right) \frac{A_{Y_j}^{\lambda_{irr},T/\lambda_{irr}}(t)}{A_{tot}^{\lambda_{irr},T/\lambda_{irr}}(t)} \rho_a \, 10^{-\rho_b \, l_{irr}} \quad (3.49)$$

So, the higher ρ_b and/or the lower ρ_a, the higher the attenuation of the incident light. For transparent matrices, $\rho_b = 0$, an assumption generally used for non-viscous solutions. And for non-reflective material of the sample holder, $\rho_a = 1$. These assumptions ($\rho_a = 1$ and $\rho_b = 0$) are adopted in the following chapters.

There is another effect of the matrix that does not relate to the attenuation of light but to the reactivity of the molecule (not purely catalysis). The reaction can proceed more quickly or more slowly depending on the matrix due, for instance, to a volume restriction and/or geometric hindrance of the structural transformation of the initial species into its photoproduct (e.g. in cyclodextrin complexes). This directly affects the photochemical quantum yield of the reaction and can be formalised as positive multiplicative factors, $\delta_{Y_j} \to Y_{j'}$ or $\delta_{Y_{j'} \to Y_j}$, associated with the quantum yields of the species in the rate-law.

Now, let us review a few circumstances where the general equations of the rate-laws (Eqs. (3.45) and (3.48)) are not applicable. For instance, in the above series of equations, it is implicitly supposed that the irradiation is performed using a purely monochromatic light beam of wavelength λ_{irr}. These equations are simply not consistent for polychromatic light sources. In such a situation, we must incorporate the variation of the quantities involved in Eq. (3.45) with excitation wavelength over the spectral range of the radiation, between λ_a and λ_b ($\Delta\lambda = \lambda_b - \lambda_a$). This means that an integration over $d\lambda$ must be introduced to account for the wavelength variation. The resulting integro-differential equations will be studied in Chap. 19. In such equations, it is evident that the absorbances and the incident light intensity depend on the wavelength. It is also important to consider that the photochemical quantum yield is, too, dependent on the excitation wavelength unless experimentally proven otherwise (see Chaps. 4 and 17).

Furthermore, strictly speaking, the formalism presented above is only valid when collimated light is used. That is when the light beam direction is perpendicular to the sample's surface (Sect. 2.6). This means that experimentally, the sample should be totally covered except for the area that allows the collimated light from the light source to reach the sample. Obviously, in the case of non-collimated light, one should amend Eq. (3.45) to integrate over all possible incidence angles of the light reaching the sample.

Finally, for samples studied in solutions, the equations suppose that the non-viscous medium is continuously stirred so that the concentration of the species is homogeneous at all times within the volume irradiated. In non-stirred solutions, the light attenuation inside the sample depends on the depth of penetration of the light as well as on the varying composition of the medium at each depth. This at least affects the value of ρ_b factor in Eq. (3.49) but may also change the formulation of the rate-equation altogether. The descriptions given in the subsequent chapters consider only vigorously stirred solutions.

3.7 Variation of the Photokinetic Factor

For illustration, let us model the variation of the photokinetic factor (Eq. (3.50)) as a function of a monotonical decrease of the total absorbance with reaction time. This case gives an idea of the exponential-like behaviour of $\text{PKF}^{\lambda_{irr},T}(t)$.

$$\text{PKF}^{\lambda_{irr},T}(t) = \frac{1 - 10^{-A_{tot}^{\lambda_{irr},T/\lambda_{irr}}(t)}}{A_{tot}^{\lambda_{irr},T/\lambda_{irr}}(t)} \tag{3.50}$$

Figure 3.1 shows the variation of the product, $P_0 \times \text{PKF}^{\lambda_{irr},T}(t)$, where the incident P_0 is constant (hence, the observed variation is due to that of PKF with time). The absorbance is considered here to decrease with time. The variation of the incident P_0-value affects the magnitude but does not change the evolution pattern of the plot. Since the expression of $A_{tot}^{\lambda_{irr},T/\lambda_{irr}}(t)$ is not defined here, Fig. 3.1 might either correspond to the data of a single species or a set of species absorbing at λ_{irr}.

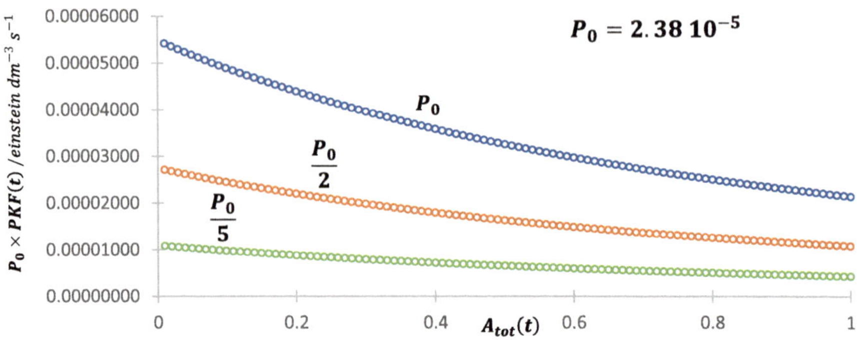

Fig. 3.1 Three $PKF^{\lambda_{irr},T}$ curves as functions of total absorbances, the latter change with irradiation time. The curves are considered at three different radiation intensities of the monochromatic light ($P_0 = 2.38\ 10^{-5}$ einstein $dm^{-3}\ s^{-1}$ (blue), 0.5 P_0 (orange), and 0.2 P_0 (green)

3.8 Power Series Expansion

The expansion in power series is a mathematical means by which a function, such as exponentials and logarithms, can be transformed into an infinite sum of polynomial terms. A number of definitions exist for the power series, among which Taylor, Mac Laurin and Laurent series. Because the Mac Laurin series converges for the exponential function, e^x, for all x, it is more suitable for the kind of problem we deal with in photokinetics.

Examples of Mac Laurin series expansions are given here for the exponential and logarithmic functions [2].

$$e^x = \sum_{n=0}^{\infty} \frac{x^n}{n!} = 1 + x + \frac{x^2}{2!} + \frac{x^3}{3!} + \ldots \tag{3.51}$$

and

$$Ln(1-x) = -\sum_{n=1}^{\infty} \frac{x^n}{n} = -x - \frac{x^2}{2} - \frac{x^3}{3} + \ldots \tag{3.52}$$

The number of terms in the expansion is theoretically infinite, leading to a very good approximation of the value of the function at hand. However, for practical reasons, most often, only a few of these terms are considered. The expansion is then said to be of 1st-order, 2nd-order … and so on when it respectively stops at the term with a power of x equal to 1, 2… and so on. Practicality in this context will, unfortunately, result in a loss of accuracy, i.e. a substantially big error is found between the value of the original function (e.g. e^x) for a given value of x and the one obtained (for the same value of x) from the calculation of the truncated power series expansion.

For instance, let us consider the power expansion of an exponential for $x = 1.3$.

Table 3.2 Variation of the estimated value of an exponential function for $x = 1.3$ with various orders of the Mac Laurin expansion

Expansion order	1st-order	2nd-order	3rd-order	4th-order	5th-order	6th-order	7th-order
$e^{1.3}$	2.3	3.145	3.511	3.630	3.661	3.668	3.669
% error	37.32	14.29	4.310	1.07	0.22	0.04	0.006

Table 3.3 Rocketing %-errors of the 2nd-order Mac Laurin expansions of e^x when the value of x increases

x value	e^x	Mac Laurin 2nd-order expansion	% error
0.001	1.001001	1.001001	1.7×10^{-8}
0.01	1.01005	1.01005	1.7×10^{-5}
0.1	1.105171	1.105	0.0155
0.2	1.221403	1.220	0.1150
0.3	1.349859	1.345	0.3612
0.4	1.491825	1.480	0.7990
0.5	1.648721	1.625	1.4597
0.6	1.822119	1.780	2.3662
0.7	2.013753	1.945	3.5348
0.8	2.225541	2.120	4.9783
0.9	2.459603	2.305	6.7073
1	2.718282	2.500	8.7313
1.5	4.481689	3.625	23.633
2	7.389056	5	47.781
3	20.08554	8.5	136.30
4	54.59815	13	319.98

$$e^{1.3} = 1 + 1.3 + \frac{(1.3)^2}{2!} + \frac{(1.3)^3}{3!} + \frac{(1.3)^4}{4!} + \frac{(1.3)^5}{5!} + \frac{(1.3)^6}{6!} + \frac{(1.3)^7}{7!} \quad (3.53)$$

The value of the exponential, $e^{1.3} = 3.66929$, is ever closely approached when the expansion order increases (Table 3.2).

The estimation of the function is also dependent on the value of x. The higher the x value, the lower the percentage difference between the true value and the approximated Mac Laurin series value (i.e. the accuracy). Table 3.3 gives an illustration of the 2nd-order expansion of the exponential function for various values of x.

In order to compensate for the loss of accuracy, the power series expansion order must increase. We notice that more accurate estimations are obtained when the values of the variable, of the function expanded, are relatively small. Table 3.4 shows an important increase of the error between the value of e^x and its values are calculated by a 5th-order Mac Laurin expansion. The % errors recorded for this case are lower compared to those obtained with a 2nd-order Mac Laurin expansions for

Table 3.4 5th-order Mac Laurin expansions of e^x compared to the true values of this function

x value	e^x	Mac Laurin 5th-order expansion	% error
0.001	1.001001	1.0010005	0
0.01	1.01005	1.010050167	0
0.1	1.105171	1.105170917	1.27×10^{-7}
0.2	1.221403	1.221402667	7.49×10^{-6}
0.3	1.349859	1.34985775	7.83×10^{-5}
0.4	1.491825	1.491818667	0.0004
0.5	1.648721	1.648697917	0.0014
0.6	1.822119	1.822048	0.0039
0.7	2.013753	2.013571	0.009
0.8	2.225541	2.2251307	0.018
0.9	2.459603	2.4587582	0.034
1	2.718282	2.7166667	0.059
1.5	4.481689	4.4617187	0.445
2	7.389056	7.2666667	1.656
3	20.08554	18.4	8.391
4	54.59815	42.866667	21.48

the same x values (Table 3.3), but, likewise, accuracy is gradually lost as the value of x increases.

Hence, any choice of the order of the Mac Laurin expansion series (with $n \ll \infty$) will result in an approximation.

The latter discussion has importance for the degree of reliability that would be reached by kinetic equations expanded by power series. A particular attention should be directed to the condition of a relatively low value of x in order to obtain acceptable accuracy of the results. This might be crucial as, most often, the expanded rate-laws, proposed in the literature [3], used high experimental values of either concentration or absorbance. It is not rare that values of absorbances exceeding 1 have been experimentally considered for a 1st-order Mac Laurin expansion. No doubt that such a practice will induce significant errors in the kinetic model so that the conclusions reached in this manner cannot be reliable.

Therefore, even though this approach has been, sometimes, recommended, it could not be endorsed for photokinetic analysis, and any experimental results generated by this method have to be considered with caution.

Furthermore, it is interesting to notice that, once the approximated rate-law is obtained by applying the truncated Mac Laurin expansion, the equation then undergoes a second approximation by which different orders of the reaction are predicted for different values of the initial concentration of the reactant. This is how a 0th-order kinetic is predicted for photoreactions when the initial concentration is high and a 1st-order kinetics for low concentrations. This double approximation does not constitute a coherent method.

When the purpose is to use the expansion for the defined goal of integrating a function that lacks a known antiderivative, a decision must be made on where to stop the expansion, i.e. what the order of expansion should be. As discussed above and as clearly indicated by the data in Tables 3.2, 3.3, and 3.4, the approximation is better when the order of the Mac Laurin expansion is high.

However, dealing with this choice relative to the best order of the expansion reveals another unsuspected problem relative to what extent a factorisation of the derived expression of the expanded function is possible. The integration of a quadratic or higher-order polynomial is not straightforward. It requires writing these polynomials, obtained by series expansion, into simpler integrable forms. In order to achieve such a simplification, we proceed in general by factorisation of the polynomial function. However, a simple factorisation is not always possible. Examples of such issues are presented in Chap. 13.

3.9 Numerical Integration Methods (NIMs)

One of the excellent tools used in photokinetics for data treatment consists of numerically integrating the system of differential equations describing the photoreaction under study. So much so that it could be stated that the numerical integration methods (NIMs) have, up to now, represented the only known general method that can be applied to any photoreaction irrespective of the photomechanism in play. We will, hereafter, review succinctly some general aspects of NIMs and hint at their possible advantages and disadvantages in relation to photokinetics.

Numerical integration methods (NIMs) are mathematical methods primarily developed for calculus problems. They are specifically useful for the evaluation of the output values of the solution of a given differential equation or a system of differential equations. That is, NIMs provide a numerical approximation to the solution of the differential equations system even though the analytical solution of the differential equations is actually not known (i.e. the antiderivatives of the integrals are not known). Hence, their name of numerical integration basically refers to a computation of integrals. In kinetics, NIMs provide the kinetic traces ($C_{Y_j}^{\lambda_{irr},T}(t)$) of species Y_j from computing the system of the rate-laws of the reaction.

A variety of approaches have been proposed for NIM formalisms, including Newton, Simpson, Euler, trapezoidal or Runge–Kutta methods. In essence, such mathematical formalisms allow the calculation of the next value of the integral of a given differential equation system at the nth value of the independent variable, t_n, when the value of the previous point, at t_{n-1}, is already known. This stepwise procedure allows the construction of the values of the integral on the whole range ($[t_0, t_n]$) considered for the independent variable t. The procedure hence needs that the initial conditions are well defined (e.g. the value of $C_{Y_j}(t_0)$). In addition, it is usually preferable to divide the selected range $[t_0, t_n]$ of the independent variable into equidistant intervals, Δt, (which defines the iteration step-size, h) for which the values of $C_{Y_j}(t)$ will be calculated.

Practically, for kinetics, this means that NIMs produce the same data as those recorded by the experiment (e.g. concentration values against reaction time). The error on the NIM-calculated data can be controlled, where the level of precision increases with a reduction of the step-size value (h) and the performance of the numerical method employed.

In subsequent chapters, the NIM used for the calculations of the photokinetic traces of photoreactions is the most widely known member of the Runge–Kutta family, namely the standard fourth-order Runge–Kutta method (*RK*4). The latter has the following formalism.

For a differential equation of the form encountered in kinetics

$$\frac{dC}{dt} = f(t, C) \tag{3.54}$$

With the initial boundary condition (for kinetics, usually $t_0 = 0$)

$$C(t_0) = C_0 \tag{3.55}$$

$C(t)$, the integrated rate-law, is an unknown function of time, which will be approximated by *RK*4. The rate-equation of C (Eq. (3.54)) is a known function of both time and C itself. At the initial time, t_0, C has the given value C_0 (Eq. (3.55)). Let the step-size be

$$\Delta t = h = \frac{t_n - t_0}{n} \tag{3.56}$$

where h is defined for the time range $[t_0, t_n]$. Hence, the iteration proceeds as

$$t_{n+1} = t_n + h \tag{3.57}$$

The *RK*4 procedure is based on the calculation of four slopes, k_1 to k_4, obtained by the following formulae.

$$k_1 = f(t_n, C_n) \tag{3.58}$$

$$k_2 = f\left(t_n + \frac{h}{2}, C_n + k_1 \frac{h}{2}\right) \tag{3.59}$$

$$k_3 = f\left(t_n + \frac{h}{2}, C_n + k_2 \frac{h}{2}\right) \tag{3.60}$$

$$k_4 = f(t_n + h, C_n + k_3\, h) \tag{3.61}$$

These will serve the successive calculation of the values of C at the n intervals, by

$$C_{n+1} = C_n + \frac{h}{6}\,(k_1 + 2\,k_2 + 2\,k_3 + k_4) \tag{3.62}$$

The major advantage of NIMs remains their ability to obtain the values of the integrated function without actually knowing the algebraic expression of the integrated function itself (e.g. $C(t)$). This is particularly appreciated when the differential equation cannot be solved easily or is impossible to solve analytically.

The main drawback of NIMs is precisely the unavailability of the integrated function, i.e. the handy algebraic mathematical formula that offers a description of the system studied. Also, practically, it is required to know the mechanism of the reaction and to fully run the calculation for each new situation (e.g. change of the reaction mechanism, the initial concentration...). In kinetics, the application of NIMs does not provide any information relative to the order of the reaction under study and does not allow the calculation of the rate-constant of the reaction.

Furthermore, there is an additional risk inherent to NIMs calculations. It is intimately related to the amount of information available on the system. Indeed, a lack of information translates into a higher number of parameters to be estimated by the operator to feed the NIMs. The power of these methods resides in the fact that a solution (for instance, fitting experimental data with high accuracy) will almost always be found by NIMs, through a judicious choice of the values for the parameters. What becomes evident from such a situation is the possibility of perfectly fitting the experimental data using different sets of values for the equations' parameters, feeding NIMs calculations. This may lead to both distinguishability and identifiability issues (Chaps. 6 and 9). Then, a good fit of the experimental data by a NIM-calculated curve ceases to be an argument for the validity of any particular set of parameters' values that allow such a fit. A delicate care should be directed towards establishing a strategy to prove the uniqueness of the results, i.e. the set of parameters' values that would be retained. Singling out the *true kinetic solution* (the true set of the parameter values of the reaction) is a real issue that, unfortunately, most often is ignored when NIMs are employed for kinetic problems.

3.10 Limitations of NIMs in Kinetic Data Analysis

It is a general understanding that the treatment of kinetic data can almost always be performed by using numerical integration methods. Such a consideration is strong enough for it to rise to the level of a concept. Consequently, details such as those presented in the previous sections might be looked at as a vain exercise that does not bring anything really new to the subject. So, is it vain? And is it true that NIMs are magical tools, overwhelming all kinds of problems in the field of photokinetics?

In order to help answer the latter questions, it would perhaps be useful to propose an informal classification of the kinetic cases. That is, we can consider three different types.

The first group consists of the kinetic cases where it is possible to achieve an analytical solution to yield the integrated rate-laws of the reactants and products involved in the photoreaction at hand. Here, the use of NIMs is redundant. Such a statement is arguably true since, for instance, no one proceeds with using NIMs for the treatment of the data of zero- or first-order thermal reactions. The reason is precisely the availability of integrated rate-laws (e.g. Eqs. (3.31), (3.34) and (3.36)) that were established through closed-form integrations of the respective rate-laws describing each reaction. Hence, the latter principle should always apply to the reaction kinetics that can be solved analytically through a robust mathematical procedure which yields the analytical expressions of the integrated rate-laws. From this perspective, it is imperative to seek to define the integrated rate-laws of any studied reaction before using NIMs.

The rest of the reaction kinetic cases (other than those discussed above making the first group) are those for which the integration of the rate-laws of the reaction is mathematically impossible to achieve in a closed form. It is then possible to divide such kinetics into two groups (representing here our second and third groups of this classification).

To the second group will belong all the kinetic cases for which approached integrated rate-laws have been already established on the basis of a strict methodology that is built on the NIMs generated data and which allow a precise indication of the limits of applicability of such formulae. The methodology proposed in Chap. 14, the semi-empirical method, could be considered one of the possible templates to solve similar problems. However, it is essential to keep in mind that the practical development of such formulae might become tedious for complex kinetics. Perhaps an input from computerised methods (such as the implementation of artificial intelligence) might help to ease the development of such approaches. In any case, once the integrated rate equations have been set, there is no further need for NIMs.

The third and last group involves all the kinetic cases that cannot be fully solved for the unknown reaction attributes, irrespective of whether or not NIMs are applied. This situation applies to cases faced by insolvable distinguishability and/or identifiability problems. Unfortunately, most of the photokinetic cases fall within this last group.

3.11 Conclusion

In this chapter, we have gone over some of the most important mathematical tools required for photokinetic investigations and the description of reactions driven by light. They will be employed throughout this book for different purposes.

References

1. Chicone C (2006) Ordinary differential equations with applications, vol 34. Springer, Cham
2. Abrmowitz M, Stegun IA (1970) Handbook of mathematical functions with formulas, graphs and mathematical tables. Dover Publications, New York. Available at https://personal.math.ubc.ca/~cbm/aands/abramowitz_and_stegun.pdf
3. Blatchley ER (2023) Photochemical reactors. Theory, methods, and applications of ultraviolet radiation. Wiley, Hoboken

Part II
Unsettled Concepts

Chapter 4
Kasha's Rule and Photochemistry

4.1 Introduction

The dependence of photochemistry on excitation wavelength is not a recently observed phenomenon; nonetheless, it has, surprisingly enough, been largely ignored in the field. The reasons for this situation are not fully understood but might be related to a provisional extension of Kasha's rule to photochemistry or perhaps to a difficulty in justifying the kind of short timescales implied in such photochemistry, which challenges the usually held view giving predominance to fast internal conversion and vibrational relaxation. Regardless of the reasons, it is still a matter of fact that a complete and satisfactory interpretation for experimentally proven wavelength-dependent photochemistry is not yet available, and the community endeavours to build a holistic understanding and a comprehensive view of the phenomenon. The present chapter compiles a non-exhaustive overview of the published data in the field, reporting on some of the most prominent features, issues and interpretations. Part of this chapter has been reused from Maafi [1] under an Open Access CC BY 4.0 license, https://creativecommons.org/licenses/by/4.0/.

4.2 An Overview

Despite the impact of excitation wavelength on photochemical reactivity (λ-effect), which has been acknowledged since the mid-twentieth century (around 1950), only recently have the effects of activation wavelengths become of genuine interest [2]. It has been argued that experimental observation relative to wavelength-dependent photochemistry has remained fragmentary, and the regularities of the phenomenon are still neither well understood nor predictable [3, 4]. Nonetheless, the literature advocates that reactions originating from higher excited-state are much more common than anticipated [5, 6].

M. Maafi, *Photokinetics*, https://doi.org/10.1007/978-3-031-98985-8_4

Much of the published work denotes such phenomena as anti-Kasha, in reference to Kasha's rule [7], which is supposed to have pre-eminence in this area. However, as we shall see, this denomination (i.e. anti-Kasha) might not be appropriate because Kasha's rule does not necessarily apply to photochemical events [8]. A complete interpretation of irradiation-wavelength variation of photoreactivity is not yet available, but one can separate the observed data into roughly two groups. The first one corresponds to irradiation at different electronic absorption bands of the reactant, whereas the second concerns sequential irradiation, at narrow wavelength-intervals, of a single absorption band of the reactant. The former type of irradiation would readily involve different excited states, whereas the latter most likely operates within the same excited-state. The reaction timescales involved in each might differ. The former might invoke competition of photochemistry with internal conversion (*IC*), while reactions induced by the latter type of irradiation would proceed at an even shorter timescale (i.e. competing with vibrational relaxation (*VR*)). This analysis indicates rather a fast to an ultrafast photochemistry, with time spans of 10^{-11} *s* (for *IC*) or shorter [9], which falls beyond the time frame corresponding to luminescence (~10^{-9} *s* or longer), and hence, does not corroborate an interpretation that suggests the applicability of Kasha's rule to photochemistry.

Evidence of the variation of photoactivity with irradiation wavelength implies the measurement of the photochemical quantum yield as one, if not the best metric for this purpose. A change of quantum yield values with irradiation requires necessarily experimental/reaction data to evidence the occurrence of the phenomenon. Alternatives to experimentally determined quantum yield values are not available, since theoretical calculation of photochemical quantum yields is a rather nascent field of research. We shall see that different experimental ways of determining the quantum yield have been adopted in different studies. A fact that underlines a lack of consensus in the community on this particular matter.

This chapter reports on some of the most prominent results on the variation of photochemical processes (i.e. composition of end-products, photoreactivity and/or change of the quantum yield values) when subjected to different irradiation lights. The selected cases presented here, although limited in number, strongly support a genuine occurrence of an excitation-wavelength-dependent photochemistry.

4.3 The Historical Kasha's Rule

In 1950, Kasha published his seminal paper describing emissive and possible stages of deactivation of a light-excited molecule. Later, this work became known as Kasha's rule [7]. The theoretical derivation, according to Kasha's rule, can be summarised by the classical Jablonski diagram [10, 11]. After a photon absorption, the excited molecule tends, in a first step, to lose some of its excitation energy by vibrational relaxation (*VR*), e.g. $S_{n,\ \nu} \rightarrow S_{n,\ 0}$, that occurs between any higher vibrational state ($0 \leq \nu \leq m$) of the *n*th singlet excited-states ($S_{n,\ \nu}$ where $n > 1$) and the lowest vibrational state of that excited-state. The new molecular state will

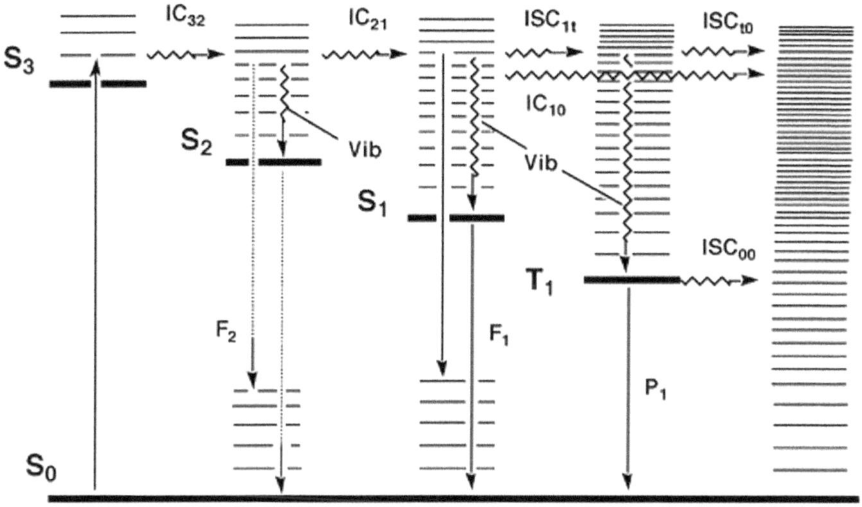

Fig. 4.1 Possible radiative and non-radiative processes of an organic molecule after absorption of a photon ($S_{0,0} \rightarrow S_{3,\nu'}$). The non-radiative IC_{ij} indicate $S_i \rightarrow S_j$ internal conversions, whereas ISC_{1t}, ISC_{t0}, and ISC_{00}, correspond to intersystem crossings from $S_{1,\ \nu}$ to $T_{1,\nu'}$, $T_{1,\ \nu}$ to $S_{0,\ \nu}$, and $T_{1,\ 0}$ to $S_{0,\ \nu}$, respectively. F_1, F_2, and P_1 indicate radiative processes (fluorescence and phosphorescence) from S_2, S_1, and T_1, respectively, and Vib is the vibrational relaxation (VR) in singlet and triplet manifolds. This modified Jablonski diagram includes an example of a non-Kasha emission (F_2) as an $S_{2,\nu} \rightarrow S_{0,\nu'}$ radiative transition. Reprint with permission from [12], *Chemical Reviews*

further lose energy by internal conversion (*IC*) followed by vibrational relaxation (*VR*), to ultimately reach the first excited-state (as $S_{n,\ 0} \ldots \rightarrow S_{2,\ \nu} \rightarrow S_{2,\ 0} \rightarrow S_{1,\ \nu} \rightarrow S_{1,\ 0}$). From that first ($S_{1,\ 0}$) excited singlet-state, the molecule can subsequently deactivate to (1) the singlet ground state ($S_{0,\ \nu}$) via a non-radiative *IC* process, (2) by emitting a photon of light ($S_{1,\ 0} \rightarrow S_{0,\ \nu}$, fluorescence), or (3) proceeds to a change of spin multiplicity by an "intersystem crossing" (*ISC*), leading it to a triplet-state ($S_{1,\ 0} \rightarrow T_{n,\ \nu}$). From the latter state, the molecule relaxes by *VR* to $T_{n,\ 0}$, then to the first excited triplet state ($T_{n,\ 0}$ — *IC* → $T_{1,\ \nu}$ — *VR* → $T_{1,\ 0}$), and from there can either phosphoresce ($T_{1,\ 0} \rightarrow S_{0,\ \nu}$) or relax, through a non-radiative *ISC*, to the ground state ($S_{0,\ \nu}$) (Fig. 4.1).

As observed from such photophysical processes, Kasha's rule can be stated as "the emitting electronic level of a given multiplicity is the lowest excited level of that multiplicity ($S_{1,\ 0}$ or $T_{1,\ 0}$)" [7, 13], but has also been stated as "in solution, emission occurs from $S_{1,\ 0}$ or $T_{1,\ 0}$, independently of the state which is initially excited" ($S_{n,\ m}$ or $T_{n,\ m}$) [14, 15]. This rule is also known as Kasha-Vavilov's rule, with Vavilov's rule stating that "the fluorescence quantum yield is independent of the excitation wavelength" [14–16].

The concept captured by Kasha-Vavilov's rule is generally valid and practically observed in a large majority of luminescence experiments, but exceptions do exist, as indicated by Fig. 4.1 and stressed out, for instance, by the IUPAC's 'glossary of terms used in photochemistry' [15], as shall be further discussed below.

4.4 The Unfounded Extension of Kasha's Rule Application to Photochemistry

Although both Kasha's and Vavilov's rules were originally deemed to apply to molecular photophysics in condensed media, and neither author suggested their applicability beyond that context [8], it was recurrently considered that the applicability of Kasha's rule "naturally" extends to photochemistry [4, 13, 17, 18]. In substance, Kasha's rule is presumably thought to also stipulate that photoreactions only proceed from the same molecular energy levels from which emissions normally arise. According to this assumption, for any system, $S_{1,\ 0}$ and $T_{1,\ 0}$ are the only energetic levels from which molecular transformations, reactions and measurable chemical changes will arise (since Kasha's rule predicts that, irrespective of the irradiation wavelength (i.e. excitation to $S_{n,\ v}$), the molecule will quickly decay to the lowest vibrational level of the first-excited electronic state ($S_{1,\ 0}$) from which it will be able to undergo photophysical processes, with this lowest excited-state of the molecule apparently also representing the starting point for photochemistry).

There is enough experimental evidence in photophysics literature to suggest that the emitted-light energy (e.g. the emission wavelength maximum), by either fluorescence or phosphorescence, is generally not disturbed by varying the excitation wavelength (as long as the remaining measurement conditions are kept unchanged). Such a simple experiment provides strong evidence supporting the argument that, for a majority of emitting species, fluorescence emission naturally originates from the lowest vibrational state of the first excited singlet-state ($S_{1,\ 0}$). It is also the best indication of the validity of Kasha's rule in photophysics for a large group of molecules. Exceptions to this norm will be reviewed in the following section.

Let us now consider the likelihood of an eventual extension of Kasha's rule to photochemistry. Such an extension might be thought coherent with the view considering that within polychromatic molecules, rotational and even vibrational levels in reactants may be so "crowded" with the populations present at the ambient temperature that selection or excitation of individual levels is impossible [19].

In this context, it is perhaps also useful to observe that no such simple experiment, equivalent to that available to photophysics (i.e. proving the invariance of the emission maximum with excitation wavelength), exists today to provide such irrefutable evidence on the occurrence of photochemistry from the lowest excited energy levels irrespective of excitation. For instance, such a straightforward experiment, if it existed and from which a wavelength-invariant photoreaction quantum yield could be inferred, would indicate Kasha's rule applicability to photochemistry. It is important to stress here that supplying such an experimental proof of the invariance of the photochemical reactivity with excitation remains key to backing up an extension to and validating Kasha's rule for photochemistry.

From a qualitative viewpoint, and keeping the spirit of Kasha's rule for photochemistry, photochemical reactions should most likely proceed from $S_{1,\ 0}$ (for singlet-state photoreactions and from $T_{1,\ 0}$ for triplet-state reactions), and so, photochemistry must compete with photophysical luminescent processes (fluorescence or phosphorescence). If we assume that the above hypothesis stands, the time-span of photochemistry has to be at most equal to that of fluorescence (~10^{-9} s) for singlet reactions proceeding from $S_{1,0}$, and obviously much longer for the long-lived triplet state, $T_{1,0}$. In this framework, there is difficulty in reconciling such timescales with those emerging from many time-resolved experiments that assign picosecond and even shorter femtosecond (10–100 fs) time spans for various molecular photoreactions [20–22].

Therefore, how to interpret this body of data? One can suggest that photochemistry seems to be well accommodated in sub-nanosecond timescales, not directly competing with fluorescence (as would be recommended by the hypothesis of an applicable Kasha's rule to photochemistry). One is then forced to conclude that photochemistry is a much faster process than fluorescence for those cases, and that it belongs to a different time domain (much shorter than those generally assigned to fluorescence and phosphorescence).

Within this framework, the photochemical transformation of such molecules in their excited states takes place well before fluorescence. The observed reactant emission is then procured by photochemically unreactive species that relax to the first single-state (i.e. the proportion of excited molecules that did not photoreact). For proposed triplet-state fast reactions, we can postulate a similar pathway, where photoreaction proceeds well before the emergence of phosphorescence that, if occurring, would be due to unreactive species. In this hypothesis, the intersystem crossing, leading to the formation of the triplet-state, seems to effectively be a much faster process than supposed by the timescales measured for phosphorescence, as was documented for some molecules [23, 24]. However, generally, the data available on triplet photoreactions tend to show relatively slow *ISC* kinetics.

An interpretation of such phenomena can perhaps, at least in part, be provided by the conical intersection process. However, it is a matter of fact that there is no holistic explanation available today that is able to rationally connect the complete set of available data.

Overall, the above qualitative analysis, which is undoubtedly a valuable hint, needs however to be supported by tangible experimental data. One of the ways to approach this problem is to look into the effect of varying excitation on the photochemical quantum yield (which is only determined experimentally). Indeed, according to an extended Kasha's rule photochemistry, a given reaction is expected to proceed from a unique excited-state (let it be the singlet-state, $S_{1,0}$). Therefore, the photochemical quantum yield of a given reaction-step, such as $X \rightarrow Y$, should not be affected by the excitation wavelength, i.e. whatever the excitation wavelength, the electrons first relax down to the first excited single-state, and only then the photoreaction can proceed. Either way, experimentally testing the invariance of the quantum yield is a good approach to bringing evidence in support or against the supposed validity of Kasha's rule for photochemistry. In essence, if the quantum yield is constant irrespective of the excitation wavelength, Kasha's rule stands (as for photophysical processes); otherwise, photochemistry has to be considered faster than emission pathways, such as fluorescence.

4.5 Anti-Kasha Emission

Before we look into fast photochemistry, let us ask whether fluorescence may compete with *IC* or *VR* and whether such a behaviour was experimentally observed.

Within Kasha's rule remit, photochemistry competes with fluorescence, and therefore, if fluorescence cannot compete with *IC* or *VR*, then photochemistry may well only proceed from the lowest vibrational level of the first excited singlet-state ($S_{1,\ 0}$). However, if emission can proceed from either a higher electronic excited-state (e.g. S_2) or a vibronic-state (e.g. $S_{1,\ \nu > 0}$), as suggested by Fig. 4.1 (e.g. the F_2 arrow), a possibility then opens for speculation on whether photochemistry is allowed from higher energy excited-states or vibrational levels.

The so-called anti-Kasha emission is a terminology relating to a molecular emission of light that does not obey Kasha's rule. That is, different excitation wavelengths induce emission spectra with distinct features. For a while, excitation-wavelength-dependent fluorescence emission spectra were thought to be rather rare. However, the number of research groups claiming to have observed anti-Kasha emissions has become relatively large, even if the reported cases represent a relatively small fraction of light emitters, for which Kasha's rule stands.

The phenomenon of wavelength-dependent fluorescence has been reported since the early 1950s [25]. Thiophosgene ($S = C(Cl)_2$) is a particularly interesting case that exhibits, in the gas phase, an S_2 fluorescence that depended on the excitation wavelength [8, 26, 27]. Fluorescence emission was selectively observed from excitation of this molecule into the three first vibronic levels of S_2, but not from higher vibronic levels where a fast internal conversion to the ground state, $S_2 \rightarrow S_0$, occurs. The authors suggested that the rate of internal conversion was dependent on the vibronic levels reached on excitation. The peculiar fluorescence of thiophosgene hence proves two extra possibilities for emission rates in addition to the usual $S_1 \rightarrow S_0$: emission may compete not only with *IC* but also with *VR*. It is one of the first fluorescence emission cases that undeniably proved a dual limitation of Kasha's rule for photophysical processes [28]. Thiophosgene emits from the second electronically excited singlet-state (not from S_1), and this emission originates from different vibrational levels of S_2, i.e. $S_{2,\ 0}$, $S_{2,\ 1}$, and $S_{2,\ 2}$. Similarly, in the gas phase, methyl salicylate [29], and three of its derivatives, methyl-2-hydroxy-3-naphthoate, methyl-1-hydroxy-2-naphthoate, and methyl-2-hydroxy-1-naphthoate exhibited a variation of the fluorescence quantum yield with the excitation wavelength [30].

Azulene [25, 31–33], pyrene [34, 35], benzopyran [36] and some of their derivatives are notoriously known examples of S_2-fluorescence in gas and condensed phases. Pyrene has been reported to emit from three singlet states, S_3 (π,π^*), S_2 (π,π^*), and S_1 (π,π^*) [34, 37, 38]. Historically, these organic molecules were thought to be exceptional anomalies, but more and more examples of wavelength-dependent fluorescence in condensed media were reported in subsequent years to enrich the documentation in this area [28, 39, 40].

Similar to the behaviour of thiophosgene, rotationally cooled but electronically excited *m*-xylyl radical, showed, in the gas phase, visible vibronic emission spectra originating from the $1^2A_2 \rightarrow 1^2B_2$ transition [39]. Several thiocarbonyl compounds showed $S_2 \rightarrow S_0$ fluorescence, from singlet π, π^* in solution [28]. In toluene, *meso*-tetraphenylporphyrin and other *meso*-substituted porphyrin derivatives have revealed a dual fluorescence emission process. The first emission was attributed to the usual $^1(\pi \leftarrow \pi^*)$ transition, whereas the second emission process was assigned to the electronic $S_0 \leftarrow S_1$ transition at the $Q_x(0,0)$ band, following excited-state tautomerism of the inner macrocyclic hydrogen atoms via an atom tunnelling process [40].

Interestingly enough, the anti-Kasha phenomenon of organic molecules in condensed media has been reported in two relatively recent reviews. One covering publications up to 2011 [12], and the other, reporting on articles published between 2011 and 2021 [41]. They described excitation-wavelength-dependent emission by experimental data and quantum-chemical computational calculations for a wide variety of organic molecules, including azulenes, aromatic acenes, polyenes, thioketones, metalloporphyrins, aromatic carbonyl compounds, quinones, halogenated aromatic compounds, cyclic triimidazoles, 1,2-diphenylphenanthroimidazole derivatives, cyanines, thiophenes and carbazoles.

The authors outlined the circumstances favourable for the occurrence of anti-Kasha's rule behaviour in condensed media. Mainly, two mechanisms were proposed to explain the fluorescence emission from S_2. Either (*ia*) when both the oscillator strength (f) of the $S_0 - S_2$ transition and the energy gap between S_1 and S_2 are large, or (*ib*) when f of $S_0 - S_1$ transition and the $S_2 - S_1$ energy gap are very small, but with a concomitant large f of the $S_0 - S_2$ transition. Whereas, fluorescence from higher ($S_{n>2}$) electronic levels were assigned to a relatively large energy gap between the electronic levels, combined with a prohibited internal conversion between S_n and S_1 (due to symmetry reasons). Phosphorescent anti-Kasha emission from T_n (with $n > 1$) were attributed to either (*iia*) an efficient *ISC* to T_2, whose symmetry differs from that of T_1, or (*iib*) to a thermal population of T_2 from T_1.

It is perhaps also useful to mention here some red-edge effect examples. The so-called "red-edge effects" were related to the existence of excited-state distribution of fluorophores in regard to their interaction energy with the environment and the slow rate of dielectric relaxation of this environment (hence, red-edge emission is not really anti-Kasha). The spectra were observed to be shifted to shorter wavelengths at the 'blue' edge of the emission spectrum [42]. However, because of a usual relatively small spectral variation, this effect is not extensively investigated.

In a polar solvent, a giant "red-edge effect" was observed on the fluorescence of graphene oxide, which was strongly dependent on the excitation wavelength [43]. The increase of the excitation-wavelength value caused up to 140-nm bathochromic shift of the fluorescence maximum of several graphene oxide sheets (e.g. measured $\frac{\lambda_{exc}}{\lambda^{max}_{emi}}$ were 350/440, and 500/580), and a broadening of the emission spectra. An $n \leftarrow \pi^*$ and a $\pi \leftarrow \pi^*$ singlet transitions were assigned to the emission of an –OH-rich graphene oxide derivative. Conversely, none of these features, including the excitation wavelength effect, was observed for graphene derivatives when a non-polar solvent was used [43]. The fluorescence spectra of many aromatic fluorophores embedded into different rigid and highly viscous media can depend on excitation wavelength, and the excited-state energy transfer, if present, fails at the 'red' excitation edge [6, 42, 44–46]. Static inhomogeneous broadening was also observed for phosphorescence spectra, where it causes long-wavelength shifts of phosphorescence maxima at the red-edge excitation in motion-restricting media [42, 47, 48].

An excitation wavelength-dependent room temperature phosphorescence was also reported in liquid and solid samples [49, 50]. For instance, the colour tunability of a series of phosphinoamines (phenylbenzothiazoles bearing phosphine groups [49]), in both solid state and solution, was due to a dual band emission of hundreds of microseconds lifespan, whose intensity depends on the excitation wavelength of ultraviolet light. *DFT* calculations supported an interpretation for such a nonconventional radiative process, by involving a combination of pathways, including a locally excited-state (the transition between excited states with different molecular geometry), a charge-transfer state, and/or an excited-state intramolecular proton transfer.

4.6 Distribution of the Literature in Photochemistry

It seems appropriate, for the purposes of the present chapter, to divide the literature relative to the possible dependence of photochemistry on excitation into two groups: one encompasses those publications where the authors considered that Kasha's rule applies to photoreactions without providing further discussion about the matter or experimentally investigating its validity. One can stipulate that this first group of research considers Kasha's rule as a principle for photoreactions. Because the latter is not yet proven with certainty, we will not be including such studies here.

The second type of documentation relates to research papers that sought to obtain experimental facts on whether the excitation wavelength has an impact on photoreactivity and/or on the values of the photochemical quantum yield. Some of these will be reviewed here. Of particular interest is the literature that employed monochromatic beams for the quantum yield measurements since polychromatic light has an averaging effect that might not allow critically resolving differences between quantum yield values measured at individual wavelength. Accordingly, it seems important to particularly review the literature that provides tangible experimental evidence on the effects of varying the monochromatic irradiation wavelength on the absolute values of the photochemical quantum yield of photoreactions. In this context, one must acknowledge the variability of the methods and equations used for the determination of the quantum yield, even when monochromatic light is employed.

4.7 Early Observations

Within less than three decades of Kasha's publication [7], Turro and his co-workers compiled what perhaps represents the first comprehensive review on the effects of excitation wavelength on photoreactions [8]. It covered a substantial amount of experimental data (more than 180 papers over the 1955–1977 period), spanning a series of phototransformations in molecules belonging to a variety of chemical families and involving an array of different photomechanisms.

Most of the described compounds showed a variation in the composition of the photoproducts when the irradiation wavelength impinging on the reactive medium was changed, i.e. different sets of end-products were generated under irradiation with different excitation wavelengths (whose beams were mostly provided by unfiltered or filtered lights of various lamps). Whereas, only a few studies reported a difference in the quantum yield with irradiation wavelength but with no alteration of the photoproduct set generated by the reaction.

The alteration of a reaction outcome in response to a change of the excitation wavelength (λ-effects), was attributed to phenomena arising from either excited-state or ground-state properties of the reactive species. The authors of the review proposed to rank this type of reaction into four classes: (a) photoreactions for which the wavelength dependence is due to the ground-state properties only; (b) photoreactions starting from higher electronic excited-states ($n > 1$), S_n^* or T_n^*; (c) photoreactions starting from higher vibrational states ((hot) excited-states, labelled by the symbol, ‡) of the lowest excited electronic states $S_1^{*\ddagger}$ or $T_1^{*\ddagger}$; and (d) reactions from T_1 through the population of the latter state from S_n^* or $S_n^{*\ddagger}$. That review focused more on differentiating spurious λ-effects due to ground-state properties from λ-effects implying higher excited-state or vibronic photochemistry, than on assigning the observed reactions to the latter two categories. The authors have, however, attempted to account for reactions with fast or slow internal conversion features.

The four classes listed above turned out to be, a priori, sufficient to explain the cases reported in the review [8], but otherwise, make a solid ground for a general classification.

4.8 Photoreactions of Species in Equilibrium

Here, a rapid thermal equilibrium is envisaged between the starting materials. A pertinent situation is whenever a relatively modest thermal barrier (not exceeding ~17 kcal/mol [8]) exists between the interconvertible molecular structures (e.g. $X \leftrightharpoons Y$). In this instance, we are dealing with ground-state species (X and Y, where Y is a product of X), obtained from thermal equilibration (Scheme 4.1). In general, the irradiation of the solution may concern two or, concomitantly, several

Scheme 4.1 Photoreactions of two species in thermal equilibrium

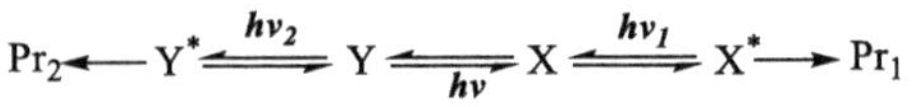

Scheme 4.2 Possible photoreactions of two species in their photostationary state

species in thermal equilibrium. In these circumstances, different excited states are populated by the excitation light (e.g. X^* and Y^*).

Each excited species might undergo an independent primary photoreaction, so that, when subjected to different irradiation wavelengths ($h\nu_1$ and $h\nu_2$), the photoreaction of one species might yield a different set of end-products (e.g. Pr_1) to those (e.g. Pr_2) obtained from another species (Scheme 4.1). Another mechanism that is overall analogous to that presented in Scheme 4.1 corresponds to the situation where a photostationary state, *pss,* (instead of the thermal equilibration) is established between *X* and *Y*. Here, the reactant (*X*) is first exposed to a light, $h\nu$, until *pss* is reached. Then, lights of different wavelengths ($h\nu_1$ and $h\nu_2$, where $h\nu_1$ might be equal to $h\nu$) are shown on the solution (now containing both thermally stable *X* and *Y*). Each irradiation light will favour one end-product (Scheme 4.2). This mechanism can also concern the central photoreversible reaction only (in which case $Pr_1 = Y$ and $Pr_2 = X$).

In the general case, the electronic absorption spectra of species *X* and *Y* might or might not overlap, but they are likely to have different absorptivities. The accumulation of Pr_1 and Pr_2 in the solution will hence depend on both their respective production quantum yields ($\Phi_{X \to Pr_1}$ and $\Phi_{Y \to Pr_2}$), absorptivities (ε_X and ε_Y, assuming that Pr_1 and Pr_2 do not, themselves, absorb) at the irradiation wavelengths, the irradiation time, and/or the intensities of the incident light beams. In these conditions, such processes, as given in Schemes 4.1 and 4.2, can readily be explained without invoking the features of the excited-states. Indeed, if one of the species is characterised by either $\varepsilon = 0$, or $\Phi = 0$ at the selected irradiation wavelength, and/or when the ratio $\varepsilon_X/\varepsilon_Y$ is very small or very high, then only the other species will be observed in solution (or at least in the largest proportion).

The benzene oxide reaction is an example of the mechanism provided by Scheme 4.1. Exposure of an ether solution of benzene oxide (**1**) at $-80\ ^\circ C$ to a 254-nm light resulted in the formation of benzene and phenol. However, at room temperature under a beam of light of wavelength higher than 310 *nm*, only (**3**) was obtained [8, 51]. The authors explained these results by the occurrence of a thermal equilibrium between (**1**) and (**2**). At low temperature, the equilibrium strongly favours (**1**) and hence only the photoreaction of that species is observed. Whereas, at room temperature, where (**1**) and (**2**) coexist, the reaction of (**2**) is predominant because its absorptivity for wavelengths >310 *nm* greatly exceeds that of (**1**).

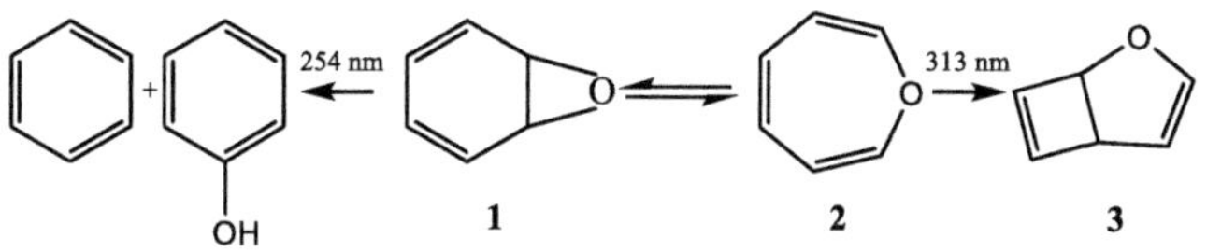

1 2 3

Bicyclo[4,3,0]nona-2,4-diene] (**4**) irradiation, in pentane, with a 300-nm light-beam leads to (**6**), whereas its exposure to a 254-nm light leads to (**5**) [52]. In fact, a *pss* is established between (**4**) and (**5**), with this ring-closure process ((**4**)→(**6**)) being much faster than that of the ring-opening ((**4**)→(**5**)). This means that over a very long exposure time with either irradiation wavelengths only (**6**) is expected to remain in solution.

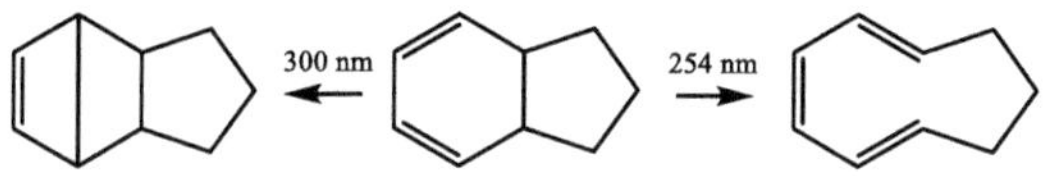

6 4 5

The experimental observation of a wavelength dependence of the end-product (**4**) for relatively short exposure times has been linked to the relative absorptivities of the three species at each irradiation wavelength. At 254 nm, the ratio $\varepsilon_{(4)}/\varepsilon_{(5)} \approx 30$, but at 300 *nm* it drops to ~0.02. Therefore, the reaction undertaken at 300 nm favors the formation of the less efficient (**6**), whereas at 254 nm, the *pss* is displaced toward (**5**). A valid interpretation of the experimental observation without invoking excited-states.

Even though we observe a measurable effect of the irradiation wavelength, still Kasha's rule cannot be tested by the systems shown in Schemes 4.1 and 4.2, due to the occurrence of a hypothetical *quasi*-independent reaction of *X* from that of *Y*. The photoreactions of *X* and *Y* might or might not, separately, obey Kasha's rule. The experiment using such systems (Schemes 4.1 and 4.2) cannot provide the useful information to reach an answer if only one irradiation is performed. Therefore, to identify whether wavelength has an impact on the reactivity of each individual species (*X* and *Y*), it is required to perform a systematic screening of the sample under several irradiation wavelengths, and the data relative to each individual species is treated separately.

The earliest known example of such a screening is that of azobenzene photoisomerisation [53]. The wavelength dependence of the photoreversible azobenzene forward and reverse quantum yields was measured at 254, 313, 365, 405, 436, 546 and 578 *nm*. The *trans* (**7**) to *cis* (**8**) quantum yield almost doubled, ranging from 0.13 to 0.27, respectively, between 254 and 436 *nm*, but decreases between 436 and 578 *nm* to a final value of 0.233. The *cis-trans* (**8** → **7**) efficiencies showed a zig-zag pattern in the region 254–578 *nm*, varying in the 0.40–0.55 range.

Table 4.1 Photoisomerisation quantum yields of azobenzene in various solvents and under different irradiation wavelengths (mercury lamp lines) [54]

Irradiation wavelength (nm)	Azobenzene concentration (M)	Medium temperature (°C)	Solvent	$\Phi_{trans\text{-}cis}$	$\Phi_{cis\text{-}trans}$
254	2.5×10^{-5}	25	Isooctane	0.13	0.14
254	6×10^{-4}	RT	Methanol	0.26	0.31
280	6×10^{-4}	RT	Methanol	0.12	0.34
313	6×10^{-4}	RT	Methanol	0.13	0.3
313	$0.04–2 \times 10^{-3}$	RT	Methanol	0.14	–
313	$0.2–5 \times 10^{-5}$	20	Methanol	0.37	–
313	10^{-5} to 10^{-4}	RT	Ethanol	0.12	0.31
313	$0.01–2 \times 10^{-3}$	15–25	Ethanol	0.22	0.69
313	$0.04–2 \times 10^{-3}$	RT	Isopropanol	0.1	0.5
313	$0.2–5 \times 10^{-4}$	20	H_2O-Ethanol	–	0.4
313	$0.2–5 \times 10^{-4}$	20	MeCN	–	0.35
313	$0.2–5 \times 10^{-4}$	20	THF	–	0.4
313	$0.025–1 \times 10^{-3}$	25	Isooctane	0.1	0.41
313	$0.04–2 \times 10^{-3}$	RT	Cyclohexane	0.1	0.42
313	$0.2–5 \times 10^{-5}$	20	Cyclohexane	0.4	–
313	$0.2–5 \times 10^{-5}$	20	n-hexane	–	0.44
334	6×10^{-4}	RT	Methanol	0.15	0.3
365	6×10^{-4}	RT	Methanol	0.15	0.35
365	$0.01–2 \times 10^{-3}$	15–20	Ethanol	0.2	0.6
365	$0.025–1 \times 10^{-3}$	25	Isooctane	0.12	0.48
365	$0.04–2 \times 10^{-3}$	RT	Cyclohexane	0.14	–
405	6×10^{-4}	RT	Methanol	0.2	0.57
405	$0.025–1 \times 10^{-3}$	25	Isooctane	0.23	0.55
436	$0.04–2 \times 10^{-3}$	RT	Methanol	0.28	–
436	6×10^{-4}	RT	Methanol	0.22	0.63
436	$0.01–2 \times 10^{-3}$	15–20	Ethanol	0.36	0.45
436	10^{-5} to 10^{-4}	RT	Ethanol	0.24	53
436	$0.04–2 \times 10^{-3}$	RT	Isopropanol	0.26	0.42
436	$0.025–1 \times 10^{-3}$	25	Isooctane	0.28	0.55
436	$0.2\text{-}5 \times 10^{-5}$	20	Cyclohexane	0.28	0.55
436	1×10^{-3}	25	Benzene	0.26	0.46

RT room temperature

7 **8**

Several subsequent studies have confirmed both the variability of azobenzene quantum yield with excitation wavelength and the lower *trans-cis* quantum yield value compared to that of its counterpart reverse reaction (Table 4.1) [54, 55]. Lower quantum yields of azobenzene were also reported for shorter UV wavelengths

[55, 56]. The latter results were corroborated by a multi-configuration calculation of azobenzene along the *NN* bond twisting that identified an S_3 singlet-state of the *trans* geometry (as the lowest excited singlet-state), on the basis of the doubly excited configuration $n^2\pi^{*2}$ [57]. The authors found that the latter state provides an explanation for azobenzene quantum yield wavelength-dependence.

The quantum yields of azobenzene isomers were determined experimentally on the basis of a differential equation for the rate processes that assumes monochromatic light, vigorous stirring, validity of Beer-Lambert's law, and that the quantum yields in both directions are independent of both the light flux provided to the reaction medium, and the initial concentration [53]. The non-linear rate-law was simplified by discarding a term involving the thermal reaction rate-constant, and its linearisation was achieved by expanding the photokinetic factor in a power series for high initial concentrations (~10^{-4} to 10^{-3} M). The *trans-cis* quantum yield was, hence, obtained from the gradient of the linear relationship of the light flux with time. This information facilitated the subsequent determination of the *cis* → *trans* quantum yield from the steady-state equation. It is also interesting to remark that a substantial decrease in the photoisomerisation rate with increasing initial concentration was experimentally shown in this study (despite the working hypothesis adopted above). The variation of the quantum yield with wavelength was later confirmed in methanol by numerically solving the system of differential equations of the reaction and the mass balance for the forward and reverse quantum yield values [58].

The *E*/*Z* photoisomerisation quantum yield of quantum dots' grafted azobenzene in chloroform (7.3 10^{-7} M) solution [59], almost tripled when monochromatically irradiating the $\pi \rightarrow \pi^*$ transition at 365 nm (0.35) and 436 nm (0.97). The reverse *Z*/*E* quantum yield was 0.39 at a 436-nm irradiation. The photokinetic method used for the determination of the quantum yields was based on an analytically derived equation [60, 61].

Solving for the quantum yields of photoreversible systems such as azobenzene, diarylethenes, spiropyrans, and fulgides through using numerical integration was also recommended [62, 63]. This contrasts with the approaches calculating the quantum yield of photoreversible systems such as azobenzenes and stilbenes, whereby the formulae of each quantum yield were worked out from a mathematical method using the equations corresponding to irradiations at two different wavelengths but assuming that the quantum yield is wavelength independent [64]. The latter strategy of evaluating the quantum yield using numerical calculation and irradiation at multiple wavelengths while considering an invariant quantum yield was also adopted for photochromes' photokinetic data, such as spiropyrans and spirooxazines [65, 66]. Unfortunately, such a mathematical strategy is not very helpful since the validity of Kasha's rule for photochemistry is, de facto, accepted and translated into the equations.

An attempt to approximate the wavelength-dependent quantum yield values of both isomers was performed by the formula, $(1 - e/E)^{\nu}$, where E is the excess vibrational energy immediately after internal conversion, e, is the height of the potential energy barrier, and ν the number of classically excited modes of vibration [53]. It was found that not only does this model predict a monotonic decrease of the

trans ⇋ *trans** ⇋ *cis** ⇋ *cis*

Scheme 4.3 Separate cis and trans excited-state species of the cis/trans photoreaction

quantum yield with increasing excitation wavelength, but also the quantum yield estimated values were less than 0.011 for all wavelengths. Both characteristics are not conforming with the experimental measurements. The authors also concluded that the latter findings are not consistent with previously proposed models by Lewis [67] and Olson [68] for *cis-trans* isomerisation. Furthermore, in the case of azobenzene, it was observed that the experimental data do not point to the presence of an indistinguishable common intermediate state, regardless of which isomer, *trans* or *cis*, was excited by the radiation, because the sum of the individual quantum yields, measured at each wavelength, do not add up to 1 [69, 70], i.e. $\Phi_{trans} + \Phi_{cis} \neq 1$ at a given irradiation wavelength. For the authors [53], the results on the quantum yield dependence of wavelength strongly suggested the existence of two separate configurations of the excited states (*trans** and *cis**) of the two individual isomers *trans* and *cis*, according to Scheme 4.3.

The latter hypothesis was somewhat corroborated by a CASSCF computation of the excited-state relaxed surfaces of *trans*-azobenzene, which indicates that the photoisomerisation mechanism can both proceed by rotation from a cold S_1 species, and by a concerted inversion when excitation in S_2 produces a hot S_1 species. These two excited-state species may explain the much lower reaction quantum yield from S_2 than that observed from S_1 excitation [71]. Incidentally, a sigmoid-like trend of the quantum yield *vs.* temperature was observed for *trans*-azobenzene-modified *DNA* [72].

DNA-psoralen monoadducts photoreaction quantum yield values ranged between 0.012 and 0.08 when the excitation varied between 248 and 365 nm, in a sort of a convex pattern (with 0.08 and 0.04 quantum yield values recorded at the short and long wavelengths of the curve) [73].

The light-driven *cis* (**10**) to *trans* (**9**) isomerisation of the *cis*-retinal chromophore in the *G*-protein coupled receptor rhodopsin was, for a long time, assumed to be wavelength independent [74] but later, in a more accurate measurement [75, 76], was found to be constant between 450 and 480 nm, and falls significantly as the irradiation wavelength was increased to 500 nm ($\Phi = 0.67$). A further 5% reduction of the quantum yield values was obtained at 570 nm. The quantum yield value at 500 nm and its wavelength dependence were corroborated by both calculation and experiment, in later studies [77, 78].

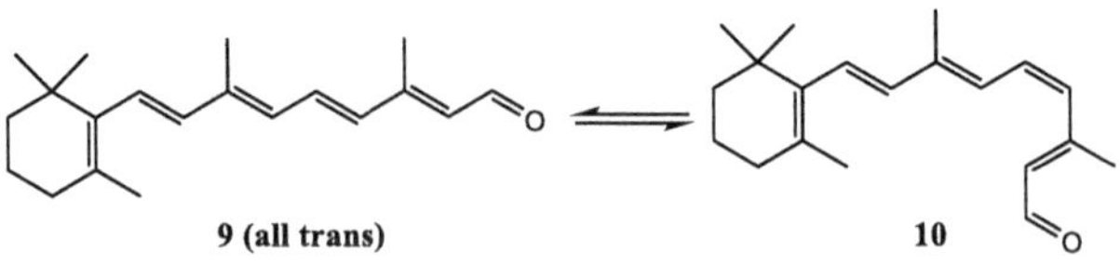

9 (all trans) **10**

Kim and co-authors hence corrected the long-held presumption that the quantum yield in vision was wavelength-independent and supported the hypothesis that the 200 *fs* photoisomerisation reaction that initiates vision is dictated by non-stationary excited-state vibrational wave packet dynamics, which corroborates the previous conclusions of an earlier study on *all-trans*-retinal, whose excitation-dependent fluorescence was assigned to a competition between photochemistry and internal conversion and between the vibrational levels of its first-excited singlet-state [79]. The quantum yield of *cis*-retinal chromophore in the *G*-protein was calculated by Eq. (4.1):

$$\Phi^{\lambda_{irr}} = \frac{N_A \frac{\partial f}{\partial t}}{P_0^{\lambda_{irr}}\ 2.303\ \varepsilon^{\lambda_{irr}}\ l\ m^{\lambda_{irr}}} \tag{4.1}$$

with N_A is Avogadro's number, $P_0^{\lambda_{irr}}$ is the incident radiation intensity, $\varepsilon^{\lambda_{irr}}$ is the molar decadic extinction coefficient (M^{-1} cm^{-1}) at the irradiation wavelength, l is the optical path length (cm), $m^{\lambda_{irr}}$ is a correction factor due to other absorbing species, and f is Dartnall's function (Eq. (4.2)) of the absorbance ($A^{\lambda_{irr}}$) [80].

$$f^{\lambda_{irr}} = -2.303\left(A^{\lambda_{irr}}(t) - A^{\lambda_{irr}}(\infty)\right) - ln\left(1 - e^{-2.303\left(A^{\lambda_{irr}}(t) - A^{\lambda_{irr}}(\infty)\right)}\right) \tag{4.2}$$

The findings of Kim et al. [75] were further compared to the predictions provided by a Landau-Zener model for dynamic curve crossing [76]. The treatment of the non-statistical internal conversion process revealed that absorption by high-frequency unreactive modes increases as the excitation wavelength shifts from 570 to 500 nm, whereas relatively less energy is deposited into reactive low-frequency modes (indicating the importance of delocalised, tortional modes in the reactive pathway of rhodopsin). However, the variation of the quantum yield of rhodopsin was assigned to a ground-state heterogeneity of a mixture of *cis* and *all-trans* retinal absorbing around 480 nm and 540 nm [81] (but unfortunately, no data from analytical chemistry techniques were shown to identify the ground-state mixture).

Irradiation in different absorption bands of *trans*-4-nitro-4′-dimethylamino-stlibene (**11**) in cyclohexane was reported to produce (a) a little variation of the isomerisation quantum yields (e.g. $\Phi_{trans}^{405} = 0.16$, and $\Phi_{cis}^{405} = 0.42$), (b) a value well below unity for the sum of the individual forward and reverse reactions' quantum yields, but, at the same time, (c) a significant change of the percentage of the *cis*-isomer in the *pss* composition, ranging between 19 and 64%, for irradiations performed respectively between 313 and 436 nm [82]. The study also demonstrated the occurrence of both temperature and concentration effects on the quantum yields. For instance, the Φ_{trans}^{366} of 4-nitro-3′-methoxy-stilbene in 1-methyl-naphthalene reduced from 0.32 to 0, as the concentration gradually increases from 10^{-5} M to 1 M.

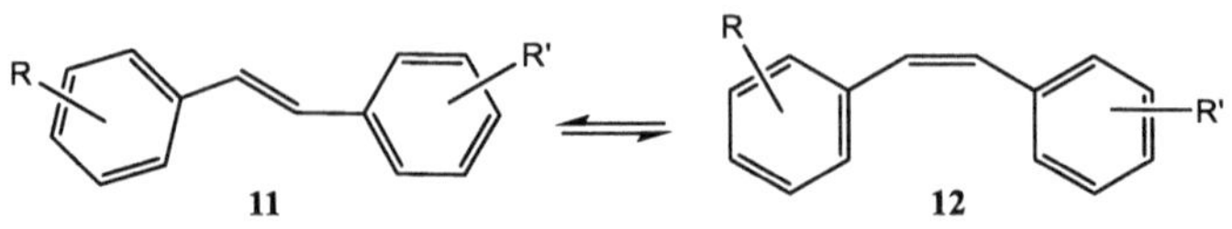

11 12

A variation of the quantum yield of *trans* to *cis* stilbene in *n*-hexane was recorded between 313 nm (0.27) and 254 nm (0.47), whereas a relatively constant quantum yield (0.29) of the *cis* to *trans* reaction was measured for the same irradiation-wavelengths [83]. The authors acknowledged differences in the way the quantum yields were determined in the individual different studies from which the wavelength effect was reported, and in general, a similar trend was observed for a number of stilbene derivatives.

We will see, in subsequent chapters (for instance, Chap. 18), more examples of several photosystems endowed with characteristic quantum yield variation with irradiation. The mathematical analyses provided therein offer a set of equations for the quantum yield that substantially differ from those presented in the present chapter. Also, they introduce one general result indicating that the photoreaction rate-constant does not depend exclusively on the absorption coefficient of the reactant, as might have been supposed, such as the reaction would be faster when the absorption is higher (the rate depends on absorption). In fact, the rate-constant of, for instance, the photoisomerisation, is a much more inclusive function involving all the parameters of the reaction at the irradiation wavelength (including the irradiated volume and area of the reactive medium [61, 84]).

4.9 Photoreactions of Multichromophoric Species Yielding Different End-Products

Some of the interpretations proposed in the previous section do not discard the fact that these reactions derive primarily from excited-state processes. If features of the latter were not really required for the interpretation of some of those reactions, the excited-state information might, however, be relevant to other situations.

For instance, and conversely to the previous cases involving two (or more) chemically independent species (Schemes 4.1 and 4.2), photochemistry can develop from a single reactant whose reactivity might well be dependent on the irradiation wavelength. The analysis now needs to rely on the excited-state properties. In the simplest case (Scheme 4.4), two different end-products (Pr_1 and Pr_2) are obtained by exposing the reactant X to lights of two different wavelengths. The most likely

$$Pr_1 \longleftarrow X^{*'} \underset{}{\overset{h\nu_1}{\rightleftharpoons}} X \underset{}{\overset{h\nu_2}{\rightleftharpoons}} X^{*''} \longrightarrow Pr_2$$

Scheme 4.4 Dual excitation path of a reactant *X*

interpretation adopted for this type of reaction considers the occurrence of two distinct excited states of X (e.g. $X^{*'}$ and $X^{*''}$). The latter, non-equivalent excited states, may have either singlet or triplet character (S_1, S_2, …, T_1, T_2,…etc.), most generally, of π,π^*, n,π^*, or n,σ^* type. The model shown in Scheme 4.4 implicitly implies that photochemistry from the excited-states $X^{*'}$ and $X^{*''}$ must occur before the redistribution of the excitation energy within the molecule [8, 23].

The most probative example of such a mechanism can stem from the excitation of different chromophores (e.g. μ_1 and μ_2 associated with the excited states $X^{*\prime}_{\mu_1}$ or $X^{*\prime\prime}_{\mu_2}$, respectively), both belonging to the molecular structure of X. The specific reactions leading to the end-products Pr_1 and Pr_2 (or more), represent a tangible proof of a difference between the primary photoprocesses, presumably proceeding from μ_1 and μ_2, respectively.

A typical and one of the earliest photoreactions of this type, concerned molecules endowed with a spatially separated C–S bond and a carbonyl group [85, 86]. For the 4,4,6,6-tetramethyl-1-thiaspiro[2,3]hexan-5-one (**13**), in methanol and under a 280-nm light, the carbonyl is excited to an n,π^* singlet (S_1), which results in the expected ring expansion to the acetals (**14**) and (**15**), without perturbation of the other cyclic moiety of the sulfur atom [86]. But a 254-nm beam (through quartz) induced specifically an n,σ^* transition of the C–S bond, corresponding to the molecular S_2-state, that leads to cyclobutanone (**16**) and the removal of the sulfur atom.

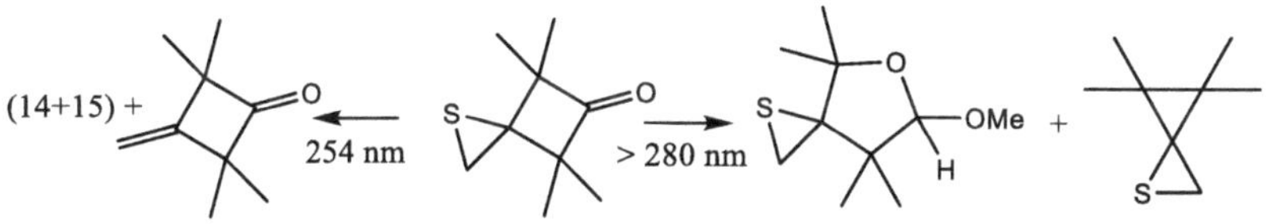

16 **13** **14** **15**

In ethanol at room temperature, a 405-nm irradiation of 1,4-dihydronaphtho [1,8-de] [1,2]diazepine (**17**) [87, 88], excites the $N = N$ bond, corresponding to an n,π^* transition that exclusively gives the highly strained *trans*-isomer (**18**). The latter reconverts into (**17**) in the dark at room temperature, or by light excitation of its n,π^* band situated at *ca.* 436 *nm*. However, irradiation of the naphthalene-absorption bands at λ_{irr}~313 nm, populating mainly the π,π^* states of the individual isomers (**17**) and (**18**), led to the exclusive formation of acenaphthene (**19**). The authors estimated the ratio of the *cis* → *trans* and *trans* → *cis* quantum yields to be 1/10, as derived from the wavelength-dependence of the isomers *pss* compositions [**17**]/[**18**]. Notice that each of (**17**) and (**18**) can individually be considered as the reactant (X) in Scheme 4.4. This kind of reaction is a process that was sometimes labelled *chromatic orthogonality* [89–91].

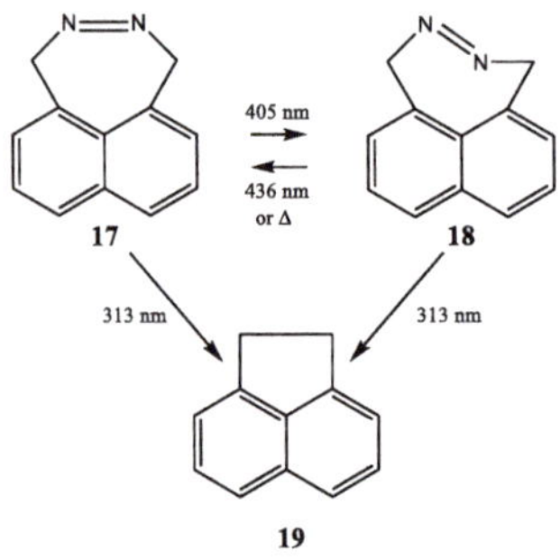

When polychromatic light is used on reactions of Scheme 4.4, such as the two chromophores, might be excited (e.g. $X^{*'}$ (as an n,π^*), and $X^{*''}$ (as a π,π^*)), selectivity can then be achieved by employing an absorption competitor (or spectator molecule) whose absorption spectrum covers one of the electronic transitions (e.g. the n,π^* responsible of $X^{*'}$), so that only the product generated from the other transition (e.g. the π,π^* of $X^{*''}$) is obtained as the major (or the only) species in the solution. A rationalisation of this phenomenon through Φ-order kinetics has been proposed [84]. This approach has been widely used for molecules encompassing a variety of dual chromophores that were envisaged for various applications. Notably, it is adopted in chromatic orthogonality and exploited in releasing photolabile protecting groups [92–103]. As an illustration, a 3D printing process requires an irradiation consisting of both blue and green lights, which should alternatively be reduced to either green or blue by the use of an adequate absorber (mask) of the unwanted light colour. The use of blue light irradiation induced both radical and cationic photopolymerisation, which produced the stiff, glassy material (**20**). Conversely, green light-absorbing photosensitizer led only to radical photopolymerisation of acrylates, producing a soft, rubbery material (**21**). Based on the chemical nature of the different domains, the *3D* printed multi-material object displayed interesting and tunable mechanical properties, including increased toughness and resistance to crack propagation compared to single-component materials [102, 104]

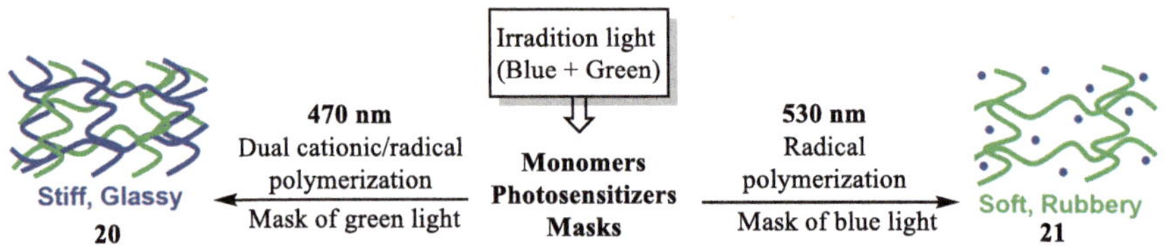

An analogous process of sequence-independent wavelength-selective (-λ-orthogonal) photoreactivity can be achieved by a selective interplay between the absorption properties of the reactive species, the quantum yields of the separate reactions in different solvents, and the irradiation wavelength [101, 105]. For instance, when *o*-methylbenzaldehyde thioether (**22**), *N*-ethylmaleimide (**23**), and methoxy-substituted diphenyltetrazole (**24**) are all present in the reaction medium, it was demonstrated that in specific solvents, different reactions are induced by LED lights of different wavelengths.

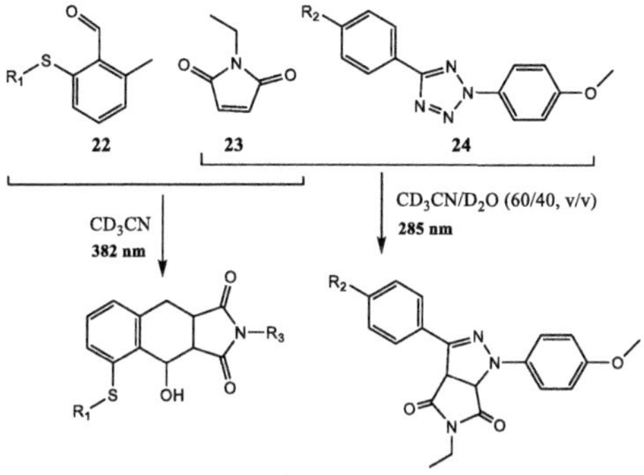

In terms of interpretation, in cases where $X^{*'}$ and $X^{*''}$ happen to be both singlet states, the process can eventually be explained by considering that vibrational relaxation within a given state can be fast, but not the internal conversion from higher to lower energy states, $S_n \rightarrow S_{n-1}$. In this perspective, the latter internal conversion must be slow enough for photochemistry to proceed from the individual distinct states. Turro et al. [8] proposed an interpretation based on the theoretical expression of the rate of internal conversion (k_{IC}) [106] from the $\nu = 0$ vibrational level of an upper excited-state ($S_{2,0}$) to a lower excited-state ($S_{1,\nu}$), that is given by the following equation:

$$k_{IC} \propto |\langle \Psi_2 | J | \Psi_1 \rangle|^2 \sum_m |\langle \chi_{20} \mid \chi_{1m} \rangle|^2 = C_{12}\, F \tag{4.3}$$

with Ψ_1 and Ψ_2 being the electronic wave functions of S_1 and S_2, respectively, and J the nuclear kinetic energy operator. χ_{1m} and χ_{20} are the nuclear wavefunctions for the appropriate vibrational levels of S_1 and S_2, respectively. The electronic factor C_{12}, results from a coupling of the electronic wave functions, and F is the Franck-Condon factor between the two states.

Equation (4.3) clearly predicts that internal conversion may be slow if either or both C_{12} and F have small values. The former situation (poor C_{12}) applies when the excited-states (Ψ_1 and Ψ_2) are orthogonal, and the latter (low F) corresponds to a large energy gap between S_2 and S_1. Even though this analysis is useful, a general, more extended description would most likely include non-adiabatic and spin-orbit coupling terms that affect transfer among states.

4.10 Photoreactions of Multi-chromophoric Species Yielding a Unique Product Irrespective of the Excitation Wavelength

One more situation (Scheme 4.5) embodies a photoconversion, with non-zero quantum yield, of one reactant to one set of end-products irrespective of the absorption band targeted by the irradiation.

Photoreaction of a reactant may well be characterised by the selective reactivity from one excited-state but not from the other [107]. The selectivity of a unique electronic excited-state to channel reaction toward the end-product is easily identified in practice since it only requires to perform irradiations at excitation wavelengths falling on the absorption band promoting the excited-state channeling reaction.

In general, however, the reactions represented by Scheme 4.5 might mean that either there is a unique reaction channel irrespective of the excitation, as when one of the excited states relaxes into the other (e.g. $S_2 \rightarrow S_1$ by internal conversion, or $S_n \rightarrow T_n \rightarrow T_1$ by ISC then IC), or the presumably different excited-states relaxing down, through specific channels, to the same set of end-products.

It is reasonable to predict that for the former case, such a process should not, a priori, show any variation of the quantum yield when the excitation light takes different wavelengths that may induce excited-states of different characters, e.g. π,π^*, n,π^* or n,σ^*. If this hypothesis stands, it may be postulated that the reactive system is deemed compatible with Kasha's rule, in the sense that the variation of the irradiation wavelength does not affect the output (quantum yield) of the reaction due to the existence of a unique intermediate after excitation. This view will still hold when the irradiation is performed under polychromatic irradiation, since such light usually covers more than one absorption band of the reactive molecule. An irradiation condition that is very often encountered in experimental photochemistry, as it predominantly uses lamps. In fact, this combination of reactions obeying Scheme 4.5 and irradiation using a lamp represents a non-negligible section of the studied photoactive reactants. Therefore, the qualitative conclusion (since no quantum yields are measured) that might be advanced here is that this behaviour fits well with Kasha's rule (the end-product(s) is independent of the excitation wavelength). The ubiquity of such reactions and behaviour in photochemistry may perhaps explain the propensity to adopt an extension of Kasha's rule to photochemistry.

The second mechanistic case (multiple reaction channels) will rather accept the possibility of a difference of the quantum yields for different excitations owing to the supposed difference of the pathways leading to the end-product(s). However, and in any case, such hypotheses will only be validated if confirmed by a detailed experimental investigation of the irradiation-wavelength effect on the quantum yield.

$$X \xrightarrow{hv} Pr$$

Scheme 4.5 Reaction generating a unique product from *X* irrespective of excitation

A short wavelength excitation (at 254 nm, S_2) was found to enhance the photodegradation of estrone (**25**) under direct photolysis by individual beams in the 254–320 nm wavelength range [108].

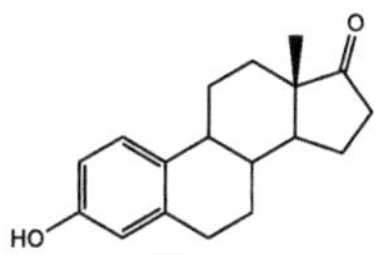

25

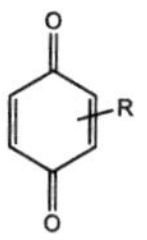

26

Flash photolysis of a series of quinone derivatives (**26**) [109] was performed at four wavelengths (440, 400, 350, 260 and 220 nm, with the beams delivered through specific filters placed between the lamp and the sample). The selected wavelength-set covers the major part of the absorption spectra of these compounds, which correspond to one large envelope spanning the 220–480 nm region. The study showed that, except for the weak long-wavelength $n \rightarrow \pi^*$ absorption band, a decrease of the quantum yield values with decreasing wavelength was manifest. Irradiation of the shortest absorption band in ethanol (220 and 260 nm) was characterised by the lowest quantum yield values (0.01–0.1) compared to those (0.07–0.98) recorded for longer-wavelength (e.g. 350 nm) irradiations.

Several α,β-unsaturated γ-dimethoxymethyl cyclohexenones (**27**) were found to undergo certain unimolecular reactions when excited to the S_2 (π,π^*) state. The latter competes with $S_2 \rightarrow S_1$ internal conversion and $S_2 \rightarrow T_n$ intersystem crossing. The unimolecular reactions do not, however, occur from the S_1 (n,π^*) or the lowest triplet (π,π^* and n,π^*) states. A very low quantum yield (<0.001) of (**27**) resulted from irradiation to its S_1 (n,π^*) state, compared to the quantum yield (0.02) recorded for its phototransformation, from a π,π^* S_2-state, into (**28**) [110].

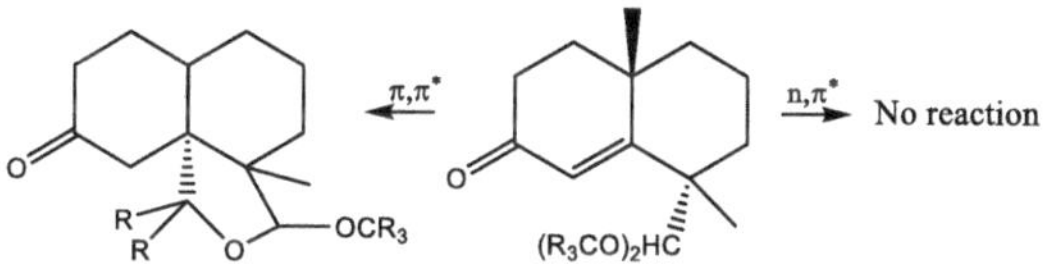

27 **28**

This type of photoreaction is widely exploited in the photoremoval of protecting groups and photocages. The molecules combining a protecting group (such as coumarin (**29**)) covalently linked to bioactive compounds (i.e. the caged species, morphine (**30**)), are totally inactive biologically. Irradiation with poly- or mono-chromatic light removes the protecting group and uncages the bioactive molecule [111].

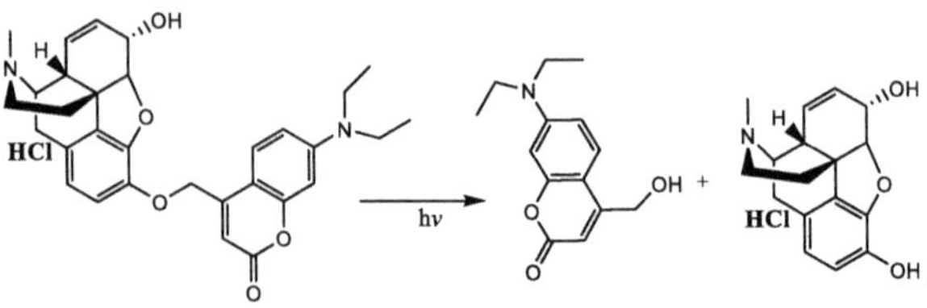

29 30

The photocleavage, leading to the production of an oxygen molecule, in a series of endoperoxides was found to exhibit a wavelength-dependent quantum yield. For instance, the quantum yields of (**31**) and (**32**) were constant (respectively, 0.13 and 0.18) between 290 and 313 nm ($S_{n \geq 2}$ excited-state), but for the S_1-state populated with irradiation at 365 nm, the quantum yields reduced to less than 2% of the constant values recorded above (respectively, 0.0026 and 0.0015). Similarly, O_2 release from Cu^{II}-superoxide compounds [112] depended on the excitation wavelength, recording more than eightfold decrease between 436 nm (0.29) and 683 nm (0.035).

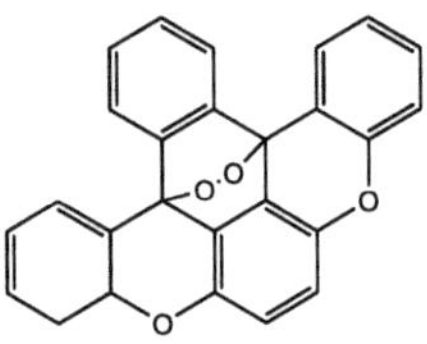

31

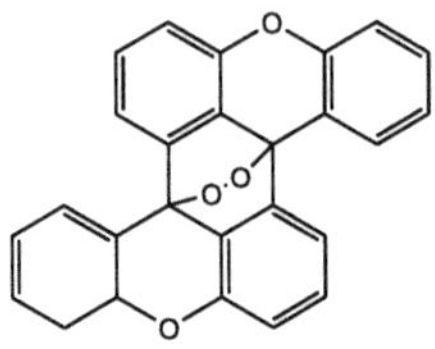

32

The dimethyldihydropyrene (*trans*-10b,10c-dimethyl-10b,10c-dihydropyrene, closed-form isomer), exhibits a negative photochromism, where the photoconversion of the coloured (**33**) into the colourless, cyclophanediene, open-form isomer (**34**), is triggered by monochromatic visible light. Nonetheless, the quantum yield of the forward (**33** → **34**) reaction has a low but positive value (0.08%) at 470 nm (corresponding to an $S_{n > 1}$ state), whereas its value is zero for an excitation of the $S_0 \rightarrow S_1$ transition at 660 nm [113].

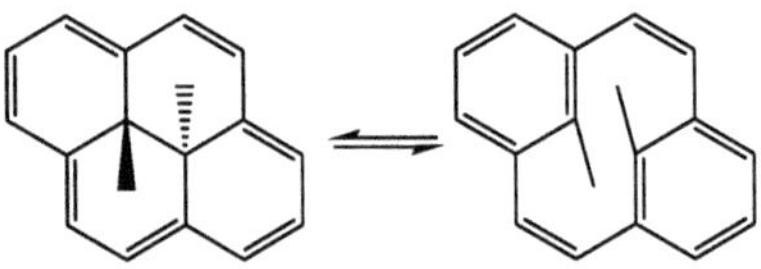

33 34

Overall, the reactions described, thus far, are transition-specific, whether the result is a unique set of end-products for different excitation wavelengths or the end-products are obtained for excitation of one band, but not for the excitation of other transitions. The description might also be considered relative to the quantum yield values, whereby, for the type of species (**27**), $\Phi \neq 0$ only for irradiation of a given absorption band, but $\Phi \cong 0$ for irradiation of the other bands.

Photodegradation of anticancer dacarbazine and cardiovascular nifedipine and nisoldipine drugs exhibit typical Scheme 4.5 reactions, with a sigmoidal variation of the quantum yield with irradiation wavelength (Chap. 18).

The photostationary state (*pss*) composition and the quantum yields of photoisomers of a series of diarylethenes-containing benzothiazolium species (**35**) were found to strongly vary with irradiation wavelength, when using mercury lamps (365/405 nm, and 405/436 nm) to induce the photoisomerisation [114]. For this series, the shorter the wavelength, the lower the *pss* percentage of the *Z*-isomer, and Φ_{Z-E} were smaller than Φ_{E-Z} at different excitation wavelengths, which has discouraged postulating a common excited-state intermediate.

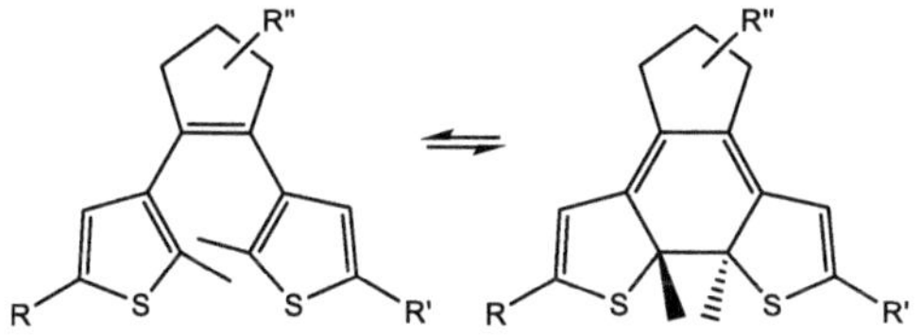

35 **36**

The cyclisation quantum yield of 1,2-bis(5-(4-ethynylphenyl)-2-methylthiophen-3-yl) perfluorocyclopentene was characterized by significantly higher values (>0.7) compared to those (<0.16) recorded for the ring-opening reaction [115]. The quantum yields were determined by the mathematical expression, Eq. (4.4).

$$\Phi = \frac{\Delta A \ V}{P_0 \ \Delta t \left(1 - 10^{-A_\lambda}\right) \varepsilon} \qquad (4.4)$$

with ΔA being the increase or decrease of the absorbance at 600 *nm* recorded under exposure of the solution of volume V to a light flux P_0 over the period Δt, A_λ being the absorbance of the solution at the specific irradiation wavelength, and ε being the absorption coefficient at 600 nm.

Both quantum yields showed dependence on the monochromatic irradiation wavelength. The ring-closing (**35** → **36**) quantum yield varied between 0.71 and 0.92 in the *UV* range (334-254 nm) but was constant ($\Phi_{o \rightarrow c} = 0.7$) between 334 and 313 nm, whereas that of ring-opening (**36** → **35**) showed a ten-times increase (from 0.0015 to 0.016) when visible-range irradiation decreased from 609 to 404 nm. A negative linear relationship was obtained in acetonitrile for the ring-opening quantum yield ($\Phi_{c \rightarrow o} = -6.94 \ 10^{-5} \lambda + 0.044$). Even though the authors stated that no explanation can be advanced for the differences in quantum yields, they nonetheless estimated that such a behaviour might be due to the existence of an excited-state energy barrier for the closed-form isomer but not for its counterpart. The ring-closure reaction of a benzoyl-phenyl-ethynyl substituted 1,2-bis [2-methyl-thien-3-yl] perfluorocyclopentene takes place with a high quantum yield through a conical intersection, which is also a relaxation funnel for the closed isomer S_1-excited-state preceded by an energy barrier [116]. The closed form was proven to be a mixture of three isomers (*cis-cis*, *cis-trans*, and *trans-trans*) with different quantum yields. For solutions of the diaryethene in CH_2Cl_2 (5×10^{-4} to 5×10^{-5} M), the quantum yields were wavelength-dependent, varying between 0.9×10^{-5} and 2.5×10^{-4} for irradiation wavelengths situated in the 514–690 nm region. The quantum yield of the ring-closing reaction was reported as being dependent on both excitation wavelength and the doses of irradiation.

The choice of the excitation wavelength (297, 313, and 334 nm) has no effect on the quantum yield (~0.03) of ring-opening of perfluoro- and perhydroxy-cyclopentane diaryethene derivatives in acetonitrile (~2×10^{-5} M at 25 ° C), whereas under the same wavelength series, a 15% variation for the ring-closure (0.51–0.60) reaction quantum yield was recorded [117]. The quantum yields were determined at non-isosbestic wavelengths located in a region where the isomers' electronic spectra overlap. They were calculated by a method exempt from simplifying assumptions using pure photokinetic data, requiring the irradiation to be collimated and monochromatic but free from imposing any other special irradiation conditions [118], a method that shall be described in Chap. 17.

Monochromatic irradiation over the 410–535 nm range of low-concentrated ethanol solution of (1,2-bis[2-methylbenzo[*b*]-thiophen-3,3,4,4,5,5-hexafluoro-1-cyclopentene) is found (Chap. 18) to have a quantum yield of the ring-opening reaction varying with a distinctive sigmoid pattern [119]. A linear relationship ($\Phi^{\lambda_{irr}}_{C\rightarrow O} = -8\ 10^{-5} + 0.472$) was also obtained for this derivative in acetonitrile [120].

Three derivatives, (**37**) to (**39**), bearing two or three chromophoric diazonaphthoquinone units were investigated in acetonitrile (*ca.* 2 10^{-5} M) under various monochromatic radiations at 254, 350, 399, and 423 *nm* at 290 and 80 ° *K*. It was found that the photoconversion quantum yields, measured by two different methods, varied between 11 and 55% for the individual molecules over the selected series of irradiation wavelengths [121].

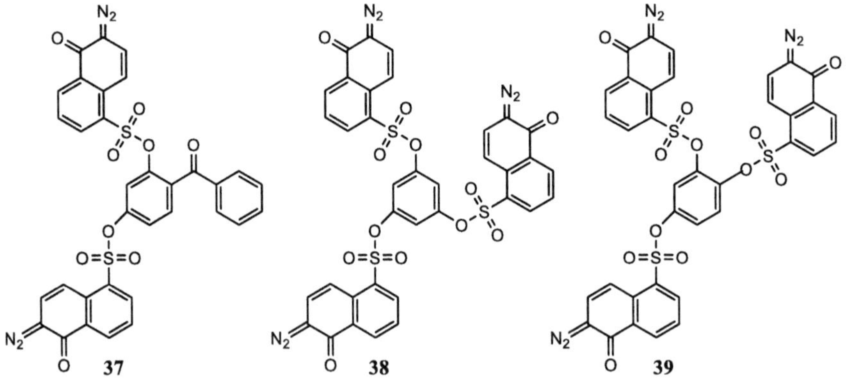

37 **38** **39**

By irradiating the $S_0 \rightarrow S_1$ transition of a spirooxazine derivative (**40**), 1,3-dihydro-3,3-dimethyl-1-isobutyl-6′-(2,3-dihydro-1*H*-indol-1-yl)spiro(2*H*-

indole-2,3′-3*H*-naphthop[2,1-*b*] [1, 4] oxazine), it was observed that the quantum yield of the photoconversion decreased with wavelength (0.22 and 0.24 at 410 and 355 nm, respectively). Whereas, excitation to a higher state ($S_0 \rightarrow S_2$), the quantum yield value first decreased to 0.1 at 316 nm, then increased to 0.53 at 295 nm (in a *V*-shaped pattern). These findings were interpreted as the photochemical yield being mainly determined by the vibrational mode of the excited singlet-state where the molecule is prepared [122].

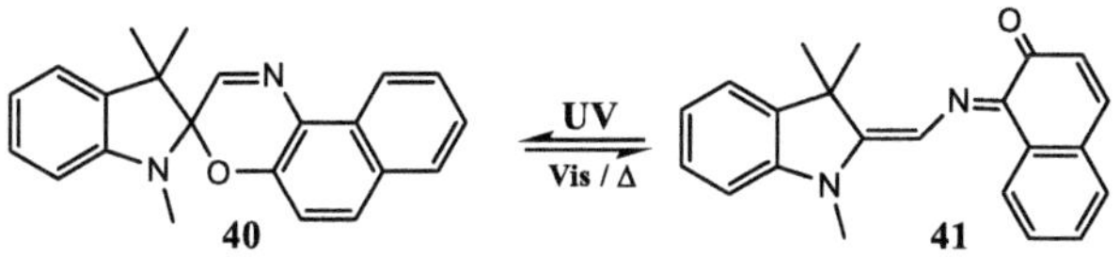

The ratios of photobleaching and photocolouration quantum yields of a series of spiro-naphthoxazines [123] were unequivocally proven to depend on the excitation wavelength at 352 and 414 nm on the basis of using a modified Fischer's method [124]. (That is, in this study, the expressions calculating the individual quantum yields were obtained from Fisher's method, assuming that the ratio of bleaching/colouration quantum yields was constant, and that the rates of the thermal reactions of ring-opening were not negligible).

The *pss* compositions as well as the quantum yields of a series of water-soluble spiropyrans (**42**) were shown to be excitation-wavelength sensitive. The ring-opening (**42** → **43**) reaction efficiencies underwent between 26% and 96% decrease for, respectively, irradiations occurring at 260 and 365 nm. A mixed increase and decrease of the quantum yield values at 530 nm was observed for different molecules [125].

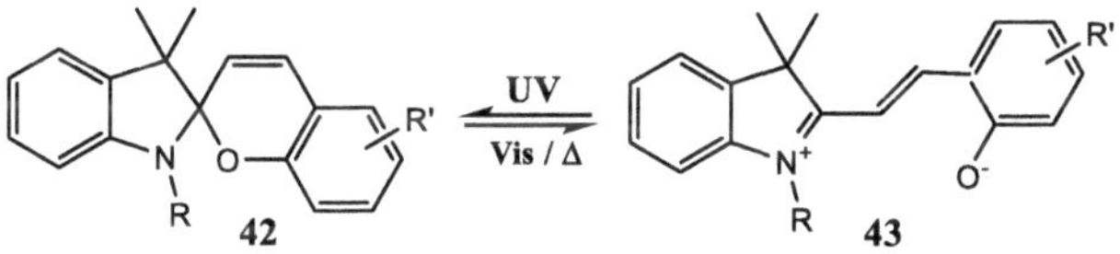

The $S_0 \rightarrow S_2$ (at 260 nm) excitation results in a more efficient reaction pathway than the $S_0 \rightarrow S_1$ (at 350 nm) excitation. The quantum yield was calculated by the ratio of the number of molecules that actually underwent phototransformation (N) to the number of photons absorbed by those molecules (N_λ), as

$$\Phi = \frac{N}{N_\lambda} \tag{4.5}$$

The photocolouration quantum yield of an oligothiophene substituted naphthopyran derivative decreased from 0.32 to 0.045 when monochromatic and continuous irradiation increased from 318 to 410 nm [126]. Similarly, the photochemical study of a biphotochromic naphthaopyran was carried out at diluted

(6×10^{-6} M) and highly (3×10^{-3} M) concentrated toluene solutions [127]. The quantum yield decreased by a third (from 0.32 to 0.11) with increasing wavelength of monochromatic beams spanning the 330–410 nm range, where measurements were performed at a 10-nm interval. Integration, initial rate, and differential methods were used for the estimation of the quantum yield at each irradiation wavelength. The dependence of the reaction quantum yield on wavelength within the S_1-excitation band, for both studies, was thought to indicate that the photoreaction occurs at a vibrational excited level $S_{1,\nu}$ in competition with vibrational relaxation.

Photoswitchable fulgides also represent a family of interest from the point of view of quantum yield variation with excitation. The experimental results showed, in general, that the quantum yield of the open- to closed-form isomer was slightly lower than that corresponding to the ring-opening quantum yield, and both recorded higher values for the reaction induced by visible light (λ_{irr} over 500 nm). These values varied between 0.15 and 0.20 (at *ca.* 400 nm, for ring-closing), ranged between 0.1 and 0.26 (in the region 300 and 400 nm, for ring-opening), and between 0.03 and 0.07 for irradiation in the visible range (500–600 nm) [128–131]. The data of quantum yield values measured at 15–20 nm intervals over the 340–700 nm spectral region has served to promote a trifluoromethyl-indolyl-fulgide derivative as a UV to NIR actinometer [132]. The quantum yield was calculated by Eq. (4.6) for concentrated solutions exposed to monochromatic beams (where the absorption of the reactive medium at the irradiation wavelength was $A_{\text{medium}}^{\lambda_{\text{irr}}} \sim 3$).

$$\Phi = \frac{\Delta n}{n_{\text{aborbed}}} = \frac{\Delta n}{P_0\, t} \tag{4.6}$$

where Δn is the molar amount (in mol) of the photo-species that reacted in time t, n_{aborbed} is the molar amount (in einsten) of photons absorbed, and P_0 is the photon flux (in einsten s^{-1}) of the light source that is absorbed by the sample.

The quantum yield of the ring-closing reaction was attributed to the $Z \rightarrow C$ (**45** → **46**) reaction-step, as a spectral unmixing of the irradiation series showed only a minor content of the E-isomer (**44**) in the reactive medium. The proposed mechanism included independent photoisomerisation of the excited-state species into their isomers ($Z^* \rightarrow C$ and $C^* \rightarrow Z$) as well as a thermal equilibrium of the ground-state isomers ($Z \leftrightharpoons C$). The latter process has been neglected for this particular molecule since both its forms were found thermally stable, as this reaction mechanism involves only photochemical reaction-steps.

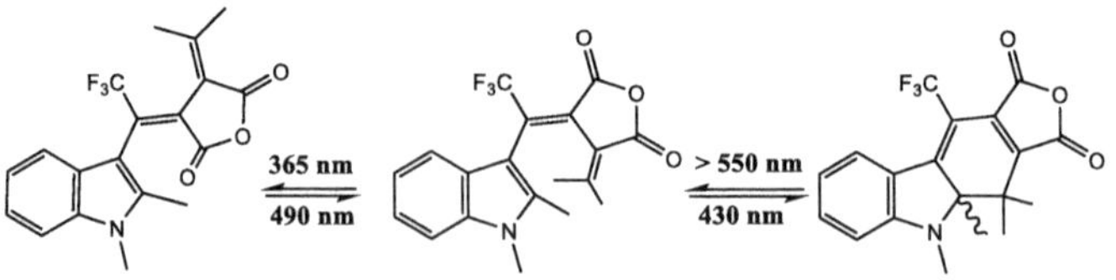

44 (E) **45 (Z)** **46 (C)**

Substantial photochemical production of the *E*-isomer, as well as thermal reactions between the ground-state species, have been observed for a number of fulgide derivatives [133, 134]. These features render the determination of the specific quantum yield values for each photoreaction-step of the reaction mechanism a tedious task. A consideration that must raise caution in regard to the quantum yield values reported in the literature for fulgides.

In general, the literature in the field reports on the determination of the quantum yields for thermally stable reactants but not for every photochemical step of the reaction mechanism (such as those leading to a photo-stationary-state, e.g. fulgides (**44**)–(**46**)). Some of such considerations are treated in Chaps. 16 and 18.

4.11 Vibronic States Photochemistry

Becker and his co-worker were perhaps the first to lay down, in the late 1960s [135], an interpretation directly linking photochemistry to vibronic states. The fluorescence excitation and absorption spectra of 2,2-diethylchromene were found to be very different [136], and the quantum yields of fluorescence ($\Phi_F^{\lambda_{irr}}$) of this compound as well as that of 5,6-benzochromene showed up to a 10-fold variation with the particular nature of which state was excited, including the electronic state, the vibrational mode and the vibrational level. The photoconversion of such molecules seemed to compete with fluorescence emission but not with phosphorescence (which was not observed, and the estimated quantum yield of *ISC* was low ~0.1) [137, 138]. A kinetic model (Eq. (4.7)) was proposed for the fluorescence quantum yield as a result of a competition between vibration relaxation (with a rate-constant, k_V) and photoreaction (with a rate-constant, k_{PC}) at the excited vibrational level, n [135].

$$\Phi_F(n) = \left(\frac{k_V}{k_V + k_{PC}}\right)^n \tag{4.7}$$

The analysis was later carried forward [139] to establish an expression (Eq. (4.8)) for the photoreaction quantum yield in relation to features of both photophysical and photochemical processes, in addition to, and for the first time in the field, taking into account a quantum yield of vibrational relaxation of a given excited vibrational level within a given mode of any electronic excited-state (Fig. 4.2).

$$\Phi_{PC}(n) = \left(\frac{k_{PC}}{k_V + k_{PC}}\right)^n \tag{4.8}$$

It was concluded that photochemistry ought to be competitive with vibrational relaxation at any excited level (electronic or vibrational/vibronic). A number of subsequent studies of various molecular structures have led to similar correlations

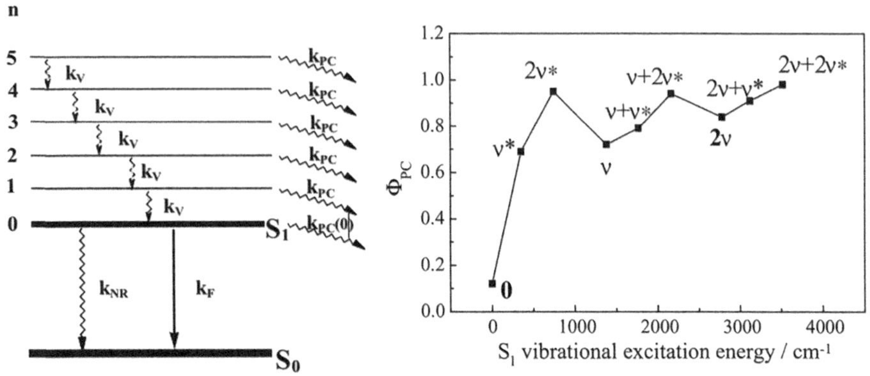

Fig. 4.2 (Left) Representation of possible photophysical and photochemical processes in the first excited state with the rate-constants for *VR* from $S_{1,\,n}$ to $S_{1,\,n-1}$ (k_V), *IC* (k_{NR}), fluorescence (k_F), and photoconversion (k_{PC}). (Right) Illustration of the variation of the photoconversion quantum yield of a non-fluorescent photochrome when excited at different vibrational levels of the first electronic excited-state. Reprinted with permission from [140], *Journal of Photochemistry and Photobiology. C: Photochemistry Reviews*

between photochemical quantum yield values and excited vibrational levels, as was described in a review [140]. It has been noted that when polychromatic light is used, the determined quantum yield of a given reaction is rather an average yield over several wavelengths, and therefore, the published data on quantum yield should be considered with caution. It is also worth mentioning that the model proposed by Becker and his co-workers always predicts higher quantum yield values for shorter excitation wavelengths due to the additive nature of the model. The latter point is shared with the model discussed above in Sect. 4.8 [53] that predicts a monotonic decrease of the quantum yield with increasing excitation wavelength. As such, these models do not seem suitable for many observed cases where the quantum yield value increased with wavelength.

A recent systematic review and analysis of reactions involving intersystem crossing, photoisomerisation, bond-breaking, and electron, proton and energy transfer has concluded to a competition of photoreaction with *VR*, radiative and non-radiative processes occurring in upper excited-states [141]. Analysis of experimental data suggested that both *VR* and *IC* to S_1 proceed at slower or comparable rates to those of photoconversion. Direct coupling of specific vibrations and photoconversion has been observed in many experiments [142, 143], and vibrational coherence has been detected for isomerisation [144], electron transfer [145, 146], and proton transfer [147]. A reversed *V*-shaped variation of the quantum yield with excitation wavelength (525, 550, 585, 625 and 600 nm) was recorded for 6-methyl-azulene-2-carboxilic acid (**47**) where the quantum yield value steeply decreased from ~0.8 at 525 nm to ~0.2 at 585 nm then gradually to ~0.1 at 600 nm [148].

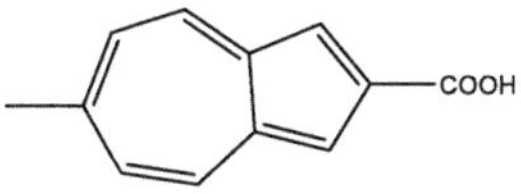

47

Evidence of the contribution of vibrational levels to photochemistry has been laid out for many other systems. The small chlorine dioxide (ClO_2) molecule undergoes, in solution, a vibrational mode-dependent photochemistry [149–151]. An interesting review focused on some photochemical reactions in the atmosphere that are driven not only by *UV*-light but also by low-energy visible light as observed at high solar zenith angles [152]. Absorbed near-IR and visible regions photons, promoting $X-H$ stretching vibrational overtones, contribute to atmospheric chemistry and play a role in the fate of certain organic compounds [153–156]. Three mechanisms were envisaged for these photoreactions. The excited OH overtone redistributes its vibrational energy within the molecules, such that enough energy is transferred into a different weaker bond of the molecule that ultimately breaks [157, 158]. A second pathway involves a concerted reaction whereby a strong coupling of the excited OH overtone with the reaction coordinate leads to the excitation of a low-frequency mode, causing the OH bond to break at much lower energies than the dissociation energy of the bond at its equilibrium geometry [154, 159]. Third, a two-photon absorption sequentially brings the overtone to an intermediate state from which the system is promoted to a repulsive dissociative first excited electronic state [160]. Similarly, the initial vibrational state preparation was found to strongly influence the photodissociation of HCNO by excitation of NH stretching overtone that enhances the breaking of the NH bond in its electronically excited-state [161].

A vibrational-mode-dependent excited-state lifetime was investigated for phenol under the S_1/S_2 conical intersection, where the O–H bond-rupture dynamics was interpreted by a possible topographical link to a lower energy minimum on the multidimensional potential energy surface [162].

The formaldehyde (CH_2O) photolysis depended on the excitation wavelength and the variation of the quantum yield of the radical channel (Fig. 4.3), was modelled by an expression (Eq. (4.9)) involving a sum of sigmoid-like functions [163, 164].

$$\Phi_{\mathrm{rad}}^{\lambda_{\mathrm{irr}}} = \frac{0.72}{1+e^{-\left(10^5/5.2\right)\left(1/\lambda_{irr}\ -\ 1/328\right)}} - \frac{0.38}{1+e^{-\left(10^5/4.7\right)\left(1/\lambda_{irr}\ -\ 1/278\right)}} \tag{4.9}$$

The excitation of vibrational states for this molecule is implicitly suggested by the narrow wavelength intervals used for irradiation (~5 nm), and the continuous variation of the measured quantum yield values (which increase then decrease with increasing wavelength).

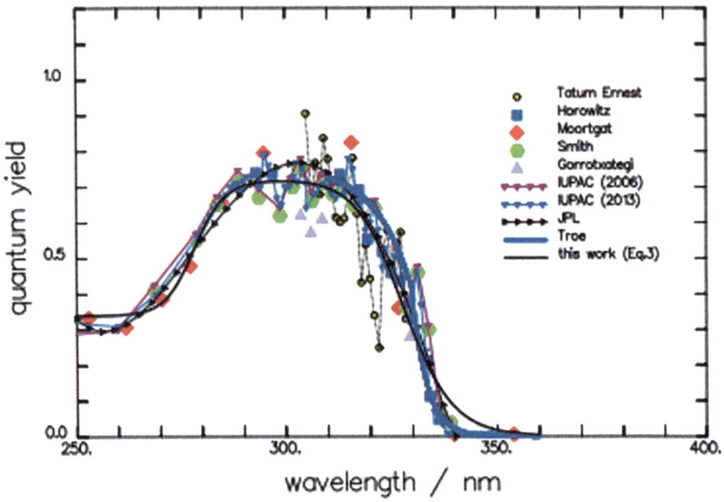

Fig. 4.3 Experimental variation of the radical channel quantum yield of the formaldehyde photolysis at room temperature, modelled by Eq. (14). Reprinted with permission from [163], *Atmospheric Chemical Journal*

4.12 Some Inorganic Actinometers

The variation of the quantum yield with excitation has also been reported for inorganic systems, including actinometers. Two illustration examples are shown here, where the reported data correspond to photonic yields rather than absolute quantum yield values. The patterns observed are different but surely convey an effect of excitation.

Nitrate and nitrite photoreactions are monitored by the photochemical productions of salicylic acid (Fig. 4.4a) and *p*-hydroxybenzoic acid, by the reaction of the photoproduced hydroxyl radical with benzoic acid [165].

Also, a series of studies revisited the properties of the ferrioxalate photoreaction that was originally proposed, for the first time in 1956 [166], for chemical actinometry. The latter inspiring work established the wavelength sensitivity of the quantum yield of the ferric/ferrous ions photoconversion. A compilation of quantum yield values from several studies, gathered in the handbook of photochemistry [54], clearly shows that the photoreduction quantum yield of the ferrioxalate ferric ions evolves within its dynamic 200–550 nm range, in a similar pattern to that observed for formaldehyde photolysis (Fig. 4.3), but with a majority of the values exceeding 1 (Fig. 4.4b). The high values of the quantum yield (exceeding unity) were attributed to the occurrence of thermal processes.

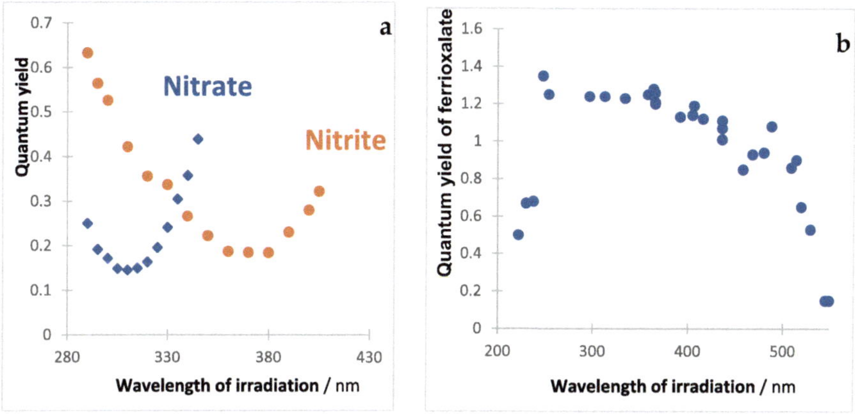

Fig. 4.4 (**a**) Interesting patterns of salicylic acid production quantum yields with irradiation wavelength, from the photoreaction of either nitrate or nitrites. Raw data taken from [165]. (**b**) Evolution of the photochemical quantum yield of $Fe^{3+} \rightarrow Fe^{2+}$ with irradiation wavelength in the ferrioxalate actinometer. Raw data taken from [54]

4.13 Miscellaneous

There are cases where reactivity variation was recorded, but quantum yields were not generally measured. The reactivity changes were due to changes in reaction/reactant features. Examples are found in industrial and technological applications. In this context, recent discussions on the quantum yield definitions and their respective equations, differentiated between the experimentally measured quantities and their actual interpretation [61, 84], and a more detailed description is available in Chaps. 5, 21 and 22. In essence, it is useful to remember that the quantum yield, a quantity measured at a given wavelength using strictly monochromatic light, is insensitive to reaction conditions such as initial concentration, radiation intensity, temperature and absorption competitors. However, the photonic yield (also labelled photonic efficiency or apparent quantum yield) is a relative quantity that is necessarily sensitive to a variation of such reaction conditions, and the only one obtained when the employed irradiation light is polychromatic. The relative character of the photonic yield meant that it is possible to observe its variation with some reaction conditions (e.g. medium temperature) while it may remain unaffected by a change of the excitation wavelength, as was observed for photoreactions of both 9- and 9,10-bromoantracenes [167].

Despite light-based stimulus has gained a noticeable popularity in drug delivery systems, long exposure to light remains a negative feature that might lead to phototoxicity (such as mutations, degradation of endogenous molecules, photocarcinogenesis, etc.). In this field, it has been recognised that the wavelength of the light employed is one of the important factors to reduce toxicity to acceptable levels or to suppress it altogether [168]. The sensitivity of DNA as a function of the irradiation wavelength (corresponding to the sun's spectrum reaching the surface of the earth) smoothly diminished, in a sigmoid pattern, between 260 and 360 nm [169]. The data was used to express the carcinogenic effectiveness of the light as a function of the light's wavelength and also to deduce a quantitative evaluation of the dose response in relation to *UV*-induced skin cancer.

In photocatalysis, semiconductor photocatalysts are increasingly characterised by the variation of the photonic yield with excitation wavelength (the so-called *action spectrum*), which ideally has the same shape as that of the fraction of radiation absorbed by the semiconductor photocatalyst with wavelength [170–172]. The action spectrum indicates a clear correlation between irradiation wavelength and photoreactivity. Similarly, photocurrent action spectra were used in order to study the influence of the geometrical structure with respect to the efficiency of thin film devices [173].

Also, reaction/reactant feature effects have long been observed on both forward and reverse quantum yields of photoreversible systems that depended on several reaction parameters such as viscosity, temperature, additives, initial concentration and excitation-wavelength [83].

For instance, a direct photolysis of polychlorinated biphenyls (*PCB*), in neutral 2-propanol at a series of irradiation wavelengths (254, 264, 294 and 300 nm) in a Rayonet photoreactor, showed a wavelength-dependence of the apparent de-chlorination quantum yield (ranging between 0.15 and 0.56) [174]. The apparent quantum yield was also found to increase with either increasing the initial concentration of the PCB or the intensity of the irradiation light. The findings of this study were proposed as a template for the remediation of PCBs. Similarly, photoproduct-formation (apparent) quantum yields of a series of fluorinated pesticides and pharmaceuticals were evaluated using LEDs emitting at 255, 275, 308, 365 and 405 nm. The fastest degradations were observed for irradiations with 255 or 275-nm compared to the other LEDs, and a different set of end-products was obtained with 365-nm irradiation [175].

Another interesting aspect in this area relates to the quantification of the quantum yield variation with structural changes of the molecule studied (e.g. in a series of derivatives) and with the consequent changes occurring on the ground-state absorption features of the individual derivatives. For instance, the photoinduced quantum yields of ring-closure and ring-opening reactions of a series of 1,2-dithienylperfluorocyclopentene derivatives followed a general trend where a lowering of the derivative transition energy led to a decrease of the quantum yield of the photoreaction of both open or closed isomers, as well as an increase of the quantum yield with increasing temperature [176]. The ring-opening reaction quantum yield increased roughly exponentially with increasing transition energy. A similar behaviour was also observed for a series of dithienylethenes with polyenic and carotenoid-like chromophores, where the quantum yields of the photochromic forward and back reactions sharply decreased with increasing polyenic chain length [177]. Such a behaviour was rationalised by the decrease of the excitation density at the central photoreactive unit and the concomitant shortening of the excited-state lifetime in the carotenoid compounds with increasing chain length. Also, an analogous dramatic decrease of the ring-opening quantum yields in a series of thiophene

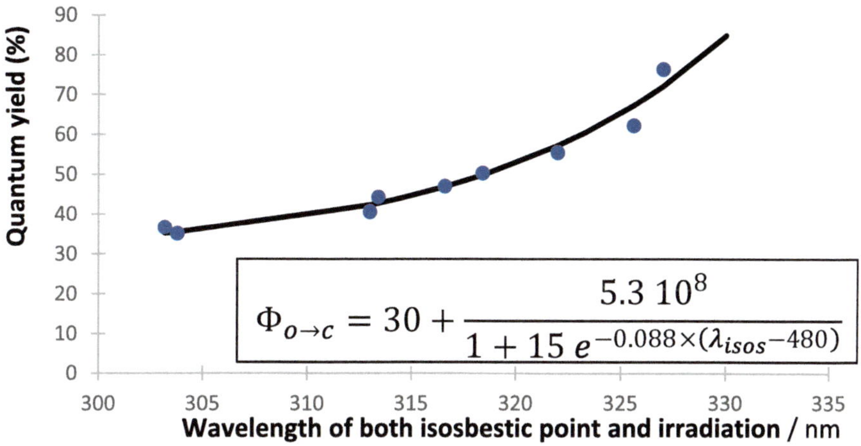

Fig. 4.5 A good sigmoid correlation (line) of the isosbestic point wavelength and the quantum yield values (dots, in %) at that wavelength for a series of bis(thienyl)ethene derivatives. The sigmoid function is indicated in the inset. Raw data taken from [179] with permission from *Materials Advances*. Reproduced with permission from [1] from *Photochem*

oligomers (including up to three thiophene rings) was observed when increasing the number of thiophene rings, whereas only a little variation was recorded for the ring-closure reaction [178]. The efficiency of the photoreactions was correlated to a threshold of the $S_0 \rightarrow S_1$ energy.

A quantification of such an effect was delivered for a series of six bis(thienyl)-ethenes, studied in THF organic solutions and *PMMA* films under irradiation from lasers or LED lights [179]. The quantum yield of the photo-colouring reaction was determined specifically at the isosbestic point of each derivative (located in the region 300–330 nm), and those of the photobleaching reactions were measured by irradiation at the maximum of each coloured isomer (in the region 574–605 nm). These quantum yields were determined by a spectrophotokinetic method [118, 180] based on an analytical integration of the rate-law of the reaction subjected to a monochromatic beam whose wavelength exactly corresponds to that of an isosbestic point.

A plot of the isosbestic point wavelength against the corresponding quantum yield value for the bis(thienyl)ethene series shows a good correlation as a sigmoid function (Fig. 4.5). It is an interesting structure/photoactivity feature.

Photoproduced diarylethene isomers are generally thermally stable, but a few exceptions were reported. The ring-opening reaction of hexafluoro- and hexahydro-diarylethene derivatives, in organic solutions, was strongly temperature-dependent [181]. The *close* → *open* thermal-rates at different temperatures were worked out from the initial linear portion of a logarithmic scale plot of the kinetic data. The observed rate of the photoinduced ring-opening was dependent on radiation intensity but not on the exponential component of its fitted traces. The authors attributed this feature to the presence of an excited-state thermal barrier.

4.14 Amending the Jablonski Diagram

The various features reported on in the present chapter point to a need to amend the Jablonski diagram in line with the recurrent findings in the fields of photophysics and photochemistry. The diagram ought to include the diversity of the possible pathways open to the excited molecule. Based on the data available, Fig. 4.6 attempts to capture this complexity, albeit in a synoptic manner. Mainly, it shows that luminescence and photochemistry might well proceed from molecular excited-states other than $S_{1,\,0}$ and $T_{1,\,0}$.

4.15 Concluding Remarks

The number of different chemical families exhibiting a photochemical quantum yield variation with excitation wavelength is arguably in support of an underlying fundamental process, and this rather challenges the view considering such variation as an exotic or rare phenomenon. A focus on the examination of the occurrence of such a phenomenon is recommended.

The cases displayed in this chapter make it obvious that the conjecture stating that the reaction quantum yield must be constant irrespective of the excitation

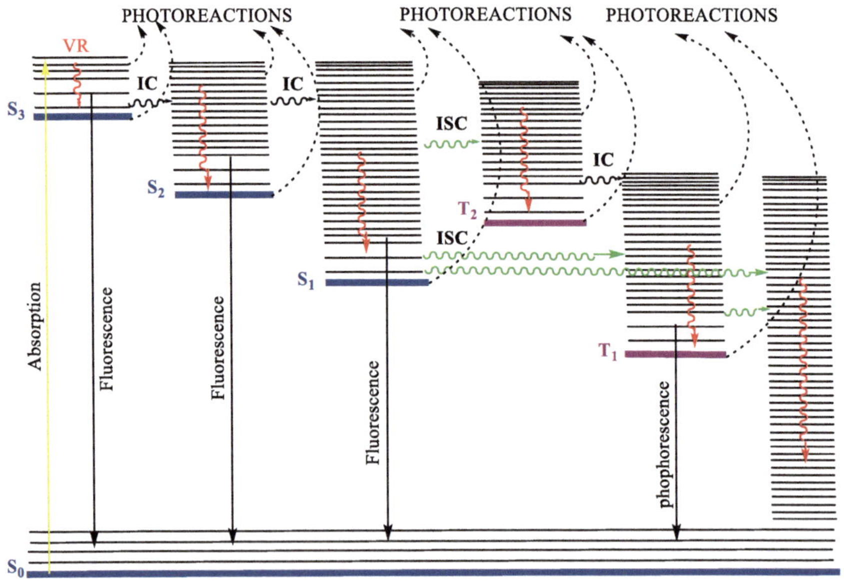

Fig. 4.6 An amended Jablonski diagram with pathways for photophysics and photochemistry emerging from hot and cold excited states, conveying possible competition of these processes with internal conversion and/or vibrational relaxation. Reprinted with permission from [1], *Photochem*

wavelength is proven false by an abundance of literature. Evidence of reactions proceeding from higher excited states and/or from hot excited-states has been mounting since the 1950s. The same has also been established for photophysical radiative processes. Competition of both luminescence and photochemistry with IC and/or VR opens new avenues of research and offers the possibility of selecting conditions suitable to specific needs and applications. As a result, Kasha's rule does not necessarily always hold for all emitters and is not automatically extendable to photochemistry. Such an observation also meant that the Jablonski diagram cannot remain unchanged and must be inclusive of IC and/or VR possible competitions with fluorescence, phosphorescence and photoreaction.

The community agrees, on one hand, that experiment is the only way to reveal which pattern will be taken by photoreactivity when changing the excitation wavelength (prediction of such pattern is not yet reliably accessible by any other means). The best metric for that purpose remains undoubtedly a measure of the photochemical quantum yield. Here, a distinction must be made between the quantum yield and the photonic yield. The former is an intrinsic feature of the molecule at a given wavelength that can only be determined by using strictly monochromatic light. Whereas, the latter is a relative quantity that can also be obtained for photoreactions subjected to polychromatic light. The photonic yield varies with changes in the experimental conditions of the reaction.

On the other hand, however, there is an obvious lack of consensus in the community on the standard way, formulae, and photokinetic methods by which the quantum yield of a photoreaction is experimentally determined. This is shown by the large number of approaches that have been adopted for the evaluation of the reaction quantum yield (a sample of expressions and methods has been provided in the present chapter). Standardisation, hence, becomes necessary. An issue that needs to be addressed by the community in order to strengthen the reliability of the data and to improve the quality of the conclusions that can be reached for the effects of excitation on reactivity. A contribution to the standardisation of these concepts will be provided in subsequent chapters.

The data available in the literature on the effect of excitation wavelength most often relate to a few (2 or 3) measurements, whereas a systematic investigation over wider wavelength ranges and larger number of actual measurements (preferably covering the whole absorption spectrum of the species studied, with measurements taken at narrow wavelength intervals) will provide a better understanding of the reaction quantum yield behaviour (as monotonical increase or decrease, or as a variable change with increasing irradiation wavelength).

Finally, a full explanation of the quantum yield dependence on wavelength is not yet available despite the many attempts to describe it. This is a future task that will certainly open up new subtleties of the excited-state.

References

1. Maafi M (2024) Excitation-wavelength dependent photochemistry. Photochemistry 4:233–270. https://doi.org/10.3390/photochem4020015
2. Kislyak A, Frisch H, Gernhardt M, Van Steenberge PHM, D'Hooge DR, Barner-Kowollik C (2020) Time-dependent differential and integral quantum yields for wavelength-dependent [4+4] photocycloadditions. Chem Eur J 26:478–484. https://doi.org/10.1002/chem.201903641
3. Ermolaev VL (2001) Ultrafast nonradiative transitions between higher excited states in organic molecules. Russ Chem Rev 70:471–490. https://doi.org/10.1070/RC2001v070n06ABEH000657
4. Tseng H-W, Shen J-Y, Kuo T-Y, Tu T-S, Chen Y-A, Demchenko AP, Chou P-T (2016) Excited-state intramolecular proton-transfer reaction demonstrating anti-Kasha behavior. Chem Sci 7:655–665. https://doi.org/10.1039/c5sc01945a
5. Bogdanova A, Popik VV (2003) Wavelength-dependent photochemistry of diazo meldrum's acid and its spirocyclic isomer, diazirino meldrum's acid: Wolff rearrangement versus isomerization. J Am Chem Soc 125(6):1456–1457. https://doi.org/10.1021/ja029528p
6. Demchenko AP (2002) The red-edge effects: 30 years of exploration. Luminescence 17:19–42. https://doi.org/10.1002/bio.671
7. Kasha M (1950) Characterization of electronic transitions in complex molecules. Discuss Faraday Soc 9:14–19. https://doi.org/10.1039/DF9500900014
8. Turro NJ, Ramamurthy V, Cherry W, Farneth W (1978) The effect of wavelength on organic photoreactions in solution. Reactions from upper excited states. Chem Rev 78:125–145. https://doi.org/10.1021/cr60312a003
9. Shi L, Xie X, Troisia A (2022) Rapid calculation of internal conversion and intersystem crossing rate for organic materials discovery. J Chem Phys 157:134106. https://doi.org/10.1063/5.0102857
10. Jabłoński A (1933) Efficiency of anti-stokes fluorescence in dyes. Nature 131:839–840. https://doi.org/10.1038/131839b0
11. Atkins P, de Paula J (2006) Atkins' physical chemistry. Oxford University Press, Oxford
12. Itoh T (2012) Fluorescence and phosphorescence from higher excited states of organic molecules. Chem Rev 112:4541–4568. https://doi.org/10.1021/cr200166m
13. Del Valle JC, Catalán J (2019) A reappraisal of Kasha's rule. Phys Chem Chem Phys 21:10061–10069. https://doi.org/10.1039/C9CP00739C
14. Vavilov SI (1922) The dependence of the intensity of the fluorescence of dyes upon the wavelength of the exciting light. Philos Mag J Sci 43(254):307–320. https://doi.org/10.1080/14786442208565217
15. Braslavsky SE (2007) Glossary of terms used in photochemistry (IUPAC). Pure Appl Chem 79:293–465. https://doi.org/10.1351/pac200779030293
16. Lewis GN, Kasha M (1944) Phosphorescence and the Triplet State. J Am Chem Soc 66:2100–2116. https://doi.org/10.1021/ja01240a030
17. Klan P, Wirz J (2009) Photochemistry of organic compounds: from concepts to practice. Wiley, Hoboken. https://doi.org/10.1002/9781444300017
18. Pianowski ZL (ed) (2022) Molecular photoswitches: chemistry, properties, and applications. Wiley, Hoboken
19. Wayne RP (1988) Principles and applications of photochemistry. Oxford University Press, Oxford
20. Rosspeintner A, Lang B, Vauthey E (2013) Ultrafast photochemistry in liquids. Annu Rev Phys Chem 64:247–271. https://doi.org/10.1146/annurev-physchem-040412-110146
21. Kumpulainen T, Lang B, Rosspeintner A, Vauthey E (2017) Ultrafast elementary photochemical processes of organic molecules in liquid solution. Chem Rev 117:10826–10939. https://doi.org/10.1021/acs.chemrev.6b00491

22. Livshits MY, Wang L, Vittardi SB, Ruetzel S, King A, Brixner T, Rack JJ (2020) An excited state dynamics driven reaction: wavelength-dependent photoisomerization quantum yields in $[Ru(bpy)_2(dmso)_2]^{2+}$. Chem Sci 11:5797–5807. https://doi.org/10.1039/d0sc00551g
23. Turro NJ, Ramamurthy V, Scaiano JC (2010) Modern molecular photochemistry of organic molecules. University Science Books, Melville
24. Yu J, Ma H, Huang W, Liang Z, Zhou K, Lv A, Li X-G, He Z (2021) Purely organic room-temperature phosphorescence endowing fast intersystem crossing from through-space spin-orbit coupling. J Am Chem Soc 1:1694–1699. https://doi.org/10.1021/jacsau.1c00290
25. Viswanath G, Kasha M (1956) Confirmation of the anomalous fluorescence of azulene. J Chem Phys 24:574–577. https://doi.org/10.1063/1.1742548
26. Levine SZ, Knight AR, Steer RP (1974) Fluorescence from the second excited singlet state of thiophosgene vapour. Chem Phys Lett 29:73–76. https://doi.org/10.1016/0009-2614(74)80137-5
27. Oka T, Knight AR, Steer RP (1975) A spectroscopic study of fluorescence from the second excited singlet state of thiophosgene vapor. J Chem Phys 63:2414–2420. https://doi.org/10.1063/1.431670
28. Steer RP (1981) Structure and decay dynamics of electronic excited states of thiocarbonyl compounds. Rev Chem Intermed 4:1–41. https://doi.org/10.1007/bf03052411
29. Acuña AU, Catalán J, Toribio F (1981) Photon energy relaxation and thermal effects on gas-phase electronically excited methyl salicylate. J Phys Chem 85:241–245. https://doi.org/10.1021/j150603a006
30. Catalán J, Del Valle JC, Palomar J, Díaz C, De Paz JLG (1999) The six-membered intramolecular hydrogen bond position as a switch for inducing an excited state intramolecular proton transfer (ESIPT) in esters of o-hydroxynaphthoic acids. J Phys Chem A 50:10921–10934. https://doi.org/10.1021/jp992631p
31. Binsch G, Heilbronner E, Jankow R, Schmidt D (1967) On the fluorescence anomaly of azulene. Chem Phys Lett 1:135–138. https://doi.org/10.1016/0009-2614(67)85008-5
32. Birks JB (1972) The photophysics of azulene. Chem Phys Lett 17:370–372. https://doi.org/10.1016/0009-2614(72)87099-4
33. Prlj A, Begušić T, Zhang ZT, Fish GC, Wehrle M, Zimmermann T, Choi S, Roulet J, Moser J-E, Vaníček J (2020) Semiclassical approach to photophysics beyond Kasha's rule and vibronic spectroscopy beyond the condon approximation. The case of azulene. J Chem Theory Comput 16:2617–2626. https://doi.org/10.1021/acs.jctc.0c00079
34. Geldof PA, Rettschnick RPH, Hoytink GJ (1969) Fluorescence from the second excited singlets of pyrene and 3,4-benzpyrene. Chem Phys Lett 4:59–61. https://doi.org/10.1016/0009-2614(69)8506
35. Peterson DL, Lytle FE, Laurendeau NM (1986) Determination of flame temperature using the anomalous fluorescence of pyrene. Opt Lett 11:345–347. https://doi.org/10.1364/OL.11.000345
36. Easterly CE, Christophorou LG, Carter JG (1973) Fluorescence from the second excited π-singlet state of aromatic hydrocarbons in solution. J Chem Soc Faraday Trans 69:471–483. https://doi.org/10.1039/F29736900471
37. Nakajima A, Baba H (1970) Fluorescence spectrum of pyrene vapor: emission from the second excited singlet state. Bull Chem Soc Jpn 43:967–967. https://doi.org/10.1246/bcsj.43.967
38. Baba H, Nakajima A, Aoi M, Chihara KJ (1971) Fluorescence from the second excited singlet state and radiationless processes in pyrene vapor. J Chem Phys 55:2433–2438. https://doi.org/10.1063/1.1676429
39. Choi IS, Lee SK (1996) Analysis of rotationally cooled vibronic emission spectra ($1^2A_2 \rightarrow 1^2B_2$) of m-Xylyl radical. Bull Korean Chem Soc 17:749–753
40. Uttamlal M, Holmes-Smith AS (2008) The excitation wavelength dependent fluorescence of porphyrins. Chem Phys Lett 454:223–228. https://doi.org/10.1016/j.cplett.2008.02.012

41. Malpicci D, Lucenti E, Giannini C, Forni A, Botta C, Cariati E (2021) Prompt and long-lived anti-Kasha emission from organic dyes. Molecules 26:6999. https://doi.org/10.3390/molecules2622699
42. Pavlovich VS (1976) Dependence of excitation spectra of solutions of dipolar molecules on registration wavelength. Zh Prikl Spektr 25:480–487
43. Cushing SK, Li M, Huang F, Wu N (2014) Origin of strong excitation wavelength dependent fluorescence of graphene oxide. ACS Nano 8:1002–1013. https://doi.org/10.1021/nn405843d
44. Samanta A (2006) Dynamic Stokes shift and excitation wavelength dependent fluorescence of dipolar molecules in room temperature ionic liquids. J Phys Chem B 110:13704–13716. https://doi.org/10.1021/jp060441q
45. Khara DC, Samanta A (2010) Solvation dynamics and red-edge effect of two electrically charged solutes in an imidazolium ionic liquid. Indian J Chem 49:714–720
46. Józefowicz M, Heldt JR (2011) Excitation-wavelength dependent fluorescence of ethyl 5-(4-aminophenyl)-3-amino2,4-dicyanobenzoate. J Fluoresc 21:239–245. https://doi.org/10.1007/s10895-010-0711-4
47. Galley WC, Purkey RM (1970) Role of heterogeneity of the solvation site in electronic spectra in solution. Proc Natl Acad Sci USA 67:1116–1121. https://doi.org/10.1073/pnas.67.3.1116
48. Milton JC, Purkey RM, Galley WC (1978) The kinetics of solvent reorientation in hydroxylated solvents from the exciting-wavelength dependence of chromophore emission spectra. J Chem Phys 68:5396–5404. https://doi.org/10.1063/1.435704
49. Khisamov RM, Ryadun AA, Sukhikh TS, Konchenko SN (2021) Excitation wavelength-dependent room-temperature phosphorescence: unusual properties of novel phosphinoamines. Mol Syst Des Eng 6:1056–1065. https://doi.org/10.1039/D1ME00117E
50. Xiao G, Ma Y-J, Qi Z, Fang CT, Yan D (2024) A flexible ligand and halogen engineering enable one phosphor-based full-colour persistent luminescence in hybrid perovskitoids. Chem Sci 15:3625–3632. https://doi.org/10.1039/D3SC06845E
51. Holovka JM, Gardner PD (1967) Photolysis and photoisomerization of the benzene oxide-oxepin system. J Am Chem Soc 89:6390–6391. https://doi.org/10.1021/ja01000a092
52. Dauben WG, Kellog MS (1971) Wavelength dependency in the photochemistry of 1,3-cyclohexadienes cis-Bicyclo [4.3.0]nona-2,4-diene to cis,cis,trans-1,3,5-cyclononatriene valence tautomerism. J Am Chem Soc 93:3805–3807. https://doi.org/10.1021/ja00744a062
53. Zimmerman G, Chow L-Y, Paik U-J (1958) The photochemical isomerization of azobenzene. J Am Chem Soc 80:3528–3531. https://doi.org/10.1021/ja01547a010
54. Montali M, Credi A, Prodi L, Gandolfi MT (2006) Handbook of photochemistry.3rd edn. Taylor and Francis, Boca Raton
55. Rau H (1995) Azo compounds. In: Neckers DC, Volman DH, von Bunau G (eds) Advances in photochemistry, vol 19. Wiley, New York
56. Rau H (2003) Photochromism: molecules and systems, vol 1. Elsevier, Amsterdam
57. Gagliardi L, Orlandi G, Bernardi F, Cembran A, Garavelli M (2004) A theoretical study of the lowest electronic states of azobenzene: the role of torsion coordinate in the cis-trans photoisomerization. Theor Chem Accounts 111:363–372. https://doi.org/10.1007/s00214-003-0528-1
58. Ladányi V, Dvořák P, Al-Anshori J, Vetráková L, Wirz J, Heger D (2017) Azobenzene photoisomerization quantum yields in methanol redetermined. Photochem Photobiol Sci 16:1757–1761. https://doi.org/10.1039/C7PP00315C
59. Ziani Z, Bellatreccia C, Battaglia FP, Morselli G, Gradone A, Ceroni P, Villa M (2024) Copper indium sulfide quantum dots enabling quantitative visible light photoisomerisation of (E)-azobenzene chromophores. Nanoscale. https://doi.org/10.1039/D4NR01997K
60. Maafi M, Brown RG (2007) The kinetic model for AB(1Φ) systems. A closed-form integration of the differential equation with a variable photokinetic factor. J Photochem Photobiol A Chem 187:319–324. https://doi.org/10.1016/j.jphotochem.2006.10.030
61. Maafi M (2023) On photokinetics under monochromatic light. Front Chem 11:1233151. https://doi.org/10.3389/fchem.2023.1233151

62. Mauser H, Gauglitz G, Compton RG, Hancock G (1998) Comprehensive chemical kinetics, photokinetics: theoretical fundamentals and applications, vol 36. Elsevier, Amsterdam
63. Stranius K, Börjesson K (2017) Determining the photoisomerization quantum yield of photoswitchable molecules in solution and in the solid state. Sci Rep 7:41145. https://doi.org/10.1038/srep41145
64. Gade R, Porada T (1997) Determination of quantum yields and product spectra by studying reversible photoisomerizations in solution. J Photochem Photobiol A Chem 107:27–34. https://doi.org/10.1016/S1010-6030(96)04594-7
65. Deniel MH, Lavabre D, Micheau JC (1999) Photokinetics under continuous irradiation. In: Crano JC, Guglielmetti RJ (eds) Organic photochromic and thermochromic compounds. Kluwer Academic/Plenum Publishers, New York. https://doi.org/10.1007/b114211
66. Delbaere S, Vermeersh G, Micheau JC (2011) Quantitative analysis of the dynamic behaviour of photochromic systems. J Photochem Photobiol C Chem Photochem Rev 12:74–105. https://doi.org/10.1016/j.jphotochemrev.2011.05.004
67. Lewis GN, Milgel TT, Lipkin D (1940) The absorption and re-emission of light by cis- and trans-stilbenes and the efficiency of their photochemical isomerization. J Am Chem Soc 62: 2973–2980. https://doi.org/10.1021/ja01868a024
68. Olson AR (1931) The study of chemical reactions from potential energy diagrams. Trans Faraday Soc 27:69–76. https://doi.org/10.1039/TF9312700069
69. Hartley GS (1938) The cis-form of azobenzene and the velocity of the thermal cis→trans-conversion of azobenzene and some derivatives. J Chem Soc 1938:633–642. https://doi.org/10.1039/JR9380000633
70. Dhammika Bandara HM, Burdette SC (2012) Photoisomerization in different classes of azobenzene. Chem Soc Rev 41:1809–1825. https://doi.org/10.1039/C1CS15179G
71. Diau EW-G (2004) A new trans-to-cis photoisomerization mechanism of azobenzene on the $S_1(n,\pi^*)$ surface. J Phys Chem A 108:950–956. https://doi.org/10.1021/jp031149a
72. Samai S, Bradley DJ, Choi TLY, Yan Y, Ginger DS (2017) Temperature dependent photoisomerization quantum yield for azobenzene-modified DNA. J Phys Chem C 121(12): 6997–7004. https://doi.org/10.1021/acs.jpcc.6b12241
73. Shi Y-B, Hearst JE (1987) Wavelength dependence for the photoreactions of DNA-psoralen monoadducts. 2: photo-cross-linking of monoadducts. Biochemist 26:3792–9798. https://doi.org/10.1021/bi00387a009
74. Schneider EE, Goodeve CF, Lythgoe RJ (1939) Proc Natl Acad Sci USA 170:102–112
75. Kim JE, Tauber MJ, Mathies RA (2001) Wavelength-dependent cis-trans isomerization in vision. Biochemistry 40:13774–13778. https://doi.org/10.1016/S0006-3495(03)75054-1
76. Kim JE, Tauber MJ, Mathies RA (2003) Analysis of the mode specific excited-state energy distribution and wavelength-dependent photoreaction quantum yield in rhodopsin. Biophys J 84:2492–2501. https://doi.org/10.1016/S0006-3495(03)75054-1
77. Hahn S, Stock G (2000) Quantum mechanical modelling of the femtosecond isomerization of rhodopsin. J Phys Chem 104:1146–1149. https://doi.org/10.1021/jp992939g
78. Tscherbul TV, Brumer P (2014) Escitation of biomolecules with incoherent light: quantum yield for the photoisomerization of model retinal. J Phys Chem 118:3100–3111. https://doi.org/10.1021/jp501700t
79. Balke DE, Becker RS (1968) Relationship between the absorption and excitation spectra and relative quantum yields of fluorescence of all-trans-retinal. J Am Chem Soc 90:6710–6711. https://doi.org/10.1021/ja01026a026
80. Dartnall HJA (1968) The photosensitivities of visual pigments in the presence of hydroxylamine. Vis Res 8:339–358. https://doi.org/10.1016/0042-6989(68)90104-1
81. Stensitzki T, Muders V, Schlesinger R, Heberle J, Heyne K (2015) The primary photoreaction of channelrhodopsin-1: wavelength dependent photoreactions induced by ground-state heterogeneity. Front Mol Biosci 2:41. https://doi.org/10.3389/fmolb.2015.00041
82. Schulte-Frohlinde D, Blume H, Gusten H (1962) Photochemical cis-trans-isomerization of substituted stilbene. J Phys Chem 66:2486–2491. https://doi.org/10.1021/j100818a039

83. Gorner H, Kuhn HJ (1995) (1995). Cis-trans isomerization of stilbenes and stilbene-like molecules. In: Neckers DC, Volman DH, von Bunau G (eds) Advances in photochemistry, vol 19. Wiley, New York
84. Maafi M (2024) On photokinetics under polychromatic light. Front Chem 12:1367276. https://doi.org/10.3389/fchem.2024.1367276
85. Gardner C, Moer JF (1969) Photochemische Reaktionen. 50. Mitteilung [1]. Zur Photochemie von gesättigten β-Ketosulfiden II. Synthese von 6-Hydroxy-2-thia-7-oxa-isotwistan. Helv Chim Acta 52:725–737. https://doi.org/10.1002/hlca.19690520322
86. Pacifici JG, Diebert C (1969) Wavelength effect in photolysis of 4,4,6,6-tetramethyl-1-thiaspiro [2.3]hexan-5-one. J Am Chem Soc 91:4595. https://doi.org/10.1021/ja01044a075
87. Gisin M, Wirz J (1976) Photolysis of the azo-precursors of 2,3- and 1, s-naphtoquinodimethane. Helv Chim Acta 59:2273–2277. https://doi.org/10.1002/hlca.19760590644
88. Pagni RM, Burnett MN, Dodd JR (1977) Photochemistry of 1,4-Dihydronaphtho [1,8- de] [1,2] diazepine. Preparation and electron spin resonance observation of the uinsubstituted 1,8-Naphthoquinodimethane. J Am Chem Soc 99:1972–1973. https://doi.org/10.1021/ja00448a048
89. Bochet CG (2004) Chromic orthogonality in organic synthesis. Synlett 13:2268–2274. https://doi.org/10.1002/chin.200508293
90. Bochet CG (2006) Photochemical release of various functional groups. Pure Appl Chem 78: 241–247. https://doi.org/10.1351/pac200678020241
91. Protti S, Ravelli D, Fagnoni M (2019) Wavelength-dependence and wavelength-selectivity in photochemical reactions. Photochem Photobiol Sci 18:2094–2101. https://doi.org/10.1039/C8PP00512E
92. Bochet CG (2000) Wavelength-selective cleavage of photolabile protecting groups. Tetrahedron Lett 41:6341–6346. https://doi.org/10.1016/S0040-4039(00)01050-9
93. Bochet CG (2001) Orthogonal photolysis of protecting groups. Angew Int Ed 40:2071–2073. https://doi.org/10.1002/1521-3773(20010601)40:11<2071::AID-ANIE2071>3.0.CO;2-9
94. Stegmaier P, Alonso JM, del Campo A (2008) Photoresponsive surfaces with two independent wavelength-selective functional levels. Langmuir 24(20):11872–11879. https://doi.org/10.1021/la802052u
95. Kotzur N, Briand B, Beyermann M, Hagen V (2009) Wavelength-selective photoactivatable protecting groups for thiols. J Am Chem Soc 25:16927–16931. https://doi.org/10.1021/ja907287n
96. Kantevari S, Matsuzaki M, Kanemoto Y, Kasai H, Ellis-Davies GCR (2010) Nat Methods 7: 123–125. https://doi.org/10.1038/nmeth.1413
97. San Miguel V, Bochet CG, del Campo A (2011) Wavelength-selective caged surfaces: How many functional levels are possible? J Am Chem Soc 133:5380–5388. https://doi.org/10.1021/ja110572j
98. Schafer F, Joshi KB, Fichte MAH, Mack T, Wachtveitl J, Heckel A (2011) Wavelength-selective uncaging of dA and dC residues. Org Lett 13:1450–1453. https://doi.org/10.1021/ol200141v
99. Cui J, San Miguel V, del Campo A (2013) Light-triggered multifunctionality at surfaces mediated by photolabile protecting groups. Macromol Rapid Commun 34:310–329. https://doi.org/10.1002/marc.201200634
100. Garcia-Fernández L, Herbivo C, San Miguel Arranz V, Warther D, Donato L, Specht A, del Campo A (2014) Dual photosensitive polymers with wavelength-selective photoresponse. Adv Mater 26:5012–5017. https://doi.org/10.1002/adma.201401290
101. Hansen MJ, Velema WA, Lerch MM, Szymanski W, Feringa BL (2015) Wavelength-selective cleavage of photoprotecting groups: strategies and applications in dynamic systems. Chem Soc Rev 2015(44):3358–3377. https://doi.org/10.1039/C5CS00118H

102. Corrigan N, Ciftci M, Jung K, Boyer C (2021) Mediating reaction orthogonality in polymer and materials science. Angew Chem Int Ed 60:1748–1781. https://doi.org/10.1002/anie.201912001
103. Kaufmann J, Sinsel F, Heckel A (2023) Chromatic selectivity of coumarin-caged oligonucleotides. Chem Eur J 29:e202204014. https://doi.org/10.1002/chem.202204014
104. Dolinski ND, Page ZA, Callaway EB, Eisenreich F, Garcia RV, Chavez R, Bothman DP, Hecht S, Zok FW, Hawker CJ (2018) Solution mask liquid lithography (SMaLL) for one-step, multimaterial 3D printing. Adv Mater 30:1800364. https://doi.org/10.1002/adma.201800364
105. Menzel JP, Feist F, Tuten B, Weil T, Blinco JP, Barner-Kowollik C (2019) Light-controlled orthogonal covalent bond formation at two different wavelengths. Angew Chem Int Ed 58: 7470–7474. https://doi.org/10.1002/anie.201901275
106. Birks JB (ed) (1973) Organic molecular photophysics, vol I. Wiley-Interscience, New York
107. Nobs F, Burger U, Schaffner K (1977) Specifically ($\pi \rightarrow \pi^*$)-induced cyclohexenone reactions: 4a-(Z-1-propenyl)-bicyclo [4.4.0]dec-1(8a)-en-2-one and 4a-Z-1-propenyl-bicyclo [4.4.0]deca-1(8a), 7-dien-2-one. Helv Chim Acta 60:1607–1628. https://doi.org/10.1002/hlca.19770600517
108. Adriano N, Ahearn C, Black C, Cracchiolo M, Ghere D, Nuñez A, Olivan L, Patel R, Saner S, Smith KR, Watkins B, Hare PM (2022) Solvent- and wavelength-dependent photolysis of estrone. Photochem Photobiol 98:783–797. https://doi.org/10.1111/php.13542
109. Bridge NK (1960) Effect of wavelength on the flash photolysis of quinones. Trans Faraday Soc 56:1001–1007. https://doi.org/10.1039/TF9605601001
110. Gloor J, Schaffner K (1974) Spezifisch $\pi \rightarrow \pi^*$-induzierte Reaktionen von γ-Dimethoxymethylcyclo hexen-2-onen: 1,3-umlagerung und wasserstoffabstraktion durch das α-kohlenstoffatom. Helv Chim Acta 57:1815–1845. https://doi.org/10.1002/hlca.19740570631
111. Dunkel P (2022) Photoremovable protecting groups. Encyclopedia 2:1225–1236. https://doi.org/10.3390/encyclopedia2030082
112. Saracini C, Liakos DG, Zapata Rivera JE, Neese F, Meyer GJ, Karlin KD (2014) Excitation wavelength dependent O2 release from copper(II)-superoxide compounds: laser flash-photolysis experiments and theoretical studies. J Am Chem Soc 136:1260–1263. https://doi.org/10.1021/ja4115314
113. Ziani Z, Loiseau F, Lognon E, Boggio-Pasqua M, Philouze C, Cobo S, Royal GC (2021) Synthesis of a negative photochrome with high switching quantum yields and capable of singlet-oxygen production and storage. Chem Eur J 27:1–13. https://doi.org/10.1002/chem.202103003
114. El-Hendawy MM, Fayed TA, Awad MK, English NJ, Etaiw SEH, Zaki AB (2015) Photophysics, photochemistry and thermal stability of diarylethene-containing benzothiazolium species. J Photochem Photochem Chem A 301:20–31. https://doi.org/10.1016/j.jphotochem.2014.12.015
115. Ribeiro A, Ballardini R, Belser P, Gandolfi MT, Iyer VM, Moggi L (2009) Photochemical investigation of photochromic diaryethene compound that can be used as a wide range actinometer. Photochem Photobiol Sci 8:1734–1742. https://doi.org/10.1039/b9pp00037b
116. Ern J, Bens AT, Martin H-D, Kuldova K, Trommsdorff HP, Kryschi C (2002) Ring-opening and -closure reaction dynamics of a photochromic dithienylethene derivative. J Phys Chem A 2002(106):1654–1660. https://doi.org/10.1021/jp012614
117. Herder M, Schmidt BM, Grubert L, Patzel M, Schwarz J, Hecht S (2015) Improving the Fatigue Resistance of Diarylethene Switches. J Am Chem Soc 137:2738–2747. https://doi.org/10.1021/ja513027s
118. Maafi M, Brown RG (2008) Kinetic analysis and elucidation options for AB(1k,2Φ) systems. New spectrokinetic methods for photochromes. Photochem Photobiol Sci 7:1360–1372. https://doi.org/10.1039/B807556E
119. Maafi M, Al-Qarni MA (2022) Mono- and polychromatic light diarylethene-actinometer for the visible range. Dyes Pigments 198:109942. https://doi.org/10.1016/j.dyepig.2021.109942

120. Sumi T, Takagi Y, Yagi A, Morimoto M, Irie M (2014) Photoirradiation wavelength dependence of cycloreversion quantum yields of diarylethenes. Chem Commun 50:3928–3930. https://doi.org/10.1039/c4cc00396a
121. Cipolloni M, Gentili PL, Ortica F, Becker RS, Favaro G (2011) Effects of solvent, excitation wavelength, and concentration on the photobehavior of some diazonaphthoquinones. Arkivoc 9:205–220. https://doi.org/10.3998/ark.5550190.0012.915
122. Di Nunzio MR, Gentili PL, Romani A, Favaro G (2008) Photochromic, thermochromic, and fluorescent spirooxazines and naphthopyrans: a spectrokinetic and thermodynamic study. ChemPhysChem 9:768–775. https://doi.org/10.1002/cphc.200700789
123. Wilkinson F, Hobley J, Naftaly M (1992) Photochromism of spiro-naphthoxazines : molar absorption coefficients and quantum efficiencies. J Chem Soc Faraday Trans 88:1511–1517. https://doi.org/10.1039/FT9928801511
124. Fischer E (1967) Calculation of photostationary states in systems A → B when only A is known. J Phys Chem 71:3704–3706. https://doi.org/10.1021/j100870a063
125. Kaiser C, Halbritter T, Heckel A, Wachtveitl J (2017) Thermal, photochromic and dynamic properties of water-soluble spiropyrans. ChemistrySelect 2:4111–4123. https://doi.org/10.1002/slct.201700868
126. Ortica F, Smimmo P, Favaro G, Mazzucato U, Delbaere S, Venec D, Vermeersch G, Frigoli M, Moustrou C, Samat A (2004) Effect of oligothiophene substituents on the photophysical and photochemical properties of a naphthopyran. Photochem Photobiol Sci 3: 878–885. https://doi.org/10.1039/B406801G
127. Ortica F, Moustrou C, Berthet J, Favaro G, Samat A, Guglielmetti R, Vermeersch G, Mazzucato U (2003) Comprehensive photokinetic and NMR study of a biphotochromic supermolecule involving two naphthopyrans linked to a central thiophene unit through acetylenic bonds. Photochem Photobiol 78:558–566. https://doi.org/10.1562/0031-8655(2003)0780558CPANSO2.0.CO2
128. Yokoyama Y, Tekahashi Y (1996) Trifluoromethyl-substituted photochromic indolylfulgide. A remarkably durable fulgide towards photochemical and thermal treatments. Chem Lett 25: 1037–1038. https://doi.org/10.1246/cl.1996.1037
129. Wolak MA, Gillespie NB, Thomas CJ, Birge RR, Lees WJ (2001) Optical properties of photochromic fluorinated indolylfulgides. J Photochem Photobiol A Chem 144:83–91. https://doi.org/10.1016/S1010-6030(01)00550-0
130. Wolak MA, Gillespie NB, Thomas CJ, Birge RR, Lees WJ (2003) Thermolysis of fluorinated cycloalkylidene fulgides yields a new class of photochromic compounds. Chem Commun 2003:992–933. https://doi.org/10.1039/B300010A
131. Cordes T, Malkmus S, DiGirolamo JA, Lees WJ, Nenov A, de Vivie-Riedle R, Braun M, Zinth W (2008) Accelerated and efficient photochemistry from higher excited electronic states in fulgide molecules. J Phys Chem A 112:13364–13371. https://doi.org/10.1021/jp807097w
132. Reinfelds M, Hermanns V, Halbritter T, Wachtveitlz J, Braun M, Slanina T, Heckel A (2019) A robust, broad-absorbing fulgide derivative as a universal chemical actinometer for UV to NIR region. ChemPhotoChem 3:441–449. https://doi.org/10.1002/cptc.201900010
133. Durr H, Bounas-Laurent H (eds) (2003) Photochromism: molecules and systems, vol 1. Elsevier, Amsterdam
134. Tian H, Zhang J (eds) (2016) Photochromic materials: preparation, properties and applications. Wiley, Hoboken
135. Becker RS, Dolan E, Balke DE (1969) Vibronic effects in photochemistry. Competition between internal conversion and photochemistry. J Chem Phys 50:239–245. https://doi.org/10.1063/1.1670783
136. Becker RS, Michl J (1966) Photochromism of synthetic and naturally occurring 2H-chromenes and 2H-pyrans. J Am Chem Soc 88:5931–5933. https://doi.org/10.1021/ja00976a044
137. Lenoble C, Becker RS (1986) Photophysics, photochemistry and kinetics of photochromic 2H-pyrans and chromenes. J Photochem 33:187–197. https://doi.org/10.1016/0047-2670(86)87033-2

138. Becker RS, Favaro G, Romani A, Gentili PL, Dias FMB (2005) Vibronic effects in pathways of photochemistry and vibrational relaxation. Chem Phys 316:108–116. https://doi.org/10.1016/j.chemphys.2005.05.008
139. Becker RS, Pellicciolo AP, Romani A, Favaro G (1999) Vibronic quantum effects in fluorescence and photochemistry. Competition between vibrational relaxation and photochemistry and consequences on photochemical control. J Am Chem Soc 121:2104–2109. https://doi.org/10.1021/ja982933p
140. Becker RS, Favaro G (2011) New concepts in photochemistry and photophysics: photochromic and other type molecules. J Photochem Photobiol C: Photochem Rev 12:167–176. https://doi.org/10.1016/j.jphotochemrev.2011.08.003
141. Demchenko AP, Tomin VI, Chou P-T (2017) Breaking the Kasha rule for more efficient photochemistry. Chem Rev 117:13353–13381. https://doi.org/10.1021/acs.chemrev.7b00110
142. Zewail AH (2000) Femtochemistry: atomic-scale dynamics of the chemical bond using ultrafast lasers (Nobel Lecture). Angew Chem Int Ed 39:2586–2631. https://doi.org/10.1002/1521-3773(20000804)39:15<2586::AID-ANIE2586>3.0.CO;2-O
143. Tamai N, Miyasaka H (2000) Ultrafast dynamics of photochromic systems. Chem Rev 100:1875–1890. https://doi.org/10.1021/cr9800816
144. Briand J, Bram O, Rehault J, Leonard J, Cannizzo A, Chergui M, Zanirato V, Olivucci M, Helbing J, Haacke S (2010) Coherent ultrafast torsional motion and isomerization of a biomimetic dipolar photoswitch. Phys Chem Chem Phys 12:3178–3187. https://doi.org/10.1039/B918603D
145. Wynne K, Hochstrasser RM (1999) Coherence and adiabaticity in ultrafast electron transfer. Adv Chem Phys 107:263–309. https://doi.org/10.1002/9780470141656.ch3
146. Rather SR, Scholes GD (2019) From fundamental theories to quantum coherences in electron transfer. J Am Chem Soc 141:708–722. https://doi.org/10.1021/jacs.8b09059
147. Chudoba C, Riedle E, Pfeiffer M, Elsaesser T (1996) Vibrational coherence in ultrafast excited state proton transfer. Chem Phys Lett 263:622–628. https://doi.org/10.1016/S0009-2614(96)01268-7
148. Myahkostupov M, Pagba CV, Gundlach L, Piotrowiak P (2013) Vibrational state dependence of interfacial electron transfer: hot electron injection from the S_1 state of azulene into TiO2 nanoparticles. J Phys Chem C 117:20485–20493. https://doi.org/10.1021/jp406662n
149. Bishenden E, Donaldson DJ (1993) Mode-specific chemical branching ratios in the photodissociation of OClO. J Chem Phys 99:3129–3132. https://doi.org/10.1063/1.465167
150. Fidder H, Tschirschwitz F, Dühr O, Nibbering ETJ (2001) Vibrational mode-specific photochemical reaction dynamics of chlorine dioxide in solution. J Chem Phys 114:6781–6794. https://doi.org/10.1063/1.1357202
151. Wallace P, Bollinger C, Hayes S, Reid A (2003) On the actinic wavelength dependence of OClO photochemistry in solution. J Chem Phys 116:1883–1890. https://doi.org/10.1063/1.1531613
152. Vaida V, Feierabend KJ, Rontu N, Takahashi K (2008) Sunlight-initiated photochemistry: excited vibrational states of atmospheric chromophores. Int J Photoenergy 2008:138091. https://doi.org/10.1155/2008/138091
153. Reynard LM, Donaldson DJ (2002) Overtone-induced chemistry of trifluoroacetic acid: an experimental and theoretical study. J Phys Chem A 106:8651–8657. https://doi.org/10.1021/jp021084w
154. Staikova M, Oh M, Donaldson DJ (2005) Overtone-induced decarboxylation: a potential sink for atmospheric diacids. J Phys Chem A 109:597–602. https://doi.org/10.1021/jp046141v
155. Young CJ, Donaldson DJ (2007) Overtone-induced degradation of perfluorinated alcohols in the atmosphere. J Phys Chem A 111:13466–13471. https://doi.org/10.1021/jp075607h
156. Manzanares CE, Delgado YP, Sunuwar S (2024) Phase shift cavity ring down and FT-NIR spectra of CH vibrational transitions and band strengths of ethyl acetate. J Quant Spectr Rad Transf 312:108810. https://doi.org/10.1016/j.jqsrt.2023.108810

157. Donaldson DJ, Tuck AF, Vaida V (2003) Atmospheric photochemistry via vibrational overtone absorption. Chem Rev 103:4717–4729. https://doi.org/10.1021/cr0206519
158. Miller Y, Chaban GM, Finlayson-Pitts BJ, Gerber RB (2006) Photochemical processes induced by vibrational overtone excitations: dynamics simulations for cis-HONO, trans-HONO, HNO_3, and HNO_3-H_2O. J Phys Chem A 110:5342–5354. https://doi.org/10.1021/jp0559940
159. Miller Y, Gerber RB (2006) Dynamics of vibrational overtone excitations of H_2SO_4, H_2SO_4-H_2O: hydrogen-hopping and photodissociation processes. J Am Chem Soc 128:9594–9595. https://doi.org/10.1021/ja062890+
160. Goss LM, Vaida V, Brault JW, Skodje RT (2001) Sequential two-photon dissociation of atmospheric water. J Phys Chem A 105:70–75. https://doi.org/10.1021/jp001828a
161. Brown SS, Metz RB, Berghout HL, Crim FF (1996) Vibrationally mediated photodissociation of isocyanic acid (HNCO): Preferential N–H bond fission by excitation of the reaction coordinate. J Chem Phys 105:6293–9303. https://doi.org/10.1063/1.472483
162. Lai HY, Jhang WR, Tseng CM (2018) Mode-dependent excited-state lifetime of phenol under the S1/S2 conical intersection. J Chem Phys 149:031104. https://doi.org/10.1063/1.5041992
163. Roth, Ehhalt DH (2015) A simple formulation of the CH_2O photolysis quantum yields. Atmos Chem Phys 15:7195–7202. https://doi.org/10.5194/acp-15-7195-2015
164. Troe J (2007) Analysis of quantum yields for the photolysis of formaldehyde at $\lambda > 310$ nm. J Phys Chem A 111(19):3868–3874. https://doi.org/10.1021/jp066886w
165. Jankowski JJ, Kieber DJ, Mopper K, Neale PJ (2000) Development and intercalibration of ultraviolet solar actinometer. Photochem Photobiol 71:431–440. https://doi.org/10.1562/0031-8655(2000)0710431DAIOUS2.0.CO2
166. Hatchard CG, Parker CA (1956) A new sensitive chemical actinometer. II. Potassium ferrioxalate as a standard chemical actinometer. Proc R Soc Lond Ser A Math Phys Sci 235(1203):518–536. https://doi.org/10.1098/rspa.1956.0102
167. Favaro G, di Nunzio MR, Gentili PL, Romani A, Becker RS (2007) Effects of the exciting wavelength and viscosity on the photobehavior of 9- and 9,10-bromoanthracenes. J Phys Chem A 111:5948–5953. https://doi.org/10.1021/jp0714015
168. Silva JM, Silva E, Reis RL (2019) Light-triggered release of photocaged therapeutics - where are we now? J Control Release 298:154–176. https://doi.org/10.1016/j.jconrel.2019.02.006
169. Setlow RB (1974) The wavelengths in sunlight effective in producing skin cancer: a theoretical analysis. Proc Natl Acad Sci USA 71:3363–3366. https://doi.org/10.1073/pnas.71.9.3363
170. Kowalska E, Abe R, Ohtani B (2009) Visible light-induced photocatalytic reaction of gold-modified titanium(iv) oxide particles: action spectrum analysis. Chem Commun 2009:241–243. https://doi.org/10.1039/B815679D
171. Lee S-K, Mills A, O'Rourke C (2017) Action spectra in semiconductor photocatalysis. Chem Soc Rev 46:4877–4894. https://doi.org/10.1039/C7CS00136C
172. Datcu A, Mendoza ML, Pérez del Pino A, Logofatu C, Luculescu C, György E (2019) UV–vis light induced photocatalytic activity of TiO_2/graphene oxide nanocomposite coatings. Catal Today 321–322:81–86. https://doi.org/10.1016/j.cattod.2018.02.026
173. Pettersson LAA, Roman LS, Inganäs O (1999) Modeling photocurrent action spectra of photovoltaic devices based on organic thin films. J Appl Phys 86:487–496. https://doi.org/10.1063/1.37075
174. Langford CH, Achari G, Izadifard M (2011) Wavelength dependence of luminescence and quantum yield in dichlorination of PCBs. J Photochem Photobiol A Chem 222:40–46. https://doi.org/10.1016/j.jphotochem.2011.04.028
175. Bhat AP, Pomerantz WCK, Arnold WA (2023) Wavelength-Dependent UV-LED Photolysis of Fluorinated Pesticides and Pharmaceuticals. Environ Sci Technol 57:5327–5336. https://doi.org/10.1021/acs.est.3c00627
176. Kuldova K, Tsyganenko K, Corval A, Trommsdorff HP, Bens AT, Kryschi C (2000) Photoswitchable dithienylethenes: threshold of the photoreactivity. Synth Met 115:163–166. https://doi.org/10.1016/S0379-6779(00)00327-1

177. Bens AT, Frewert D, Kodatis K, Kryschi C, Martin H-D, Trommsdorff HP (1998) Coupling of chromophores: carotenoids and photoactive diarylethenes – photoreactivity versus radiationless deactivation. Eur J Org Chem 11:2333–2338. https://doi.org/10.1002/(SICI)1099-0690(199811)1998:11<2333::AID-EJOC2333>3.0.CO;2-1
178. Irie M, Eriguchi T, Takada T, Uchida K (1997) Photochromism of diarylethenes having thiophene oligomers as the aryl groups. Tetrahedron 53:12263–12271. https://doi.org/10.1016/S0040-4020(97)00558-9
179. Nagorny S, Schewe M, Weingartz T, Eitzeroth A, Adams J, Rembe C, Schmidt A (2024) Stabilities of bis(thienyl)ethenes in polymethyl methacrylate (PMMA) coatings as absorbance modulation layers for nanoscale imaging. Mater Adv 5:159–170. https://doi.org/10.1039/D3MA00791J
180. Maafi M (2008) Useful spectrokinetic methods for the investigation of photochromic and thermo-photochromic spiropyrans. Molecules 13:2260–2302. https://doi.org/10.3390/molecules13092260
181. Dulić D, Kudernac T, Pužys A, Feringa BL, van Wees BJ (2007) Temperature gating of the ring-opening process in diarylethene molecular switches. Adv Mater 19:2898–2902. https://doi.org/10.1002/adma.200700161

Chapter 5
The Photochemical Quantum Yield Formula

5.1 Introduction

The quantum yield definitions and formulae have been discussed succinctly in the previous *chapters* (e.g. Sect. 1.9). The importance of determining the correct value of the quantum yield of a reaction, or more precisely of a photoactive species (reactant or product), is essential in photokinetics not only as a means to evidence a possible excitation-wavelength effect (Chap. 4) but also to characterise the reactive species, to appropriately model photoreactivity and to predict the evolution of the reaction. We have seen, in the previous chapter, that several equations were proposed by a number of studies for the photochemical quantum yield. This sample is by no means exhaustive of similar formulae found in the related literature. A situation that raises a question about which equation is suitable for the purpose or whether a series of different equations can safely be used to determine the quantum yield of a species. These matters are looked into here with the aim to lift, at least partially, the confusion that might exist around the photochemical quantum yield equation.

5.2 A Notoriously Used Formula

It is almost a matter of principle that in photochemistry, the quantum yield of photoreactions is calculated using a unique set of two formulae. An equation for the integrated quantum yield (Eq. (1.4), or Eq. (5.1)), is the most often employed, and one conveying the differential quantum yield (Eq. (1.5)).

M. Maafi, *Photokinetics*, https://doi.org/10.1007/978-3-031-98985-8_5

$$\Phi = \frac{\text{number of reacted molecules}}{\text{number of quanta absobed}} = \frac{n_{\text{tot.rect.mol.}}}{n_{\text{tot.abs.h}\nu}} \tag{5.1}$$

Even though the two equations are supposed to be equivalent and interchangeable, a rationale linking them mathematically is missing in the literature.

Furthermore, despite the occurrence of the former equation (Eq. (5.1)) in photochemistry textbooks and research publications, its derivation from physical chemistry principles is still enigmatic. A review is then needed to assess its applicability and validity.

5.3 A Succinct Background

A century ago, Warburg [1–4] defined the specific photochemical effect, the first quantitative measure in photochemistry, as the ratio of moles reacted or formed to the energy absorbed (expressed in gram calories) [5]. Not long after, and using Einstein's postulate stipulating that one photon absorbed excites one molecule, Warburg amended the definition to the ratio of the molecules (or moles) reacted or formed divided by the photons absorbed, i.e. constituting our modern photochemical quantum yield [6]. Soon after, in 1925, Allmand [5, 7] added a specification to the concept in that the ratio must involve the number of moles of the starting material that have reacted, dividing the precise number of quanta that have been absorbed by the starting material, a quantity now known as the primary quantum yield. Currently, the primary photochemical process is defined [6] as '*any elementary chemical process undergone by an electronically excited molecular entity and yielding a primary photoproduct*', which is quantitatively evaluated by the quantum efficiency that is identical, in this case, to the quantum yield. However, the quantum yield, in general, bears a somewhat wider definition in the sense that primary as well as secondary processes might be involved, formulated as '*the number of defined events occurring per photon absorbed by the system*'.

The importance of the quantum yield is evident from its occurrence in a large number of publications over the last century, and its definition (as given by Eq. (5.1) or equivalent) is found in almost all photochemistry textbooks (as well as in some of the chapters dealing with photochemistry in other speciality books). The formula of the quantum yield, ubiquitously used, has become not only a beacon in photochemistry but also, perhaps, an unwavering and self-standing concept.

5.4 Quantum Yield and Quantum Efficiency Definitions

In Eq. (5.1), the number of reacted molecules (or $n_{\text{tot. rect. mol.}}$ in Eq. (1.4), expressed in mol/L) is obtained by an analytical technique, and the number of quanta absorbed (or $n_{\text{tot. abs. h}\nu}$ in Eq. (1.4)) is determined by actinometry [6, 8]. The procedure might be carried out by placing the actinometer in front of and behind the sample so that the difference between the incident (P_0) and transmitted (P_{tr}) measured light intensities for the irradiation time interval corresponds to the light absorbed by the sample ($P_{abs} = P_0 - P_{tr}$). The light intensity of the light source [8] is here expressed in units of einstein $dm^{-3}\ s^{-1}$ (Sect. 2.18), and Φ is dimensionless [6, 8].

The formulation of the quantum yield (Eq. (5.1)) is equivalent to a probability, as it is expressed as the ratio of the positive/realised events (the reacted molecules of the species, e.g. X), to the total number of events (all the molecules that have absorbed a photon). This definition complies with the two fundamental laws of photochemistry [8], stipulating that only absorbing molecules may photo-react and a molecule can only absorb one photon. Therefore, Eq. (5.1) is correct only and only if these two numbers ($n_{\text{tot. rect. mol.}}$ and $n_{\text{tot. abs. h}\nu}$) are accurately determined.

The second formula of the quantum yield given by Eq. (1.5) [6], defining the differential quantum yield (under low radiation field and homogeneous media), as the ratio (dx/dt) of the rate of change of a measurable quantity (e.g. concentration in mol/(L s) or mol/(dm^3 s)), and the amount of incident photons (prior to absorption) per volume and per time interval, $q^0_{n,p,\lambda}\left(1 - 10^{-A(\lambda)}\right)$, (expressed einstein/($dm^3$ s) or equivalent mol/(L s); $q^0_{n,p} = P_0$). $A(\lambda)$ is the total absorbance at the excitation wavelength. The quantum efficiency is identical to the quantum yield for a primary photochemical process [5, 7].

The discussion that follows will mainly focus on the primary photochemical process $XY(1\Phi)$, a phototransformation of species X into a photoproduct Y (X —$h\nu$→ Y). The quantum yield (quantum efficiency) of this unimolecular photoreaction (Eqs. (1.4) and (1.5)) will, in the following, be evaluated against five considerations.

5.5 Consideration Relative to the Quantum Yield Formulae

5.5.1 Consideration I: The Absorbed Number of Photons

Implicitly, Eq. (1.4) (or Eq. (5.1)) requires that the absorbing species in the reaction medium is itself the one that reacts (i.e. the reactant X in $XY(1\Phi)$). This condition is necessary since the quantum yield (which here is equivalent to the quantum efficiency) is specific to the phototransformation of one species (X).

Accordingly, Eq. (5.1) should be applicable to those reactions (1) for which X is the unique initial reactant, (2) that undergo a direct photoreaction, (X —$h\nu$→ $Y_1 \cdots \rightarrow Y_n$), and (3) whose photoproducts ($Y_1 \cdots \rightarrow Y_n$) do not absorb over the wavelength span of the excitation light responsible for the phototransformation of reactant X.

Equation (5.1) is exclusive to measuring photochemical quantum yields of reactants (point (1) above) as specified by Allmand [5, 7]. Technically, Eq. (5.1) cannot be applied to determine the photochemical quantum yields of species that might be involved in secondary photoreactions occurring after excitation of the reactant (e.g. species Y_1 in the consecutive photoreaction $X— h\nu \rightarrow Y_1 — h\nu \rightarrow Y_2$). Similarly, point (2) excludes concurrent, reversible and cyclic mechanisms that involve species *X*, from the domain of applicability of Eq. (5.1). Also, since Eq. (5.1) does not explicitly take into account thermal reactions, that start or end at species *X*, these types of mechanisms are also excluded from the reactions covered by the equation.

Accordingly, Eq. (5.1) is only applicable to the group of reactions for which *X* undergoes a single forward photochemical step. The latter set is reduced even more by point (3), to include only the photoreactions for which the electronic spectra of *X* do not overlap, over the wavelength domain of the excitation light, those of the products of the reaction.

If points (1) and (2) have usually been implicitly respected in the literature, the last point (3) has very rarely been considered. The impact of overlooking point (3) is an increase in the absorbed quanta encompassed by the value of $n_{\text{tot. abs. h}\nu}$, thus unavoidably leading to an incorrect estimation of the reaction quantum yield. Of course, the situation is more acute if the photoproducts' absorptivities were higher than that of *X*.

The contribution of the products to the total absorption, $A(\lambda)$, represents then a manifest source of inaccuracy in the determination of the quantum yield value when using Eq. (5.1). From a spectrophotometric viewpoint, it is usually the case that the electronic absorption spectra of most photochemical species (reactants) overlap, partly or fully, with those of their photoproducts (a situation often observed for photoreactive organic materials). Yet, the general experimental photochemistry literature does not consider this particular aspect. In fact, there are no established methods for the use of Eq. (5.1) that quantify and separate the number of photons absorbed by the photoproducts from that absorbed by *X*. Hence, it is not currently possible to bring the necessary correction of this effect to Eq. (5.1) ($n_{\text{tot. abs. h}\nu}$ measured by actinometry corresponds to all absorbing species, including *X*). This also means that the values of the photochemical quantum yields measured according to the procedure using Eq. (5.1) are usually plausible since they should be less than unity ($\Phi(\lambda_{irr}) < 1$). However, this plausibility is not by itself a proof of the validity of the results obtained. This set of reasons considerably reduces the field of application and the claimed universality attributed to Eq. (5.1).

5.5.2 Consideration II: The Light Type

In principle, Eq. (5.1) should not be considered for application to systems subjected to polychromatic light, according to the caveat stipulated in the *IUPAC* recommendations, as '*Strictly, the term quantum yield applies only for monochromatic excitation*' [6]. Also, it shouldn't because its mathematical formulation does not involve any reference to an irradiation over a wavelength range. In the case of polychromatic light, Eq. (1.5) should be a function of two variables (i.e. dt and $d\lambda$, so that the solution of the rate-equation should be achieved by a double integration). It is striking to notice that Eq. (5.1) is routinely employed in the literature for polychromatic irradiation (delivered by filtered or unfiltered fluorescent and discharge lamps, LEDs and/or daylight). To date, the photokinetics community has not yet reached consensus on uniquely establishing a closed-form integrated solution of the rate-law corresponding to polychromatic irradiation (see Chap. 19).

5.5.3 Consideration III: The Quantum Yield Invariance

In order to justify the practicality of extending the application of Eq. (5.1) to reactions under polychromatic-light irradiation (ranging between λ_a and λ_b), it might be argued that the quantum yield (or quantum efficiency) is wavelength invariant on the provision that Kasha's rule [9] applies to photochemistry processes (which is an unproven assumption, see Chap. 4). In support of that argument, one can consider that the quantum yield can be written as a sum of probabilities (Eq. (5.2)) at each wavelength of irradiation impinging on the reaction.

As the incident light, in practical terms, is described by a profile (variable values, $P_0^{\lambda_{irr}}$) over the excitation-wavelength range, the number of absorbed photons, as well as the amount of reacted molecules of X should also be variable with wavelength.

Here, we suppose that only X absorbs the light (not Y), the total number of absorbed photons ($n_{tot.\ abs.\ h\nu}$) encompasses photons absorbed at all wavelengths, and the number of photons absorbed by X at a given wavelength λ_i ($n_{X.abs.h\nu}^{\lambda_i}$), its reacted molecules ($n_{X.rect.mol.}^{\lambda_i}$), due to that absorption of light, define the quantum yield at that wavelength (as $\Phi_{X\to Y}^{\lambda_i} = n_{X.rect.mol.}^{\lambda_i}/n_{X.abs.h\nu}^{\lambda_i}$), and the quantum yield is wavelength-independent ($\Phi_{X\to Y}^{\lambda_i} = \Phi_{X\to Y}$). Under these conditions, we can write,

$$\Phi_{X\to Y} = \frac{n_{X.rect.mol.}}{n_{tot.abs.h\nu}} = \sum_{\lambda_i=\lambda_a}^{\lambda_b} \frac{n_{X.rect.mol.}^{\lambda_i}}{n_{tot.abs.h\nu}} = \sum_{\lambda_i=\lambda_a}^{\lambda_b} \frac{n_{X.abs.h\nu}^{\lambda_i}}{n_{tot.abs.h\nu}} \frac{n_{X.rect.mol.}^{\lambda_i}}{n_{X.abs.h\nu}^{\lambda_i}} = \sum_{\lambda_i=\lambda_a}^{\lambda_b} \frac{n_{X.abs.h\nu}^{\lambda_i}}{n_{tot.abs.h\nu}} \Phi_{X\to Y}^{\lambda_i}$$

$$= \Phi_{X\to Y} \frac{\sum_{\lambda_i=\lambda_a}^{\lambda_b} n_{X.abs.h\nu}^{\lambda_i}}{n_{tot.abs.h\nu}} = \Phi_{X\to Y} \tag{5.2}$$

Under this hypothesis, Eqs. (5.1) and (5.2) are supposedly equivalent. However, such an equality is disapproved by the fact that the sum in the last ratio of Eq. (5.2), $\sum_{\lambda_i=\lambda_a}^{\lambda_b} n_{X.abs.h\nu}^{\lambda_i}$ is not equal to $n_{tot.\ abs.\ h\nu}$, as implicitly used in Eq. (5.2), because it does not mathematically stand owing to $\sum_{\lambda_i=\lambda_a}^{\lambda_b} n_{X.abs.h\nu}^{\lambda_i} = \sum_{\lambda_i}^{\lambda_n} P_0^{\lambda_{irr}}\left(1-10^{-A_X^{\lambda_{irr}}}\right) \neq P_0^{\lambda_{irr}}\left(1-10^{-\sum_{\lambda_i}^{\lambda_n} A_X^{\lambda_{irr}}}\right) = n_{tot.abs.h\nu}$.

Furthermore, the whole argument can be seen as flawed since we have seen in Chap. 4 that the claim about the necessary condition relative to the invariance of $\Phi_{X\to Y}^{\lambda_{irr}}$ might not be always satisfied, unless the wavelength independence of the photochemical quantum yield is experimentally proven (i.e. Kasha's rule stands). Therefore, for Eq. (5.1) (or equivalent equations) to be applicable to polychromatic light irradiation, there must be tangible experimental evidence, obtained from monochromatic irradiations, to support the invariance of $\Phi_{X\to Y}^{\lambda_{irr}}$ before considering applying it. Such a corroboration has neither been stipulated in the recommendations set out in the quantum yield definition nor provided in the literature employing Eq. (5.1) to polychromatic light irradiation.

5.5.4 Consideration IV: The Irradiation Time

The current formulation of Eq. (5.1) suggests that it is applicable irrespective of the irradiation time interval selected (as long as this time allows enough reaction progress to accurately measure the variation in concentrations). This means that the same value of $\Phi_{X\to Y}^{\lambda_{irr}}$ should be obtained for two similar samples when, for instance, the first sample is irradiated during a time interval t_1, and the other one is exposed to the same light for a longer period t_2 (e.g. $t_2 = 2\, t_1$).

In order to make this point clear, let's consider again the simple unimolecular photoreaction $XY(1\Phi)$, and assume that its absorbance (or concentration, C_X) changes fairly linearly over a period of reaction time equal to or longer than $2t_1$. Such a behaviour is similar to that observed for a 0th-order reaction. So, for the considered interval of time, we can write an equation for the line (Eq. (5.3), with the slope, k, supposedly representing the 0th-order overall rate-constant of the photoreaction). Incidentally, the concentration of the product formed is given by $C_Y(t) = kt$.

$$C_X(t) = C_X(0) - kt \tag{5.3}$$

Let's perform this photoreaction on the above samples by subjecting them, respectively, to irradiation time intervals, t_1 and t_2. Therefore, according to Eq. (5.3), the concentration at t_2 is equal to

$$\begin{aligned} C_X(t_2) &= C_X(0) - kt_2 = C_X(0) - kt_1 + kt_1 - kt_2 = C_X(t_1) - k(t_2 - t_1) \\ &= C_X(t_1) - kt_1 \end{aligned} \tag{5.4}$$

The number of absorbed photons is defined as the product of the incident light intensity ($P_0^{\lambda_{irr}}$) at λ_{irr}, the fraction of light absorbed $\left(1 - 10^{-\varepsilon_X^{\lambda_{irr}}\, l\, C_X(t)}\right)$ at that wavelength (Eq. (2.20)), and the period of irradiation time (t). Hence, we can write

$$n_{tot.abs.h\nu}(t_1) = n_{X.abs.h\nu}(t_1) = P_0^{\lambda_{irr}}\left(1 - 10^{-\varepsilon_X^{\lambda_{irr}}\ l\ C_X(t_1)}\right)t_1 \tag{5.5}$$

and for t_2 using Eq. (5.4),

$$\begin{aligned} n_{tot.abs.h\nu}(t_2) &= n_{X.abs.h\nu}(t_2) = P_0^{\lambda_{irr}}\left(1 - 10^{-\varepsilon_X^{\lambda_{irr}}\ l\ C_X(t_2)}\right)t_2 \\ &= P_0^{\lambda_{irr}}\left(1 - 10^{-\varepsilon_X^{\lambda_{irr}}\ l\ C_X(t_1)}\ 10^{\varepsilon_X^{\lambda_{irr}}\ l\ k\ t_1}\right)t_2 \end{aligned} \tag{5.6}$$

The number of photons is expressed here relative to the volume of the irradiated solution as it has the unit of ($P\ t$), i.e. mol dm^{-3} (or einstein dm^{-3}).

We can now write Eq. (5.1) for the two experiments, which are supposed to give the same value for $\Phi_{X \to Y}$, as

$$\begin{aligned} \Phi_{X \to Y}^{\lambda_{irr}} &= \frac{C_Y(t_1)}{n_{tot.abs.h\nu}} = \frac{C_Y(t_1)}{P_0^{\lambda_{irr}}\left(1 - 10^{-\varepsilon_X^{\lambda_{irr}}\ l\ C_X(t_1)}\right)t_1} \\ &= \frac{C_Y(t_2)}{P_0^{\lambda_{irr}}\left(1 - 10^{-\varepsilon_X^{\lambda_{irr}}\ l\ C_X(t_1)}\ 10^{\varepsilon_X^{\lambda_{irr}}\ l\ k\ t_1}\right)t_2} \end{aligned} \tag{5.7}$$

Or, by replacing C_Y by its equation, we obtain

$$\begin{aligned} \Phi_{X \to Y}^{\lambda_{irr}} &= \frac{k}{P_0^{\lambda_{irr}}} \frac{1}{\left(1 - 10^{-\varepsilon_X^{\lambda_{irr}}\ l\ C_X(t_1)}\right)} \\ &= \frac{k}{P_0^{\lambda_{irr}}} \frac{1}{\left(1 - 10^{-\varepsilon_X^{\lambda_{irr}}\ l\ C_X(t_1)}\ 10^{\varepsilon_X^{\lambda_{irr}}\ l\ k\ t_1}\right)} \end{aligned} \tag{5.8}$$

For the equality in Eq. (5.8) to stand, the second fractions in each of the last two terms of the equation must be equal. This means that

$$10^{-\varepsilon_X^{\lambda_{irr}}\ l\ C_X(t_1)} = 10^{-\varepsilon_X^{\lambda_{irr}}\ l\ C_X(t_1)}\ 10^{\varepsilon_X^{\lambda_{irr}}\ l\ k\ t_1} \tag{5.9}$$

This identity would only be true if $t_1 = 0$ (i.e. $10^{\varepsilon_X^{\lambda_{irr}}\ l\ k\ t_1} = 1$), that is, when there is no reaction. This situation leads to the conclusion that the quantum yield value in Eq. (5.1) is necessarily dependent on irradiation time. The case demonstrated here for illustration (a linear kinetic trace) can easily be generalised to any pattern of the kinetic trace. Simulated examples show that $\Phi_{X \to Y}^{\lambda_{irr}}$ undergoes a decreasing trend with reaction time (as it would be expected from the increase of the factor $10^{\varepsilon_X^{\lambda_{irr}}\ l\ k\ t_1}$ in Eq. (5.8)).

This means that the value of $\Phi_{X \to Y}^{\lambda_{irr}}$ obtained from Eq. (5.1) must always be associated with a specific irradiation time. This condition has not been a prerequisite that experimentalists impose when Eq. (5.1) is employed. Anyway, an irradiation-time-dependent quantum yield would not make physical sense. Overall, the status of Eq. (5.1) is weakened by the present consideration and renders its validity questionable.

Assuming Eq. (5.1) to be valid, one can summarise the important features required for its application to the determination of the quantum yield of a given primary photoreaction. This practically necessitates, at least, nine conditions to be satisfied.

(a) A single and well-defined reactant X must be identified.
(b) The photochemical quantum yield of only X is determined in the process.
(c) Reactant X must be thermally stable.
(d) The reaction of X must not be photoreversible.
(e) X must not be involved in a concurrent or cyclic photo- or thermal reaction.
(f) The photoproducts generated by the phototransformation of X must not absorb at any of the wavelengths covered by the irradiation beam.
(g) Other species occurring in the reaction medium (exposed to the irradiation beam) must not absorb the excitation light.
(h) The initial concentration employed should lay within the linearity range of the spectrophotometric calibration graph of X (otherwise the Beer-Lambert law does not apply (i.e. Eq. (5.5) is not accurate).
(i) The irradiation duration must be specified.

The wider experimental photochemistry literature does not state that the actually reported values of the photochemical quantum yields comply with the above-listed conditions; with points (f), (g) and (h) often representing obvious limitations.

The serious applicability issues of Eq. (5.1), as laid out above, raise a few fundamental concerns about its identity as discussed in the following section.

5.5.5 Consideration V: The Identity of the Basic Rate-Law

Since Eq. (5.1) involves kinetic parameters (i.e. $\Phi_{X \to Y}$, $C_X(t)$ or $C_Y(t)$, and $n_{tot.\ abs.\ h\nu}$), it is only legitimate to seek a relationship between this equation and the rate-law equation of the photoreaction. In other words, the equation expressing the integral photochemical quantum yield should conceptually originate from the differential equation of the studied photoreaction. Unfortunately, to the question 'which differential equation generates Eq. (5.1)?', there is, as far as we are aware, no answer to be found in the photochemistry literature, including Warburg papers (the equation proposed by Warburg could be considered as a postulate) [1–4].

If we assume that such a rate-law exists, it would automatically mean that all photoreactions, irrespective of their features or irradiation conditions/nature of the employed light (e.g. mono- or polychromatic), have de facto the same rate-law because they all have effectively the same formula for the quantum yield. Such a general statement cannot be true in absolute kinetic terms, as different photo- and photothermal reactions are expected to be described by different rate-laws. It is relevant here to remember the basic mathematical principle stating that different differential equations will have different solutions and therefore the output formulation for the quantum yield equation should not be expected to be unique. According to this observation, the single formula proposed for the integral quantum yield (Eq. (5.1)) is certainly not universal.

A further mathematical analysis of Eq. (5.1) shows that if this equation (in its actual form or rearranged) was a solution of a rate-law, then it would necessarily include a term corresponding to the initial condition (i.e. at $t_0 = 0$, $C_X(0) \neq 0$) for the factor in Eq. (5.5) $\left(1 - 10^{-\varepsilon_X^{\lambda_{irr}}\, l\, C_X(t)}\right)$; the initial condition term should presumably read $\left(1 - 10^{-\varepsilon_X^{\lambda_{irr}}\, l\, C_X(0)}\right)$. The fact that the boundary condition term does not appear in Eq. (5.1) raises strong concerns on its mathematical coherence. The equation, however, takes into account the initial conditions of time and concentration (second ratio on the right-hand side of Eq. (5.10)).

$$\Phi_{X\to Y}^{\lambda_{irr}} = \frac{1}{P_0^{\lambda_{irr}}\left(1 - 10^{-\varepsilon_X^{\lambda_{irr}}\, l\, C_X(t)}\right)} \frac{C_X(0) - C_X(t)}{t - t_0} \tag{5.10}$$

Looking into the identity of the rate-law behind Eq. (5.1), the mathematical formulation of Eq. (5.10) can be better appreciated if rearranged (with, $\Delta C_X = C_X(t) - C_X(0)$), as

$$-\frac{\Delta C_X}{\Delta t} = \Phi_{X\to Y}^{\lambda_{irr}}\, P_0^{\lambda_{irr}}\left(1 - 10^{-\varepsilon_X^{\lambda_{irr}}\, l\, C_X(t_1)}\right) \tag{5.11}$$

If we consider that Eq. (5.11) is also valid for infinitesimal variation of time, a reconstitution of a differential equation (Eq. (5.12)) can be attempted, as

$$\frac{dC_X}{dt} = -\Phi_{X\to Y}^{\lambda_{irr}}\, P_0^{\lambda_{irr}}\left(1 - 10^{-\varepsilon_X^{\lambda_{irr}}\, l\, C_X(t)}\right) \tag{5.12}$$

Interestingly enough, Eq. (5.12) is precisely the rate-equation for the unimolecular photoreaction,

X — $h\nu \rightarrow Y$, with Y transparent to the irradiation light (as we shall see in Chap. 11). This might lead us to think that an answer to our earlier question relative to the rate-law has been found, i.e. Eq. (5.12) generates Eq. (5.1). Unfortunately, it is easily proven that Eq. (5.12) does not accept Eq. (5.1) as a solution. There is, though, a single situation where Eq. (5.1) could be the mathematical solution of Eq. (5.12), that is, when the absorbance, $\varepsilon_X^{\lambda_{irr}}\, l\, C_X(t)$, is invariant with time (and $dC_X/dt = constant$). This must correspond to a monochromatic and isosbestic irradiation, which is impossible to perform on the reaction studied ($\varepsilon_Y^{\lambda_{irr}} = 0$) since an isosbestic point cannot show on the electronic absorption spectra of X at the irradiation wavelength.

There is another interesting fact about Eq. (5.12): rearranging it again can lead to the equation of the differential quantum yield, as defined by Eq. (1.5), but notice that the minus sign is omitted in that definition (in which case, *dx*/*dt*, will correspond specifically to the rate of the product formation).

$$\Phi_{X \to Y}^{\lambda_{irr}} = \frac{-\frac{dC_X}{dt}}{P_0^{\lambda_{irr}} \left(1 - 10^{-\varepsilon_X^{\lambda_{irr}}\, l\, C_X(t)}\right)} \tag{5.13}$$

The differential quantum yield Eq. (5.13) (or Eq. (1.5)) is supposed to be interchangeable with Eq. (5.1), the integral quantum yield. The two definitions appear one after the other under the same entry '*quantum yield, Φ*' of the glossary of terms used in photochemistry [6] with no additional information on their possible connection or relationship. In light of the conclusion reached above (i.e. Eq. (5.13) (or Eq. (1.5)) and Eq. (5.1) are not related), such a dual definition increases opacity towards understanding the physical chemistry origin and photokinetic meaning of Eq. (5.1). However, Eq. (5.13) (which, might incidentally justify our attempted Eq. (5.12)), coherently roots the differential quantum yield to the rate-equation of the unimolecular photoreaction with a transparent product (and not any other reaction).

5.6 Conclusion

The assessment, undertaken in this chapter, of the equations adopted in photochemistry for the experimental measurement of photochemical quantum yields (Eqs. (1.4) and (1.5)), can be considered as an attempt to answer four basic questions: (a) Is Eq. (1.4) universal? (b) Is it kinetically and mathematical sound? (c) What is the differential equation from which Eq. (1.4) (i.e. Eq. (5.1)) could rationally be derived? (d) Is there a relationship between the formulae of the integral (Eq. (1.4)) and the differential (Eq. (1.5)) quantum yields?

It is obvious from the above discussion that a negative answer should be assigned to every one of these questions. As a matter of fact, Eq. (5.1) (or Eq. (1.4)) cannot possibly be the unique or a universal formula for the determination of the photochemical quantum yield of photoreactions. Its coherence with mathematical and kinetic principles seems unclear, as are its generating rate-law and its relationship to Eq. (1.5).

Such findings call for a tighter definition of quantum yield formulae. The subsequent chapters will tackle this important matter in great detail.

References

1. Warburg E (1917) Über die Anwendung der Quantenhypothese auf die Photochemie. Naturwissenschaften 5:489–494. https://doi.org/10.1007/BF02450059
2. Warburg E (1920) Electrochemistry 26:54–59
3. Warburg E (1924) Die Quantenregeln in der Photochemie. Naturwissenschaften 1924(12): 1058–1063. https://doi.org/10.1007/BF01452119
4. Warburg E, Rump W (1929) Über die Photolyse der Lösungen von Schwefelwasserstoff in Hexan und in Wasser und über die Photolyse von Lösungen im allgemeinen. Z Phys 58:291–305. https://doi.org/10.1007/BF01340381
5. Rubin MB, Braslavski SE (2010) Quantum yield: the term and the symbol. A historical search. Photochem Photobiol Sci 9:670–674. https://doi.org/10.1039/B9PP00182D
6. Braslavski SE (2007) Glossary of terms used in photochemistry, 3rd edition (IUPAC Recommendations 2006). Pure Appl Chem 2007(79):293–465. https://doi.org/10.1351/pac200779030293
7. Allmand AJ (1926) Part I.—Einstein's law of photochemical equivalence. Introductory address to Part I. Trans Faraday Soc 21:438–452. https://doi.org/10.1039/TF9262100438
8. Kuhn HJ, Braslavski SE, Schmidt R (2004) Chemical actinometry (IUPAC Technical Report). Pure Appl Chem 76:2105–2146. https://doi.org/10.1351/pac200476122105
9. Kasha M (1950) Characterization of electronic transitions in complex molecules. Discuss Faraday Soc 9:14–19. https://doi.org/10.1039/DF9500900014

Chapter 6
Hidden Hurdles in Chemical Kinetics: Distinctability, Distinguishability, and Identifiability Concepts

6.1 Introduction

Distinguishability, identifiability, and distinctability concepts lie at the heart of kinetic analysis despite the fact that they have not been described in relation to photokinetic investigation. The three concepts represent the issue of degeneracy of the kinetic solution. This chapter gives their definitions and, more importantly, demonstrates that a kinetic analysis of any photoreaction cannot be considered complete until these three concepts are fully addressed and the issues they may pose are consistently resolved. They should naturally represent the tripod pillar of a full and consistent kinetic analysis. Not addressing any of them exposes the kinetic analysis, results, and conclusions to serious mistakes, misinterpretation, and/or invalidation.

6.2 Primary Purposes, Tools, and Methods of Chemical Kinetic Analysis

Chemical kinetic analysis' primary concerns are either to define the reaction mechanism undergone by the reactive system or, when the mechanism is known, to define the values of the kinetic parameters and attributes of the reaction under investigation. Sometimes, the kinetic analysis attempts to determine both the unknown mechanism and kinetic parameters.

It is given that, in general, chemical analysis cannot be performed without analytical chemistry techniques. The latter quantitatively monitors the progress of the reaction. Usually, one technique is employed (e.g. spectrophotometry), but often, the data obtained by more than one analytical chemistry technique are needed. The specificity for each kinetic problem decides which and how many techniques are required for a full kinetic analysis.

M. Maafi, *Photokinetics*, https://doi.org/10.1007/978-3-031-98985-8_6

The kinetic analysis must also involve, at one point or another, some degree of mathematical description of the reactive system. As for thermally driven kinetics, generally, an equation expressing the variation of the concentration (one for each species involved in the reaction) is established as the integrated rate-law. Such integrated rate-laws must be built according to the adequate rate-law of the specific mechanism undergone by the photoreaction under investigation. The integrated rate-law equation itself or any of its rearranged forms is used to fit the experimental data. It is hence clear that setting up the correct rate-law and appropriately solving it have a preponderant role in the methodology of kinetic analysis.

As briefly introduced in Chap. 2, if the mechanism is unknown, it is imperative that a mechanism is preconceived for the reaction at hand in order to be able to derive its rate-law and, hence, its integrated rate-law. This is always possible. Whether the mechanism is known with certainty or preconceived, it is usual to consider in kinetic practice that, if the derived integrated rate-law equation happens to fit the experimental data well, it is enticing to accept the good fit as strong proof for the validity of the mechanism for the system studied. The conceptual hypothesis here assumes that *only the integrated rate-law equations belonging to the true mechanism of the reactive system are able to fit, with high precision, its experimental data.* In simple words, a good fit is a confirmation of the mechanism proposed. This represents the main evaluation procedure adopted in kinetics [1].

Once the mechanism and an integrated rate-law are known, the parameters of the reaction will be determined from the fitting data. In general, one set of values is derived. These are accepted as long as they are physically meaningful, plausible, and do neither conflict with the chemical nature of the species or the information available on the process nor with the mechanism used for the reaction. An interpretation follows from these parameters on the basis that the kinetic analysis has established the true mechanism of the reaction and the true set of parameter values.

The above procedure is conceptually applied for chemical and photochemical kinetics, irrespective of the specific details of each type of reaction.

When numerical integration methods are employed, reaction data are generated from the system of differential equations set out for the considered mechanism. Parameters' values are fed to numerical integration methods (NIMs) to produce calculated traces, which are then fitted to the experimental kinetic traces. A good fit of these data is a confirmation of the consistency of the kinetic analysis. Overall, this approach remains fundamentally similar to the one discussed above, so the same conclusions relative to the validity of the mechanism and/or the parameters' values also apply here.

From there, it is possible to formulate this concept as "*the kinetics of any reactive system, is always solvable if the mechanism of the reaction was known (or pre-conceived), and the corresponding integrated rate-law or data obtained by numerical integration delivered a good fitting of the experimental data.*" Conventional wisdom probably confirms this statement. However, we will still ask: is there room for a question on the reliability of the above kinetic analysis methodology? Can one explore whether this approach is universal and tight enough to lead with certainty to the true mechanism and/or the true set of the parameters' values (within the experimental error)?

It turns out that there are some issues that must be clarified before a kinetic analysis is deemed complete. We will review three of them in the following section.

6.3 The Identifiability Problem, or When Do the True Values of the Set of Parameters Become Elusive?

The concept of *identifiability* can be understood as a description of the degeneracy of the kinetic solution. What is meant here by the kinetic solution is a set of values, "S_r," attributed to the set of reaction parameters. S_r includes a particular value for each parameter relevant to the dynamics of the process investigated (e.g. the intrinsic parameters, ε_{Y_j}, $\Phi_{Y_j \rightarrow Y_{j'}}$, …, etc.). The found solution S_r is a solution if and only if, when using the S_r values in the reaction equations, the experimental or NIMs' synthetic kinetic traces of the reaction are reproduced with high precision.

We have seen in the previous section that the good fit of the experimental data by the theoretical model implemented with the S_r values supposes that this set of parameters' values is unique. However, the identifiability becomes a challenging issue when, for instance, a reaction of known mechanism accepts more than one (possibly an infinity) of the kinetic solutions S_{r1}, S_{r2}, …, S_{rn}, all faithfully reproducing the kinetic traces of the reaction. It is obvious that in such a situation, we are faced with a serious difficulty, in that it is not possible to decide with certainty which of the possible S_{r1}, S_{r2}, …, S_{rn} sets is the true parameter set ($S_{r(\text{true})}$) that represents the reaction at hand.

The identifiability issue has been described in relation to reaction kinetics as early as the 1940s [2]. One of the prominent studied systems in this area is the trimolecular reaction involving two consecutive irreversible first-order thermal reactions ($X \rightarrow Y_1 \rightarrow Y_2$), whose intermediate species Y_1 is not possibly isolated and/or separately investigated, and the only experimental data available on this reaction consist of the absorbance trace. In these conditions, the kinetic solution, S_r, corresponds to the simultaneous determination of the values of three parameters, namely, the absorptivity of Y_1 (ε_{Y_1}) and the rate-constants of the reaction-steps ($k_{X \rightarrow Y_1}$ and $k_{Y_1 \rightarrow Y_2}$). It has been shown that S_r is not unique because of the slow–fast ambiguity, proving the non-uniqueness in parameter values, and the existence of several possible (S_{r1}, S_{r2}, …, S_{rn}) solutions [3–5]. Using either Laplace transforms or the state space similarity transformation, some limits were set for what can be achieved in terms of elucidating the trimolecular system kinetics [5]. The latter study has underlined both the close dependency between the performance of working out solutions and the number of unknowns and the need to be able to measure and/or define additional quantities in order to achieve the true solution ($S_{r(\text{true})}$). The same authors followed up their investigation with a more general case of identifiability in chemical kinetics for reactions other than the trimolecular $XY_2(2k)$ [6].

Two methods employing least squares–based numerical integration and the analysis of numerically evaluated derivatives have been applied to derive rate-constants of steroidal imidazoles' reactions using the experimental absorbance time data of the reaction, whose rate-laws cannot be integrated in closed form. A significant variation between sets of rate-constant values was observed, indicating the occurrence of an identifiability problem [7]. The author used the two methods for deriving rate-constants from absorbance time data, and their use was illustrated. One method uses least squares fitting to the curve obtained by numerical integration of the rate-law. The other involves analysis of the numerically evaluated derivative with respect to time of the absorbance time curve. This allows calculation of rate constants in a direct fashion, either from the initial and minimum values of a function of the derivative or by fitting the decrease in the value of this function to a first-order rate equation when this is suitable. The advantages and disadvantages of the methods are discussed. The illustrations are from recent studies of the reactions of steroidal imidazoles with aryl esters. The derivative method appears to be superior and of potentially wider utility.

Identifiability has also been assessed with relation to first- or pseudo-first-order photophysical processes of excimers and/or exciplexes, when monitored by time-resolved fluorescence spectroscopy [8]. The investigated two-state models involved a set of eight parameters, including thermal rate constants (k_1 and k_{-1}), fluorescence intensity functions ($a(\nu)$ and $b(\nu)$), lifetimes (τ_a and τ_b), and initial species concentrations ($C_a(0)$ and $C_b(0)$). The recorded decay profiles were fitted by double exponential functions of the eight parameters. The fitting of the curves could be achieved by a multitude of solutions, characterised by different sets of values of the parameters. The authors explored further the conditions of solving the unidentifiability problem by introducing special conditions, allowing some of the unknown parameters to become available. It was demonstrated that the solvable cases must have three unknowns or less.

Beyond chemical kinetics, unidentifiability has also been observed for a wide range of fields, such as catalytic methanation [9], fluorescence decays in aromatic polymers [10], non-linear pharmacokinetic [11] and parametric models [12], mechanical [13] and polynomial state space systems [14], Volmer-Heyrovsky electrochemical reactions [15], enhancing machine perception for robots' autonomy [16], chain polymerisation [17], MurC reaction [18], biochemical reaction networks [19], analysis of state space models [20], infectious disease transmission models [21], linear compartmental models [22], differential models with parameters [23], and statistical inference [24].

From a kinetic point of view, the identifiability will be sought for systems for which ($a_{id.}$) the reactant, products, and mechanism (number and nature of reaction-steps) are known, and ($b_{id.}$) the reaction obeys given kinetic orders, but ($c_{id.}$) some or all the parameters' values are not known. In this case, identifiability translates to the existence of different (more than one) sets of reaction parameters that can explain (e.g. by a good fit of) the experimental data of the reaction.

The occurrence of an identifiability issue means that a good fit of the experimental data by a model equation based on one of the value sets of the parameters is only a necessary condition but cannot be considered sufficient to accept, as valid, the set of parameters' values used. Therefore, the classical procedure of kinetic analysis, as discussed in Sect. 6.2, must include clear evidence on whether an identifiability issue exists and a clear procedure leading to the resolution of this problem. The resolution of the identifiability problem is achieved by singling out the true values of the parameter set, $S_{r(\text{true})}$, that truthfully correspond to the physical system and the reaction at hand. Such a procedure can be dubbed "*the elucidation method*" (see an example in Chap. 10).

6.4 Distinguishability: Does It Pose a Problem in Kinetic Analysis?

Distinguishability arises for reactions whose kinetic traces can be interpreted according to several (more than one) mechanistic options. This situation defines what is called the *distinguishability* problem when different reaction schemes (mechanisms) reproduce the experimental data with equal goodness of fit. The kinetics of the system is then considered to be non-distinguishable.

According to the literature listed in the previous section, the distinguishability issue is generally concurrent with the identifiability problem (but this is not a rule).

Let us consider, for illustration, the absorption traces of a thermal reaction, established to involve three species $XY_2(nk)$, but the number (n) and/or the disposition of the first-order reaction-steps within the reaction mechanism are not known with certainty. If the absorptivity of Y_1 is not known, the reaction traces could be fitted, with high accuracy, to several equation models that correspond to six distinct reaction mechanistic options [4, 25], as shown in Scheme 6.1, for a $XY_2(3k)$ sequence.

A systematic method proposed later [5], considering five possible parameters (four first-order rate-constants and the absorptivity ε_{Y_1}, i.e. $XY_2(4k)$) and a series of constraints, theoretically established a large set of reaction schemes that are indistinguishable for the trimolecular system.

In practice, the kinetic experimental data of such a reaction would perfectly be mapped out to more than one mechanistic option but without the possibility to single out the true mechanism of the reaction. The kinetic analysis remains, therefore,

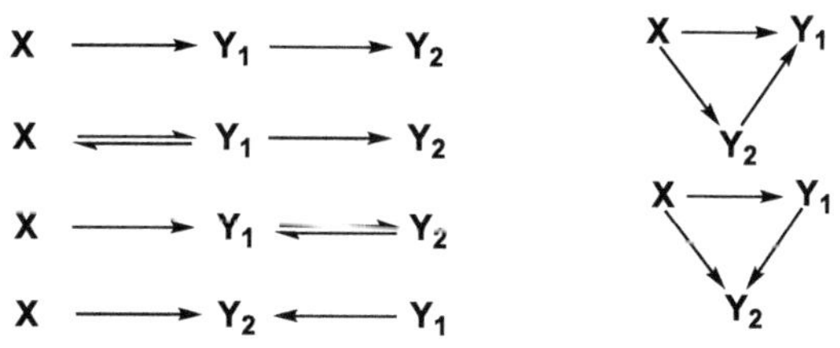

Scheme 6.1 Six mechanistic options for the thermal $XY_2(3k)$ reaction

incomplete as long as this distinguishability problem is not fully solved (otherwise, the conclusions reached for the reaction kinetics are not pertinent).

The general conclusion that stands out from the consideration of the distinguishability problem is that the fitting of the data cannot, on its own, be considered a sufficient means to reach a definitive conclusion about the identity of the true mechanism undergone by the reaction. The good fit of the traces becomes only a necessary but not a sufficient condition to decide whether the selected reaction mechanism at the origin of the equations fitting the data is the true one for the reaction investigated. In other words, the only certainty the analyst may have is that when the experimental data is not faithfully fitted to the model equation of a proposed mechanism, then this mechanism is discarded and cannot be used to describe the reaction at hand.

Kinetically, distinguishability must be examined for reactive systems for which ($a_{dist.}$) the reactant and products are known, ($b_{dist.}$) the mechanism (the precise number, orientation, and combination of reaction-steps) is unknown, ($c_{dist.}$) the reaction-steps obey known kinetic orders, and ($d_{dist.}$) the reaction parameters' values are not known. The distinguishability problem translates to the existence of different (more than one) reaction mechanisms that can be used to interpret the experimental data of the reaction investigated.

6.5 The New Concept of Distinctability in Relation to Kinetics

It is perhaps useful to mention two other types of concepts inherent to kinetic interpretation that may lead to confusion. One is general in kinetics, and the other is more specific to photoreactions.

Intrinsic distinctability refers to the possible occurrence of distinct kinetic orders for the investigated reaction due to the nature of the reactive system itself and the impossibility of foreseeing or predicting the kinetic order of a reaction from only examining its mechanism. We have a good example in the thermal reaction $X \rightarrow Y$, whose kinetic behaviour cannot be known prior to conducting the experiment, as this reaction may obey either 0th- or 1st-order kinetics. The situation addressed here is not to be confused with the manifestation of pseudo-reaction orders. Distinctability of the reaction order is also an issue in photokinetics. We will see in the following chapters that a photoreaction may obey distinct reaction kinetic orders depending on the irradiation type used.

The variety of irradiation types can also cause another difficulty in interpretation. We have seen in Chap. 2 that a photoreaction can be driven by a variety of light types, including ($a_{h\nu}$) isosbestic or ($b_{h\nu}$) non-isosbestic monochromatic lights or ($c_{h\nu}$) polychromatic irradiations of disparate profiles. *Extrinsic distinctability* emerges in relation to the light type, where the only analysis of the experimental kinetic traces of the reaction cannot inform about or single out the type of light that was used to

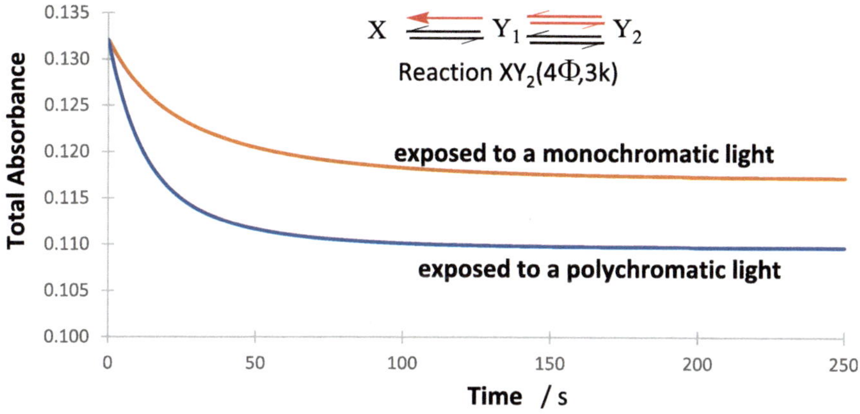

Fig. 6.1 Similarity of the shapes of the kinetic traces obtained with mono- and polychromatic light impinging on the same $XY_2(3k, 4\Phi)$ photoreaction (the thermal reaction-steps are indicated in red)

induce the photoreaction at hand. Similarly, the experimental trace informs neither of the state of collimation of the light used nor of the geometry of the irradiation setup.

Both types of distinctability are independent of the analysis of the trace data, in the sense that both are related to properties of the reactive system and the reaction conditions. The intrinsic distinctability can generally be elucidated experimentally by an appropriate mathematical analysis of the traces, whereas the extrinsic distinctability requires information that must be provided by the experimentalist who performed the experiment (i.e. indicating the light type, light setup employed, etc.). As far as we know, there is no direct means that is able to determine the light type inducing the reaction from only examining experimental traces obtained from the investigated reaction.

Therefore, in kinetics and photokinetics, the distinctability concept applies to a given photoreaction whose reactant and products, as well as its full mechanism (i.e. the number and nature of reaction-steps), are all known with certainty. There are still two pieces of information that are not accessible directly from a knowledge of the mechanism or the kinetic traces of the reaction (Fig. 6.1).

The relevance of distinctability is indicated by the recommendation that the order of the reaction is a feature that must be experimentally determined and not postulated (e.g. by analogy to or from a guessed mathematical model) and the necessity of clearly indicating the light type used to avoid incorrectly assigning an interpretation that might be valid for one type of light to another type of light, for which that interpretation is no longer valid. The latter situation is mostly encountered for the determination of the quantum yield of a reaction subjected to polychromatic light using methods and equations exclusive to photokinetics, obtained under irradiation with monochromatic light.

Accordingly, we are led to conclude that in order to avoid mistakes, it is mandatory to know the full set of distinctability options offered to a given reaction. This, as a matter of fact, must clearly represent one of the first steps of a consistent kinetic analysis. We shall see tangible examples of distinctability cases in photokinetics in the following chapters.

6.6 Occam's Razor Principle and Kinetic Solution Degeneracy

According to the description of the general procedure of kinetic analysis, as adopted in the literature (Sect. 6.2), when the fitting of the kinetic traces gives a perfect match between experimental data and the model equation, a final conclusion may be reached about the kinetic order, mechanism, and the reaction parameter values obtained as the true features for the reaction, and the kinetics can be solved. Therefore, in this case, the satisfactory outcome should represent the end of a successful and complete kinetic analysis. This conclusion might be supported by *Occam's razor* principle [26], stating that "of two competing theories, the simpler explanation of an entity is to be preferred," hence giving precedence to simplicity. The principle can also be invoked in different terms but with the same substantial meaning as "in explaining a thing no more assumptions should be made than are necessary." Hence, for the above kinetic description, since we found an explanation for the experimental data at hand, we should not progress further with more assumptions.

Occam's razor, in cases such as the one discussed here, if not carefully handled, would seem to oppose simplicity to truth. However, in the scientific method, the truth is paramount, and simplicity is an option only when the truth is preserved. In the context of kinetics and photokinetics, degeneracy makes it not possible to decide with confidence which of the solution options is the right one to explain the kinetic data and faithfully represent the physical system studied.

Once degeneracy operates, each kinetic solution obtained offers a similar full explanation of the data at hand, and therefore, Occam's razor principle's simplest option is difficult to characterise (e.g. the solution with the lowest values of the parameters).

In fact, the correct question to pose in kinetic analysis should be "Which kinetic solution is the true one?" Occam's razor, in this context, means that the simplest answer to the latter question would be preferred.

Clearly, for degenerate kinetic solutions, there are no objective indications to sustain Occam's razor principle in its classical form; that is, the information available does not allow us to confirm that, for instance, the reaction data available cannot belong to another reaction (that would have a different mechanism compared to the one we have first chosen) if the model equations of these mechanisms fit the experimental trace with high accuracy.

6.7 The Fundamental Ground for the Occurrence of Degeneracy

It is understood that the fitting of the kinetic traces by the model proposed (as the only observable on the dynamic system since the parameters of the reaction cannot be measured directly) is a necessary condition in kinetic analysis. Irrespective of the validity of the model equations, we know that it has to be based on the reaction mechanism (true or preconceived), and this model has to be expressed mathematically in terms of equations. Therefore, equations are fundamental to kinetics and kinetic analysis, whether or not they address identifiability and distinguishability issues.

The general rule imposes that, among all the equations of the kinetic system, only the set of those that are linearly independent is useful for the description and the later solution of the kinetics (the equations that are not linearly independent can just be ignored). This becomes a fundamental condition in kinetic analysis.

Let us assume that our system of equations, derived for the reactive system under investigation, is exclusively a system of linear equations; let it be five equations and four unknowns. This information will not allow us to automatically conclude that this system is solvable. In fact, we need to first check whether at least four of these equations are linearly independent. This means that not more than one of these five equations can be worked out by a linear combination of any of the other four. In case the five equations are linearly independent, then any one of them can be discarded because it is not necessary for achieving a complete solution of the system (here, we need only four linearly independent linear equations because there are four unknowns). In kinetics, the latter case, where the number of equations exceeds that of the unknowns, is very rare, as we are going to subsequently see (e.g. Chaps. 9 and 10).

When the equations describing the kinetics are non-linear, the situation is not much different from the above case of linear equations, in that the equations retained must be linearly independent. However, there is a supplementary complication, in that it is not always certain that a solution can be found, even when the number of equations is equal to the number of unknowns (obviously, when the number of equations is lower than that of the unknowns, there will be no solution possible). What number of equations is required to solve a system of non-linear but linearly independent equations depends on the system of equations at hand.

Therefore, in kinetic analysis, it is essential to identify the fundamental set of unknowns of the investigated reaction, as well as the number of linearly independent equations that describe the dynamics.

As a matter of fact, in photokinetics, such a process, most often, shows that there are fewer linearly independent equations than the number of unknowns. And in all cases, some, if not all, of these equations are non-linear.

Fundamentally, the reason behind the emergence of degeneracy is a lack of information, which mathematically translates to either or both a higher number of unknowns compared to the number of linearly independent equations in the system of equations describing the kinetics and/or the linearly independent equations being non-linear (even if, in the latter case, the number of unknowns and equations is the same).

Defining the unknowns and the linearly independent equations of a reactive system, comparing their numbers and examining the linearity of the equations, is a practical way to establish the potential occurrence of identifiability and/or distinguishability.

6.8 Constraints of Degeneracy on Kinetic Investigation

As evidenced from the previous sections, it is not judicious to ignore the possible occurrence of one or more of the kinetic solution degeneracy issues (i.e. distinctability, distinguishability, and identifiability issues) when performing analysis of kinetic data. The investigation and kinetic analysis seek to determine both the true mechanism of the reaction and/or the true values of the unknown kinetic and spectroscopic parameters of the system under study. Therefore, the kinetic analysis must seriously take these issues into account. Failing to do so will shed a significant degree of uncertainty and doubt on the veracity of the conclusions reached and/or the numerical values obtained. In both these cases, the purpose of describing the kinetics of a reaction is not achieved, even though a good fit of the experimental data is obtained and the values of the parameters all seem plausible.

In fact, the distinction between a fully correct kinetic analysis and a useless set of results may well reside in the difference between obtaining just plausible results and obtaining plausible results from a kinetic analysis that addressed and solved the distinctability, distinguishability, and identifiability issues. We can add another element to our previous conclusion, in that the plausibility of the results obtained for the parameters, together with a good fit of the experimental data, represents a necessary but not a sufficient pair of conditions for a consistent kinetic analysis. Simply stated, the constraint that kinetic analysis has to overcome can be formulated as follows: if the distinctability, distinguishability, and identifiability for any kinetic problem are not proven to be resolvable and effectively resolved by a detailed mathematical analysis, then the kinetics of the reaction at hand has not been solved. Therefore, any conclusions and/or parameter values that could be proposed for the dynamic system are useless unless the above statement is fully verified and the possibility of resolving the above issues is clearly demonstrated.

The constraint to kinetic analysis can be expressed as the need for a detailed *elucidation method* that leads to a single/unique set of the parameters' values and/or the mechanism of the reaction. The uniqueness of the solution must be proven before envisaging the interpretation of any kinetic data.

6.9 Dealing with Degeneracy

When either or both an identifiability or a distinguishability problem are established with certainty for a given kinetic situation, the only possible way to solve such issues is to increase the amount of information on the reactive system. This can be achieved by two ways: either by defining the values of one or more unknown parameters by alternative means (each specific kinetic problem will require a minimum number of these unknowns to be defined) or by increasing the number of equations but without increasing the number of unknown parameters (otherwise, if the number of parameters increases, we displace the problem without being able to solving it).

Unravelling the numerical values of some unknowns can be delivered by employing other analytical chemistry techniques, in addition to the necessary spectrophotometry. For instance, separation chromatographic techniques can identify the concentration of the species (e.g. those reached at the end of the reaction).

Increasing the number of equations will require an experimental design. The carefully selected supplementary experiments are tuned towards generating new equations to be added to the original set, with the condition stated above that the final total number of unknowns will be equal to or lower than the final total number of equations. This implies a good understanding of the kinetics and the experimental means available to the experimentalist.

It is important to underline that true identifiability and distinguishability problems cannot be solved by using common mathematical methods (e.g. they are not solvable by numerical integration methods) or sophisticated processes (such as neural networks and fuzzy logic).

6.10 Degeneracy in Photokinetic Literature

It is truly impressive to observe that photokinetic documentation is practically bare of references relative to identifiability and distinguishability problems. As far as we are aware, research papers as well as textbooks do not give these concepts the consideration they deserve.

Without trying to interpret history, this situation is probably due to the very limited number of authors who made an effort to bring these concepts to light in kinetics, so much so that the identifiability and distinguishability issues remained in the shadows and seemed to apply only to rare and very particular situations (which is far from true).

In any case, it is a fact that the vast majority of photokinetic studies do not consider these issues in their proposed treatment of kinetic data.

We will discuss such photokinetic problems in the upcoming chapters.

6.11 Conclusion

The uncertainty hovering over the concepts reviewed thus far in Part II calls for more scrutiny and necessitates a more detailed examination of the definitions used. The impact of using the current understanding regarding these essential aspects is evidently significant on the quantification of various photochemical features.

Based on the various observations seen in the above chapters, a photokinetic investigation must include some fundamentals and clearly identify:

(a) The light type used, polychromatic or monochromatic
(b) The quantity measured, either quantum yield or photonic yield
(c) The correct equation for the quantum yield
(d) The true kinetic solution by characterising and solving eventual degeneracy problems

These important points might well be summarised in a single concept, that is, the *elucidation method*. An elucidation of a photokinetic problem means that the degeneracy issues have been solved, a true set of equations has been laid out for the different (unknown) quantities, and a procedure implements the elucidation steps for specifically defined experimental conditions, including the light type used.

References

1. Lente G (2015) Deterministic kinetics in chemistry and systems biology: the dynamics of complex reaction networks. Springer, Cham. https://doi.org/10.1007/978-3-319-15482-4
2. Swain CG (1944) The kinetic analysis of consecutive irreversible first order reactions. J Am Chem Soc 66:1696–1700. https://doi.org/10.1021/ja01238a027
3. Alcock NW, Benton DJ, Moore P (1970) Kinetics of series first-order reactions. Analysis of spectrophotometric data by the method of least squares and an ambiguity. Trans Faraday Soc 66:2210–2213. https://doi.org/10.1039/TF9706602210
4. Jackson WG, Harrowfield JM, Vowles PD (1977) Consecutive, irreversible first-order reactions. Ambiguities and practical aspects of kinetic analyses. Int J Chem Kinet 9:535–548. https://doi.org/10.1002/kin.550090404
5. Vajda S, Rabitz H (1988) Identifiability and distinguishability of first-order reaction systems. J Phys Chem 92:701–707. https://doi.org/10.1021/j100314a024
6. Vajda S, Rabitz H (1994) Identifiability and distinguishability of general reaction systems. J Phys Chem 98:5265–5271. https://doi.org/10.1021/j100071a016
7. Guthrie JP (1982) Determination of rate constants when analytical integration of the rate law is not possible: approaches based on numerical differentiation and numerical integration. Can J Chem 60:765–771. https://doi.org/10.1139/v82-112
8. Sienicki K, Durocher G (1992) On the identifiability and distinguishability of first-order reaction systems by time-resolved fluorescence spectroscopy. Macromolecules 24:1102–1106. https://doi.org/10.1016/1010-6030(92)85193-X
9. Walter E, Lecourtier Y, Happel J, Kao J-Y (1986) Identifiability and distinguishability of fundamental parameters in catalytic methanation. AICHE J 32:1360–1366. https://doi.org/10.1002/aic.690320813

10. Sienicki K, Durocher G (1991) Fluorescence decays in aromatic polymers: analysis of kinetic models and their identifiability and distinguishability. J Photochem Photobiol A Chem 63:279–287. https://doi.org/10.1021/ma00005a020
11. Godfrey KR, Michael TM, Chapman J, Vajda S (1994) Identifiability and indistinguishability of nonlinear pharmacokinetic models. J Pharmacokin Biopharm 22:229–251. https://doi.org/10.1007/BF02353330
12. Walter E, Pronzato L (1996) On the identifiability and distinguishability of nonlinear parametric models. Math Comput Simul 42:125–134. https://doi.org/10.1016/0378-4754(95)00123-9
13. Zhang J-F, Yan Q-H, Zhai T (2011) Numerical approach to identifiability test of parametric models in nonlinear mechanical systems. J Dyn Syst Meas Control 133:051002. https://doi.org/10.1115/1.4004062
14. Gerbet D, Röbenack K (2021) An algebraic approach to identifiability. Algorithms 14:255–269. https://doi.org/10.3390/a14090255
15. Berthier F, Diard J-P, Pronzato L, Walter E (1996) Identifiability and distinguishability concepts in electrochemistry. Automatica 32:973–984. https://doi.org/10.1016/0005-1098(96)00031-3
16. Debus T, Dupont P, Howe RD (2005) Distinguishability and identifiability testing of contact state models. Adv Robot 19:545–566. https://doi.org/10.1163/156855305323383794
17. Rantow FS, Masoud S, Grady MC (2006) Global parametric identifiability of mechanistic models in chain polymerization. IEEE Proc Am Contr. https://doi.org/10.1109/ACC.2006.1657188
18. Hattersley JG, Pérez-Velázquez J, Chappell MJ, Bearup D, Roper D, Dowson C, Bugg T, Evans ND (2011) Indistinguishability and identifiability of kinetic models for the MurC reaction in peptidoglycan biosynthesis. Comp Methods Progr Biomed 104:70–80. https://doi.org/10.1016/j.cmpb.2010.07.009
19. Szederkényi G, Banga JR, Alonso AA (2012) CRNreals: a toolbox for distinguishability and identifiability analysis of biochemical reaction networks. Bioinformatics 28:1549–1550. https://doi.org/10.1093/bioinformatics/bts171
20. Meshkat N, Rosen Z, Sullivant S (2018) Algebraic tools for the analysis of state space models. Adv Stud Pure Math 77:171–205. https://doi.org/10.48550/arXiv.1609.07985
21. Roosa K, Chowell G (2019) Assessing parameter identifiability in compartmental dynamic models using a computational approach: application to infectious disease transmission models. Theor Biol Med Model 16:1–15. https://doi.org/10.1186/s12976-018-0097-6
22. Gross E, Meshkat N, Shiu A (2022) Identifiability of linear compartmental models: the singular locus. Adv Appl Math 133:102268. https://doi.org/10.1016/j.aam.2021.102268
23. Ovchinnikov A, Pogudin G, Thompson P (2023) Input-output equations and identifiability of linear ODE models. IEEE Trans Autom Contr 68:812–824. https://doi.org/10.1109/TAC.2022.3145571
24. Hennig C (2024) Parameters not empirically identifiable or distinguishable, including correlation between Gaussian observations. Stat Pap 65:771–794. https://doi.org/10.1007/s00362-023-01414-3
25. Milligan WO, Mullica DF, Pennington DE, Lok CKC, Kwong DWJ (1984) Application of nonlinear least squares analysis on three different consecutive, irreversible first order kinetic processes. Comput Chem 8:285–298. https://doi.org/10.1016/0097-8485(84)85021-4
26. Encyclopaedia Britannica. Available at https://www.britannica.com/topic/Occams-razor

Chapter 7
The Debated Order of a Photoreaction

7.1 Introduction

Does a photoreaction have a kinetic order? Thus far, this basic question has no clear answer in the literature. Most often, photochemical reactions were treated by the classical thermal kinetic models set out for zeroth-, first- or second-order kinetics. Even if a debate was shortly held about this issue to justify the relevance of thermal kinetic orders for photoreactions, the controversy was not definitely settled. Hence, today, whether a primary photoprocess has a particular kinetic order or whether it should simply obey a thermal order is not a fully resolved issue. This chapter will succinctly review the scarce literature around that opening basic question.

7.2 The Debate

A little more than two decades ago, an interesting discussion took place about whether a monomolecular photoreaction (Scheme 7.1) has a kinetic order [1–3]. On one side [2, 3], it was argued that the basic photoreaction should obey second-order kinetics since the reaction can be considered bimolecular, i.e. involving a photoreactive species, X, and a photon, γ (and yielding a photoproduct, Y). On the other side of the argument [1, 3], the photoreaction was supposed to have no kinetic order at all due to a change in its kinetic behaviour that depends on the initial concentration of the reactive species, $C_X(0)$, i.e. it may change from a linear-(0th-order) type kinetics for relatively high $C_X(0)$ values to a more curved temporal variation (first-order-like) profile when the initial concentration values are much lower.

M. Maafi, *Photokinetics*, https://doi.org/10.1007/978-3-031-98985-8_7

$$X + \gamma \rightarrow X^{*} \rightarrow Y$$

Scheme 7.1 Transformation of reactant X upon absorbing a photon γ

$$X + \gamma \quad \rightarrow \quad X^{*} \quad \rightarrow \quad Y$$

In response to the latter group, the supporter of the first argument claimed that if either or both $C_X(t)$ and $C_\gamma(t)$ (the photon concentration) remained approximately constant during the reaction, then the apparent order might be zero or one; however, the total order of the photoreaction must still be two [2, 3]. X and γ were set to obey first-order kinetics, as given by Eq. (7.1).

$$r_X(t) = -\frac{dC_X(t)}{dt} = k\, C_X(t)\, C_\gamma(t) \tag{7.1}$$

As far as we are aware, there has been no further debate on this issue in the literature. There are still no known standard integrated rate-laws for photosystems, and photochemistry textbooks do not usually provide any comprehensive description of photokinetics in terms of photoreaction orders. Yet 0th, 1st- or 2nd-order kinetic treatments are routinely employed in many research papers to evaluate the rate-constants of photoreactions [4–8] or reported on in textbooks [9, 10].

A further examination of the literature reveals that experimentally determined rates of a number of photoreactions, including unimolecular reactions (Scheme 7.1), whose X and Y absorbed the excitation light, were found to increase with increasing light intensity (I_0) and decreasing $C_X(0)$. This dual effect has not been mentioned in the aforementioned debate [1–3] since the discussion focused exclusively on the reaction-order variation with $C_X(0)$.

As a matter of fact, the debate remains open on whether or not a photoreaction has or should have a well-defined, standard kinetic order as thermal reactions do.

7.3 Consistency of the Rate-Law

In 1966, Shaw and Toby [11] published a short communication to warn about and prevent the spread of errors relative to the attribution of reaction order in photokinetics that were observed in some of the textbooks (available at that time). The authors pointed out that the rate-law of the unimolecular photochemical reaction ($X + h\nu \rightarrow 2\ Y$) was commonly expressed in one of three of the following forms (where Φ was assumed equal to 1 for simplification):

$$-\frac{dC_X(t)}{dt} = k_1\, I_a \tag{7.2}$$

$$= k_2\, C_X(t) \tag{7.3}$$

$$= I_a \tag{7.4}$$

The authors then evaluated the consistency of these equations. The first equation (Eq. (7.2)) was deemed unsatisfactory for two main reasons. Since the absorbed light intensity, I_a, like the incident intensity, I_0, has dimensions of einstein per area – time (i.e. electromagnetic energy impinging on unit area in unit time, according to the analysis developed in the study), then k_1 would have dimensions of m^{-1}, which makes little physical sense. Otherwise, if I_a is taken to mean the average absorbed light, $\overline{I_a}$ (assuming a uniform concentration of X in the photolysis slab reactor and considering the average of the reaction-rate over the reactor length), then k_1 is redundant.

The authors noticed that because k_2 in Eq. (7.3) was quasi-independent of temperature but should be dependent on concentration (except for very low optical densities), Eq. (7.3) is weak and inconsistent.

They, however, considered that Eq. (7.4) is the best way of describing the disappearance-rate of the reactant X. They concluded that, in the general case of the unimolecular reaction, the equation must include the quantum yield as

$$-\frac{dC_X(t)}{dt} = \Phi\, I_a \tag{7.5}$$

Accordingly, the classical first-order thermal kinetics, in its classical formulation, was proven to be inappropriate to treat photochemical data.

For the proposed second-order kinetics, as given by Eq. (7.1), one can reach the following assessment. Equation (7.1) can be debunked for the same reasons invoked for Eq. (7.2) when the initial concentration of X is high (i.e. $C_X(t) = C_X(0)$) since, in that equation, the photon concentration, $C_\gamma(t)$, must represent the absorbed light inducing the reaction, i.e. I_a, and hence Eq. (7.1) is nothing else but Eq. (7.2).

If, now, $C_\gamma(t)$ is rather considered to be the initial (or incident) light intensity, I_0, then it must be constant, and hence, Eq. (7.1) is equivalent to Eq. (7.3) and would be discarded for the same reasons given for the latter.

For Eq. (7.1) at low $C_X(0)$ values, the rate-law includes both the reactant concentration and the light intensity, as $-dC_X(t)/dt = k_3\, I_a\, C_X(t)$. Here, if $C_\gamma(t)$ is equal to I_0 (i.e. $k_3\, I_a = k_3\, I_0 =$ constant), then we go back to Eq. (7.3). Otherwise, the argument that the reaction can be described as following a second-order reaction fails because the sum of the partial order is not 2 but now unknown, the reason being that the light intensity, I_a, is neither a concentration, per se, nor is it linearly proportional to a concentration since I_a is a non-linear function of $C_X(t)$. Therefore, the assumption that in Eq. (7.1), it is consistent to proceed with the sum of the partial orders of $C_X(t)$ and $C_\gamma(t)$ (which are both equal to unity, to find a global reaction order equal to 2) cannot be deemed a valid procedure.

The discussion provided above should have at least closed the debate on whether thermal reaction kinetics could characterise photoreactions. Unfortunately, this is still a common practice in the literature. In this context, it would be relevant to draw attention to the fact that a discussion on which rate-law equation would describe a photoreaction should logically take place before the analysis of the kinetic data or which already existing thermal kinetic model better fits the experimental data of photoreactions.

Equation (7.5) was the formulation later accepted in the field [12] for the unimolecular reaction whose product is transparent to the incident light but with the expanded equation of I_a (given by Eq. (2.20)) as

$$-\frac{dC_X(t)}{dt} = I_0\left(1 - e^{-\varepsilon\, C_X(t)\, l}\right)\Phi \tag{7.6}$$

7.4 Inferences for the Reaction Order

The analysis described in the previous section [11] on defining the most consistent rate-law for the unimolecular photochemical reaction did not examine the proper question relating to the reaction order. In a later publication, Toby [3] confirmed the validity of Eq. (7.5) but concluded that a photoreaction "*lacks a meaningful order that is independent of the concentration* (magnitude)," so "*the concept of reaction order for a photoreaction is of little use*," a conclusion that falls within the scope of Logan's [1] inference that "*only the quantum yield* (Φ) *denotes how a photochemical conversion proceeds.*" Resorting to the absence of a reaction order in photochemistry seems to have been able to cross over the field, so much so that photochemistry literature, in general, does not provide a hint about what the true kinetic order of a photoreaction is.

Nonetheless, one can surely argue that since the observed kinetic behaviour of a given photoreaction is reproducible, it would be difficult to definitely dismiss the existence of a rationale that can model such kinetic behaviour (and kinetic traces) of that photoreaction.

Despite the above status quo, the concept of a reaction order for photoreactions cannot be dismissed from a purely mathematical point of view. Equation (7.5) is obviously not directly proportional to the concentration, and a rate-constant (e.g. k) does not explicitly show in this equation. Clearly, this indicates that it belongs to a kinetic system that is intrinsically and physically different from a thermal reaction. The behaviour of a photoreaction would be mapped out by a different integrated rate-law compared to those of thermal reactions, a feature that should be expected since, as we have seen in Sect. 3.5, different differential equations must have different solutions. It is obvious that Eq. (7.5) is significantly different from those assigned to the thermal reaction rate-laws (Table 7.1), and therefore, particular order and properties, unlike those of a thermal reaction, might characterise a photoreaction.

Table 7.1 Kinetic properties characterising some basic thermal reactions

Kinetic order	Rate-law $-\frac{dC_X(t)}{dt}=$	Integrated rate-law $C_X(t)=$	Half-life time $t_{\frac{1}{2}}=$	Initial velocity $-r_{0,\,X}=$	Unit of the rate-constant k
0th	k	$-kt + C_X(0)$	$\frac{C_X(0)}{2\,k}$	k	Ms^{-1}
1st	$k\ C_X(t)$	$C_X(0)\ e^{-k\,t}$	$\frac{ln\,2}{k}$	$k\ C_X(0)$	s^{-1}
2^{nd}_a $C_X(0) \neq C_{X'}(0)$[a]	$k\ C_X(t)\ C_{X'}(t)$	$\frac{(C_{X'}(0)-C_X(0))\ C_X(0)\ e^{-(C_{X'}(0)-C_X(0))\ k\ t}}{C_{X'}(0)-C_X(0)\ e^{-(C_{X'}(0)-C_X(0))\ k\ t}}$	$\frac{ln\,2}{k\ (C_{X'}(0)-C_X(0))}$	$k\ C_X(0)\ C_{X'}(0)$	$M^{-1}s^{-1}$
2^{nd}_b $C_X(0) = C_{X'}(0)$	$k\ (C_X(t))^2$	$\frac{C_X(0)}{1+k\ t\ C_X(0)}$	$\frac{1}{k\ C_X(0)}$	$k\ (C_X(0))^2$	$M^{-1}s^{-1}$

[a] X and X' are chosen such as $C_X(0) < C_{X'}(0)$

In this context, it might be useful to consider the case of an induced reaction order change by different experimental conditions of bimolecular reactions. Here, the change of the experimental conditions causes irreducible changes to the differential equation. It is interesting that 2nd-order thermal reactions actually obey two different and independent integrated rate-laws, depending on whether or not the initial concentrations of the reactants X and X' are equal (i.e. the integrated rate-law of one cannot be worked out from that of the other one; Table 7.1). Therefore, one could assume that there is a need to differentiate between these behaviours (the sum of the concentration exponents does not catch this subtlety). To bring more clarity to the matter, it would be considered that either the reaction obeys a different order for each experimental condition or there are suborders (e.g. 2_a^{nd} and 2_b^{nd}) to the kinetics of bimolecular reactions.

7.5 Conclusion

The literature provides certainty about two aspects of photoreactions: (1) there is little doubt that the rate-law in photokinetics is represented by the template formulated in Eq. (7.5), and (2) as far as we are aware, photoreactions have no standard reaction orders as the thermal reactions do. Therefore, for photoreactions, the question initially raised in the above debate, "*Does a photoreaction have a kinetic order*?" is still legitimate and standing.

Some of the subsequent chapters will explore this question in detail and propose consistent answers for it.

References

1. Logan RS (1997) Does a photochemical reaction have a reaction order? J Chem Educ 74:1303. https://doi.org/10.1021/ed074p1303
2. Hippler M (2003) Photochemical kinetics: reaction orders and analogies with molecular beam scattering and cavity ring-down experiments. J Chem Educ 80:1074–1077. https://doi.org/10.1021/ed080p1074
3. Toby S (2005) Does a photochemical reaction have a reaction order? J Chem Educ 82:37–38. https://doi.org/10.1021/ed082p37.2
4. Thoma K, Klimek R (1991) Photostabilisation of drugs in dosage forms without protection from packaging materials. Int J Pharm 67:169–175. https://doi.org/10.1016/0378-5173(91)90429-R
5. Nassar N, Trivella A, Mokh S, Al-Iskandarani M, Budzinski H, Mazellier P (2017) Photodegradation of sulfamethazine, sulfamethoxypiridazine, amitriptyline, and clomipramine drugs in aqueous media. J Photochem Photobiol A Chem 336:176–182. https://doi.org/10.1016/j.jphotochem.2016.12.008
6. Jorfi S, Barzegar G, Ahmadi M, Soltani RDC, Haghighifard NAJ, Takdastan A, Saeedi R, Abtahi M (2016) Enhanced coagulation-photocatalytic treatment of acid red 73 dye and real textile wastewater using UVA/synthesized MgO nanoparticles. J Environ Manag 177:111–118. https://doi.org/10.1016/j.jenvman.2016.04.005

7. Uher G, Pillans JJ, Hatton AD, Upstill-Goddard R (2017) Photochemical oxidation of dimethylsulphide to dimethylsulphoxide in estuarine and coastal waters. Chemosphere 186: 805–816. https://doi.org/10.1016/j.chemosphere.2017.08.050
8. Hu C, Xu M, Zhang J, Hu B, Yu G (2019) Recyclable MoO_3 nanobelts for photocatalytic degradation of Rhodamine B by near infrared irradiation. Int J Chem Kinet 51:3–13. https://doi.org/10.1002/kin.21223
9. Durr H, Bounas-Laurent H (eds) (2003) Photochromism: molecules and systems, vol 1. Elsevier, Amsterdam
10. Tonnesen HH (ed) (2004) Photostability of drugs and drug formulations. CRC Press, Boca Raton
11. Shaw H, Toby S (1966) Light absorption in photochemistry. J Chem Educ 43:408–410. https://doi.org/10.1021/ed043p408
12. Mauser H, Gauglitz G, Compton RG, Hancock G (1998) Comprehensive chemical kinetics, photokinetics: theoretical fundamentals and applications, vol 36. Elsevier, Amsterdam

Part III
Photokinetics for Monochromatic Isosbestic Irradiation

Chapter 8
The Reaction-Order Under Monochromatic and Isosbestic Irradiation

8.1 Introduction

The simplest light type from the viewpoint of a mathematical description of photoreactions is the monochromatic and isosbestic irradiation (at λ_{isos}, see Sect. 2.9). We start here from that relatively scarce irradiation condition since, if they occur, isosbestic points exist only in limited numbers for any given reaction. Isosbesticity also means that there should be more than one reactive species absorbing the incident light.

8.2 A Template Reaction

For the purpose of developing the mathematical framework, we will consider a reaction template, which is neither too simple nor too complex, as is the trimolecular photoreaction. The most extended reaction between three species is the cyclic mechanism $XY_2(6k, 6\Phi)$, involving a combination of twelve thermal and photochemical reaction-steps (Scheme 8.1).

The correct kinetic description of this general trimolecular mechanism has the advantage of providing the correct descriptions for any of the possible sub-mechanisms (more than 50) that are worked out from this reaction. We will study the kinetics of the reaction under isosbestic irradiation, whose system of differential equations can be solved. Here, we consider that, at the isosbestic point, the irradiation light is absorbed by all three species.

M. Maafi, *Photokinetics*, https://doi.org/10.1007/978-3-031-98985-8_8

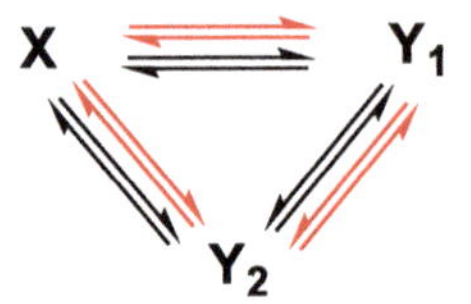

Scheme 8.1 The utmost inclusive mechanism of photothermal trimolecular $XY_2(6k, 6\Phi)$ reaction, where black and red arrows correspond, respectively, to photochemical and thermal reaction-steps

8.3 The Sub-mechanisms of $XY_2(6\Phi)$

It is useful to count the number of sub-mechanisms that can be worked out from a generic $XY_2(nk, m\Phi)$ mechanism. A reaction mechanism such as $XY_2(6k, 6\Phi)$ (Scheme 8.1) can either be used for the reaction involving six thermal and six photochemical reaction-steps or as a generic mechanism encompassing a large set of sequences (which can be worked out from that generic format, such as $XY_2(1k, 6\Phi)$, $XY_2(6k, 1\Phi)$ and $XY_2(2k, 3\Phi)$). Let us look hereafter at the relatively smaller set of sub-mechanisms (Table 8.1) belonging to the simpler $XY_2(6\Phi)$ photoreaction. This reaction is defined by eight intrinsic parameters, including the six photochemical quantum yields of the six photochemical reaction-steps and two spectroscopic parameters, i.e. the absorption coefficients of the two photoproducts ($\varepsilon_X^{\lambda_{obs}}$ is supposed to be known). Overall, 25 situations can be counted for the $XY_2(6\Phi)$ photoreaction.

The ticks in Table 8.1 indicate the unknown fundamental parameters of the corresponding reaction. Such a specification, most often, corresponds to different elucidation difficulties of each kinetic case and, therefore, helps in listing the set of cases requiring different elucidation strategies.

This rule stands for all the sub-mechanisms of Table 8.1, including the unimolecular $XY(1\Phi)$ reaction, which has two options, $SubM_1$ and $SubM_2$, that differ by whether or not the product (Y) absorbs the incident light.

8.4 The Descriptive System of Differential Equations for $XY_2(6k, 6\Phi)$ Reaction

A system of three ordinary differential equations fully describes the kinetics of $XY_2(6k, 6\Phi)$ reactions subjected to monochromatic and isosbestic light of wavelength λ_{isos} (where the absorption coefficients of the three species are equal, $\varepsilon_X^{\lambda_{isos}} = \varepsilon_{Y_1}^{\lambda_{isos}} = \varepsilon_{Y_2}^{\lambda_{isos}}$ (Sect. 2.9)). The individual quantum yields ($\Phi_{Y_j \to Y_{j'}}^{\lambda_{isos}}$) are considered to be different for the involved photochemical reaction-steps. The system of linear and homogeneous ordinary differential equations (ODEs) can be set out as the temporal variation of the individual concentrations ($C_{Y_j}(t)$) as functions of these concentrations [1](as in Eq. 3.1, dC/dt=F(C)). The ODEs are written for a slab

Table 8.1 A set of kinetic cases from the generic $XY(6\Phi)$ reaction that requires different elucidation methods (the reaction parameter to be elucidated are indicated for each sub-reaction)

Parameters → sub-mechanism ↓	Reaction	Reaction scheme	$\varepsilon^{\lambda}_{Y_1}$	$\varepsilon^{\lambda}_{Y_2}$	$\Phi^{\lambda}_{X\rightarrow Y_1}$	$\Phi^{\lambda}_{X\rightarrow Y_2}$	$\Phi^{\lambda}_{Y_1\rightarrow X}$	$\Phi^{\lambda}_{Y_2\rightarrow X}$	$\Phi^{\lambda}_{Y_1\rightarrow Y_2}$	$\Phi^{\lambda}_{Y_2\rightarrow Y_1}$
$SubM_1$	$XY(1\boldsymbol{\Phi})$	$X \rightarrow Y$			✓					
$SubM_2$			✓		✓					
$SubM_3$	$XY_2(2\boldsymbol{\Phi})$	$Y_2 \leftarrow X \rightarrow Y_1$			✓	✓				
$SubM_4$			✓		✓	✓				
$SubM_5$			✓	✓	✓	✓				
$SubM_6$	$XY(2\boldsymbol{\Phi})$	$X \leftrightharpoons Y$			✓		✓			
$SubM_7$			✓		✓		✓			
$SubM_8$	$XY_2(3\boldsymbol{\Phi})$	$Y_2 \leftarrow X \leftrightharpoons Y_1$			✓	✓	✓			
$SubM_9$			✓		✓	✓	✓			
$SubM_{10}$			✓	✓	✓	✓	✓			
$SubM_{11}$	$XY_2(4\boldsymbol{\Phi})$	$Y_2 \leftrightharpoons X \leftrightharpoons Y_1$	✓	✓	✓	✓	✓	✓		
$SubM_{12}$	$XY_2(2\boldsymbol{\Phi})$	$X \rightarrow Y_1 \rightarrow Y_2$	✓		✓				✓	
$SubM_{13}$	$XY_2(3\boldsymbol{\Phi})$	$X \leftrightharpoons Y_1 \rightarrow Y_2$	✓		✓		✓		✓	
$SubM_{14}$		$X \rightarrow Y_1 \leftrightharpoons Y_2$	✓	✓	✓				✓	✓
$SubM_{15}$		$X \rightarrow Y_1 \rightarrow Y_2 \rightarrow X$	✓	✓	✓			✓	✓	
$SubM_{16}$		$X \rightarrow Y_1 \rightarrow Y_2 \leftarrow X$	✓	✓	✓	✓			✓	
$SubM_{17}$	$XY_2(4\boldsymbol{\Phi})$	$X \leftrightharpoons Y_1 \rightarrow Y_2 \rightarrow X$	✓	✓	✓		✓	✓	✓	
$SubM_{18}$		$X \leftrightharpoons Y_1 \rightarrow Y_2 \leftarrow X$	✓	✓	✓	✓	✓		✓	
$SubM_{19}$		$X \leftrightharpoons Y_1 \leftarrow Y_2 \leftarrow X$	✓	✓	✓	✓	✓			✓
$SubM_{20}$		$X \rightarrow Y_1 \leftrightharpoons Y_2 \rightarrow X$	✓	✓	✓			✓	✓	✓
$SubM_{21}$		$X \rightarrow Y_1 \leftrightharpoons Y_2 \leftarrow X$	✓	✓	✓	✓			✓	✓
$SubM_{22}$	$XY_2(5\boldsymbol{\Phi})$	$X \leftrightharpoons Y_1 \leftrightharpoons Y_2 \rightarrow X$	✓	✓	✓		✓	✓	✓	✓
$SubM_{23}$		$X \leftrightharpoons Y_1 \leftrightharpoons Y_2 \leftarrow X$	✓	✓	✓	✓	✓		✓	✓
$SubM_{24}$		$X \leftrightharpoons Y_1 \rightarrow Y_2 \leftrightharpoons X$	✓	✓	✓	✓	✓	✓	✓	
$SubM_{25}$	$XY_2(6\boldsymbol{\Phi})$	$X \leftrightharpoons Y_1 \leftrightharpoons Y_2 \leftrightharpoons X$	✓	✓	✓	✓	✓	✓	✓	✓

reactor, where the reactive medium is vigorously stirred, maintained at a constant temperature and exposed to a collimated/isosbestic light. At all times, the concentration of excited-state species is considered negligible.

$$\frac{dC_X^{\lambda_{isos},T}(t)}{dt} = \mathrm{a}_1^{\lambda_{isos}}\, C_X^{\lambda_{isos},T}(t) + \mathrm{a}_2^{\lambda_{isos}}\, C_{Y_1}^{\lambda_{isos},T}(t) + \mathrm{a}_3^{\lambda_{isos}}\, C_{Y_2}^{\lambda_{isos},T}(t) \tag{8.1}$$

$$\frac{dC_{Y_1}^{\lambda_{isos},T}(t)}{dt} = \mathrm{a}_4^{\lambda_{isos}}\, C_X^{\lambda_{isos},T}(t) + \mathrm{a}_5^{\lambda_{isos}}\, C_{Y_1}^{\lambda_{isos},T}(t) + \mathrm{a}_6^{\lambda_{isos}}\, C_{Y_2}^{\lambda_{isos},T}(t) \tag{8.2}$$

$$\frac{dC_{Y_2}^{\lambda_{isos},T}(t)}{dt} = \mathrm{a}_7^{\lambda_{isos}}\, C_X^{\lambda_{isos},T}(t) + \mathrm{a}_8^{\lambda_{isos}}\, C_{Y_1}^{\lambda_{isos},T}(t) + \mathrm{a}_9^{\lambda_{isos}}\, C_{Y_2}^{\lambda_{isos},T}(t) \tag{8.3}$$

The constant coefficients $\mathrm{a}_{1-9}^{\lambda_{isos}}$ are expressed in units s^{-1}. They are functions of both the rate-constants ($k_{Y_j \to Y_{j'}}^{\lambda_{isos},T}$) of the thermal reaction-steps and the quantum yields of the photochemical reaction-steps.

$$\mathrm{a}_1^{\lambda_{isos}} = -k_{X\to Y_1}^{\lambda_{isos},T} - \Phi_{X\to Y_1}^{\lambda_{isos}}\,\gamma_1 - k_{X\to Y_2}^{\lambda_{isos},T} - \Phi_{X\to Y_2}^{\lambda_{isos}}\,\gamma_1 \tag{8.4}$$

$$\mathrm{a}_2^{\lambda_{isos}} = k_{Y_1\to X}^{\lambda_{isos},T} + \Phi_{Y_1\to X}^{\lambda_{isos}}\,\gamma_1 \tag{8.5}$$

$$\mathrm{a}_3^{\lambda_{isos}} = k_{Y_2\to X}^{\lambda_{isos},T} + \Phi_{Y_2\to X}^{\lambda_{isos}}\,\gamma_1 \tag{8.6}$$

$$\mathrm{a}_4^{\lambda_{isos}} = k_{X\to Y_1}^{\lambda_{isos},T} + \Phi_{X\to Y_1}^{\lambda_{isos}}\,\gamma_1 \tag{8.7}$$

$$\mathrm{a}_5^{\lambda_{isos}} = -k_{Y_1\to X}^{\lambda_{isos},T} - \Phi_{Y_1\to X}^{\lambda_{isos}}\,\gamma_1 - k_{Y_1\to Y_2}^{\lambda_{isos},T} - \Phi_{Y_1\to Y_2}^{\lambda_{isos}}\,\gamma_1 \tag{8.8}$$

$$\mathrm{a}_6^{\lambda_{isos}} = k_{Y_2\to Y_1}^{\lambda_{isos},T} + \Phi_{Y_2\to Y_1}^{\lambda_{isos}}\,\gamma_1 \tag{8.9}$$

$$\mathrm{a}_7^{\lambda_{isos}} = k_{X\to Y_2}^{\lambda_{isos},T} + \Phi_{X\to Y_2}^{\lambda_{isos}}\,\gamma_1 \tag{8.10}$$

$$\mathrm{a}_8^{\lambda_{isos}} = k_{Y_1\to Y_2}^{\lambda_{isos},T} + \Phi_{Y_1\to Y_2}^{\lambda_{isos}}\,\gamma_1 \tag{8.11}$$

$$\mathrm{a}_9^{\lambda_{isos}} = -k_{Y_2\to Y_1}^{\lambda_{isos},T} - \Phi_{Y_2\to Y_1}^{\lambda_{isos}}\,\gamma_1 - k_{Y_2\to X}^{\lambda_{isos},T} - \Phi_{Y_2\to X}^{\lambda_{isos}}\,\gamma_1 \tag{8.12}$$

with the coefficient γ_1 being a product of the reactant absorptivity ($\varepsilon_X^{\lambda_{isos}}$, expressed in M^{-1} cm^{-1}), the incident light flux per irradiated volume and area of the sample ($P_0^{\lambda_{isos}}$, in einstein dm^{-3} s^{-1}), the optical path length of the excitation light inside the sample (l_{irr}, in cm) and the dimensionless photokinetic factor ($\mathrm{PKF}^{\lambda_{isos}}$). The dimension analysis yields units of s^{-1} for the coefficient γ_1.

$$\gamma_1 = \varepsilon_X^{\lambda_{isos}}\, P_0^{\lambda_{isos}}\, l_{irr}\, PKF^{\lambda_{isos}} \tag{8.13}$$

The photokinetic factor ($PKF^{\lambda_{isos}}$) explicit formula (Eq. (8.14)) indicates that it is independent of reaction time due to the constant value of $A_{tot}^{\lambda_{isos},T/\lambda_{isos}}$. This feature has a considerable consequence on the linearity of the differential equations of the system describing the reaction.

$$\mathrm{PKF}^{\lambda_{isos}} = \frac{1 - 10^{-A_{tot}^{\lambda_{isos},T/\lambda_{isos}}}}{A_{tot}^{\lambda_{isos},T/\lambda_{isos}}} \tag{8.14}$$

with the total absorbance at λ_{isos} being

$$\begin{aligned} A_{tot}^{\lambda_{isos}/\lambda_{isos}} &= \varepsilon_X^{\lambda_{isos}} \; l_{irr} C_X^{\lambda_{isos},T}(t) + \varepsilon_{Y_1}^{\lambda_{isos}} \; l_{irr} C_{Y_1}^{\lambda_{isos},T}(t) + \varepsilon_{Y_2}^{\lambda_{isos}} \; l_{irr} C_{Y_2}^{\lambda_{isos},T}(t) \\ &= \varepsilon_X^{\lambda_{isos}} \; l_{irr}\left(C_X^{\lambda_{isos},T}(t) + C_{Y_1}^{\lambda_{isos},T}(t) + C_{Y_2}^{\lambda_{isos},T}(t)\right) \\ &= \varepsilon_X^{\lambda_{isos}} \; l_{irr}\left(C_X^{\lambda_{isos},T}(0) + C_{Y_1}^{\lambda_{isos},T}(0) + C_{Y_2}^{\lambda_{isos},T}(0)\right) \end{aligned} \tag{8.15}$$

8.5 Preparing the ODE System for Integration

If the solution of the ODE system (Eqs. (8.1)–(8.3)) is approached by a simple conventional method, the procedure leads to a third-order algebraic linear equation that is not soluble. That is, the three roots of this equation can only be determined if the fundamental parameters of the reaction are known. But the determination of the latter parameters is supposed to be achieved by kinetic analysis after solving the third-order algebraic equation.

To circumvent this hurdle, it is possible to use a number of methods that, most usually, employ sophisticated mathematical procedures and/or require computational expenditure [2]. Nonetheless, for the $XY_2(6k, 6\Phi)$ reaction, it turns out that an analytical solution is achievable by classical means.

A new approach [1] proposed to reduce the three differential equations to two by considering an observable corresponding to the cumulative absorbance (Eq. (8.16)) measured at a given non-isosbestic observation wavelength, λ_{obs}, of two species among the three involved in the reaction, e.g. Y_1 and Y_2.

$$A_{Y_1,Y_2}^{\lambda_{isos},T/\lambda_{obs}} = \varepsilon_{Y_1}^{\lambda_{obs}} \; l_{irr} C_{Y_1}^{\lambda_{isos},T}(t) + \varepsilon_{Y_2}^{\lambda_{obs}} \; l_{irr} C_{Y_2}^{\lambda_{isos},T}(t) \tag{8.16}$$

The change of variable does not cause any loss of information and leads to a differential system of two equations (Eqs. (8.17) and (8.18)) that fully describes the reactive system, as the original one (Eqs. (8.1)–(8.3)) did with three equations. It can be expressed as a function of two variables, the reactant concentration, $C_X(t)$, and the measured cumulative absorbance of products, Y_1 and Y_2, $A_{Y_1,Y_2}^{\lambda_{isos},T/\lambda_{obs}}$, as

$$\frac{dA_{Y_1,Y_2}^{\lambda_{isos},T/\lambda_{obs}}}{dt}=\mathrm{a}_{10}\,C_X^{\lambda_{isos},T}(t)+\mathrm{a}_{11}\,A_{Y_1,Y_2}^{\lambda_{isos},T/\lambda_{obs}}+\mathrm{a}_{12}=\left(A_{Y_1,Y_2}^{\lambda_{isos},T/\lambda_{obs}}\right)' \tag{8.17}$$

$$\frac{dC_X^{\lambda_{isos},T}(t)}{dt}=\mathrm{a}_{13}\,C_X^{\lambda_{isos},T}(t)+\mathrm{a}_{14}\,A_{Y_1,Y_2}^{\lambda_{isos},T/\lambda_{obs}}+\mathrm{a}_{15}=\left(C_X^{\lambda_{isos},T}(t)\right)' \tag{8.18}$$

The a_{10} to a_{15} coefficients are expressed by

$$a_{10}=\left(\varepsilon_{Y_1}^{\lambda_{obs}}\,\mathrm{a}_4^{\lambda_{isos}}+\varepsilon_{Y_2}^{\lambda_{obs}}\,\mathrm{a}_7^{\lambda_{isos}}\right)-\frac{a_{12}}{C_{tot}^{\lambda_{isos},T}} \tag{8.19}$$

$$a_{11}=\left(\frac{\varepsilon_{Y_1}^{\lambda_{obs}}}{\varepsilon_{Y_2}^{\lambda_{obs}}}\,\mathrm{a}_6^{\lambda_{isos}}-\mathrm{a}_9^{\lambda_{isos}}\right)-\frac{a_{12}}{\varepsilon_{Y_2}^{\lambda_{obs}}\,C_{tot}^{\lambda_{isos},T}} \tag{8.20}$$

$$a_{12}=\left(\mathrm{a}_5^{\lambda_{isos}}+\frac{\varepsilon_{Y_2}^{\lambda_{obs}}}{\varepsilon_{Y_1}^{\lambda_{obs}}}\,\mathrm{a}_8^{\lambda_{isos}}-\frac{\varepsilon_{Y_1}^{\lambda_{obs}}}{\varepsilon_{Y_2}^{\lambda_{obs}}}\,\mathrm{a}_6^{\lambda_{isos}}-\mathrm{a}_9^{\lambda_{isos}}\right)\frac{1}{\Delta\varepsilon}\,C_{tot}^{\lambda_{isos},T} \tag{8.21}$$

$$a_{13}=\mathrm{a}_1^{\lambda_{isos}}-\frac{a_{15}}{C_{tot}^{\lambda_{isos},T}} \tag{8.22}$$

$$a_{14}=\frac{a_3^{\lambda_{isos}}}{\varepsilon_{Y_2}^{\lambda_{obs}}}-\frac{a_{15}}{\varepsilon_{Y_2}^{\lambda_{obs}}\,C_{tot}^{\lambda_{isos},T}} \tag{8.23}$$

$$a_{15}=\left(\frac{\mathrm{a}_2^{\lambda_{isos}}}{\varepsilon_{Y_1}^{\lambda_{obs}}}-\frac{\mathrm{a}_3^{\lambda_{isos}}}{\varepsilon_{Y_2}^{\lambda_{obs}}}\,\mathrm{a}_6^{\lambda_{isos}}-\mathrm{a}_9^{\lambda_{isos}}\right)\frac{1}{\Delta\varepsilon}\,C_{tot}^{\lambda_{isos},T} \tag{8.24}$$

where

$$\Delta\varepsilon=\frac{1}{\varepsilon_{Y_1}^{\lambda_{obs}}}-\frac{1}{\varepsilon_{Y_2}^{\lambda_{obs}}} \tag{8.25}$$

and

$$C_{tot}^{\lambda_{isos},T}=C_X^{\lambda_{isos},T}(0)+C_{Y_1}^{\lambda_{isos},T}(0)+C_{Y_2}^{\lambda_{isos},T}(0)=C_X^{\lambda_{isos},T}(t)+C_{Y_1}^{\lambda_{isos},T}(t)+C_{Y_2}^{\lambda_{isos},T}(t) \tag{8.26}$$

Analytically solving a system of two differential equations is performed by a standard, well-known mathematical method (free from any approximations or simplifying assumptions). The method requires obtaining the second derivative of one of the two variables (e.g. $A_{Y_1,Y_2}^{\lambda_{isos},T/\lambda_{obs}}$, by differentiating Eq. (8.17)) and replacing the derivatives $(C_X(t))'$ and $\left(A_{Y_1,Y_2}^{\lambda_{isos},T/\lambda_{obs}}\right)'$, appearing in the new equation in their respective expressions given by Eqs. (8.17) and (8.18). Therefore, the obtained second-order differential equation (Eq. (8.27)) depends only on $C_X(t)$ and $A_{Y_1,Y_2}^{\lambda_{isos},T/\lambda_{obs}}$, as

$$\left(A_{Y_1,Y_2}^{\lambda_{isos},T/\lambda_{obs}}\right)'' = \mathrm{a}_{16}\, C_X^{\lambda_{isos},T}(t) + \mathrm{a}_{17}\, A_{Y_1,Y_2}^{\lambda_{isos},T/\lambda_{obs}} + \mathrm{a}_{18} \tag{8.27}$$

where

$$\mathrm{a}_{16} = \mathrm{a}_{10}\,(\mathrm{a}_{13} + \mathrm{a}_{11}) \tag{8.28}$$

$$\mathrm{a}_{17} = \mathrm{a}_{10}\,\mathrm{a}_{14} + \mathrm{a}_{11}^2 \tag{8.29}$$

$$\mathrm{a}_{18} = \mathrm{a}_{10}\,\mathrm{a}_{15} + \mathrm{a}_{11}\,\mathrm{a}_{12} \tag{8.30}$$

It is now useful to eliminate the variable $C_X(t)$ from Eq. (8.27) by expressing it in terms of $A_{Y_1,Y_2}^{\lambda_{isos},T/\lambda_{obs}}$ and $\left(A_{Y_1,Y_2}^{\lambda_{isos},T/\lambda_{obs}}\right)'$. Hence, the second-order differential equation becomes

$$\left(A_{Y_1,Y_2}^{\lambda_{isos},T/\lambda_{obs}}\right)'' = \mathrm{a}_{19}^{\lambda_{isos}} \left(A_{Y_1,Y_2}^{\lambda_{isos},T/\lambda_{obs}}\right)' + \mathrm{a}_{20}^{\lambda_{isos}}\, A_{Y_1,Y_2}^{\lambda_{isos},T/\lambda_{obs}} + \mathrm{a}_{21} \tag{8.31}$$

with the coefficients of Eq. (8.31) that can exclusively be expressed by the coefficients $\mathrm{a}_1^{\lambda_{isos}} - \mathrm{a}_9^{\lambda_{isos}}$ as

$$\mathrm{a}_{19}^{\lambda_{isos}} = \mathrm{a}_1^{\lambda_{isos}} + \mathrm{a}_5^{\lambda_{isos}} + \mathrm{a}_9^{\lambda_{isos}} \tag{8.32}$$

$$\mathrm{a}_{20}^{\lambda_{isos}} = \mathrm{a}_7^{\lambda_{isos}}\left(\mathrm{a}_3^{\lambda_{isos}} - \mathrm{a}_2^{\lambda_{isos}}\right) + \mathrm{a}_8^{\lambda_{isos}}\left(\mathrm{a}_1^{\lambda_{isos}} - \mathrm{a}_3^{\lambda_{isos}}\right) + \mathrm{a}_9^{\lambda_{isos}}\left(\mathrm{a}_2^{\lambda_{isos}} - \mathrm{a}_1^{\lambda_{isos}}\right) \tag{8.33}$$

$$\mathrm{a}_{21}^{\lambda_{isos}} = \left[\varepsilon_{Y_1}^{\lambda_{obs}}\left(\mathrm{a}_3^{\lambda_{isos}}\,\mathrm{a}_4^{\lambda_{isos}} - \mathrm{a}_1^{\lambda_{isos}}\,\mathrm{a}_6^{\lambda_{isos}}\right) + \varepsilon_{Y_2}^{\lambda_{obs}}\left(\mathrm{a}_2^{\lambda_{isos}}\,\mathrm{a}_7^{\lambda_{isos}} - \mathrm{a}_1^{\lambda_{isos}}\,\mathrm{a}_8^{\lambda_{isos}}\right)\right] C_X^{\lambda_{isos},T}(0) \tag{8.34}$$

The coefficients $\mathrm{a}_{19}^{\lambda_{isos}}$ and $\mathrm{a}_{20}^{\lambda_{isos}}$ are specifically defined by the isosbestic parameters and do not depend on the initial concentration. Coefficient $\mathrm{a}_{21}^{\lambda_{isos}}$, instead, is positive or equal to zero and depends on isosbestic parameters, the initial concentration of X and the absorptivities, $\varepsilon_{Y_1}^{\lambda_{obs}}$ and $\varepsilon_{Y_1}^{\lambda_{obs}}$. In the case of a reaction (Scheme 8.1) where Y_2 is not involved, coefficients $\mathrm{a}_{20}^{\lambda_{isos}}$ and $\mathrm{a}_{21}^{\lambda_{isos}}$ will take a zero value.

The final step in the mathematical transformation consists of rearranging the general form of the non-homogeneous second-order differential equation (Eq. (8.31)) into

$$\left(A_{Y_1,Y_2}^{\lambda_{isos},T/\lambda_{obs}}\right)'' - \mathrm{a}_{19}^{\lambda_{isos}} \left(A_{Y_1,Y_2}^{\lambda_{isos},T/\lambda_{obs}}\right)' - \mathrm{a}_{20}^{\lambda_{isos}}\, A_{Y_1,Y_2}^{\lambda_{isos},T/\lambda_{obs}} = \mathrm{a}_{21}^{\lambda_{isos}} \tag{8.35}$$

8.6 Complementary and Particular Solutions

Two steps are required to resolve Eq. (8.35). In the first, the equation is considered without its second term (i.e. it assumes that $a_{21}^{\lambda_{isos}} = 0$), whose solution will correspond to the *complementary function* (labelled as $\overline{A}$). This solution, together with Eq. (8.35), is solved in a second step to yield the *particular solution* (labelled as A^*). The sum of these two partial solutions (Eq. (8.36)) represents the general solution of Eq. (8.35) and, hence, a solution for the original system of differential equations (Eqs. (8.1)–(8.3)).

$$A_{Y_1,Y_2}^{\lambda_{isos},T/\lambda_{obs}} = \overline{A} + A^* \tag{8.36}$$

The complementary function, $\overline{A}$**,** can be obtained by considering that the solution of the homogeneous equation (i.e. Eq. (8.35) with $a_{21}^{\lambda_{isos}} = 0$) takes an exponential form of time, such as $\overline{A} = e^{\rho\, t}$. By introducing $\overline{A}$ in the homogeneous equation, we obtain the auxiliary or characteristic equation (Eq. (8.37)).

$$\rho^2 - a_{19}^{\lambda_{isos}}\, \rho - a_{20}^{\lambda_{isos}} = 0 \tag{8.37}$$

The auxiliary equation, Eq. (8.37), is a quadratic equation, whose discriminant (Δ_{aux}) can be positive, zero or negative.

$$\Delta_{aux} = \left(a_{19}^{\lambda_{isos}}\right)^2 + 4\, a_{20}^{\lambda_{isos}} \tag{8.38}$$

The negative values of Δ_{aux} are discarded because they lead to imaginary roots with no physical meaning in kinetics. The real roots (ρ_+ and ρ_-) of Eq. (8.37), when $\Delta_{aux} \geq 0$, have the classical formulae:

$$\rho_{\pm} = \frac{a_{19}^{\lambda_{isos}} \pm \sqrt{\Delta_{aux}}}{2} \tag{8.39}$$

For positive values of Δ_{aux}, the solution of the complementary function is a linear combination of two independent exponential terms.

$$\overline{A} = \rho_1\, e^{\rho_+\, t} + \rho_2\, e^{\rho_-\, t} \tag{8.40}$$

But when $\Delta_{aux} = 0$, only one (double) solution is derived ($\rho_{\pm} = 0.5\, a_{19}^{\lambda_{isos}}$), and the complementary function has the following form.

$$\overline{A} = \rho_3\, e^{\rho_{\pm}\, t} \tag{8.41}$$

The constants $\rho_1 - \rho_3$ will subsequently be determined by considering the initial conditions.

The experimental kinetic traces observed for (cyclic and non-cyclic) trimolecular or two-species reactions are predicted, at long reaction time, to tend asymptotically towards a constant value due to the exponential functions (the numerical values of the roots are either negative or zero).

The particular solution, $\boldsymbol{A}^*$, of the non-homogeneous equation is derived on the basis of the values of both $a_{21}^{\lambda_{isos}}$ and Δ_{aux}. There are four cases to consider:

Case I: $\Delta_{aux} > 0$, $a_{21}^{\lambda_{isos}} = 0$ and $\boldsymbol{A}^* = 0$

The particular solution here is equal to zero ($\boldsymbol{A}^* = 0$), and the solution is similar to the complementary function derived earlier (e.g. Eq. (8.40)).

$$A_{Y_1,Y_2}^{\lambda_{isos},T/\lambda_{obs}} = \rho_1\, e^{\rho_+\, t} + \rho_2\, e^{\rho_-\, t} \tag{8.42}$$

Case II: $\Delta_{aux} > 0$ and $a_{21}^{\lambda_{isos}} \neq 0$

The particular solution is a constant because $a_{21}^{\lambda_{isos}}$ is a constant, and therefore, it is a solution of the non-homogeneous equation ($\boldsymbol{A}^* = \rho_4$ and hence, $(\boldsymbol{A}^*)' = (\boldsymbol{A}^*)'' = 0$ and $\rho_4 = -a_{21}^{\lambda_{isos}}/a_{20}^{\lambda_{isos}}$). Note that ρ_4 has the same properties as $A_{Y_1,Y_2}^{\lambda_{isos},T/\lambda_{obs}}$ and, hence, must be positive, indicating that the coefficient $a_{20}^{\lambda_{isos}}$ is negative.

The general solution for the non-homogeneous equation for this case takes the form

$$A_{Y_1,Y_2}^{\lambda_{isos},T/\lambda_{obs}} = \rho_1\, e^{\rho_+\, t} + \rho_2\, e^{\rho_-\, t} + \rho_4 \tag{8.43}$$

Case III: $\Delta_{aux} = 0$ and $a_{21}^{\lambda_{isos}} = 0$

Since a unique root ($\rho_\pm$ because $\Delta_{aux} = 0$ in Eq. (8.39)) is found for the complementary function, the solution of this case cannot be derived as a linear combination of exponential functions. Instead, two linearly independent terms based on the double root ($e^{\rho_\pm\, t}$ and $t\, e^{\rho_\pm\, t}$) are used for the general solution.

$$A_{Y_1,Y_2}^{\lambda_{isos},T/\lambda_{obs}} = \rho_3\, e^{\rho_\pm\, t} + \rho_5\, t\, e^{\rho_\pm\, t} = (\rho_3 + \rho_5\, t\,)\, e^{\rho_\pm\, t} \tag{8.44}$$

Case IV: $\Delta_{aux} = 0$ and $a_{21}^{\lambda_{isos}} \neq 0$

For this situation, $\boldsymbol{A}^* = \rho_4$ so that the solution has the general form:

$$A_{Y_1,Y_2}^{\lambda_{isos},T/\lambda_{obs}} = (\rho_3 + \rho_5\, t\,)\, e^{\rho_\pm\, t} + \rho_4 \tag{8.45}$$

The general solutions obtained for *cases I and II* are quite different from those belonging to *cases III* and *IV*. The former are clearly of first-order character, whereas the latter combines first-order kinetics (i.e. $\rho\, e^{\rho_\pm\, t}$ terms) and $\rho\, t\, e^{\rho_\pm\, t}$ functions. The second type is dubbed here as *t-first-order kinetics*.

8.7 Boundary Conditions

In order to finalise the mathematical treatment, it is important to consider the boundary conditions. The first one is relative to $A_{Y_1,Y_2}^{\lambda_{isos},T/\lambda_{obs}}(0)$ that can be non-zero at $t=0$ if either Y_1 and Y_2 are initially present. Assuming $l_{irr}=l_{obs}$ at the initial time, we have

$$A_{Y_1,Y_2}^{\lambda_{isos},T/\lambda_{obs}}(0)=\varepsilon_{Y_1}^{\lambda_{obs}}\, l_{irr}C_{Y_1}^{\lambda_{isos},T}(0)+\varepsilon_{Y_2}^{\lambda_{obs}}\, l_{irr}C_{Y_2}^{\lambda_{isos},T}(0) \tag{8.46}$$

The second initial condition relates to the initial reaction-rate that can be derived from a transformation of Eq. (8.17), as

$$\left(A_{Y_1,Y_2}^{\lambda_{isos},T/\lambda_{obs}}\right)'_{t=0}=r_{A,Y_1,Y_2}^{\lambda_{isos},T/\lambda_{obs}}(0)=\left(\varepsilon_{Y_1}^{\lambda_{obs}}\,\mathrm{a}_4^{\lambda_{isos}}+\varepsilon_{Y_2}^{\lambda_{obs}}\,\mathrm{a}_7^{\lambda_{isos}}\right)C_X^{\lambda_{isos},T}(0)\, l_{irr}+$$
$$\left(\varepsilon_{Y_1}^{\lambda_{obs}}\,\mathrm{a}_5^{\lambda_{isos}}+\varepsilon_{Y_2}^{\lambda_{obs}}\,\mathrm{a}_8^{\lambda_{isos}}\right)C_{Y_1}^{\lambda_{isos},T}(0)\, l_{irr}+\left(\varepsilon_{Y_1}^{\lambda_{obs}}\,\mathrm{a}_6^{\lambda_{isos}}+\varepsilon_{Y_2}^{\lambda_{obs}}\,\mathrm{a}_9^{\lambda_{isos}}\right)C_{Y_2}^{\lambda_{isos},T}(0)\, l_{irr} \tag{8.47}$$

8.8 General *C*(*t*) Solutions of the First-Order Type

Based on the mathematical framework developed above, the analytical expressions for the integrated rate-laws ($C_{Y_j}(t)$) in terms of temporal variations of the three species' concentrations (Scheme 8.1) can now be derived [1]. The forthcoming solutions are equally valid for photothermal $XY_2(6k,6\Phi)$, photochemical $XY_2(6\Phi)$ or purely thermal $XY_2(6k)$ reactions involving three species and up to six photochemical and/or six thermal reaction-steps.

The first-order kinetic type is worked out for both *cases I* and *II,* whose $\Delta_{aux}>0$ and either $\rho_4=0$ or $\rho_4\neq 0$, respectively. The expressions of the various coefficients involved in Eqs. (8.48)–(8.50) have been provided in the previous sections. Overall, the kinetic traces for trimolecular cyclic and non-cyclic photothermal systems exposed to a monochromatic and isosbestic irradiation are characterised by bi-exponential trends.

$$C_X^{\lambda_{isos},T}(t)=\rho_1\frac{\rho_+-a_{11}}{a_{10}}\,e^{\rho_+\,t}+\rho_2\frac{\rho_--a_{11}}{a_{10}}\,e^{\rho_-\,t}+\frac{\rho_4\,a_{11}+a_{12}}{a_{10}} \tag{8.48}$$

$$C_{Y_1}^{\lambda_{isos},T}(t)=\rho_1\frac{\left(a_{11}-\rho_+\right)\varepsilon_{Y_2}^{\lambda_{obs}}-a_{10}}{\left(\varepsilon_{Y_2}^{\lambda_{obs}}-\varepsilon_{Y_1}^{\lambda_{obs}}\right)a_{10}}\,e^{\rho_+\,t}+\rho_2\frac{\left(a_{11}-\rho_-\right)\varepsilon_{Y_2}^{\lambda_{obs}}-a_{10}}{\left(\varepsilon_{Y_2}^{\lambda_{obs}}-\varepsilon_{Y_1}^{\lambda_{obs}}\right)a_{10}}\,e^{\rho_-\,t}$$

$$+\frac{\left(a_{12}+a_{10}\,C_{tot}+a_{11}\,\rho_4\right)\,\varepsilon_{Y_2}^{\lambda_{obs}}-\rho_4\,a_{10}}{\left(\varepsilon_{Y_2}^{\lambda_{obs}}-\varepsilon_{Y_1}^{\lambda_{obs}}\right)\,a_{10}} \tag{8.49}$$

$$C_{Y_2}^{\lambda_{isos},T}(t)=-\rho_1\frac{\left(a_{11}-\rho_+\right)\,\varepsilon_{Y_1}^{\lambda_{obs}}-a_{10}}{\left(\varepsilon_{Y_2}^{\lambda_{obs}}-\varepsilon_{Y_1}^{\lambda_{obs}}\right)\,a_{10}}\,e^{\rho_+\,t}-\rho_2\frac{\left(a_{11}-\rho_-\right)\,\varepsilon_{Y_1}^{\lambda_{obs}}-a_{10}}{\left(\varepsilon_{Y_2}^{\lambda_{obs}}-\varepsilon_{Y_1}^{\lambda_{obs}}\right)\,a_{10}}\,e^{\rho_-\,t}$$
$$-\frac{\left(a_{12}+a_{10}\,C_{tot}+a_{11}\,\rho_4\right)\,\varepsilon_{Y_2}^{\lambda_{obs}}-\rho_4\,a_{10}}{\left(\varepsilon_{Y_2}^{\lambda_{obs}}-\varepsilon_{Y_1}^{\lambda_{obs}}\right)\,a_{10}} \tag{8.50}$$

8.9 General Solutions of the Type: t-First-Order Kinetics

The second type of kinetic behaviour under isosbestic irradiation is observed for *cases III* and *IV* (where $\Delta_{aux}=0$ and $\rho_4=0$ or $\rho_4\neq 0$), including reactions whose integrated rate-laws depend on a mix of two different time functions.

$$C_X^{\lambda_{isos},T}(t)=\left(\rho_5+\left(\rho_\pm-a_{11}\right)\,\left(\rho_3+\rho_5\,t\right)\right)\,\frac{e^{\rho_\pm\,t}}{a_{10}}-\frac{\rho_4\,\,a_{11}+a_{12}}{a_{10}} \tag{8.51}$$

$$C_{Y_1}^{\lambda_{isos},T}(t)=-\left(\rho_5+\left(\rho_\pm-a_{11}+\frac{a_{10}}{\varepsilon_{Y_2}^{\lambda_{obs}}}\right)\,\left(\rho_3+\rho_5\,t\right)\right)\,\frac{e^{\rho_\pm\,t}}{a_{10}\,\varepsilon_{Y_1}^{\lambda_{obs}}\,\Delta\varepsilon}$$
$$-\frac{a_{10}\,C_{tot}+\rho_4\,\left(a_{11}-\frac{a_{10}}{\varepsilon_{Y_2}^{\lambda_{obs}}}\right)+a_{12}}{a_{10}\,\varepsilon_{Y_1}^{\lambda_{obs}}\,\Delta\varepsilon} \tag{8.52}$$

$$C_{Y_2}^{\lambda_{isos},T}(t)=\left(\rho_5+\left(\rho_\pm-a_{11}+\frac{a_{10}}{\varepsilon_{Y_2}^{\lambda_{obs}}}+a_{10}\Delta\varepsilon\right)\,\left(\rho_3+\rho_5\,t\right)\right)\,\frac{e^{\rho_\pm\,t}}{a_{10}\,\varepsilon_{Y_2}^{\lambda_{obs}}\,\Delta\varepsilon}$$
$$+\frac{-a_{10}\,C_{tot}+\rho_4\,\left(a_{10}\,\varepsilon_{Y_2}^{\lambda_{obs}}\,\Delta\varepsilon-a_{11}+\frac{a_{10}}{\varepsilon_{Y_2}^{\lambda_{obs}}}\right)-a_{12}}{a_{10}\,\varepsilon_{Y_2}^{\lambda_{obs}}\,\Delta\varepsilon} \tag{8.53}$$

In practice, the kinetic traces obeying first-order kinetics (described by bi-exponential integrated rate-laws) seem to be more experimentally observed than those behaving along t-first-order kinetics.

8.10 General Solutions of $XY(mk, n\Phi)$ Reactions

The procedure described above can also give analytically derived general solutions for photothermal systems of two molecules, i.e. $XY(2k, 2\Phi)$ and their sub-mechanisms, when exposed to an isosbestic irradiation.

In this case, both in the equations representing the coefficients $a_4^{\lambda_{isos}}$ to $a_9^{\lambda_{isos}}$ (Eqs. (8.4)–(8.12)) and in the subsequent coefficients and equations, the properties involving or corresponding to the second product, Y_2, are all set to zero (e.g. $k_{Y_2\to X}=\Phi_{Y_2\to X}^{\lambda_{isos}}=k_{Y_1\to Y_2}=\Phi_{Y_1\to Y_2}^{\lambda_{isos}}=\varepsilon_{Y_2}^{\lambda_{obs}}=0$).

In fact, the solution given below can readily be generalised to any numbers, "*m*" and "*n*," of reaction-steps occurring between the reactant and its photoproduct, i.e. $XY(mk, n\Phi)$, as long as the $a_i^{\lambda_{isos}}$ coefficients are modified accordingly.

Because the reaction $XY(mk, n\Phi)$ has a unique positive root ($\rho_\pm\neq 0$) and a coefficient $a_{20}^{\lambda_{isos}}$ equal to zero, this particular reactive system (under isosbestic irradiation) accepts only one form for the integrated rate-laws, i.e. the mono-exponential function (*t*-first-order models do not exist for this type of reaction). The exponential argument, $\rho_\pm$ (or $a_{19}^{\lambda_{isos}}$), is equivalent to the negative global rate-constant ($-k_r^{\lambda_{isos},T}$) of the photothermal $XY(2k, 2\Phi)$ reaction, which includes the rate-constants, $k_{X\to Y}^T$ and $k_{Y\to X}^T$, of the individual thermal reaction-steps.

$$k_r^{\lambda_{isos},T}=\Phi_{X\to Y}^{\lambda_{n-isos}}\gamma_1+\Phi_{X\to Y}^{\lambda_{n-isos}}\gamma_1+k_{X\to Y}^T+k_{Y\to X}^T \tag{8.54}$$

The integrated rate-laws assigned to the reactant and the photoproduct are

$$C_X^{\lambda_{isos},T}(t)=C_X^{\lambda_{isos},T}(0)+\left(\frac{a_4^{\lambda_{isos}}}{a_{19}^{\lambda_{isos}}}C_X^{\lambda_{isos},T}(0)+\frac{a_5^{\lambda_{isos}}}{a_{19}^{\lambda_{isos}}}C_Y^{\lambda_{isos},T}(0)\right)\left(1-e^{\rho_\pm t}\right) \tag{8.55}$$

$$C_{Y_1}^{\lambda_{isos},T}(t)=C_{Y_1}^{\lambda_{isos},T}(0)+\left(\frac{a_4^{\lambda_{isos}}}{a_{19}^{\lambda_{isos}}}C_X^{\lambda_{isos},T}(0)+\frac{a_5^{\lambda_{isos}}}{a_{19}^{\lambda_{isos}}}C_Y^{\lambda_{isos},T}(0)\right)\left(e^{\rho_\pm t}-1\right) \tag{8.56}$$

The equation relative to the total absorbance is

$$A_{tot}^{\lambda_{isos},T/\lambda_{obs},T}(t)=A_{tot}^{\lambda_{isos},T/\lambda_{obs},T}(\text{initial})+\frac{r_{0,A}^{\lambda_{isos},T/\lambda_{obs}}}{k_r^{\lambda_{isos},T}}\left(1-e^{-k_r^{\lambda_{isos},T}t}\right) \tag{8.57}$$

with the label "*initial*" indicating the initial time of the considered reaction, e.g. *initial* = 0 or *pss*, and the total initial rate is the sum of the individual species' initial rates (Eq. (8.47)). The observation wavelength is equivalent here to a non-isosbestic wavelength (i.e. $\lambda_{obs}=\lambda_{n-isos}$).

$$r_{0,A}^{\lambda_{isos},T/\lambda_{obs}} = \left(\varepsilon_Y^{\lambda_{n-isos}} - \varepsilon_X^{\lambda_{n-isos}}\right) l_{obs} \times$$
$$\left(\left(\Phi_{X\to Y}^{\lambda_{n-isos}} \gamma_1 + k_{X\to Y}^T\right) C_X^{\lambda_{isos},T}(\text{initial}) - \left(\Phi_{Y\to X}^{\lambda_{n-isos}} \gamma_1 + k_{Y\to X}^T\right) C_Y^{\lambda_{isos},T}(\text{initial})\right) \tag{8.58}$$

When the reaction is performed in the dark ($P_0^{\lambda_{isos}} = 0 = \gamma_1$) after a period of irradiation (e.g. when the reaction reaches *pss*, this time becomes the starting point of the thermal reaction; hence, *initial* = *pss*) and if the reactant is thermally stable ($k_{X\to Y}^T = 0$), the initial rate is expressed for the thermal reaction as

$$r_{0,A}^{\lambda_{obs},T} = -\left(\varepsilon_Y^{\lambda_{n-isos}} - \varepsilon_X^{\lambda_{n-isos}}\right) l_{obs}\, k_{Y\to X}^T\, C_Y^{\lambda_{isos},T}(\text{initial}) \tag{8.59}$$

By setting the missing parameters to zero, the above equations will deliver the integrated rate-laws for the sub-mechanisms of $XY(2k, 2\Phi)$. The equations can be applied to the primary photoreaction (X — $h\nu$ → Y) and other unimolecular reactions, whose photoproduct absorbs the incident light at λ_{isos}, as well as to reversible reactions, such as $XY(2k)$, $XY(2\Phi)$, $XY(1k, 1\Phi)$ and $XY(2\Phi, 1k)$. It is useful to indicate that Eqs. (8.55) and (8.56) yield exactly the same equations found in the literature for $XY(2k, 2\Phi)$ and its sub-mechanisms, though those equations were derived analytically using different solving methods compared to the one presented here.

8.11 A General Template for $XY_q(mk, n\Phi)$ Reactions

The preceding analysis leads us to consider a general template for the integrated rate-law of a universal photothermal reaction $XY_q(mk, n\Phi)$ (which include q products interlinked by m thermal and n photochemical reaction-steps). The concentration of a species Y_j of the reaction mechanism (with $0 \le j \le q$ and $X = Y_0$) is, hence, expressed as a sum of p mono-exponential functions for the reaction obeying a first-order kinetics (Eq. (8.60)) or a dual sum of mono-exponential and the product of time and mono-exponential functions (Eq. (8.61)).

$$C_{Y_j}^{\lambda_{isos},T}(t) = \alpha_j + \sum_{i=1}^{p_j} \beta_{ij}\, e^{-\gamma_{ij}\, t} \tag{8.60}$$

or

$$C_{Y_j}^{\lambda_{isos},T}(t) = \delta_j + \sum_{i=1}^{p_j} \sigma_{ij}\left(\theta_{ij}\, e^{-\mu_{ij}\, t} + t\, e^{-\mu_{ij}\, t}\right) \tag{8.61}$$

The Greek letters in the above equations are real coefficients, and the number of terms in the sum depends on both the species Y_j considered and the mechanism of the reaction.

8.12 Conclusion

The analytical derivation presented in this chapter clearly shows that exposing photothermal reactions to a monochromatic and isosbestic irradiation necessarily leads to a time invariance of the photokinetic factor that makes, from a mathematical viewpoint, photokinetics analogous to thermal kinetics.

As a direct consequence, the order of the photothermal reaction governed by $XY_2(6k, 6\Phi)$ mechanism or any of its sub-mechanisms is one of two possible options: either the reaction obeys a classical first-order kinetics, or its behaviour corresponds to a combination of first-order and t-first-order kinetics.

Considering the large number of mechanistic options included in the studied $XY(6k, 6\Phi)$, this reaction can be viewed as an acceptable template for the general $XY_q(mk, n\Phi)$ photothermal reactions. Therefore, we can draw a general statement about photothermal reactions, that is, when subjected to monochromatic and isosbestic irradiation, these reactions behave according to the kinetics of thermal reactions, obeying either first- or first- and t-first-order kinetics.

References

1. Brown RG, Maafi M (2005) General analytical solutions for the kinetics of AB(k,Φ) and ABC (k,Φ) systems. Int J Chem Kinet 37:162–174. https://doi.org/10.1002/kin.20061
2. Mauser H, Gauglitz G, Compton RG, Hancock G (1998) Comprehensive chemical kinetics, photokinetics: theoretical fundamentals and applications, vol 36. Elsevier, Amsterdam

Chapter 9
Degeneracy of the Kinetic Solution: Distinguishability and Identifiability in Relation to $XY(mk, n\Phi)$ Photothermal Reactions and Naphthopyran (NPY) Thermal Bleaching

9.1 Introduction

We will explore the occurrence of degeneracy in kinetics due to distinctability, distinguishability, and/or identifiability, in relation to the analyses of kinetic equations and analytical solutions that were proposed in the previous chapter. Some sub-mechanisms of the $XY_2(mk, n\Phi)$ reactions are studied hereafter for illustration of the degeneracy of the kinetic solution.

9.2 Distinctability

The intrinsic distinctability of isosbestically irradiated $XY_2(mk, n\Phi)$ reactions arises from the two effective kinetic options offered to such reactions, i.e. following either a first-order or a first-order/t-first-order kinetics. This means that there is no reasonable way that allows us to know which kinetics is likely to apply to any of these reactions by an examination of its reaction mechanism, the exception being the $XY(mk, n\Phi)$ reactions that are not concerned by this type of distinctability issues (the option t-first-order is not possible for these reactions).

The extrinsic distinctability can be envisaged for $XY_2(mk, n\Phi)$ and $XY(mk, n\Phi)$ reactions subjected to one or several concomitant monochromatic irradiations, whose wavelengths correspond to one or several isosbestic points (i.e. mono- or poly-isosbestic irradiation). Irrespective of the irradiation type, the resulting traces will conform to the general templates stated in Eq. 8.60 or 8.61. Therefore, it will not be possible to identify which of a mono- or a poly-isosbestic irradiation was used to induce the reaction by an examination of the kinetic traces of the reaction at hand. Incidentally, if a poly-isosbestic irradiation is possible, it remains relatively hard to implement experimentally.

M. Maafi, *Photokinetics*, https://doi.org/10.1007/978-3-031-98985-8_9

9.3 The Case of the *XY*(2*k*, 2*Φ*) Reactions

The true mechanistic option governing a specific *XY* reaction investigated under isosbestic irradiation cannot be known by only considering the kinetic traces of the reaction. Both reversible and non-reversible *XY*(2*k*, 2*Φ*) reaction options have a first-order character, generating mono-exponential curves of the form given by Eqs. 8.55, 8.56 and 8.57. The uniqueness of the latter equations (and their applicability to the *XY*(2*k*, 2*Φ*) sequence and the set of its sub-mechanisms) is, by itself, a proof of the occurrence of a distinguishability problem. If the traces available correspond to the temporal variation of the total absorbance, then the kinetic traces of the set of reversible *XY*(2*k*, 2*Φ*) reactions, characterised by an equilibrium state, will be indistinguishable from the one-step reactions (i.e. *XY*(1*Φ*) and *XY*(1*k*)) if the products of the latter reactions absorb at λ_{obs}.

Most often, the total absorbance traces obtained experimentally for photothermal reactions, including the reversible *XY*(2*k*, 2*Φ*), are obtained in the form of experimental $A_{tot}^{\lambda_{isos},T/\lambda_{obs}}(t)$ that is measured through l_{obs} (with $A_{tot}^{\lambda_{isos},T/\lambda_{isos}}(t)$ being a constant) due to the facility of measurement offered by spectrophotometry. The treatment of such traces is done by the corresponding equation (Eq. 9.1), which is worked out from adding the absorbancies of the individual species (*X* and *Y*) using Eqs. 8.55 and 8.56 (as given in Eq. 8.57).

$$A_{tot}^{\lambda_{isos},T/\lambda_{obs}}(t)=A_{tot}^{\lambda_{isos},T/\lambda_{obs}}(0)+\left(\frac{k_{X\to Y}^{T}+\Phi_{X\to Y}^{\lambda_{isos}}\gamma_1}{k_r^{\lambda_{isos},T}}C_X(0)-\frac{k_{Y\to X}^{T}+\Phi_{Y\to X}^{\lambda_{isos}}\gamma_1}{k_r^{\lambda_{isos},T}}C_Y(0)\right)$$
$$\times l_{obs}\left(\varepsilon_Y^{\lambda_{obs}}-\varepsilon_X^{\lambda_{obs}}\right)\left(1-e^{-k_r^{\lambda_{isos},T}t}\right) \tag{9.1}$$

This equation can be reduced to a general form, Eq. 9.2, where $\omega^{\lambda_{isos},T/\lambda_{obs}}$ can take positive or negative values, but it is equal to zero if there is no reaction, i.e. when $A_{tot}^{\lambda_{isos},T/\lambda_{obs}}(t)=A_{tot}^{\lambda_{isos},T/\lambda_{obs}}(\infty)=A_{tot}^{\lambda_{isos},T/\lambda_{obs}}(0)$.

$$A_{tot}^{\lambda_{isos},T/\lambda_{obs}}(t)=A_{tot}^{\lambda_{isos},T/\lambda_{obs}}(0)+\omega^{\lambda_{isos},T/\lambda_{obs}}\left(1-e^{-k_r^{\lambda_{isos},T}t}\right) \tag{9.2}$$

Here, the total absorbance is measured at a selected observation wavelength, $\lambda_{obs}\neq\lambda_{isos}$, and across the observation optical path length, l_{obs}, with its value, $A_{tot}^{\lambda_{isos},T/\lambda_{obs}}(\infty)$, being recorded when the reaction reaches equilibrium (or the photostationary state or completion at $t=\infty$).

$$A_{tot}^{\lambda_{isos},T/\lambda_{obs}}(t)=\varepsilon_X^{\lambda_{obs}}\,l_{obs}C_X^{\lambda_{isos},T}(t)+\varepsilon_Y^{\lambda_{obs}}\,l_{obs}C_Y^{\lambda_{isos},T}(t) \tag{9.3}$$

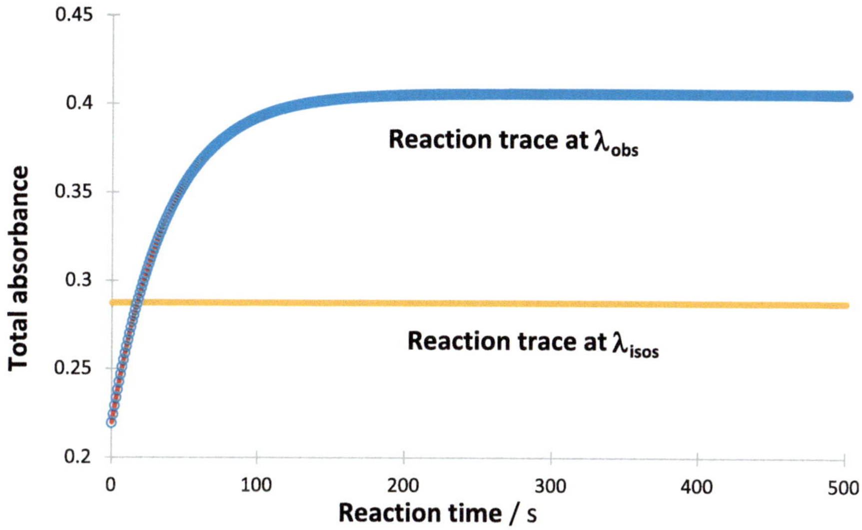

Fig. 9.1 An example of $XY(2\Phi, 1k)$ reaction trace (blue circles) and its fitting with Eq. 9.2 (red line), whose parameters are $A_{tot}^{\lambda_{isos},T/\lambda_{obs}}(0)$, $k_r^{\lambda_{isos},T}$ and $\omega^{\lambda_{isos}/\lambda_{obs}}$. The values of the fitting parameters are $A_{tot}^{\lambda_{isos},T/\lambda_{obs}}(0) = 0.2198$, $\omega^{\lambda_{isos}/\lambda_{obs}} = 0.1862$ and $k_r^{\lambda_{isos},T} = 0.0263\ s^{-1}$. The trace measured at the isosbestic point at λ_{isos} ($A_{tot}^{\lambda_{isos},T/\lambda_{isos}}$, yellow line) is constant with time

Practically, the experimental trace of the reaction is fitted to Eq. 9.2 in order to determine both the values of the reaction rate-constant, $k_r^{\lambda_{isos},T}$, and the coefficient, $\omega^{\lambda_{isos},T/\lambda_{obs}}$ (that depends on both λ_{isos} and λ_{obs}, Eqs. 9.1 and 9.2).

Let us use the example of the reversible photothermal reaction $XY(2\Phi, 1k)$ to illustrate the identifiability issues. Let the trace shown in Fig. 9.1 be numerically calculated (using *RK*4) for the reaction when exposed to an isosbestic/monochromatic irradiation. The trace corresponds to the temporal variation of the total absorbance recorded at an observation (λ_{obs}) wavelength. The absorptivities of X at the isosbestic point ($\varepsilon_X^{\lambda_{isos}} = 13625\ M^{-1}\ cm^{-1}$) and at the observation wavelength ($\varepsilon_X^{\lambda_{obs}} = 10415\ M^{-1}\ cm^{-1}$), as well as its initial concentration ($C_X(0) = 2.11\ 10^{-5}\ M$) are supposed to be known. $C_X(0)$ is taken to fall within the linearity range of the calibration graph of the reactant X. The optical lengths of irradiation and observation have different values, $l_{irr} = 2.38\ cm$ and $l_{obs} = 1\ cm$.

The values of $k_r^{\lambda_{isos},T}$ and $\omega^{\lambda_{isos}/\lambda_{obs}}$ can be related to different values of the four fundamental reaction parameters, which are unknown at this stage, i.e. the forward and reverse quantum yields at the irradiation wavelength, $\Phi_{X\to Y}^{\lambda_{isos}}$ and $\Phi_{Y\to X}^{\lambda_{isos}}$, the thermal rate-constant ($k_{Y\to X}$), and the absorption coefficient of the photoproduct Y at the observation wavelength, $\varepsilon_Y^{\lambda_{obs}}$. The rate-constant of the reaction has a known formula ($k_r^{\lambda_{isos},T} = -\rho_{\pm} = -a_{19}^{\lambda_{isos}}$, Eq. 8.39, expressed in s^{-1}). Its numerical value is obtained from the fitting the trace with Eq. 9.2.

$$k_r^{\lambda_{isos},T} = k_{Y\to X}^T + \left(\Phi_{X\to Y_1}^{\lambda_{isos}} + \Phi_{Y\to X}^{\lambda_{isos}}\right)\gamma_1 = 0.0263\ s^{-1} \tag{9.4}$$

with the rate-constant of the thermal reaction-step experimentally obtained from the fitting of the relaxation trace obtained in the dark after the reaction has reached its stationary state ($k_{Y\to X}^T = 0.013\ s^{-1}$). The coefficient γ_1 (Eq. 8.13) being

$$\gamma_1 = \varepsilon_X^{\lambda_{isos}}\, P_0^{\lambda_{isos}}\, l_{irr}\, \frac{1 - 10^{-\left(A_{tot}^{\lambda_{isos},T/\lambda_{isos}} \frac{l_{irr}}{l_{obs}}\right)}}{A_{tot}^{\lambda_{isos},T/\lambda_{isos}} \frac{l_{irr}}{l_{obs}}} = 1.1915\ s^{-1} \tag{9.5}$$

The pre-exponential factor in Eq. 9.2 has the following expression (with $C_Y(0) = 0$). Its value is obtained from the fitting of the trace to Eq. 9.2.

$$\omega^{\lambda_{isos}/\lambda_{obs}} = -\frac{\Phi_{X\to Y}^{\lambda_{isos}}\gamma_1}{k_r^{\lambda_{isos},T}}\, C_X(0)\, l_{obs}\left(\varepsilon_X^{\lambda_{obs}} - \varepsilon_Y^{\lambda_{obs}}\right) = 0.1862 \tag{9.6}$$

When treating this kinetics in the conditions set out above it becomes quickly evident that there is a set of solutions (S_r, $r > 1$) that all fit the purpose of reproducing the trace (Fig. 9.1) with high precision. The identifiability is easily evidenced when considering the rearranged Eqs. 9.4 and 9.6 into Eqs. 9.7 and 9.8 whereby each given value to the forward quantum yield ($\Phi_{X\to Y}^{\lambda_{isos}}$ in Eq. 9.7) results in generating the values of both the reverse quantum yield ($\Phi_{Y\to X}^{\lambda_{isos}}$ from Eq. 9.7) and the absorption coefficient of the product at λ_{obs} ($\varepsilon_Y^{\lambda_{obs}}$ from Eq. 9.8).

$$\Phi_{Y\to X}^{\lambda_{isos}} = \frac{k_r^{\lambda_{isos},T} - k_{Y\to X}^T}{\gamma_1} - \Phi_{X\to Y}^{\lambda_{isos}} \tag{9.7}$$

$$\varepsilon_Y^{\lambda_{obs}} = \varepsilon_X^{\lambda_{obs}} + \frac{\omega^{\lambda_{isos}/\lambda_{obs}}\, k_r^{\lambda_{isos},T}}{\Phi_{X\to Y}^{\lambda_{isos}}\gamma_1\, C_X(0)\, l_{obs}} \tag{9.8}$$

Table 9.1 shows a few examples of the parameter sets for the kinetic solutions (this set is by no means exhaustive). Good correlation coefficient (r^2), sums of squared errors (SSE), and root-mean-square errors (RMSE) were obtained for the traces ($S_1 - S_7$) when fitted to Eq. 9.2 that use the parameters of Table 9.1. The values inputted for $\Phi_{X\to Y}^{\lambda_{isos}}$ are always positive and less than unit, but negative values can be obtained for the two other reaction parameters ($\Phi_{Y\to X}^{\lambda_{isos}}$ and $\varepsilon_Y^{\lambda_{obs}}$), in which case such solutions must be discarded. The values reported in the last two columns of Table 9.1 have been recalculated by Eqs. 9.4 and 9.6 and the values of their reaction parameters, as given therein. It is clear that the errors recorded on these parameters ($\omega^{\lambda_{isos}/\lambda_{obs}}$ and $k_r^{\lambda_{isos},T}$) are very small when compared with the fitting values (as shown in Eqs. 9.4 and 9.6). The true solution that has fed the RK4 calculation is given by S_3 (Table 9.1), which is shown here to signify that it could be obtained, but it is not possible at this stage to identify it as the true solution among the selected

Table 9.1 A sample of kinetic solutions for the *XY*(2*Φ*, 1*k*) reaction represented by the traces of Fig. 9.1

	Parameters[a]				
Kinetic solution	$\Phi_{X\to Y}^{\lambda_{isos}}$	$\Phi_{Y\to X}^{\lambda_{isos}}$	$\varepsilon_Y^{\lambda_{obs}}$	$k_r^{\lambda_{isos},T}$	$\omega^{\lambda_{isos}/\lambda_{obs}}$
S_1	0.01021	9.532 10^{-3}	29492.77	0.0263	0.1862
S_2	0.00953	1.632 10^{-3}	30854.03	0.0263	0.1862
S_3	0.00841	2.752 10^{-3}	33576.00	0.0263	0.1862
S_4	0.00367	4.792 10^{-3}	40993.33	0.0263	0.1862
S_5	0.00421	6.952 10^{-3}	56681.98	0.0263	0.1862
S_6	0.00186	9.302 10^{-3}	115137.6	0.0263	0.1862
S_7	0.00084	1.032 10^{-2}	242300.7	0.0263	0.1862

[a]: the values of the remaining relevant parameters are: $l_{irr} = 2.38\ cm$, $l_{obs} = 1\ cm$, $C_X(0) = 2.11\ 10^{-5}\ M$, $\varepsilon_X^{\lambda_{isos}} = 13625\ M^{-1}\ cm^{-1}$, $\varepsilon_X^{\lambda_{obs}} = 10415\ M^{-1}\ cm^{-1}$, and $k_{Y\to X}^{T} = 0.013\ s^{-1}$

set $S_1 - S_7$. The last two solutions provide absorption coefficient higher than $10^5\ M^{-1}\ cm^{-1}$, which might be considered unrealistic.

The differences between the parameter sets of Table 9.1 in the various kinetic solutions exclude any interpretations of these differences based on experimental errors or a variability within the data making up the kinetic trace. These differences are objective and lead, for each set, to a consistent fitting of the trace (Fig. 9.1). Also, such a degeneracy does not conceivably allow us to single out the true set (the values of the unknowns are all perfectly plausible). The kinetics is truly unidentifiable, and the true set of the parameters' values is simply impossible to know within the framework and conditions set out above. Any deliberate choice of a particular set of parameters among the possible solutions is merely an unfounded guess that does not provide any consistent information about the kinetics of the reaction but certainly would lead to erroneous conclusions.

Following up from this simple reaction, which is characterised by only four unknown parameters, it is evident that identifiability issues are most likely to emerge for kinetics involving a larger set of unknown parameters. We will see another example in the next section.

It is also interesting to mention that, as previously invoked for the distinguishability issue, the degeneracy of the solution at the origin of identifiability represents a limit for the applicability of Occam's razor (Sect. 6.6). The latter principle would rather apply to the ways by which the true solution can be found (i.e. to the ways in which the identifiability is resolved). The simpler means to get to the true solution (i.e. the elucidation method) should be considered the best (so that, according to Occam's razor, there is no need to find other elucidation methods). Examples of elucidation methods will be reviewed in the next chapter for photokinetics under isosbestic irradiation.

9.4 The Bleaching of Coloured Naphthopyrans

Naphthopyrans are a class of photothermochromes whose colourless, ring-closed parent species absorbs in the UV (reactant $X = NPY$), whereas the coloured, ring-opened isomers (the Y_j products) absorb around the blue end of the visible spectrum [1–4]. The mechanism proposed for this family of photochromes is supposed to involve up to four photoproducts in a $XY_4(n\Phi, mk)$ reaction (Scheme 9.1).

However, in practice, spectrophotometric measurements can only detect the long-living and most stable ($Y_1 = CT$ and $Y_2 = TT$) photoproducts, hence reducing the reaction to a trimolecular dynamics $XY_2(n\Phi, mk)$. The long-range conjugation, spreading over the fused benzene rings, the oxygen sp^2 atom, and the chain of double bonds, confers to the open-ring isomer structures greater planarity and rigidity, which, in turn, contributes to lowering the energies of their π, π^* and n, π^* transitions and causing their emerging absorption in the visible spectrum [5]. The resulting *CT* and *TT* broad absorption bands (Fig. 9.2) are practically indistinguishable, appearing as a single band (showing, during both forward and reverse reactions, no emerging new attributes, shoulders, and/or changes in the recorded

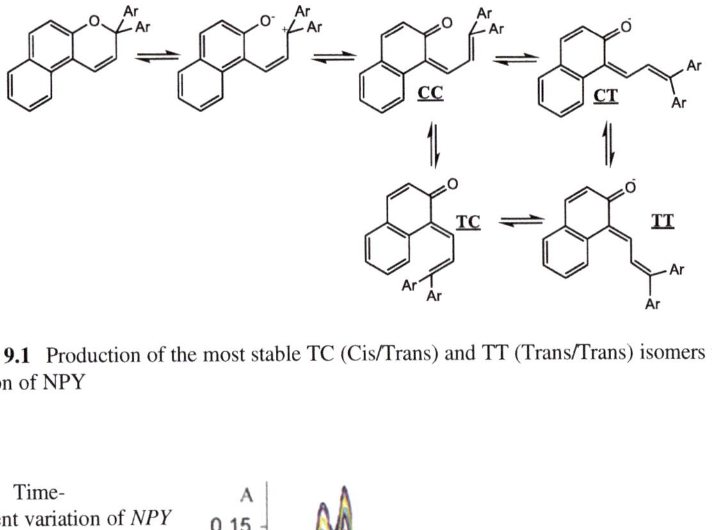

Scheme 9.1 Production of the most stable TC (Cis/Trans) and TT (Trans/Trans) isomers from the excitation of NPY

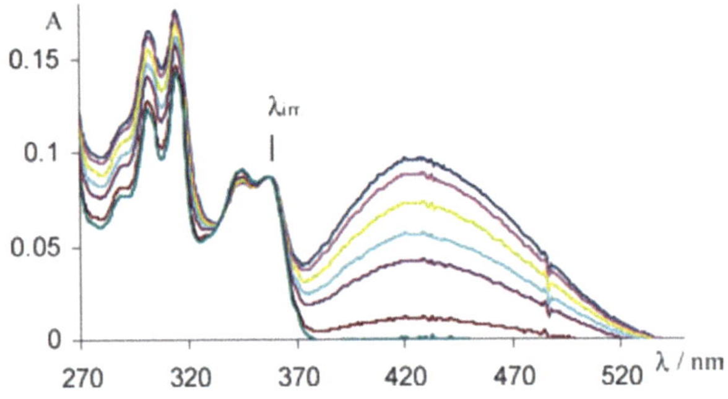

Fig. 9.2 Time-dependent variation of *NPY* ($2.4 \times 10^{-5} M$) absorption spectra in acetonitrile under a 358-*nm* isosbestic irradiation and 24 ° *C*. (Reproduced with permission [4], *International Journal of Chemical Kinetics*)

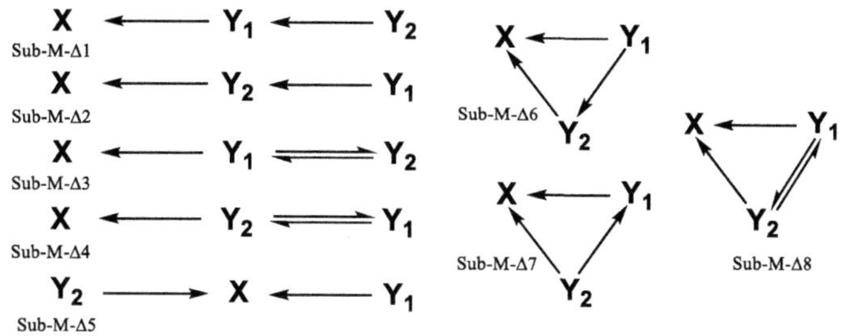

Scheme 9.2 A set of eight plausible mechanistic routes for the thermal bleaching of $XY_2(3k)$ *NPY*

maximum position). This feature might indicate that either the relative amounts of isomers remain constant over reaction time, or the differences in their *E*/*Z* conformations have, in fact, little impact on the respective spectral shapes of their electronic absorption bands. In any case, the new band is red-shifted by ~65 *mn* (with a maximum around 425 *mn*) compared to the parent compound (*NPY*).

Both the colouring and fading reactions, respectively, conducted under irradiation and in the dark, produce a set of smooth spectra, characterised by a series of four isosbestic points recorded at 230, 258, 338, and 357 *nm*. Both dynamics, hence, proceed without side reactions.

Based on qualitative evidence, the overall reaction dynamics of naphthopyrans is known to involve a thermally stable but photochemically active reactant and both thermally and photochemically reactive ring-opened isomers. The precise combination of photochemical and thermal reaction-steps responsible for the interconversion of *NPY* and its *CT* and *TT* isomers is still unknown today. The generic $XY_2(n\Phi, mk)$ option remains most likely [6–11], even though much simplified sequences might have been proposed, e.g. $XY(1\Phi, 1k)$ [12].

The $XY_2(n\Phi, mk)$ option underlines an obvious and complicated distinguishability problem due to the high number of possible mechanistic variations that are expressed by Eq. 8.61 (for first-order kinetics under isosbestic irradiation). Thus far, the exact mechanism governing the photothermal reaction of naphthopyrans has, unfortunately, not been solved and is still not fully identified. The values of the coefficients m and n and the positions of the respective reaction-steps *vis-à-vis* the three reactive species are not yet established and remain subjects of speculation. Nonetheless, we know that $n > 1$ as *NPY* is photoactive (but n can take any value up to 6), and $2 \leq m \leq 4$ since the coloured species regenerate, spontaneously and thermally, the thermally stable reactant, *NPY*.

Since no particular conditions can be imposed on the thermal bleaching reaction mechanism, any possible mechanism should be considered. According to the observation that $2 \leq m \leq 4$, eight plausible reaction routes can theoretically be postulated (Scheme 9.2), including sequences assigning consecutive, opposing, divergent, and cyclic reactions ($Sub - M - \Delta 1$ to $Sub \quad M - \Delta 8$).

Owing to the particular spectroscopic features of *NPY* photoisomers (i.e. the similarity of the absorption bands of *CT* and *TT* (Y_1 and Y_2) isomers), some of the sub-mechanisms laid out in Scheme 9.2 become redundant. Therefore, $Sub - M - \Delta 2$, 4, and 6 are indistinguishable from their respective homologues $Sub - M - \Delta 1$, 3, and 7, so they can be ignored. Hence, the effective set of genuinely different mechanistic options is reduced to five, i.e. $Sub - M - \Delta 1$, 3, 5, 7, and 8.

The literature does neither provide tangible experimental proof on whether or not an isothermal isomerisation proceeds between the geometric *cis/trans* conformations (processes $k_{Y_2 \to Y_1}$ and $k_{Y_1 \to Y_2}$) occurring between *CT* and *TT*, nor on the improbable direct conversion of *TT* into *NPY*. The latter, presumably, necessitates isomerisation of the first exocyclic double bond that should then probably proceed through a short-living intermediate species (that should not be observed at the observation wavelength (440 *nm*), and/or $k_{Y_2 \to X}$ must be the limiting step). Accordingly, for the five mechanistic schemes remaining, $k_{Y_1 \to X}$ values may be expected to be higher than $k_{Y_2 \to X}$. However, if one speculates that the pathway producing *NPY* directly from *TT* is not reasonable due to a probable existence of a high energy barrier between the isomers, then only two mechanisms, $Sub - M - \Delta 1$ and $Sub - M - \Delta 3$, remain for consideration. Unfortunately, despite a substantial reduction in the number of options, the distinguishability problem cannot be fully solved for the thermal reaction. A much less reducible set of photochemical mechanistic options can similarly and easily be constructed. A complexity that explains the up-to-date failure to unravel the true operating mechanism in *NPY* reaction. In this respect, any preconceived mechanism that seems to conform to the experimental data is only speculative and can easily be replaced by another one that provides similar conformity with experiment.

Along with the distinguishability issue, there is also an identifiability problem of the same proportions and equally with no currently available methods to solve it. To give an idea of the extent of the identifiability problem, let us look at the thermochromic reaction of *NPY* (Scheme 9.2). It is understood that the equations will now correspond to the trace of the thermochromic reaction only. This supposes that *NPY* is irradiated by the isosbestic/monochromatic light until, for instance, the *pss* is reached (with its composition $C_X^{\lambda_{isos},T}(pss)$, $C_{Y_1}^{\lambda_{isos},T}(pss)$, and $C_{Y_2}^{\lambda_{isos},T}(pss)$), and then, from there, the light is switched off, and the trace, $A_{tot}^{T,\lambda_{obs}}(t)$, of the thermal reaction is collected by observing (monitoring) the reactive medium at a wavelength, e.g. in the visible range, λ_{obs}, where only the photoproducts Y_1 and Y_2 absorb. During the thermal reaction, the species concentration evolves from their initial *pss* values to $C_X^T(\infty) = C_X^T(0)$ and $C_{Y_1}^T(\infty) = C_{Y_2}^T(\infty) = 0$.

The previous mathematical formalism (Eq. 8.61, Sect. 8.8) can be presented for the sub-reactions of Scheme 9.2 by their first-order kinetics character. The thermal relaxation traces, $A_{tot}^{T,\lambda_{obs}}(t)$, are given by the following bi-exponential model equations.

$$A_{tot}^{T,\lambda_{obs}}(t) = -\frac{r_{0,A}^{T,\lambda_{obs}} - \rho_- A_{tot}^{T,\lambda_{obs}}(0)}{\rho_- - \rho_+} e^{\rho_+ t} + \frac{r_{0,A}^{T,\lambda_{obs}} - \rho_+ A_{tot}^{T,\lambda_{obs}}(0)}{\rho_- - \rho_+} e^{\rho_- t} \tag{9.5}$$

with $r_{A,0}^{\lambda_{obs},T}$ being the initial velocity obtained from the total absorbance trace of the thermal relaxation. $r_{A,0}^{T,\lambda_{obs}}$ in Eq. 9.6 is expressed as a general function of the three species concentrations recorded at *pss*.

$$\begin{aligned} r_{0,A}^{T,\lambda_{obs}} &= \left(\varepsilon_{Y_1}^{\lambda_{obs}}\, a_4^{\lambda_{isos}} + \varepsilon_{Y_2}^{\lambda_{obs}}\, a_7^{\lambda_{isos}}\right) C_X^{T,\lambda_{isos}}(pss) \\ &+ \left(\varepsilon_{Y_1}^{\lambda_{obs}}\, a_5^{\lambda_{isos}} + \varepsilon_{Y_2}^{\lambda_{obs}}\, a_8^{\lambda_{isos}}\right) C_{Y_1}^{T,\lambda_{isos}}(pss) \\ &+ \left(\varepsilon_{Y_1}^{\lambda_{obs}}\, a_6^{\lambda_{isos}} + \varepsilon_{Y_2}^{\lambda_{obs}}\, a_9^{\lambda_{isos}}\right) C_{Y_2}^{T,\lambda_{isos}}(pss) \end{aligned} \tag{9.6}$$

For the specific reaction of *NPY* considered here, the term of $C_X^{\lambda_{isos},T}(pss)$ is not considered in Eq. 9.6 because its coefficient, $a_4^{\lambda_{isos}} = a_7^{\lambda_{isos}} = 0$ and the actual values of the other *pss* concentrations are not known. Also, by replacing the rest of the $a^{\lambda_{isos}}$ factors in the above equation by their respective expressions (Eqs. 8.8, 8.9, 8.11 and 8.12), $r_{0,A}^{T,\lambda_{obs}}$ can be rewritten as

$$\begin{aligned} r_{0,A}^{T,\lambda_{obs}} &= \left(-\left(k_{Y_1\to X}^{T} + k_{Y_1\to Y_2}^{T}\right)\varepsilon_{Y_1}^{\lambda_{obs}} + k_{Y_2\to Y_1}^{T}\,\varepsilon_{Y_2}^{\lambda_{obs}}\right) C_{Y_1}^{T,\lambda_{isos}}(pss)(pss)\, l_{obs} \\ &+ \left(k_{Y_1\to Y_2}^{T}\,\varepsilon_{Y_1}^{\lambda_{obs}} - \left(k_{Y_2\to Y_1}^{T} + k_{Y_2\to X}^{T}\right)\varepsilon_{Y_2}^{\lambda_{obs}}\right) C_{Y_2}^{T,\lambda_{isos}}(pss)\, l_{obs} \end{aligned} \tag{9.7}$$

The absorbance $A_{tot}^{T,\lambda_{obs}}(0)$ (recorded at λ_{obs}) has the form

$$A_{tot}^{T,\lambda_{obs}}(0) = \varepsilon_{Y_1}^{\lambda_{obs}}\, C_{Y_1}^{T,\lambda_{isos}}(pss)\, l_{obs} + \varepsilon_{Y_2}^{\lambda_{obs}}\, C_{Y_2}^{T,\lambda_{isos}}(pss)\, l_{obs} \tag{9.8}$$

The formulae for the roots can be worked out from Eq. 9.9 by putting to zero the missing thermal rate-constants in accordance with each reaction sub-mechanism step (Scheme 9.2). The resulting roots, ρ_+ and ρ_-, are negative irrespective of the mechanism.

$$\rho_\pm = \frac{1}{2}\left(-\left(k_{Y_1\to X}^{T} + k_{Y_1\to Y_2}^{T} + k_{Y_2\to Y_1}^{T} + k_{Y_2\to X}^{T}\right) \pm\sqrt{\left(k_{Y_1\to X}^{T} + k_{Y_1\to Y_2}^{T} + k_{Y_2\to Y_1}^{T} + k_{Y_2\to X}^{T}\right)^2 - 4\left(k_{Y_1\to X}^{T} k_{Y_2\to Y_1}^{T} + k_{Y_1\to X}^{T} k_{Y_2\to X}^{T} + k_{Y_1\to Y_2}^{T} k_{Y_2\to X}^{T}\right)}\right) \tag{9.9}$$

The experimental trace ($A_{tot}^{440}(t)$) of the thermal relaxation of *NPY* photoproducts has been recorded in acetonitrile at $T = 24\ ^\circ$C and $\lambda_{obs} = 440\ nm$ (Fig. 9.3). The pattern of the curve clearly indicates the occurrence of at least two kinetic regimes (the one occurring first is much faster than the subsequent one). At the end of the thermal reaction and the complete depletion of the coloured photoproducts ($t \gg 500\ s$), the obtained absorption spectrum corresponds to that of the reactant

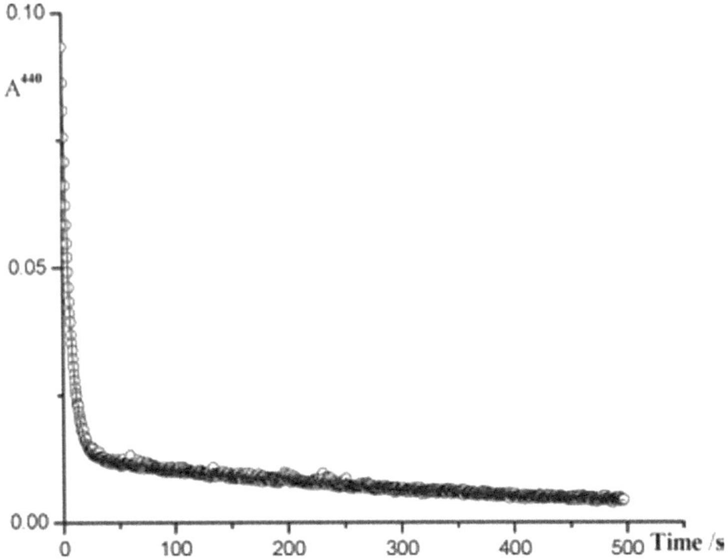

Fig. 9.3 Total absorbance against time experimental trace (circles) of the thermo-fading of NPY photoproducts and its fitting to Eq. 9.5 (red line). (Reproduced with permission (Fig. 2 p-721 from [4], *International Journal of Chemical Kinetics*)

(indicating that no side reactions took place during the photocolouring and thermo-fading processes).

In general, the fitting of such experimental traces is performed by a general bi-exponential equation (Eq. 9.10) for the extraction of the four equation parameters, where β_1 and β_2 represent the rate-constants of the two thermal reaction-steps.

$$A_{tot}^{\lambda_{obs},T}(t) = \alpha_1 \, e^{\beta_1 \, t} + \alpha_2 \, e^{\beta_2 \, t} \tag{9.10}$$

The fitting can aslo be achieved by Eq. 9.5, whose application provides the values of the rate-constants of the thermal reaction-steps (ρ_+ and ρ_-) but also those for the initial rate ($r_{A,0}^{T,\lambda_{obs}}$) and the initial absorbance ($A_{tot}^{T,\lambda_{obs}}(0)$). In addition, the shared factors between the expressions of pre-exponential and exponential terms in Eq. 9.5 add an extra constraint on the fitting (than when using Eq. 9.10) that increases the reliability of the results.

The fitting parameters of Eq. 9.5 for the trace shown in Fig. 9.3 are obtained with high precision as $A_{tot}^{24,440}(0) = 0.09244\left(\pm 6 \times 10^{-5}\right)$, $r_{0,A}^{24,440} = -0.01253\left(\pm 1.7 \times 10^{-4}\right)$, $\rho_+ = -0.15776$ ($\pm 5.2 \times 10^{-4}$), and $\rho_- = -0.00229$ ($\pm 1 \times 10^{-5}$) [4]. These parameters define four equations. The latter equations are themselves expressed by a total of eight fundamental (and a minimum of six) parameters of the reaction at hand (Eqs. 9.7 and 9.8), namely, the absorption coefficients, $\varepsilon_{Y_1}^{\lambda_{obs}}$ and $\varepsilon_{Y_2}^{\lambda_{obs}}$; the *pss* concentrations, $C_{Y_1}^{\lambda_{isos},T}(pss)$ and $C_{Y_2}^{\lambda_{isos},T}(pss)$; and up to four rate-constants, $k_{Y_1 \to X}^{T}$, $k_{Y_1 \to Y_2}^{T}$, $k_{Y_2 \to Y_1}^{T}$, and $k_{Y_2 \to X}^{T}$.

Table 9.2 Sets of kinetic solutions and their parameter values[a] for the thermal *XY*(2*k*) or *XY*(4*k*) reactions applied to the kinetic trace of *NPY* shown in Fig. 9.2

	Initial concentrations[b]		Absorptivities		Rate-constants			
	$C_{Y_1}(pss)$	$C_{Y_2}(pss)$	$\varepsilon_{Y_1}^{\lambda_{obs}}$	$\varepsilon_{Y_2}^{\lambda_{obs}}$	$k_{Y_1 \to X}$	$k_{Y_1 \to Y_2}$	$k_{Y_2 \to Y_1}$	$k_{Y_2 \to X}$
Mechanism ***XY*₂(2*k*)**, *sub-M-Δ1*								
1*a*	1.11×10^{-5}	2.03×10^{-6}	71687	6399	0.15776	–	0.00229	–
1*b*	4.44×10^{-6}	8.13×10^{-7}	17890	15973	0.15776	–	0.00229	–
1*c*	1.76×10^{-6}	3.23×10^{-7}	45000	40178	0.15776	–	0.00229	–
Mechanism ***XY*₂(4*k*)**, *sub-M-Δ8*								
8*a*	8.50×10^{-6}	1.20×10^{-8}	10842	13222	0.13970	0.01788	0.00146	0.00100
8*b*	1.15×10^{-5}	4.37×10^{-9}	7980	6000	0.12830	0.02936	0.00389	0.00199
8*c*	1.09×10^{-7}	7.21×10^{-6}	15721	12576	0.00119	0.00121	0.01563	0.14199

[a]: concentrations, absorptivities and rate-constants are, respectively, expressed in M, $M^{-1}\,cm^{-1}$, and s^{-1}

[b]: the concentration at the start of the thermal reaction corresponds to the values reached at *pss*

Qualitatively, owing to the non-linearity and the number of the (four) equations provided by the fitting of the trace, it is obvious that the occurrence of an identifiability problem is highly probable for the five plausible mechanisms ($Sub - M - \Delta 1$, 3, 5, 7, and 8). This stems from the impossibility of deriving an algebraic solution for all the parameters. Consequently, a relatively large number of kinetic solutions can be found, each offering a high accuracy of fit of the experimental trace. Therefore, irrespective of the kinetic option considered for *NPY* thermal relaxation, the kinetics is unidentifiable.

For illustration, let us consider two mechanistic options ($Sub - M - \Delta 1$ and $Sub - M - \Delta 8$, involving two and four thermal reaction-steps, respectively) to interpret the trace of Fig. 9.3. A few examples of the kinetic solutions are provided in Table 9.2 [4]. The data of the kinetic solutions are all different, confirming the occurrence of the anticipated identifiability and distinguishability problems. Similar degeneracy results are found with all five mechanisms.

9.5 Conclusion

The degeneracy of the kinetic solution and the presence of several options are complications posed on the journey of data interpretation and a proper characterisation of the kinetics. The distinctability does not allow us to guess the behaviour of the reaction from its mechanism, but it reviews the possible behaviours that could be adopted by the reaction kinetics. The identifiability and distinguishability problems will exist despite a good fit of the experimental or simulated data by the model-integrated rate-law equation(s) describing the reaction. Such degeneracy is observed for reactions involving two and three reactive species. The examples studied in this chapter make it reasonable to expect that these phenomena will also be present for the kinetics of any photo- or photothermal reaction, whether under isosbestic or non-isosbestic irradiation. Therefore, the degeneracy problems must, de facto, be part of the kinetic investigation and analysis. They must be identified and addressed. The proof of the occurrence of the phenomena was evidenced in the present chapter, and the kinetic elucidation is reviewed in the next chapter for reactions under isosbestic irradiation.

References

1. Harie G, Samat A, Guglielmetti R (1997) Comparative study of chromenes containing different spiro-carbocyclic moieties. Mol Cryst Liq Cryst 297:263. https://doi.org/10.1080/10587259708036131
2. Pozzo JL, Lokshin V, Samat A, Gugglielmetti R, Dubest R, Aubard J (1998) Effect of heteroaromatic annulation with five-membered rings on the photochromism of *2H*-[1]-benzopyrans. J Photochem Photobiol A Chem 114:185–191. https://doi.org/10.1016/S1010-6030(98)00215-9
3. Fan P, Pan G, Wei JQ, Ming Y, Zhu A, Fan MG, Hung WM (2000) Synthesis and photochromic mechanism of 3-Phenyl-3-[1,2-Dimethylindol-3-yl]-*3H*-Naphtho[2,1-b]Pyran. Mol Cryst Liq Cryst 344:283–288. https://doi.org/10.1080/10587250008023850
4. Maafi M, Brown RG (2005) Photophysics and kinetics of naphthopyran derivatives, part 2: analysis of Diarylnaphthopyran kinetics. Degeneracy of the kinetic solution. Int J Chem Kinet 37:717–727. https://doi.org/10.1002/kin.20130
5. Berlman IB (2012) Handbook of fluorescence spectra of aromatic molecules. Academic Press, New York. ISBN. 9780323161671
6. Lenoble C, Becker RS (1986) Photophysics, photochemistry and kinetics of photochromic 2H-pyrans and chromenes. J Photochem 33:187–197. https://doi.org/10.1016/0047-2670(86)87033-2
7. Luccioni-Houzé B, Campredon M, Guglielmetti R, Giusti G (1997) Kinetic analysis of Fluoro-[2h]-Chromenes at the Photostationary states. Mol Cryst Liq Cryst 297:161–165. https://doi.org/10.1080/10587259708036117
8. Pan GL, Wei JQ, Fan P (2000) Laser induced photochromic mechanism of 3-phenyl-3-[1,2-dimethylindol-3-yl]-3H-naphtho[2,1-b]pyran. Res Chem Intermed 2000(26):829–838. https://doi.org/10.1163/156856700X00732
9. Aubard J, Maurel F, Buntinx G, Poizat O, Levi G, Guglielmetti R, Samat A (2000) Femto/picosecond transient absorption spectroscopy of photochromic 3,3-diphenyl-naphtho[2,1-b]pyran. Mol Cryst Liq Cryst 345:215–220. https://doi.org/10.1080/10587250008023921
10. Brazevic S, Nizinski S, Sliwa M, Abe J, Rode MF, Burdzinski G (2020) Control of the photo-isomerization mechanism in 3*H*-naphthopyrans to prevent formation of unwanted long-lived photoproducts. Int J Mol Sci 21. https://doi.org/10.3390/ijms21217825
11. Gierczyk B, Rode MF, Burdzinski G (2022) Mechanistic insights into photochromic 3H-naphthopyran showing strong photocoloration. Sci Rep 12:10781. https://doi.org/10.1038/s41598-022-14679-9
12. McFadden ME, Barber RW, Overholts AC, Robb MJ (2023) Naphthopyran molecular switches and their emergent mechanochemical reactivity. Chem Sci 14:10041–10067. https://doi.org/10.1039/D3SC03729K

Chapter 10
Kinetic Elucidation

10.1 Introduction

The occurrence of degeneracy of the kinetic solution, as seen in the previous chapter, imposes the development of an elucidation method by which the true kinetic solution is obtained. The elucidation method must provide a set of equations that uniquely define each of the unknown parameters of the reaction at hand. In these formulae, each parameter must be expressed as a function of experimentally accessible quantities or those previously made available by the elucidation procedure.

In the present context, irradiation is isosbestic and monochromatic, and first-order kinetics applies (i.e. no distinctability issues). The reaction mechanism is considered to be known for each reactive system studied (i.e. no distinguishability problems). Hence, the elucidation will only target solving the identifiability issue.

One of the best ways to test the efficacy of an elucidation method is to apply it to numerically calculated kinetic traces.

10.2 A Procedure for Kinetic Elucidation

A general procedure can be proposed for the elucidation of identifiability issues of photothermal systems. One can follow the step-by-step procedure given below. The first two points ((*i*) and (*ii*)) indicate that the mechanism of the reaction is fully known. This supposes that other techniques and methods have been previously used to establish this information, and the distinguishability problem is, at this stage, considered fully solved. Points (*iii*) to (*viii*) lay out the irreducible set of basic equations and unknowns of the reaction under study and check the compatibility of integrated rate-laws and experimental data. An evaluation of the solvability of the identifiability problem is proposed in the subsequent points (*ix*) to (*xiii*).

M. Maafi, *Photokinetics*, https://doi.org/10.1007/978-3-031-98985-8_10

(i) Find out the number of products generated from the reactant.
(ii) Confirm the detailed mechanism of the reaction by identifying the number, nature, and position of all reaction-steps in operation.
(iii) List all the unknown parameters of the reaction that cannot be obtained directly from the experiment.
(iv) Write down all the integrated rate-law equations that describe reaction and species.
(v) Confirm the good fit of the experimental traces by the adequate integrated rate-laws.
(vi) Write down all the equations that describe the reaction.
(vii) Check if the equations are linearly-independent.
(viii) Discard the equations that are linearly-dependent.
(ix) Check whether the number of the fundamental systems of linearly-independent equations is higher or at least equal to that of the reaction's unknowns.
(x) If the number of equations is less than the number of unknowns, then the kinetics is not solvable because of an unsolvable identifiability problem.
(xi) If the number of equations is higher or at least equal to that of the unknowns, then try to derive for each unknown a specific equation expressed only as a function of experimentally accessible and/or known quantities.
(xii) If the previous point is satisfied, then the elucidation method is complete, and the kinetic solution, i.e. a set of the unknown parameters' values, can be defined in a unique way.
(xiii) If point (*xi*) cannot be achieved, then the kinetic solution cannot be reached, and extra information on the reactive system is required. If such an information is obtained, then roll the procedure again, starting from point (*iii*).

The light used must be strictly monochromatic, and its isosbestic character requires that the reactant and at least one other product absorb the excitation wavelength. The temperature of the medium must be constant throughout the reaction time. The procedure also requires that the concentrations used for the experimental work fall within the linearity range of the calibration graph of the reactant and subsequently formed species. Whenever possible, this must be experimentally established by constructing the calibration graphs. If the species cannot be physically isolated from the reactive medium and/or when they undergo thermal reactions, it is advised that the experimental work should be performed on solutions whose measured total absorbances, at all wavelengths and over reaction time, do not exceed a value of 0.5. Such a value of the total absorbance will, most likely, mean that the individual species'absorbancies belong to the linearity ranges of the respective species' calibration graphs.

10.3 Unknowns and Characteristic Equations of $XY(2\Phi, 2k)$ Reactions

The eight mechanistic options that can be deduced from the generic $XY(2\Phi, 2k)$ photothermal reaction (Table 10.1), are defined by up to six unknowns (indicated by ✓ symbol), i.e. two thermal reaction rate-constants, two quantum yields, and the absorptivities of the species.

The traces of these reactions are described by the same general set of equations, whether under irradiation or in the dark. The temperature of the reaction must be maintained at a constant value. Specific quantities occurring in the following general equation will be discarded in accordance with the specific sub-mechanism at hand.

The integrated rate-laws of reactant and photoproduct concentrations have the general forms laid out by Eqs. 8.55 and 8.56 but can also reformulated by Eqs. 10.1 and 10.2. These equations apply to both photochemical, photothermal, and thermal reactions. It is important to specify which species are present at the beginning of the reaction and the mechanism of the reaction under study, and then, work out the different quantities of the equations accordingly.

$$C_X^{\lambda_{isos},T}(t) = C_X^{\lambda_{isos},T}(0) + \frac{r_0^{\lambda_{isos},T}}{k_r^{\lambda_{isos},T}}\left(1 - e^{-k_r^{\lambda_{isos},T}\,t}\right) \tag{10.1}$$

$$C_Y^{\lambda_{isos},T}(t) = C_Y^{\lambda_{isos},T}(0) + \frac{r_0^{\lambda_{isos},T}}{k_r^{\lambda_{isos},T}}\left(e^{-k_r^{\lambda_{isos},T}\,t} - 1\right) \tag{10.2}$$

with the initial velocity of the reaction expressed as

$$r_0^{\lambda_{isos},T} = \left(k_{X\to Y}^{T} + \Phi_{X\to Y}^{\lambda_{isos}}\,\gamma_1\right) C_X^{\lambda_{isos},T}(0) - \left(k_{Y\to X}^{T} + \Phi_{Y\to X}^{\lambda_{isos}}\,\gamma_1\right) C_Y^{\lambda_{isos},T}(0) \tag{10.3}$$

and the reaction rate-constant ($k_r^{\lambda_{isos},T} = -\rho_{\pm} = -a_{19}^{\lambda_{isos}}$) is

Table 10.1 All sub-mechanisms of $XY(2\Phi, 2k)$ reaction with their respective unknowns

	Parameters						
Sub-mechanism	Reaction	ε_X^{λ}	ε_Y^{λ}	$\Phi_{X\to Y}^{\lambda}$	$k_{X\to Y}^{T}$	$\Phi_{Y\to X}^{\lambda}$	$k_{Y\to X}^{T}$
$SubM_1$	$XY(1\Phi)$		✓	✓			
$SubM_2$	$XY(1\Phi, 1k)$		✓	✓	✓		
$SubM_3$	$XY(1\Phi, 1k)$		✓	✓			✓
$SubM_4$	$XY(2\Phi)$		✓	✓		✓	
$SubM_5$	$XY(2k)$	✓	✓		✓		✓
$SubM_6$	$XY(2\Phi, 1k)$		✓	✓	✓	✓	
$SubM_7$	$XY(2\Phi, 1k)$		✓	✓		✓	✓
$SubM_8$	$XY(2\Phi, 2k)$	✓	✓	✓	✓	✓	✓

$$k_r^{\lambda_{isos},T} = (k_{X\to Y}^T + k_{Y\to X}^T) + (\Phi_{X\to Y}^{\lambda_{isos}} + \Phi_{Y\to X}^{\lambda_{isos}})(\varepsilon_X^{\lambda_{isos}}\ P_0^{\lambda_{isos}}\ l_{irr}\ PKF^{\lambda_{isos}}) \tag{10.4}$$

At all times, the species‘concentrations obey the mass balance equation,

$$C_X^{\lambda_{isos},T}(0) + C_Y^{\lambda_{isos},T}(0) = C_X^{\lambda_{isos},T}(t) + C_Y^{\lambda_{isos},T}(t) = C_X^{\lambda_{isos},T}(pss) + C_Y^{\lambda_{isos},T}(pss) \tag{10.5}$$

The total absorbance of the medium, measured at the observation wavelength λ_{obs}, corresponds to the sum,

$$A_{tot}^{\lambda_{isos},T/\lambda_{obs}}(t) = \varepsilon_X^{\lambda_{obs}}\ C_X^{\lambda_{isos},T}(t)\ l_{obs} + \varepsilon_Y^{\lambda_{obs}}\ C_Y^{\lambda_{isos},T}(t)\ l_{obs} \tag{10.6}$$

where the concentrations of the species in Eq. 10.4 take the forms given by Eqs. 10.1 and 10.2. They also combine in the formula of the total absorbance to take the mono-exponential form,

$$A_{tot}^{\lambda_{isos},T/\lambda_{obs}}(t) = A_{tot}^{\lambda_{isos},T/\lambda_{obs}}(0) + \frac{r_{0,A}^{\lambda_{isos},T/\lambda_{obs}}}{k_r^{\lambda_{isos},T}}\left(1 - e^{-k_r^{\lambda_{isos},T}\ t}\right) \tag{10.7}$$

The initial velocity of the reaction, corresponding to the total absorbance trace, is worked out as

$$\left(A_{tot}^{\lambda_{isos},T/\lambda_{obs}}(t)\right)'_{t=0} = r_{0,A}^{\lambda_{isos},T/\lambda_{obs}} = \left(\varepsilon_Y^{\lambda_{obs}} - \varepsilon_X^{\lambda_{obs}}\right)\ l_{obs}\ r_0^{\lambda_{isos},T} \tag{10.8}$$

For reversible reactions, the thermodynamic constant is expressed as the ratio of the species concentrations reached of the equilibrium ($t = \infty$).

$$K_{eq} = \frac{C_Y^{\lambda_{isos},T}(\infty)}{C_X^{\lambda_{isos},T}(\infty)} \tag{10.9}$$

where the included concentrations have the following formulae.

$$C_X^{\lambda_{isos},T}(\infty) = -\frac{a_5^{\lambda_{isos}}}{k_r^{\lambda_{isos},T}} = \frac{k_{Y\to X}^T + \Phi_{Y\to X}^{\lambda_{isos}}\,\gamma_1}{k_r^{\lambda_{isos},T}}\ C_X^{\lambda_{isos},T}(0) \tag{10.10}$$

$$C_Y^{\lambda_{isos},T}(\infty) = \frac{a_4^{\lambda_{isos}}}{k_r^{\lambda_{isos},T}} = \frac{k_{X\to Y}^T + \Phi_{X\to Y}^{\lambda_{isos}}\,\gamma_1}{k_r^{\lambda_{isos},T}}\ C_X^{\lambda_{isos},T}(0) \tag{10.11}$$

The fitting of the traces can be performed by the integrated rate-laws Eqs. 10.1, 10.2, or 10.7. It can also be achieved by a generic equation (Eq. 10.12) if some of the reaction parameters present in the previous equations are not available. The numerical values of the parameters $\gamma_2^{\lambda_{isos},T}$, $\gamma_3^{\lambda_{isos}}$, and $k^{\lambda_{isos},T}$, obtained from the fitting of a given trace, will be equated to their algebraic expressions in the theoretical equation of the considered trace (e.g. Eqs. 10.1, 10.2, or 10.7).

$$C_{X\ or\ Y}^{\lambda_{isos},T}(t)\ or\ A_{tot}^{\lambda_{isos},T/\lambda_{obs}}(t)=\gamma_2^{\lambda_{isos},T}+\gamma_3^{\lambda_{isos},T}\ e^{-k_r^{\lambda_{isos},T}\ t} \tag{10.12}$$

For elucidation purposes, it is important to use only the linearly-independent equations. There are six of such equations, e.g. Eqs. 10.3, 10.4, 10.5, 10.6, 10.8, and 10.9. However, this number can be reduced depending on the information available on the reaction and its kinetics.

10.4 The Kinetic Solution for *XY*(1*Φ*)

The photoproduct in this unimolecular reaction absorbs the incident light at the isosbestic wavelength ($\varepsilon_Y^{\lambda_{isos}}\neq 0$) but is unreactive ($\Phi_{Y\rightarrow X}^{\lambda_{isos}}=0$). Obviously, its initial concentration is equal to zero. The forward quantum yield ($\Phi_{X\rightarrow Y}^{\lambda_{isos}}$) and the absorptivity of Y at the observation wavelength ($\varepsilon_Y^{\lambda_{obs}}$) are the two unknowns of the reaction ($SubM_1$ in Table 10.1). The absorptivity spectrum of the reactant X, $\varepsilon_X^{\lambda_{obs}}$, is worked out from its absorption spectrum and its known initial concentration, $C_X^{\lambda_{isos},T}(0)$.

The integrated rate-law of the reactant is derived from Eq. 10.1, where $k_{X\rightarrow Y}^T=k_{Y\rightarrow X}^T=\Phi_{Y\rightarrow X}^{\lambda_{isos}}=C_Y^{\lambda_{isos},T}(0)=0$ and $k_r^{\lambda_{isos},T}=-\Phi_{X\rightarrow Y}^{\lambda_{isos}}\gamma_1$.

$$C_X^{\lambda_{isos},T}(t)=C_X^{\lambda_{isos},T}(0)\ e^{-\Phi_{X\rightarrow Y}^{\lambda_{isos}}\ \varepsilon_X^{\lambda_{isos}}\ P_0^{\lambda_{isos}}\ l_{irr}\ PKF^{\lambda_{isos}}\ t} \tag{10.13}$$

The integrated rate-law of Y in terms of concentration is worked out as

$$\begin{aligned}C_Y^{\lambda_{isos},T}(t)&=C_X^{\lambda_{isos},T}(0)-C_X^{\lambda_{isos},T}(t)\\&=C_X^{\lambda_{isos},T}(0)\ (1-e^{-\Phi_{X\rightarrow Y}^{\lambda_{isos}}\ \varepsilon_X^{\lambda_{isos}}\ P_0^{\lambda_{isos}}\ l_{irr}\ PKF^{\lambda_{isos}}\ t})\end{aligned} \tag{10.14}$$

Using the latter equations, together with Eq. 10.6, we define the integrated rate-law in terms of total absorbance, either depending on both initial concentration and species absorbancies,

$$\begin{aligned}A_{tot}^{\lambda_{isos},T/\lambda_{obs}}(t)&=\varepsilon_Y^{\lambda_{obs}}\,C_X^{\lambda_{isos},T}(0)\,l_{obs}\\&+(\varepsilon_X^{\lambda_{obs}}-\varepsilon_Y^{\lambda_{obs}})\,l_{obs}\,C_X^{\lambda_{isos},T}(0)\,e^{-\Phi_{X\rightarrow Y}^{\lambda_{isos}}\,\varepsilon_X^{\lambda_{isos}}\,P_0^{\lambda_{isos}}\,l_{irr}\,PKF^{\lambda_{isos}}\,t}\end{aligned} \tag{10.15}$$

or depending on the measured absorbancies

Scheme 10.1 Chemical structures for three reactants obeying an *XY*(1*Φ*) sequence (*NIF*, *NIS*, and *DBZ*)

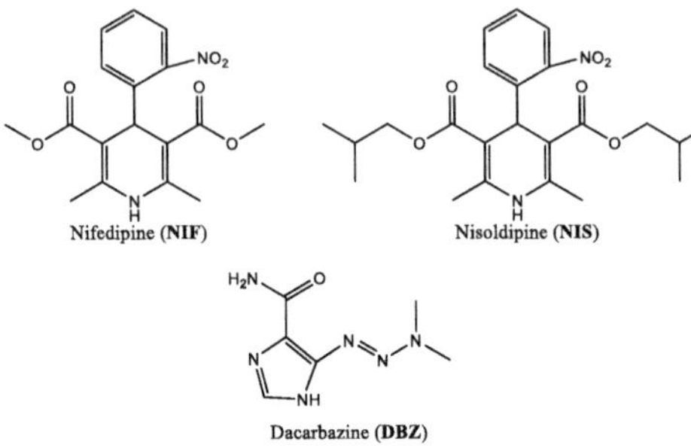

$$A_{tot}^{\lambda_{isos},T/\lambda_{obs}}(t) = A_{tot}^{\lambda_{isos},T/\lambda_{obs}}(\infty) + (A_{tot}^{\lambda_{isos},T/\lambda_{obs}}(0) - A_{tot}^{\lambda_{isos},T/\lambda_{obs}}(\infty))\, e^{-\Phi_{X\to Y}^{\lambda_{isos}}\, \varepsilon_X^{\lambda_{isos}}\, P_0^{\lambda_{isos}}\, l_{irr}\, PKF^{\lambda_{isos}}\, t} \quad (10.16)$$

For the cases where absorbance measurements are accessible, the absorptivity spectrum of the photoproduct is calculated by

$$\varepsilon_Y^{\lambda_{obs}} = \frac{A_{tot}^{\lambda_{isos},T/\lambda_{obs}}(\infty)}{l_{obs}\, C_X^{\lambda_{isos},T}(0)} \quad (10.17)$$

Fitting one of the reaction traces ($C_X^{\lambda_{isos},T}(t)$, $C_Y^{\lambda_{isos},T}(t)$, or $A_{tot}^{\lambda_{isos},T/\lambda_{obs}}(t)$) to the appropriate mono-exponential model equation (Eqs. 10.13, 10.14, or 10.16) will deliver a value for the fitting parameter, corresponding to the reaction rate-constant, $k_r^{\lambda_{isos},T}$.

$$k_r^{\lambda_{isos},T} = \Phi_{X\to Y}^{\lambda_{isos}}\, \varepsilon_X^{\lambda_{isos}}\, P_0^{\lambda_{isos}}\, l_{irr}\, PKF^{\lambda_{isos}} \quad (10.18)$$

The elucidation method is then completed by calculating the quantum yield of the reaction using Eq. 10.19 since the quantities on the right-hand side of that equation are experimentally accessible.

$$\Phi_{X\to Y}^{\lambda_{isos}} = \frac{k_r^{\lambda_{isos},T}}{\varepsilon_X^{\lambda_{isos}}\, P_0^{\lambda_{isos}}\, l_{irr}\, PKF^{\lambda_{isos}}} \quad (10.19)$$

The photodegradation of three drugs, nisoldipine, nifedipine, and dacarbazine (Scheme 10.1), are governed by the typical *XY*(1*Φ*) mechanism with an absorbing photoproduct at λ_{isos}. The spectrophotometric spectra are characterised by the disappearance of the absorption band possessing the longest maximum wavelength (Fig. 10.1). The photoreaction, i.e. the depletion of *X* and the production of *Y* in the medium, is a smooth process, as indicated by the occurrence of one or several isosbestic points on the overlaid spectra.

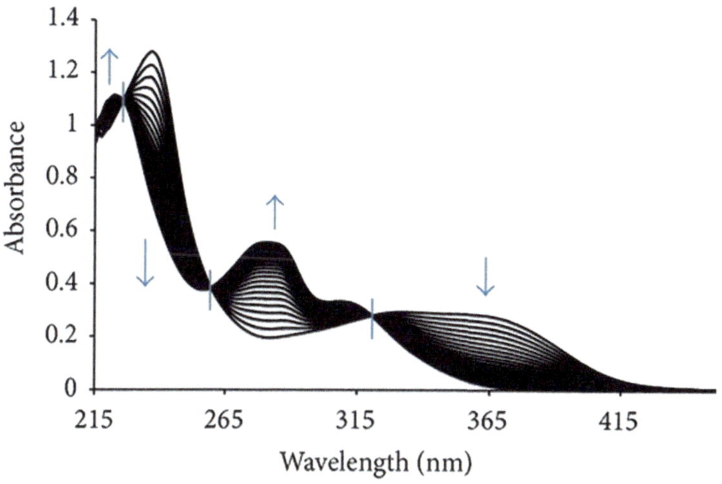

Fig. 10.1 Time variation of the electronic absorption spectra of an ethanolic solution of nisoldipine (*NIS*, 5.88 × 10^{-5} M). The arrows indicate the direction of the peaks' evolution during the photoreaction, and the vertical lines cross the spectra at the isosbestic points (232, 259, and 320 nm). (Reproduced with permission (Fig. 1 p-4 from [1], *International Journal of Photoenergy*)

Table 10.2 Some experimental and fitting features of the photodegradation of $XY(1\Phi)$ drugs, performed under isosbestic irradiation

Drug	λ_{isos} /nm	λ_{obs} /nm	$C_X^{\lambda_{isos},T}(0)$ /M	$A_{tot}^{\lambda_{isos},T/\lambda_{obs}}(0)$	$P_0^{\lambda_{isos}} \times 10^7$ /einsten dm^{-3} s^{-1}	$k_r^{\lambda_{isos},T} \times 10^3$ /s^{-1}	$\Phi_{X \to Y}^{\lambda_{isos}}$
Nifedipine	321	390	5.14 × 10^{-5}	0.0261	3.18	2.36	0.267
Nisoldipine	259	390	5.88 × 10^{-5}	0.0222	2.64	0.33	0.028
Nisoldipine	320	390	5.88 × 10^{-5}	0.0198	3.53	2.80	0.304
Dacarbazine	221	325	5.47 × 10^{-6}	0.0518	15.3	0.13	0.001

Depending on the observation wavelength, the absorbance gradually increases or decreases with irradiation time (when irradiation is performed with a monochromatic light of an isosbestic wavelength). The acquisition at a given λ_{obs} of the temporal evolution of the total absorbance represents the experimental kinetic trace of the reaction. Irrespective of λ_{obs}, the trace is fitted with high accurracy by Eq. 10.16. The fitting parameter, $k_r^{\lambda_{isos},T}$, serves the determination of the value of forward quantum yield by the application of Eq. 10.19. The values of $\Phi_{X \to Y}^{\lambda_{isos}}$ for the three drugs at the respective irradiation wavelengths are shown in Table 10.2.

10.5 The Kinetic Solution for the Photothermal Reversible $XY(1\Phi, 1k)$

Many spiropyran photochromes have been assigned an $XY(1\Phi, 1k)$ mechanism [2–4], where the forward reaction is initiated photochemically from the closed-ring reactant ($C_Y^{\lambda_{isos},T}(0) = k_{X\to Y}^T = 0$) and reversed thermally from the open-ring product ($\Phi_{Y\to X}^{\lambda_{isos}} = 0$). Time-resolved, computational, and mechanistic studies have established that the photo-colouration process of spiropyrans involve a triplet excited-state and generates several short-living intermediates that convert into the much more stable trans-merocyanine that is observed by spectrophotometry [4–6].

Three unknown parameters characterise the $XY(1\Phi, 1k)$ kinetics, namely, the quantum yield of the forward reaction-step, the rate-constant of the thermal reaction-step, and the absorptivity of Y. The remaining parameters of the reaction are all experimentally accessible.

The kinetics of 1′,3′-dihydro-1′,3′,3′-trimethyl-6-nitrospiro[2*H*-1-benzopyran-2,2′-(2*H*)-indole derivative was studied in ethyl acetate at a constant temperature of 14 °C (Scheme 10.2). The reactant was subjected to a continuous and monochromatic irradiation at the wavelength of the isosbestic point situated in the ultraviolet, $\lambda_{isos} = 339\ nm$.

The progress of the reaction is indicated by a smooth increase of a new absorption band emerging in the visible region of the spectrum ($450 - 650\ nm$). The observation wavelength corresponds to the maximum of the latter transition at $\lambda_{obs} = 580\ nm$ (Fig. 10.2). After the reaction reaches the photostationary state ($t = 600\ s$), the excitation light is switched off, and the reactive medium is allowed to evolve in the dark. The reverse reaction ($Y \to X$), corresponding to a thermal bleaching of the solution, is attested by a continuous decrease of the absorbance at λ_{obs} until a complete depletion of Y ($A_{tot}^{\lambda_{isos},T/\lambda_{obs}}(\infty) = A_{tot}^{339,14/580}(\infty) = 0$).

The experimental kinetic traces (Fig. 10.2) were well-fitted to the theoretical model equations of the total absorbance with high accuracy. These are defined for the separate photo-colouring and thermo-bleaching reactions taking place under irradiation and in the dark, respectively. Therefore, Eq. 10.8 is adapted to the two situations.

The data treatment proceeds first by fitting the trace of the bleaching isothermal ($T = 14$ °C) reaction with Eq. 10.20. The equation parameters extracted from the fitting have the following values for the absorbance at *pss*, $A_{tot}^{\lambda_{isos},T/\lambda_{obs}}(pss) = 1.131$ (which, with $A_{tot}^{\lambda_{isos},T/\lambda_{obs}}(0)$, are directly available from the experimental spectra or

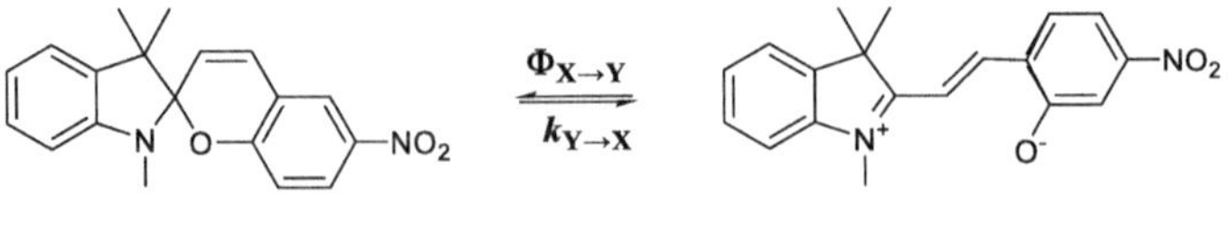

Scheme 10.2 A proposed reversible photothermal $XY(1\Phi, 1k)$ reaction for a nitro-spiropyran

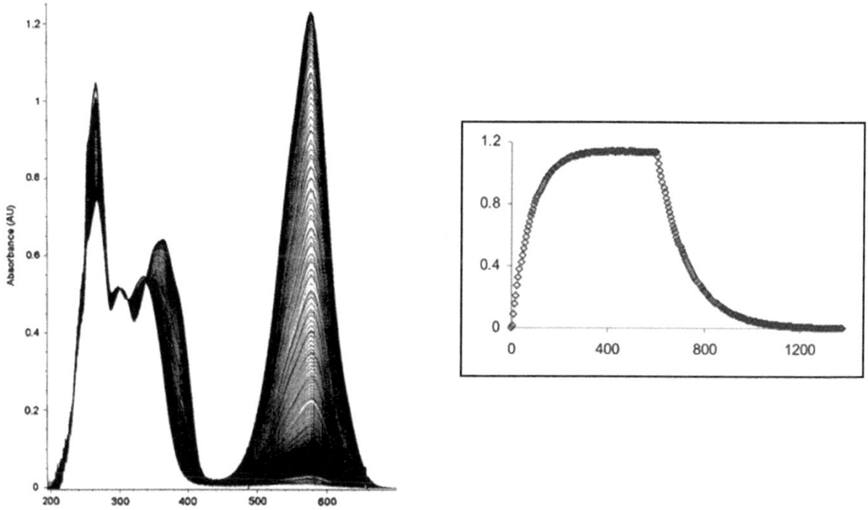

Fig. 10.2 _Left_: Time evolution of the electronic absorption spectra of the nitro-spiropyran derivative (Scheme 10.2) in ethyl acetate. The initial concentration of the ring-closed isomer was $C_X(0) = 5.83\ 10^{-5}$ M, the isosbestic irradiation wavelength was $\lambda_{isos} = 339\ nm$, the constant temperature was 14 °C, and the irradiation and observation path lengths were $l_{irr} = 2.1\ cm$ and $l_{obs} = 1cm$, respectively. The irradiated area and volume of the sample were $S_{irr} = 1\ cm^2$ and $V_{irr} = 2.1\ mL$. _Right_: Photo-colouring (up to 600 s) and thermo-bleaching (600 − 1300 s) kinetic traces of the reaction medium, monitored as the variation of the total absorbance measured at $\lambda_{obs} = 580\ nm$. The experimental data (squares) are fitted to the adequate equation models (lines). (Reproduced with permission from [7], *International Journal of Chemical Kinetics*)

trace), the rate-constant $k_r^T = -a_{19}^T = k_{Y\to X}^T = 7.69 \times 10^{-3}\ s^{-1}$, and the initial rate $r_{0,A}^{T,\lambda_{obs}} = -8.73 \times 10^{-3}\ s^{-1}$.

$$A_{tot}^{\lambda_{obs},T}(t) = A_{tot}^{\lambda_{isos},T/\lambda_{obs}}(pss) + \frac{r_{0,A}^{T,\lambda_{obs}}}{k_r^T}\left(1 - e^{-k_r^T\, t}\right) \tag{10.20}$$

The fitting of the photocolouring trace ($T = 14$ °C and $\lambda_{isos} = 339\ nm$) with Eq. 10.21 yields the two parameters, $k_r^{\lambda_{isos},T} = k_{Y\to X}^T + \Phi_{X\to Y}^{\lambda_{isos}}\, \varepsilon_X^{\lambda_{isos}}\, P_0^{\lambda_{isos}}\, l_{irr}\, PKF^{\lambda_{isos}} = 12.27 \times 10^{-3}\ s^{-1}$ and $r_{0,A}^{\lambda_{isos}/\lambda_{obs},T} = -14.06 \times 10^{-3}\ s^{-1}$.

$$A_{tot}^{\lambda_{isos},T/\lambda_{obs}}(t) = A_{tot}^{\lambda_{isos},T/\lambda_{obs}}(0) + \frac{r_{0,A}^{\lambda_{isos}/\lambda_{obs},T}}{k_r^{\lambda_{isos},T}}\left(1 - e^{-k_r^{\lambda_{isos},T}\, t}\right) \tag{10.21}$$

The quantum yield value is calculated from the reactions' rate constants as

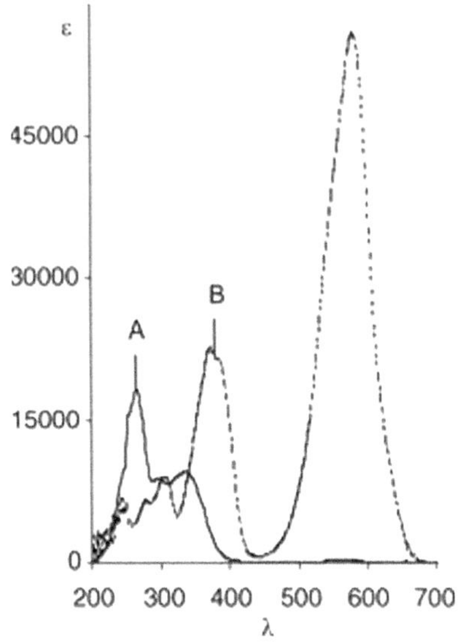

Fig. 10.3 Electronic absorption spectra of *X* (solid line) and *Y* (dashed line). The spectrum of *Y* has been reconstructed using Eq. 10.25. (ε is expressed in $M^{-1}\ cm^{-1}$, and λ is in *nm*). (Reproduced with permission [8], *International Journal of Chemical Kinetics*)

$$\Phi_{X\rightarrow Y}^{\lambda_{isos}} = \frac{k_r^{\lambda_{isos},T} - k_{Y\rightarrow X}^T}{\varepsilon_X^{\lambda_{isos}}\ P_0^{\lambda_{isos}}\ l_{irr}\ PKF^{\lambda_{isos}}} \tag{10.22}$$

and found to be equal to 4.58×10^{-3}. The *pss* concentrations are obtained from Eqs. 10.10 and 10.11 as functions of the reaction's rate-constants,

$$C_X^{\lambda_{isos},T}(pss) = \frac{k_{Y\rightarrow X}^T}{k_r^{\lambda_{isos},T}}\ C_X^{\lambda_{isos},T}(0) \tag{10.23}$$

$$C_Y^{\lambda_{isos},T}(pss) = C_X^{\lambda_{isos},T}(0) - C_X^{\lambda_{isos},T}(pss) = \frac{k_r^{\lambda_{isos},T} - k_{Y\rightarrow X}^T}{k_r^{\lambda_{isos},T}}\ C_X^{\lambda_{isos},T}(0) \tag{10.24}$$

with their values being $C_X(pss) = 3.65 \times 10^{-5}\ M$ and $C_Y(pss) = 2.18 \times 10^{-5}\ M$.

The last unknown, $\varepsilon_Y^{\lambda_{obs}}$ (for $200 < \lambda_{obs} < 700\ nm$), is worked out using equation Eq. 10.6, as

$$\varepsilon_Y^{\lambda_{obs}} = \frac{A_{tot}^{\lambda_{isos},T/\lambda_{obs}}(pss) - \varepsilon_X^{\lambda_{obs}}\ l_{obs}\ C_X^{\lambda_{isos},T}(pss)}{C_Y^{\lambda_{isos},T}(pss)\ l_{obs}} \tag{10.25}$$

By applying Eq. 10.25 to the overall absorption region, the spectrum of species *Y* can be fully reconstructed (Fig. 10.3).

The elucidation method described above allows an easy and quick determination of the absolute values of both fundamental reaction parameters, i.e. the quantum yield and the absorption coefficient spectrum of the thermally unstable species *Y* for an $XY(1\Phi, 1k)$ system under an isosbestic irradiation. The individual species spectra are determined, irrespective of whether they partially or fully overlap. It also allows us to determine the composition of the *pss* without the need for a physical separation of the species (which is, in fact, not possible due to the thermal instability of Y).

We can also discuss the concept of *colourability,* which is defined as "the ability of a colourless or a slightly coloured (pale yellow) photochromic material to develop colouration" [9, 10]. The colourability has, therein, been quantified in dilute solutions, by expressing the initial absorbance, $A_{tot}^{\lambda}(0)$, as proportional to the concentration of the colourless species, e.g. C_X^{λ}; the quantum yield (labelled as the colouration quantum yield); and the absorptivity of the coloured species, e.g. ε_Y^{λ}, with δ being the proportionality constant (Eq. 10.26).

$$A_{tot}^{\lambda}(0) = \delta\, \Phi_{col}^{\lambda}\, \varepsilon_Y^{\lambda}\, C_X^{\lambda} \tag{10.26}$$

The presence of the kinetic and spectroscopic quantities in the latter equations supposes that the kinetics has to be first elucidated, which did not facilitate the application of Eq. 10.26. Furthermore, Eq. 10.26 applies to reactions where the reactant is thermally stable [11]. Owing to that difficulty, it was suggested that the colourability may be defined by exclusively experimentally available quantities for less restricted conditions on the reaction properties [8]. This alternative can be expressed by a positive quantity (Eq. 10.27) for the reversible $XY(1\Phi, 1k)$, which also applies to $XY(1\Phi)$ reactions.

$$COL_{XY}^{\lambda_{isos},T} = \varepsilon_X^{\lambda_{isos}}\, \Phi_{X\to Y}^{\lambda_{isos}} = \frac{r_{0,A}^{\lambda_{isos},T/\lambda_{obs}}}{\gamma_1\, C_X^{\lambda_{isos},T}(0)} \tag{10.27}$$

Nonetheless, both of the above colourability definitions might not fit our purpose because they relay the properties of the reactant only, whereas the final colour of the solution depends on the properties of both the reactant and the product at *pss*. This is also the case for any photochromic system, whatever the number of its products or the mechanism of its reaction may be.

10.6 Elucidation of the Photoreversible $XY(2\Phi)$ Reaction Kinetics

The $XY(2\Phi)$ reaction is one of the most ubiquitously studied photosystems in photochemistry literature. It is also an interesting kinetic case that deserves to be investigated in some detail.

10.6.1 The Unknowns and Trace-Fitting Equations

The kinetic solution for $XY(2\Phi)$ consists of the values of three unknown parameters of this reaction, namely, the absorptivity of the photoproduct Y and the forward and reverse quantum yields.

The integrated rate-laws for the reactant and the product of a photoreversible reaction performed under isosbestic irradiation have mono-exponential forms. The experimental traces of this reaction for the total absorbance are fitted to Eq. 10.21, and those for concentrations can be fitted to the following reformulated equations from Eqs. 10.1 and 10.2.

$$C_X^{\lambda_{isos},T}(t) = C_X^{\lambda_{isos},T}(pss) + \left(C_X^{\lambda_{isos},T}(0) - C_X^{\lambda_{isos},T}(pss)\right) e^{-k_r^{\lambda_{isos},T}\ t} \tag{10.28}$$

$$C_Y^{\lambda_{isos},T}(t) = C_Y^{\lambda_{isos},T}(pss) - C_Y^{\lambda_{isos},T}(pss)\ e^{-k_r^{\lambda_{isos},T}\ t} \tag{10.29}$$

10.6.2 The Descriptive Equations

The equations that describe the photoreversible system under isosbestic irradiation are extracted from its integrated rate-law for the total absorbance (Eq. 10.21), the values of its fitting parameters, the characteristics at the start and end (pss) of the reaction, and the mass balance equation.

The two particular absorbances of the trace are

$$A_{tot}^{\lambda_{isos},T/\lambda_{obs}}(0) = \varepsilon_X^{\lambda_{obs}}\ l_{obs} C_X^{\lambda_{isos},T}(0) \tag{10.30}$$

$$A_{tot}^{\lambda_{isos},T/\lambda_{obs}}(pss) = \varepsilon_X^{\lambda_{obs}}\ l_{obs} C_X^{\lambda_{isos},T}(pss) + \varepsilon_Y^{\lambda_{obs}}\ l_{obs} C_Y^{\lambda_{isos},T}(pss) \tag{10.31}$$

with the expressions of the species concentrations at pss given by

$$C_X^{\lambda_{isos},T}(\mathrm{pss}) = \frac{\Phi_{Y\to X}^{\lambda_{isos}}}{\left(\Phi_{Y\to X}^{\lambda_{isos}} + \Phi_{X\to Y}^{\lambda_{isos}}\right)}\ C_X^{\lambda_{isos},T}(0) \tag{10.32}$$

$$C_Y^{\lambda_{isos},T}(\mathrm{pss}) = \frac{\Phi_{X\to Y}^{\lambda_{isos}}}{\left(\Phi_{Y\to X}^{\lambda_{isos}} + \Phi_{X\to Y}^{\lambda_{isos}}\right)}\ C_X^{\lambda_{isos},T}(0) \tag{10.33}$$

which obey the following mass balance:

$$C_X^{\lambda_{isos},T}(0) = C_X^{\lambda_{isos},T}(pss) + C_X^{\lambda_{isos},T}(pss) \tag{10.34}$$

The thermodynamic equilibrium constant (Eq. 10.9) is

$$K_{eq}^{\lambda_{isos}} = \frac{C_Y^{\lambda_{isos},T}(pss)}{C_X^{\lambda_{isos},T}(pss)} = \frac{\Phi_{X\to Y}^{\lambda_{isos}}}{\Phi_{Y\to X}^{\lambda_{isos}}} \tag{10.35}$$

And finally, we have the equations corresponding to the fitting parameters.

$$k_r^{\lambda_{isos},T} = \left(\Phi_{Y\to X}^{\lambda_{isos}} + \Phi_{X\to Y}^{\lambda_{isos}}\right) \varepsilon_X^{\lambda_{isos}} P_0^{\lambda_{isos}} l_{irr} PKF^{\lambda_{isos}} \tag{10.36}$$

and

$$r_{0,A}^{\lambda_{isos},T/\lambda_{obs}} = \left(\varepsilon_Y^{\lambda_{obs}} - \varepsilon_X^{\lambda_{obs}}\right) l_{obs} \Phi_{X\to Y}^{\lambda_{isos}} \varepsilon_X^{\lambda_{isos}} P_0^{\lambda_{isos}} l_{irr} PKF^{\lambda_{isos}} C_X^{\lambda_{isos},T}(0) \tag{10.37}$$

Therefore, these eight equations represent the complete set of equations available for the kinetic investigation of the *XY*(2*Φ*) reaction.

10.6.3 The Experimentally Available Quantities

The numerical values of the following quantities are all experimentally accessible and are fixed (invariable) for a given trace. These are $C_X^{\lambda_{isos},T}(0)$; $\varepsilon_X^{\lambda_{obs}}$; $\varepsilon_X^{\lambda_{isos}}$; l_{obs}; l_{irr}; $P_0^{\lambda_{isos}}$; $A_{tot}^{\lambda_{isos},T/\lambda_{obs}}(0)$; $A_{tot}^{\lambda_{isos},T/\lambda_{obs}}(pss)$; $k_r^{\lambda_{isos},T}$ (the fitting parameter of the trace); the temperature of the medium, *T*; and $PKF^{\lambda_{isos}}$ (obtained using the constant value of $A_{tot}^{\lambda_{isos},T/\lambda_{isos}}$ in Eq. 8.14).

10.6.4 The Irreducible Set of Equations

Let us now explore Eqs. 10.30, 10.31, 10.32, 10.33, 10.34, 10.35, 10.36 and 10.37 with the aim of identifying the ones that can be worked out from a linear combination of any of the others or those that are redundant. The ultimate purpose is to find a set of equations that cannot be reduced further. This irreducible set corresponds to the linearly-independent equations that will serve later in the development of the elucidation method.

In this perspective, we first notice that one of the equations (Eq. 10.30) is not useful for kinetic elucidation because it does not involve any of the three unknowns of the system; however, it serves well in the determination of the reactant absorptivity, $\varepsilon_X^{\lambda_{obs}}$. Also, it is obvious that Eq. 10.34 can easily be worked out from the sum of the *pss* concentrations (Eqs. 10.32 and 10.33). Therefore, these two equations (Eqs. 10.30 and 10.34) can be discarded from being part of the irreducible set.

Now, if on one hand, the formulae of the *pss* concentrations are introduced in the expression of the total absorbance measured at *pss* (Eq. 10.31), and the obtained formula is rearranged, we get

$$A_{tot}^{\lambda_{isos},T/\lambda_{obs}}(pss)\frac{\left(\Phi_{Y\to X}^{\lambda_{isos}}+\Phi_{X\to Y}^{\lambda_{isos}}\right)}{C_X^{\lambda_{isos},T}(0)\,l_{irr}}=\varepsilon_X^{\lambda_{obs}}\,\Phi_{Y\to X}^{\lambda_{isos}}+\varepsilon_Y^{\lambda_{obs}}\,\Phi_{X\to Y}^{\lambda_{isos}} \tag{10.38}$$

and, on the other hand, rearranging the equation of the overall reaction rate-constant (Eq. 10.37) as

$$\frac{k_r^{\lambda_{isos},T}\,\varepsilon_X^{\lambda_{obs}}}{\varepsilon_X^{\lambda_{isos}}\,P_0^{\lambda_{isos}}\,l_{irr}PKF^{\lambda_{isos}}}=\varepsilon_X^{\lambda_{obs}}\,\Phi_{Y\to X}^{\lambda_{isos}}+\varepsilon_X^{\lambda_{obs}}\,\Phi_{X\to Y}^{\lambda_{isos}} \tag{10.39}$$

and subtracting it from Eq. 10.38, we reconstitute the expression of the initial rate as

$$\begin{aligned} r_{0,A}^{\lambda_{isos},T/\lambda_{obs}} = & \left(A_{tot}^{\lambda_{isos},T/\lambda_{obs}}(pss)\frac{\left(\Phi_{Y\to X}^{\lambda_{isos}}+\Phi_{X\to Y}^{\lambda_{isos}}\right)}{C_X^{\lambda_{isos},T}(0)}-\frac{k_r^{\lambda_{isos},T}\,\varepsilon_X^{\lambda_{obs}}}{\varepsilon_X^{\lambda_{isos}}\,P_0^{\lambda_{isos}}\,PKF^{\lambda_{isos}}}\right) \\ & \times l_{obs}\,\varepsilon_X^{\lambda_{isos}}\,P_0^{\lambda_{isos}}\,PKF^{\lambda_{isos}}\,C_X^{\lambda_{isos},T}(0) \end{aligned} \tag{10.40}$$

Hence, one of the latter equations can be discarded. Let it be the total absorbance at *pss* (Eq. 10.33). Similarly, the expression of $k_r^{\lambda_{isos},T}$ can be derived from one of the equations of the *pss* concentrations as

$$k_r^{\lambda_{isos},T}=\Phi_{Y\to X}^{\lambda_{isos}}\varepsilon_X^{\lambda_{isos}}P_0^{\lambda_{isos}}\,l_{irr}PKF^{\lambda_{isos}}\frac{C_X^{\lambda_{isos},T}(0)}{C_X^{\lambda_{isos},T}(\mathrm{pss})}=\Phi_{X\to Y}^{\lambda_{isos}}\varepsilon_X^{\lambda_{isos}}P_0^{\lambda_{isos}}\,l_{irr}PKF^{\lambda_{isos}}\frac{C_X^{\lambda_{isos},T}(0)}{C_Y^{\lambda_{isos},T}(\mathrm{pss})} \tag{10.41}$$

This makes the expressions (Eqs. 10.32 and 10.33) of the *pss* concentrations redundant and not useful for the solution of the problem.

In summary, we end up with a set of only three linearly-independent equations Eqs. 10.35, 10.36, and 10.37 (none of these equations can be worked out from a linear combination of the other two). This represents the irreducible set for the *XY*(2*Φ*) reaction. One of the equations, Eq. 10.35, involves three unknowns, whereas the other two are each a function of two. Two of these linearly-independent equations (Eqs. 10.35 and 10.37) are non-linear (whereas, the third one (Eq. 10.36) is linear).

The irreducible set can be presented in the following formulations where the right-hand side terms of Eqs. 10.43 and 10.44 depend exclusively on experimentally accessible quantities ($k_r^{\lambda_{isos},T}$ and $r_{0,A}^{\lambda_{isos}}$ can be obtained from the fitting of the total absorbance trace to Eq. 10.21).

$$\Phi_{X\to Y}^{\lambda_{isos}}=K_{eq}^{\lambda_{isos}}\,\Phi_{Y\to X}^{\lambda_{isos}} \tag{10.42}$$

$$\Phi_{Y\to X}^{\lambda_{isos}} + \Phi_{X\to Y}^{\lambda_{isos}} = \frac{k_r^{\lambda_{isos},T}}{\varepsilon_X^{\lambda_{isos}}\, P_0^{\lambda_{isos}}\, l_{irr}\, PKF^{\lambda_{isos}}} \tag{10.43}$$

$$\Phi_{X\to Y}^{\lambda_{isos}}\left(\varepsilon_Y^{\lambda_{obs}} - \varepsilon_X^{\lambda_{obs}}\right) = \frac{r_{0,A}^{\lambda_{isos},T/\lambda_{obs}}}{l_{obs}\, \varepsilon_X^{\lambda_{isos}}\, P_0^{\lambda_{isos}}\, l_{irr}\, PKF^{\lambda_{isos}}\, C_X(0)} \tag{10.44}$$

The irreducible set, uniquely defining the kinetics of the photoreversible reaction, is then the basis for solving the kinetics and determining its three unknowns, $\Phi_{X\to Y}^{\lambda_{isos}}$, $\Phi_{X\to Y}^{\lambda_{isos}}$, and $\varepsilon_Y^{\lambda_{obs}}$. The question is now: to what extent will this be possible?

To answer this question, let us proceed with a hypothetical case where a value for $\Phi_{X\to Y}^{\lambda_{isos}}$ is deliberately assigned (such as $0 < \Phi_{X\to Y}^{\lambda_{isos}} \leq 1$). It is then clear that both the values of $\Phi_{Y\to X}^{\lambda_{isos}}$ (from Eq. 10.43) and later, $\varepsilon_Y^{\lambda_{obs}}$ (from Eq. 10.44) can be determined. The set found can be considered the kinetic solution for the reaction, as long as the values of the three parameters are plausible.

Of course, this process can be repeated indefinitely for different chosen values of the forward quantum yield. It is, hence, certain that a different set of parameters' values ($\Phi_{X\to Y}^{\lambda_{isos}}$, $\Phi_{X\to Y}^{\lambda_{isos}}$ and $\varepsilon_Y^{\lambda_{obs}}$) will be obtained for each different starting value of $\Phi_{X\to Y}^{\lambda_{isos}}$. Accordingly, we can generate a large number of plausible values for our set of unknowns. This practically means that the experimental trace can exactly be reproduced by equations using many and significantly different sets of values of the unknowns. Therefore, the true set of the parameter values (the true kinetic solution), characterising a given $A_{tot}^{\lambda_{isos},T/\lambda_{obs}}(t)$ trace, cannot be known. It is a typical unidentifiability problem. The origin of this unidentifiability stems from the non-linearity of some of the fundamental and linearly-independent equations of the irreducible set describing the system. Such a degeneracy of the kinetic solution indicates that the system of the three fundamental equations cannot be uniquely solved for the three unknowns defining the system.

Incidentally, it is perhaps useful to remember that employing numerical integration methods has no effect on improving the overall situation, i.e. either on the occurrence of the identifiability issue for this kinetics or in singling out the true kinetic solution.

This simply means that, as a rule, it is impossible to solve the photoreversible reaction kinetics based on a single kinetic trace of the total absorbance.

This fact perfectly explains why the published methods using a single (total absorbance) trace had to include in their procedures at least one simplifying assumption. While these simplifications may lead to a unique kinetic solution, such a solution cannot be proven in any possible way to be the true kinetic solution of the reactive photosystem. Hence, it could, at most, be considered as one of the degenerate sets of solutions caused by the unidentifiability of the kinetics.

In this context, it is interesting to investigate whether increasing the number of total absorbance traces generated by different isosbestic irradiations can improve the outcome of the kinetic analysis. This is discussed in the next section.

A further discussion of the methods proposed to solve this kinetics under non-isosbestic irradiation will be provided in Chaps. 16 and 20.

10.6.5 Using Multiple Irradiations

One of the hypotheses proposes that the photoreversible kinetics would become solvable if the data of more than one irradiation are used. In the case where the reaction investigated is subjected to two independent isosbestic irradiations (λ_{isos_1} and λ_{isos_2}), two irreducible sets, each constituted of three equations (Eqs. 10.42, 10.43 and 10.44), will describe these reactions. Each of the sets is dependent of three unknowns (*i.e.*, $set_1 : \Phi_{Y\rightarrow X}^{\lambda_{isos_1}}$, $\Phi_{X\rightarrow Y}^{\lambda_{isos_1}}$, and $\varepsilon_Y^{\lambda_{obs_1}}$ and $set_2 : \Phi_{Y\rightarrow X}^{\lambda_{isos_2}}$, $\Phi_{X\rightarrow Y}^{\lambda_{isos_2}}$, and $\varepsilon_Y^{\lambda_{obs_2}}$). The quantum yields are supposed, a priori, to be wavelength-dependent, based on our discussion in Chap. 4. A kinetic solution based on the assumption that the quantum yield is wavelength-independent is not reliable.

However, it is valid to reduce by one the number of unknowns when considering a single observation wavelength, $\lambda_{obs_1} = \lambda_{obs_2}$ (i.e. $\varepsilon_Y^{\lambda_{obs_1}} = \varepsilon_Y^{\lambda_{obs_2}}$). Hence, we end up with six equations to solve for five unknowns. A review of the equations of the irreducible set indicates that only Eq. 10.44 is dependent on $\varepsilon_Y^{\lambda_{obs}}$ so that the two equations Eq. 10.44 obtained for the two isosbestic irradiations are functions of three unknowns, and therefore, they cannot be solved. Also, they can replace, by their ratio, Eq. 10.45, a non-linear equation of the type of Eq. 10.42.

$$\frac{\Phi_{X\rightarrow Y}^{\lambda_{isos_1}}}{\Phi_{X\rightarrow Y}^{\lambda_{isos_2}}} = \frac{r_{0,A}^{\lambda_{isos_1},T/\lambda_{obs}}}{r_{0,A}^{\lambda_{isos_2},T/\lambda_{obs}}} \tag{10.45}$$

The total number of linearly-independent equations describing the photoreversible reaction under two independent isosbestic irradiations is then reduced to five, i.e. two Eq. 10.42, two Eq. 10.43, and Eq. 10.45. Because Eqs. 10.42 and 10.45 are non-linear, the five equations will not suffice to solve for the five unknowns, and therefore, an analytical solution of the kinetics (the true kinetic solution) is not reachable. The duplication of the irradiation has very little input with regard to solving the kinetics.

In general, increasing the number of the isosbestic irradiations increases both the number of unknowns (by two for each extra isosbestic irradiation (i.e. the two quantum yields) if the observation wavelength is kept the same), $n_{unk} = 3 + 2\,(n_{isos} - 1)$, and the total number of equations, $n_{eq} = 3\,n_{isos} - 1$, for $n_{isos} > 1$. It is obvious that $n_{eq} > n_{unk}$ from $n_{isos} \geq 3$. Table 10.3 gives an account of n_{eq} and n_{unk} for up to six independent isosbestic irradiations.

Because of the non-linearity of Eqs. 10.42 and 10.45, the kinetics for irradiation with $n_{isos} = 3$ is most likely not solvable. The occurrence of three or more isosbestic points in the spectrum of a reaction is relatively rare, so it is experimentally relatively difficult to perform the reaction under such conditions. The discussion here is rather a theoretical case of solving this kinetics under multiple successive isosbestic irradiations, but its complete investigation falls beyond the scope of this chapter.

Table 10.3 List of the numbers of equations (n_{eq}) and unknowns (n_{unk}) for *XY*(2*Φ*) reaction subjected to multiple independent isosbestic irradiations (n_{isos})

n_{isos}	Eq. 10.42	Eq. 10.43	Eq. 10.44	Eq. 10.45	n_{eq}	n_{unk}
1	1	1	1	0	3	3
2	2	2	2	1	5	5
3	3	3	3	2	8	7
4	4	4	4	3	11	9
5	5	5	5	4	14	11
6	6	6	6	5	17	13

10.6.6 The Elucidation Method

The above discussion clearly shows the impossibility in solving the kinetics of *XY*(2*Φ*) reaction when only the total absorbance trace is available (in addition to the quantities listed in Sect. 10.6.3). Hence, there is a need for extra information on the kinetics. This could be to know the numerical value of one of the three unknowns, e.g. $\varepsilon_Y^{\lambda_{obs}}$ at a given observation wavelength, which is, generally, not affordable for this particular quantity.

A more accessible way remains the measurement of the concentration trace of the reactant using a separation method such as high performance liquid chromatography (*HPLC*). To do so, it is mandatory to first have the calibration graph of the reactant constructed by spectrophotometry and its linear range specified. The initial concentration used for the photolysis experiment and the determination of the reactant kinetic trace must conform to that linearity range found by spectrophotometry (because the linearity ranges obtained by different analytical chemistry techniques, e.g. *HPLC* and *UV-Vis*, can differ). The linearity range determined by spectrophotometry has ascendance because, in order to be valid, the absorbed light formula requires a consistent value of the absorbance (i.e. falling within the linearity range limits determined by *UV-Vis* spectroscopy). The calibration graph of *X* is still required to be separately constructed by *HPLC* so as to be used for the determination of $C_X^{\lambda_{isos},T}(t)$ values during photolysis (making up the reactant concentration trace). An alternative and perhaps easier option for gaining extra information would be to determine only the value of $C_X^{\lambda_{isos},T}(pss)$ (instead of the full trace).

In either case, the values of $C_Y^{\lambda_{isos},T}(t)$ or $C_Y^{\lambda_{isos},T}(pss)$ are worked out from the mass balance equation (Eq. 10.5), $C_X^{\lambda_{isos},T}(0)$, and the values obtained for $C_X^{\lambda_{isos},T}(t)$. In the following, we will develop the elucidation method for the simpler option of measuring $C_X^{\lambda_{isos}}(pss)$ only.

Knowing $C_X^{\lambda_{isos},T}(pss)$ and $C_Y^{\lambda_{isos},T}(pss)$ will allow the determination of $\varepsilon_Y^{\lambda_{obs}}$ from the equation (Eq. 10.5) of the total absorbance recorded at *pss*.

The absorptivity spectrum of the photoproduct *Y* can then be fully reconstructed by using the same equation and a screening over the values of λ_{obs}, as given by

$$\varepsilon_Y^{\lambda_{obs}} = \frac{A_{tot}^{\lambda_{isos},T/\lambda_{obs}}(pss) - \varepsilon_X^{\lambda_{obs}}\, l_{obs} C_X^{\lambda_{isos},T}(pss)}{C_Y^{\lambda_{isos},T}(pss)\, l_{obs}} \tag{10.46}$$

The value of $\varepsilon_Y^{\lambda_{obs}}$ facilitates the determination of the forward quantum yield value from Eq. 10.44 as

$$\Phi_{X\to Y}^{\lambda_{isos}} = \frac{r_{0,A}^{\lambda_{isos},T/\lambda_{obs}}}{l_{obs}\, \varepsilon_X^{\lambda_{isos}}\, P_0^{\lambda_{isos}}\, l_{irr}\, PKF^{\lambda_{isos}}\, C_X(0)\, (\varepsilon_Y^{\lambda_{obs}} - \varepsilon_X^{\lambda_{obs}})} \tag{10.47}$$

And finally the quantum yield of the reverse reaction can be obtained either from Eq. 10.43 as

$$\Phi_{Y\to X}^{\lambda_{isos}} = \frac{k_r^{\lambda_{isos},T}}{\varepsilon_X^{\lambda_{isos}}\, P_0^{\lambda_{isos}}\, l_{irr}\, PKF^{\lambda_{isos}}} - \Phi_{X\to Y}^{\lambda_{isos}} \tag{10.48}$$

or from Eq. 10.42 since $K_{eq}^{\lambda_{isos}}$ is calculated by Eq. 10.35.

$$\Phi_{Y\to X}^{\lambda_{isos}} = \frac{\Phi_{X\to Y}^{\lambda_{isos}}}{K_{eq}^{\lambda_{isos}}} \tag{10.49}$$

The elucidation method through Eqs. 10.47, 10.48 and 10.49 will yield the true kinetic solution for *XY*(2*Φ*) kinetics.

10.6.7 Application Examples

The elucidation method developed above for *XY*(2*Φ*) reactions has been successfully applied to several photoreactive reversible systems driven by isosbestic irradiations, including, sunitinib [12], montelukast [13], fluvoxamine [14], pinosylvin [15], and oxyresveratrol [16]. The full reactant concentration traces and *pss* concentrations were obtained by *HPLC*. The $C_X^{\lambda_{isos},T}(t)$ traces were fitted to Eq. 10.28. The optical path length for observation was $l_{obs} = 1\ cm$, whereas $l_{irr} = 2\ cm$ was used for the molecules‘ photolysis experiments, except for that of fluvoxamine ($l_{irr} = 1\ cm$). A sample of the kinetic parameters and the kinetic solutions are presented in Table 10.4.

Table 10.4 True kinetic solutions and reaction features of a few $XY(2\Phi)$ photosystems under isosbestic irradiations. Elucidation performed according to the method displayed in Sect. 10.6.6

Drug	$\lambda_{isos}/\lambda_{obs}$ $/nm/nm$	$C_X^{\lambda_{isos}}(0)$ $/M$	$\varepsilon_X^{\lambda_{isos}}$ $/M^{-1}cm^{-1}$	$P_0^{\lambda_{isos}} \times 10^6$ $/einsten\ dm^{-3}\ s^{-1}$	$k^{\lambda_{isos}} \times 10^3$ $/s^{-1}$	$\varepsilon_Y^{\lambda_{isos}}$ $/M^{-1}cm^{-1}$	$\Phi_{X\to Y}^{\lambda_{isos}}$	$\Phi_{Y\to X}^{\lambda_{isos}}$
Sunitinib	345/445	7.53×10^{-5}	4021	2.040	3.50	3863	0.0200	0.064
Montelukast	258/345	4.70×10^{-5}	23,003	0.643	1.17	23,762	0.0650	0.024
Fluvoxamine	226/245	1.37×10^{-5}	14,254	0.609	0.197	13,802	0.0038	0.00157
Fluvoxamine	285/245	2.25×10^{-5}	1774	0.470	0.480	586	0.2167	0.0844

10.6.8 Extension to XY(1k, 2Φ) and XY(2Φ, 1k) Photothermal Reactions

To distinguish between the two possible cases, let $XY(1k, 2\Phi)$ be the mechanistic option for which the thermal reaction-step corresponds to a forward reaction ($k^T_{X\rightarrow Y}$) and the notation $XY(2\Phi, 1k)$ when the thermal reaction-step is a reverse reaction ($k^T_{Y\rightarrow X}$).

We consider here as well that only the total absorbance traces are available. The numerical values of the rate-constants of the thermal and the photothermal reactions are determined from the fitting of the traces to equations that are equivalent to Eq. 10.21.

The kinetic elucidation method for $XY(1k, 2\Phi)$ reactions is straightforward if the absorptivity of the reactant is determined at a relatively low temperature by spectrophotometry, (assuming it to be temperature independent), otherwise the elucidation of this kinetic will the same as for the second case, XY(2Φ,1k) shown below, but starting from the product of XY(1k,2Φ). Firstly, the solution of the reactant is left in the dark for a long enough period at a constant temperature to lead to its total transformation into the thermally stable product (with $C_Y^{\lambda_{isos},T}(\infty) = C_X^{\lambda_{isos},T}(0)$). Hence, the spectrum of the product, $\varepsilon_Y^{\lambda_{obs}}$, can be recorded, and likewise, for different initial concentrations of *X*, its calibration graph can be constructed. Next, the *pss* equation of the total absorbance and the mass balance will be solved for the *pss* concentrations of *X* and *Y*. The ratio of the latter, together with the rate-constant of the photochemical reaction (that is better determined starting from an isosbestic irradiation of the thermally produced *Y*), is finally used to work out the values of forward and reverse quantum yields (i.e. solving the system of two linear equations, Eqs. 10.50 and 10.51).

$$\Phi^{\lambda_{isos}}_{X\rightarrow Y} + k^T_{X\rightarrow Y} = K^{\lambda_{isos}}_{eq}\,\Phi^{\lambda_{isos}}_{Y\rightarrow X} \tag{10.50}$$

$$\Phi^{\lambda_{isos}}_{Y\rightarrow X} + \Phi^{\lambda_{isos}}_{X\rightarrow Y} = \frac{k_r^{\lambda_{isos},T} - k^T_{X\rightarrow Y}}{\varepsilon_X^{\lambda_{isos}}\,P_0^{\lambda_{isos}}\,l_{irr}PKF^{\lambda_{isos}}} \tag{10.51}$$

Conversely, the kinetics of $XY(2\Phi, 1k)$ cannot be elucidated when only the total absorbance traces are available, in a similar way to that reviewed for $XY(2\Phi)$ kinetics. So let us suppose here, as we did for the photoreversible reaction, that $C_X^{\lambda_{isos}}(pss)$ can be obtained (e.g. determined at a constant temperature using an adequate analytical means, such as nuclear magnetic resonance, *NMR*). In these conditions, the elucidation method developed for $XY(2\Phi)$ is applicable to $XY(2\Phi, 1k)$ but with the equations relative to this system, i.e. that include the thermal rate-constant, $k^T_{Y\rightarrow X}$, such as

$$k_r^{\lambda_{isos},T} = \left(\Phi_{Y\rightarrow X}^{\lambda_{isos}} + \Phi_{X\rightarrow Y}^{\lambda_{isos}}\right) \varepsilon_X^{\lambda_{isos}} P_0^{\lambda_{isos}} l_{irr} PKF^{\lambda_{isos}} + k_{Y\rightarrow X}^T \tag{10.52}$$

$$\Phi_{X\rightarrow Y}^{\lambda_{isos}} = K_{eq}^{\lambda_{isos}} \left(\Phi_{Y\rightarrow X}^{\lambda_{isos}} + k_{Y\rightarrow X}^T\right) \tag{10.53}$$

10.6.9 The Case of XY(2k, 2Φ) *and* XY(2k, 1Φ) *Systems*

For these reactions, it is not possible to establish the calibration graph for the reactant due to its thermal instability. This removes the essential tool, allowing us to quantify the reactant concentration, especially at the thermal equilibrium or *pss* (unless the measurement by the analytical technique, e.g. *NMR*, is performed while the reaction is being irradiated [17]). This represents an important experimental hurdle that makes the kinetics of these reactions hard to solve.

10.7 The Trimolecular $XY_2(2\Phi, 2k)$ Kinetics

Trimolecular photothermal reactions are ubiquitous in photochemistry [18–21]. Nonetheless, little effort was conceded in the literature to establish standard methods to solve their kinetics [22–24]. The present and next sections review elucidation methods that were proposed for some trimolecular systems under monochromatic/isosbestic irradiation.

10.7.1 Descriptive Equations, Unknowns, and the Kinetic Elucidation of $XY_2(2\Phi, 2k)$ *Reactions*

The $XY_2(2\Phi, 2k)$ mechanistic options have been proposed for a number of photoactive systems [18–22]. In general terms, the kinetics is defined by seven unknown parameters, namely, two rate-constants of the thermal reaction-steps, two photochemical quantum yields, and the absorptivities of the three species. The elucidation method described below [8] applies to those mechanistic sequences that do not include a reversible thermal process. The procedure also assumes that the initial concentrations of the starting materials are known to the experimentalist.

$$C_{tot}^{\lambda_{isos},T}(0) = C_X^{\lambda_{isos},T}(0) + C_{Y_1}^{\lambda_{isos},T}(0) + C_{Y_2}^{\lambda_{isos},T}(0) \tag{10.54}$$

The total absorbance, measured at λ_{obs}, of a first-order type kinetics is worked out according to the formalism laid out in Chap. 8. The general expression given by Eq. 8.61 is used here to fit the experimental traces of both photothermal ($A_{tot}^{\lambda_{isos},T/\lambda_{obs}}(t)$) and purely thermal ($A_{tot}^{\lambda_{obs},T}(t)$) reactions.

The set of fitting parameters for the thermal traces, which generally start when the light is switched off, include the rate-constants ρ_+^T and ρ_-^T of the corresponding reaction-steps. The latter makes part of the rate-constants of the photothermal reactions.

The coefficients of $a_1 - a_{21}$ will be defined for the selected $XY_2(2\Phi, 2k)$ mechanistic option by the expressions provided in Chap. 8. The numerical values of coefficients $a_{19}^{\lambda_{isos}}$ and $a_{20}^{\lambda_{isos}}$ can be calculated by using the fitting data as

$$a_{19}^{\lambda_{isos}} = \rho_-^{\lambda_{isos}} + \rho_+^{\lambda_{isos}} \quad (10.55)$$

$$a_{20}^{\lambda_{isos}} = -\rho_-^{\lambda_{isos}} \rho_+^{\lambda_{isos}} \quad (10.56)$$

This pair of simultaneous equations can be solved for the two unknown quantum yields (since the numerical values of ρ_+^T and ρ_-^T are already known from the fitting of the total absorbance trace of the thermal reaction). If Eq. 10.56 is linear, the solution of the pair equation system is unique and easily derived (Eq. 10.55 is always linear). However, if this equation is non-linear (e.g. quadratic), several solutions might be found. In this case, all the solutions are retained for further analysis and testing.

For each of these solutions, the numerical values of the corresponding $a_1 - a_9$ coefficients are defined for both thermal and photothermal reactions. This means that, at this stage of the elucidation method, all kinetic parameters, including rate-constants and quantum yields, are numerically defined before any quantification of the spectroscopic data of the three compounds (X, Y_1, and Y_2) or the concentrations at the equilibrium state are determined.

The next step consists of defining the *pss* concentrations. The cumulative total absorption of species Y_1 and Y_2 measured at *pss*, as defined in Eq. 8.61, is

$$A_{tot}^{\lambda_{isos},T/\lambda_{obs}}(pss) = C_{Y_1}^{\lambda_{isos},T}(pss)\, \varepsilon_{Y_1}^{\lambda_{obs}}\, l_{irr} + C_{Y_2}^{\lambda_{isos},T}(pss)\, \varepsilon_{Y_2}^{\lambda_{obs}}\, l_{irr} \quad (10.57)$$

which can also be expressed as the following ratio

$$\begin{aligned} A_{tot}^{\lambda_{isos},T/\lambda_{obs}}(pss) = -\frac{a_{21}^{\lambda_{isos}}}{a_{20}^{\lambda_{isos}}} &= \frac{a_1^{\lambda_{isos}}\, a_6^{\lambda_{isos}} - a_3^{\lambda_{isos}}\, a_4^{\lambda_{isos}}}{a_{20}^{\lambda_{isos}}}\, C_{tot}^{\lambda_{isos},T}(0)\, \varepsilon_{Y_1}^{\lambda_{obs}}\, l_{irr} \\ &+ \frac{a_1^{\lambda_{isos}}\, a_8^{\lambda_{isos}} - a_2^{\lambda_{isos}}\, a_7^{\lambda_{isos}}}{a_{20}^{\lambda_{isos}}}\, C_{tot}^{\lambda_{isos},T}(0)\, \varepsilon_{Y_2}^{\lambda_{obs}}\, l_{irr} \end{aligned} \quad (10.58)$$

Therefore, by analogy between Eqs. 10.57 and 10.58, the species concentrations will be defined as

$$C_{Y_1}^{\lambda_{isos},T}(pss) = \frac{a_1^{\lambda_{isos}}\, a_6^{\lambda_{isos}} - a_3^{\lambda_{isos}}\, a_4^{\lambda_{isos}}}{a_{20}^{\lambda_{isos}}}\, C_{tot}^{\lambda_{isos},T} \quad (10.59)$$

$$C_{Y_2}^{\lambda_{isos},T}(pss) = \frac{a_1^{\lambda_{isos}}\,a_8^{\lambda_{isos}} - a_2^{\lambda_{isos}}\,a_7^{\lambda_{isos}}}{a_{20}^{\lambda_{isos}}}\,C_{tot}^{\lambda_{isos},T}(0) \tag{10.60}$$

Since, the coefficients $a_1 - a_9$ depend on quantum yields and rate-constants of thermal reaction-steps (determined above), the numerical values of all three species concentrations at *pss* can be calculated (their formulae are defined by known, exclusively kinetic quantities, i.e. not incorporating absorptivities). $C_{Y_1}^{\lambda_{isos},T}(pss)$ and $C_{Y_2}^{\lambda_{isos},T}(pss)$ are easily calculated by Eqs. 10.59 and 10.60, which, in turn, allows the *pss* concentration of *X* to be defined from the mass balance equation.

$$C_X^{\lambda_{isos},T}(pss) = C_{tot}^{\lambda_{isos},T}(0) - C_{Y_1}^{\lambda_{isos},T}(pss) - C_{Y_2}^{\lambda_{isos},T}(pss) \tag{10.61}$$

Incidentally, Eqs. 10.58, 10.59 and 10.60 are general equations that also apply to $XY_2(6\Phi, 6k)$ reactions under isosbestic irradiation. Furthermore, another general definition for equilibrium constant emerges from these equations.

$$K_{eq}^{\lambda_{isos}} = \frac{C_{Y_1}^{\lambda_{isos},T}(pss)}{C_{Y_2}^{\lambda_{isos},T}(pss)} = \frac{a_1^{\lambda_{isos}}\,a_6^{\lambda_{isos}} - a_3^{\lambda_{isos}}\,a_4^{\lambda_{isos}}}{a_1^{\lambda_{isos}}\,a_8^{\lambda_{isos}} - a_2^{\lambda_{isos}}\,a_7^{\lambda_{isos}}} \tag{10.62}$$

The final step of the elucidation concerns the determination of the species absorptivities. We assume that the specific spectroscopic features of the three species are unknown. We hence need three linear and linearly-independent equations. Such a set is available if including the two initial velocities of the thermal and the photothermal reactions (Eq. 8.47) and the total absorbance either at the initial time or at the *pss* (Eq. 10.57). These are obtained as fitting parameters of Eq. 8.61 corresponding to the appropriate kinetics. The system of these three linear equations is solved for the three unknown absorptivities.

The elucidation method is complete but the identifiability problem might not have been fully solved due to the possible occurrence of several solutions of the system of Eqs. 10.55 and 10.56. This can be addressed by further testing each solution to the fitting-determined parameters and the experimentally observed features. An example of such a situation is discussed in the following section.

10.7.2 Shedding Light on NPY Kinetics

Naphthopyrans (*NPY*) are typical trimolecular systems endowed with photochromic properties. The colouration emerges after *UV*-irradiation of the reactant. The latter is the thermally stable closed-form isomer (*X*) that photoconverts into two merocyanine-like open forms, featuring *cis* – *trans* (Y_1) and *trans* – *trans* (Y_2) conformations (*CT* and *TT*, respectively, as laid out in Scheme 9.1).

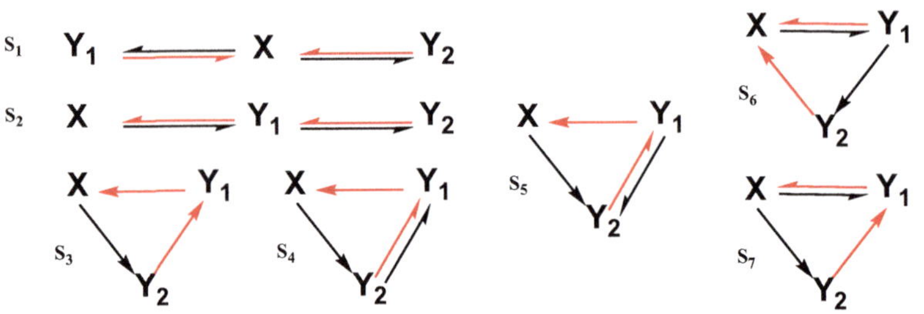

Scheme 10.3 Sets of $XY_2(1\ or\ 2\Phi, 2k)$ sequences for *NPY* (where the thermal reaction-steps appear in red)

The true mechanism governing the *NPY* photothermal reaction is still unknown despite extensive study of this family of photochromes. The only little confirmed information on the mechanism is deduced from the experimental behaviour of the reaction. In essence, the reactant only photoreacts and is thermally regenerated from the coloured species. Mechanistically, this summarises into a mechanism engaging at least three reaction-steps, such as a minimum/irreducible set, i.e. a photochemical step starting from X, and two thermal steps for the bleaching of Y_1 and Y_2 in the dark. As far as we are aware, any extra reaction-step is considered rather as a suggestion since it cannot objectively be confirmed but only proposed on the basis of a qualitative analysis. In this context, the minimum hypothesis proposed for the *NPY* reaction obeyed the $XY_2(1\Phi, 2k)$ mechanism. This minimum hypothesis can be interpreted as a type of application of Occam's razor, where there is no need to consider more reaction-steps to the above reaction option if the latter fits the experimental data of the reaction well. However, this minimum hypothesis has not been tested beyond a good fit of the traces. Because such a good fit cannot provide a definitive proof of the occurrence of the mechanistic option, as previously discussed (Sect. 6.6), the true mechanism is, in fact, still unknown.

To shed light on the validity of the above option, a careful testing is carried out against the elucidation method developed in the previous section. Furthermore, we will proceed with a similar procedure for the slightly more extended $XY_2(2\Phi, 2k)$ mechanism. The point here is to explore the possibility of whether the kinetics of *NPY* could analytically be proven to be described by the $XY_2(2\Phi, 2k)$ mechanism.

Within the general $XY_2(2\Phi, 2k)$ model, we can list seven sub-mechanistic options for *NPY*, including divergent, consecutive, and cyclic mechanisms (Scheme 10.3). The series includes only one $XY_2(1\Phi, 2k)$ sequence.

The photothermal reaction of the *NPY* derivative, 3,3-diphenyl-*3H*-naphtho [2,1-*b*]pyran [8], was experimentally studied in ethyl acetate solutions ($C_X(0) = 5.87\ 10^{-5}\ M$ and 14 °C). Separate experimental traces of photothermal ($\lambda_{isos} = 358\ nm$) and thermal reactions have been recorded at the observation wavelength $\lambda_{obs} = 423\ nm$ (Fig. 10.4). The remaining experimental conditions and set-up are similar to those used in Sect. 9.4 [25]. The analytical signal (i.e. the total absorbance) recorded in the visible region at λ_{obs} is due to the appearance and

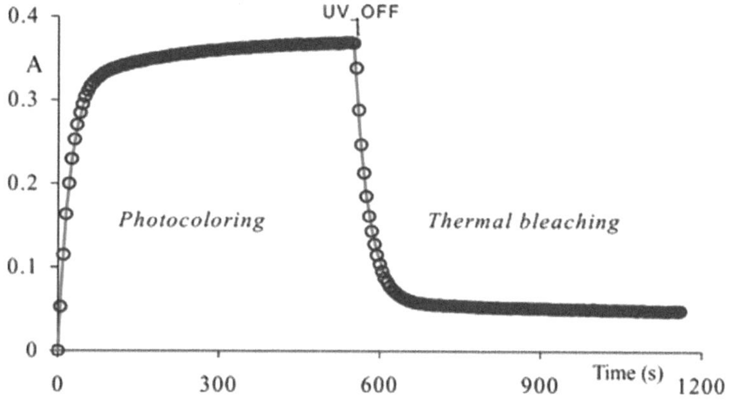

Fig. 10.4 *NPY* photo-colouring and thermo-fading experimental traces (circles) and their fitting (line) with appropriate equations. The reactions were performed in ethyl acetate solutions at 14 °C, where $C_X(0) = 5.87\ 10^{-5}\ M$, $l_{obs} = 1\ cm$, $l_{irr} = 2.1\ cm$, $\lambda_{isos} = 358\ nm$, and $\lambda_{obs} = 423\ nm$. (Reproduced with permission from [8], *International Journal of Chemical Kinetics*)

Table 10.5 Parameters of the trace fitting of *NPY* photothermal reactions and thermo-bleaching in ethyl acetate (Fig. 10.4). Photocolouring under isosbestic irradiation at $\lambda_{isos} = 357\ nm$ and thermal bleaching process undertaken at 14 °C

Drug	$r_{0,A}$ /s^{-1}	$A_{tot}(0)$	$A_{tot}(\infty)$	$-\rho_+$ /s^{-1}	$-\rho_-$ /s^{-1}	r^{2a}
Photothermal reaction	0.015180	0	0.3756	0.04606	0.00385	0.999
Thermal reaction	−0.01109	0.369	0	0.03528	0.00019	0.998

[a]: squared correlation coefficients of experimental against calculated total absorbance traces

disappearance of the coloured isomers Y_1 and Y_2 (the initial species $X = NPY$ being colourless).

The fitting of the traces has been performed by Eq. 8.61 that is adapted to the considered kinetics (i.e. both for light switched on and off). Once established, these equations apply irrespective of the sub-mechanism (Scheme 10.3). The kinetic traces are fitted with high accuracy as clearly shown in Fig. 10.4 and indicated by a value almost unity of the correlation coefficients obtained on a plot of the experimental trace and the data obtained by Eq. 8.61 (Table 10.5). The values of the rest of the fitting parameters of the equations are also given in Table 10.5.

The elucidation method laid out in the previous section is now applied separately to each of the photothermal (*PT*) sequences, S_1^{PT}–S_7^{PT} (Scheme 10.3). Using the data of Table 10.5, together with the values of the coefficients $a_{19}^{\lambda_{isos}}$ and $a_{20}^{\lambda_{isos}}$ (equal to −0.01444 and $-1.77331\ 10^{-4}$, respectively), the kinetic solution, i.e. the numerical values of the kinetic and spectroscopic parameters, are independently determined for each mechanistic option.

The findings divide the set of options into two groups. The first includes S_2^{PT}–S_4^{PT} and S_6^{PT}. At least one of the determined intrinsic parameters for these sequences has a negative numerical value, a feature that requires to rule out the mechanistic options

Table 10.6 Kinetic solutions of *NPY* photothermal S_1^{PT}, S_5^{PT}, and S_7^{PT} reactions

	$\Phi^{\lambda_{isos}}_{X\to Y_1}$	$\Phi^{\lambda_{isos}}_{Y_1\to Y_2}$	$\Phi^{\lambda_{isos}}_{X\to Y_2}$	$C_X(pss)$ /M	$C_{Y_1}(pss)$ /M	$C_{Y_2}(pss)$ /M	$\varepsilon^{423}_{Y_1}$ /M^{-1} cm^{-1}	$\varepsilon^{423}_{Y_2}$ /M^{-1} cm^{-1}
S_1^{PT}	$9.65\ 10^{-3}$	–	$4.78\ 10^{-3}$	$2.219\ 10^{-6}$	$6.072\ 10^{-7}$	$5.587\ 10^{-5}$	23,637	6347
S_5^{PT}	–	$1.07\ 10^{-2}$	$3.69\ 10^{-3}$	$2.219\ 10^{-6}$	$2.322\ 10^{-7}$	$5.625\ 10^{-5}$	70,050	6271
S_7^{PT}	$9.68\ 10^{-3}$	–	$4.76\ 10^{-3}$	$2.219\ 10^{-6}$	$9.082\ 10^{-7}$	$5.557\ 10^{-5}$	23,638	6254

of this first group since the intrinsic parameters' values must be positive. These sequences, including the $XY_2(1\Phi, 2k)$ option featuring the minimum set of reaction-steps, cannot possibly be considered to represent *NPY* dynamics.

The second group comprises the three remaining candidates, one divergent S_1^{PT} and two cyclic S_5^{PT} and S_7^{PT} mechanisms. Their kinetic solutions, involving only positive values for the intrinsic parameters, are reported in Table 10.6. Because these are unique for each sequence, the elucidation method is proven to solve the identifiability problem for each sub-mechanism (S_1^{PT}, S_5^{PT}, and S_7^{PT}). However, because the mathematical framework available for the description of the reaction does not allow us to achieve any further progress in the kinetic analysis, the three mechanistic options (S_1^{PT}, S_5^{PT}, and S_7^{PT}) remain plausible, indicating an evident insoluble distinguishability issue.

The *pss* concentrations of *X* and Y_2 have very similar values for the three options, and the values of the parameters characterising sequences S_1^{PT} and S_7^{PT} are overall quasi-identical (Table 10.6). The latter similarity stems, perhaps, from a close correspondence of the expressions defining the $a_1^{\lambda_{isos}}$ –$a_9^{\lambda_{isos}}$ coefficients of these reaction options.

To evaluate further the plausible sequences, the analysis can rely on a qualitative prospection based on the reasonable occurrence of events and/or by invoking some thermodynamic criteria.

For instance, even if the molecular structures of isomers Y_1 and Y_2 have not been attributed, it is rather unlikely that both the *trans* – *trans* and the *cis* – *trans* open-form isomers undergo identical reactions, as is suggested by S_1^{PT}, Specifically, that their thermal routes exclusively generate the closed-form *NPY*. If this pathway looks reasonable for the *cis* – *trans* isomer, it might be deemed very improbable for its *trans* – *trans* homologue. The reason is the need for the latter, open-form isomer to overcome two successive energy barriers in order to revert to the *NPY* closed-form configuration, where one of these barriers is, most likely, an isomerisation of the *trans* – *trans* isomer into the *cis* – *trans* species. Therefore, it is not realistic to consider the occurrence of a mechanism that contains the *trans* – *trans* → *NPY* transformation without simultaneously encompassing the *trans* – *trans* → *cis* – *trans* reaction-step. Such a sequence of events does not belong to the S_1^{PT} divergent mechanism, and hence, the latter may be excluded from consideration for *NPY* kinetics despite the plausibility of the values (Table 10.6) generated by the elucidation method for the unknown parameters of the reaction. Here, we can underline again that a good fit as well as plausible values of the parameters do not make the kinetic solution necessarily acceptable.

If the above hypothesis, considering that the thermal bleaching reaction must unfold according to the sequence *trans – trans* → *cis – trans* → *NPY*, is correct, then species Y_1 and Y_2 in both S_5^{PT} and S_7^{PT} must be identified as *cis – trans* and *trans – trans*, respectively.

Accordingly, it would mean that the photoreactions of both *NPY* and Y_1 (=*cis – trans*) in S_5^{PT} mechanism exclusively yield the *trans – trans* configuration. This situation is physically unlikely because it seems to prioritise the strained option (*NPY* → *trans – trans*) over much simpler ones (*NPY* → *cis – trans* or *cis – trans* → *NPY*). Therefore, S_5^{PT} is highly improbable to describe the kinetics of *NPY*.

Thus far, S_7^{PT} seems to be the only option remaining among the seven possible options (Scheme 10.3) to describe *NPY* kinetics when it is assumed to belong to an $XY_2(2\Phi, 2k)$ system. S_7^{PT} kinetics predicts that only *NPY* (i.e. *X*) photoreacts (in a divergent reaction) to produce species Y_1 twice as efficiently as Y_2 (Table 10.6). The thermal recyclisation reaction of Y_1 ($Y_1 \rightarrow NPY$) is much faster than the thermal *E*/*Z* isomerisation $Y_2 \rightarrow Y_1$. The ratio of the *pss* concentrations of the coloured species $(C_{Y_2}(pss)/C_{Y_1}(pss))$ is equal to 60, and the ratio of the species absorptivities is $\varepsilon_{Y_2}^{423}/\varepsilon_{Y_1}^{423} \approx 4$. This leads to a larger contribution of species Y_2 to the *pss* observed total absorbance than Y_1.

Although the results obtained so far for S_7^{PT} seem satisfactory, it is, however, mandatory to check the consistency of all equations describing both the photothermal and the thermal reactions before definitely accepting this option as the true *NPY* mechanism. For instance, this is achieved by comparing the values of some fitting parameters with those calculated by using both the theoretical equations of the parameters and the data obtained from the elucidation method (Table 10.6). The calculated values of initial velocity and pre-exponential terms of the fading process are found to be different from those obtained from the fitting of the traces. The explanation of such a discrepancy is the dependency of the set of values obtained for $\varepsilon_{Y_1}^{423}$ and $\varepsilon_{Y_2}^{423}$ on the pair of equations used for that purpose (there are more than two equations available, e.g. $r_{0,\,A}$ and $A_{tot}(\infty)$).

As a result, the kinetics of the *3H*-naphtho[2,1-*b*]pyran derivative is incontestably proven not to obey any of the sequences presented in Scheme 10.3, and consequently, it should be described by a higher-level reaction mechanism than $XY_2(2\Phi, 2k)$. A conclusion that is consistent with a lack of objective reasons ruling out photochemical reactivity of Y_1 and/or Y_2 as has been proven for a *2H*-naphthopyran homologue structure [26] and some spiro-naphthopyran derivatives [22].

10.7.3 Shapes of the Photo-Isomer Absorption Spectra

Perhaps the second information about the *NPY* system, apart from not being described by an $XY_2(2\Phi, 2k)$ system, that can be obtained with certainty from this study is that the absorption spectra of Y_1 and Y_2 have exactly the same overall shape but not necessarily the same absorption coefficients.

This conclusion stems from the fact that linear plots, with intercepts equal to zero (e.g. $r_{0,A}^{\lambda_{obs},T} = \rho_i \, A_{tot}^{\lambda_{isos},T/\lambda_{obs}}(pss)$), are experimentally obtained between each initial velocity ($r_{0,\,A}$) and the *pss* total absorbance ($A_{tot}(pss)$) for the individual wavelength over the $\lambda_{obs} = 380 - 580$ *nm* spectral region. These plots can still be constructed, even if the true *NPY* mechanism and the values of its species absorbancies are unknown since initial velocities and total absorbance at *pss* are available as fitting prameters of the experiemtal traces. The quantities ($Q_i^{\lambda_{obs}} = r_{0,A}$ or $A_{tot}(pss)$) used for these plots depend on the variation of the absorption coefficients of Y_1 and Y_2. The theoretical expressions of $Q_i^{\lambda_{obs}}$ each has the algebraic form of a linear combination of $\varepsilon_{Y_1}^{\lambda_{obs}}$ and $\varepsilon_{Y_2}^{\lambda_{obs}}$, as given by Eq. 10.63 (the reaction kinetic parameters and *pss* concentrations are constant for a given experiment).

$$Q_i^{\lambda_{obs}} = \alpha_i \, \varepsilon_{Y_1}^{\lambda_{obs}} + \delta_i \, \varepsilon_{Y_2}^{\lambda_{obs}} \tag{10.63}$$

The ratios of any two of $r_{0,\,A}$ and $A_{tot}(pss)$, $Q_1^{\lambda_{obs}}$, and $Q_2^{\lambda_{obs}}$ (Eq. 10.64) are constants for all wavelengths as the two quantities correlate in straight lines (with α_i and δ_i as constants).

$$\frac{Q_1^{\lambda_{obs}}}{Q_2^{\lambda_{obs}}} = \frac{\alpha_1 \, \varepsilon_{Y_1}^{\lambda_{obs}} + \delta_1 \, \varepsilon_{Y_2}^{\lambda_{obs}}}{\alpha_2 \, \varepsilon_{Y_1}^{\lambda_{obs}} + \delta_2 \, \varepsilon_{Y_2}^{\lambda_{obs}}} = \frac{\alpha_1 + \delta_1 \left(\varepsilon_{Y_2}^{\lambda_{obs}} / \varepsilon_{Y_1}^{\lambda_{obs}}\right)}{\alpha_2 + \delta_2 \left(\varepsilon_{Y_2}^{\lambda_{obs}} / \varepsilon_{Y_1}^{\lambda_{obs}}\right)} = \rho_{1,2} \tag{10.64}$$

Rearranging the latter equation leads to the ratio of the absorptivities, which is expressed by constants as

$$\frac{\varepsilon_{Y_2}^{\lambda_{obs}}}{\varepsilon_{Y_1}^{\lambda_{obs}}} = \frac{\rho_{1,2} \, \alpha_2 - \alpha_1}{\delta_1 + \rho_{1,2} \, \delta_2} = n \tag{10.65}$$

Therefore, the absorptivity spectra of *NPY* isomers Y_1 and Y_2 have the same shape over the wavelength range considered. This feature agrees with the results of the transient and permanent spectra gained from nanosecond experiments on a 2*H-NPY* analogue [26]. The present approach demonstrates that similar information on the ε-spectra of the coloured species of trimolecular systems can also be obtained from the fitting data of their traces.

10.8 $XY_2(6\Phi, 2k)$ Kinetics Yielding a Mono-Exponential Trace

Continuing with the prospection into trimolecular photothermal kinetics, especially that of naphthopyrans, which should not a priori obey any $XY_2(2\Phi, 2k)$ sub-mechanisms (under isosbestic irradiation) (see previous section) but still has no known or established reaction mechanism, we consider here the particular case when the experimental trace of the irradiated system follows a mono-exponential trend. The experimental data of *NPY* photocolouring traces usually have a bi-exponential behaviour [22], but it was observed that these traces can be described by (or approximated to) a mono-exponential function by an effect of temperature, generally when *T* exceeds 25 °C [27–29]. These investigations have also indicated the need to monitor the reaction over a relatively short period of time (typically less than 100 s) for the approximation to be valid, and *NPY* mechanism was supposed to obey $XY_2(1\Phi, 2k)$ kinetics in a first approximation to later use the data obtained to propose a more realistic proposal for the mechanism [27]. At 30 °C, the time of validity for the approximation was extended to 200 s for the 3,3-diphenyl-3*H*-naphtho[2,1-*b*]pyran derivative in toluene [30]. The kinetic analysis was supported by formulating a theoretical approach to the evaluation of the kinetics of three-component systems, along with the mechanistic $XY_2(6\Phi, 2k)$ options.

10.8.1 Basic Equations

The unknowns of the $XY_2(6\Phi, 2k)$ system are the six photochemical quantum yields, the absorption coefficients of *NPY*, Y_1 and Y_2 isomers, and the two thermal rate-constants. The kinetic traces of both photothermal and thermo-fading reactions corresponding to the cumulative absorbance of the photoproducts are measured at $\lambda_{obs} = 430\ nm$.

The elucidation of the kinetics will then require at least a minimum of ten linearly-independent equations. The basic equations are deduced from the general formalism developed in Chap. 8.

According to our conditions, the trace of the photothermal reaction is a mono-exponential curve, and the integrated rate-law takes the form,

$$A_{tot}^{\lambda_{isos},T/\lambda_{obs}}(t) = \left(A_{Y_2}^{\lambda_{isos},T/\lambda_{obs}}(0) + \frac{a_{12}^{\lambda_{isos}}}{a_{11}^{\lambda_{isos}}} \right) e^{a_{11}^{\lambda_{isos}}\, t} - \frac{a_{12}^{\lambda_{isos}}}{a_{11}^{\lambda_{isos}}} \tag{10.66}$$

where for *NPY* kinetics, the initial absorbance is equal to zero at $\lambda_{obs} = 430\ nm$, and the coefficients $a_{11}^{\lambda_{isos}}$ and $a_{12}^{\lambda_{isos}}$ are constants expressed as

$$a_{11}^{\lambda_{isos}} = a_6^{\lambda_{isos}} \frac{\varepsilon_{Y_1}^{\lambda_{obs}}}{\varepsilon_{Y_2}^{\lambda_{obs}}} + a_9^{\lambda_{isos}} - \left(a_8^{\lambda_{isos}} \frac{\varepsilon_{Y_1}^{\lambda_{obs}}}{\varepsilon_{Y_2}^{\lambda_{obs}}} + a_5^{\lambda_{isos}} - a_6^{\lambda_{isos}} \frac{\varepsilon_{Y_1}^{\lambda_{obs}}}{\varepsilon_{Y_2}^{\lambda_{obs}}} - a_9^{\lambda_{isos}} \right) \frac{\varepsilon_{Y_1}^{\lambda_{obs}}}{\left(\varepsilon_{Y_2}^{\lambda_{obs}} - \varepsilon_{Y_1}^{\lambda_{obs}}\right)} \tag{10.67}$$

$$a_{12}^{\lambda_{isos}} = \left(a_8^{\lambda_{isos}} \frac{\varepsilon_{Y_1}^{\lambda_{obs}}}{\varepsilon_{Y_2}^{\lambda_{obs}}} + a_5^{\lambda_{isos}} - a_6^{\lambda_{isos}} \frac{\varepsilon_{Y_1}^{\lambda_{obs}}}{\varepsilon_{Y_2}^{\lambda_{obs}}} - a_9^{\lambda_{isos}} \right) \frac{\varepsilon_{Y_1}^{\lambda_{obs}} \varepsilon_{Y_2}^{\lambda_{obs}}}{\left(\varepsilon_{Y_2}^{\lambda_{obs}} - \varepsilon_{Y_1}^{\lambda_{obs}}\right)} C_X^{\lambda_{isos}}(0) \tag{10.68}$$

An excellent fit of the experimental traces with Eq. 10.66 was obtained, yielding the values of the fitting parameters as $a_{11}^{\lambda_{isos}} = -0.1483\ s^{-1}$ and $a_{12}^{\lambda_{isos}} = 0.0119\ s^{-1}\ cm^{-1}$. Equations 10.67 and 10.68 form the first set of equations required for the elucidation method. A second set is provided by the coefficients $a_1^{\lambda_{isos}}$ -$a_9^{\lambda_{isos}}$ (in s^{-1}), which are linked together by the following three expressions.

$$a_1^{\lambda_{isos}} = -a_4^{\lambda_{isos}} - a_7^{\lambda_{isos}} \tag{10.69}$$

$$a_5^{\lambda_{isos}} = -a_2^{\lambda_{isos}} - a_8^{\lambda_{isos}} \tag{10.70}$$

$$a_9^{\lambda_{isos}} = -a_3^{\lambda_{isos}} - a_6^{\lambda_{isos}} \tag{10.71}$$

Conforming with a mono-exponential photocolouring kinetics, the only plausible option to represent an XY_2 system, kinetically behaving as an XY reaction (i.e. the variations of the three absorbances are mono-exponential trends with time), is when the coefficient $a_{10}^{\lambda_{isos}} = 0$. The latter feature leads to an additional equation of the elucidation set.

$$a_4^{\lambda_{isos}} \varepsilon_{Y_1}^{\lambda_{obs}} + a_7^{\lambda_{isos}} \varepsilon_{Y_2}^{\lambda_{obs}} = \left(a_8^{\lambda_{isos}} \frac{\varepsilon_{Y_1}^{\lambda_{obs}}}{\varepsilon_{Y_2}^{\lambda_{obs}}} + a_5^{\lambda_{isos}} - a_6^{\lambda_{isos}} \frac{\varepsilon_{Y_1}^{\lambda_{obs}}}{\varepsilon_{Y_2}^{\lambda_{obs}}} - a_9^{\lambda_{isos}} \right) \frac{\varepsilon_{Y_1}^{\lambda_{obs}} \varepsilon_{Y_2}^{\lambda_{obs}}}{\left(\varepsilon_{Y_2}^{\lambda_{obs}} - \varepsilon_{Y_1}^{\lambda_{obs}}\right)} \tag{10.72}$$

The last set is the series of $a_{19}^{\lambda_{isos}}$ -$a_{21}^{\lambda_{isos}}$ coefficients, defined under the working condition ($a_{10}^{\lambda_{isos}} = 0$) within the theoretical model laid out in Chap. 8 and the known values of $a_{11}^{\lambda_{isos}}$ and $a_{12}^{\lambda_{isos}}$.

$$a_{19}^{\lambda_{isos}} = a_1^{\lambda_{isos}} + a_5^{\lambda_{isos}} + a_9^{\lambda_{isos}} = 2\, a_{11}^{\lambda_{isos}} \tag{10.73}$$

$$a_{20}^{\lambda_{isos}} = a_7^{\lambda_{isos}} \left(a_3^{\lambda_{isos}} - a_2^{\lambda_{isos}}\right) + a_8^{\lambda_{isos}} \left(a_1^{\lambda_{isos}} - a_3^{\lambda_{isos}}\right) + a_9^{\lambda_{isos}} \left(a_2^{\lambda_{isos}} - a_1^{\lambda_{isos}}\right) = \left(a_{11}^{\lambda_{isos}}\right)^2 \tag{10.74}$$

$$a_{21}^{\lambda_{isos}} = \left(\left(a_3^{\lambda_{isos}} a_4^{\lambda_{isos}} - a_1^{\lambda_{isos}} a_6^{\lambda_{isos}}\right) \varepsilon_{Y_1}^{\lambda_{obs}} + \left(a_2^{\lambda_{isos}} a_7^{\lambda_{isos}} - a_1^{\lambda_{isos}} a_8^{\lambda_{isos}}\right) \varepsilon_{Y_2}^{\lambda_{obs}} \right) C_X^{\lambda_{isos}}(0) = -a_{11}^{\lambda_{isos}} a_{12}^{\lambda_{isos}} \tag{10.75}$$

The thermal relaxation trace of the *NPY* derivative under study [30] has a mono-exponential trend but does not reach a zero value at $t = \infty$. This feature was interpreted as a depletion of the product Y_1, present at the *pss*, whose thermal rate-constant is noticeably higher than that of the other Y_2 isomer so that, in contrast to the photocolouring analysis, a condition $a_{10}^T = 0$ is not applicable for the thermal kinetics because $A_{tot}^{\lambda_{obs},T}(\infty) \neq 0$ (incidentally, the latter indicates a good fatigue resistance, i.e. Y_2 is thermally stable for the time span used in that study). The appearance of species Y_2 in the reaction medium can either or both be due to the thermal conversion of Y_1 ($k_{Y_1 \rightarrow Y_2}^T$) or to either photochemical conversions $\Phi_{X \rightarrow Y_2}^{\lambda_{isos}}$ or $\Phi_{Y_1 \rightarrow Y_2}^{\lambda_{isos}}$. However, the kinetic and spectroscopic data of the reaction at hand do not allow us to know which species (X, Y_1, or both) are responsible for its production. In addition, if we suppose that the thermal rate-constant, $k_{Y_1 \rightarrow Y_2}^T = 0$, the residual absorbance recorded after the fading process of isomer Y_1 corresponds to the amount of Y_2 that has been formed at the *pss* during the photocolouring process. However, several studies have implicitly assumed $k_{Y_1 \rightarrow Y_2}^T = 0$ as a working hypothesis [27, 28, 31–33], although there is no available experimental evidence to prove that $A_{Y_2}^{\lambda_{isos},T/\lambda_{obs}}(pss) = A_{Y_2}^{\lambda_{obs},T}(\infty)$. We will adapt in the following discussion the same assumption by considering that the thermal isomerisation of *NPY* isomer Y_1 into Y_2 does not take place (i.e. $k_{Y_1 \rightarrow Y_2}^T = 0$) at 30°C. We make use of the experimental value of the absorbance recorded at the end of the thermal reaction, $A_{Y_2}^{\lambda_{isos},T/\lambda_{obs}}(pss) = A_{tot}^{\lambda_{obs},T}(\infty) = A_{Y_2}^{\lambda_{obs},T}(\infty) = 0.0165$, to establish an additional equation expressed in terms of the same coefficients used so far.

$$\left(a_2^{\lambda_{isos}} a_7^{\lambda_{isos}} - a_1^{\lambda_{isos}} a_8^{\lambda_{isos}}\right) \varepsilon_{Y_2}^{\lambda_{obs}} = a_{20}^{\lambda_{isos}} \frac{A_{Y_2}^{\lambda_{obs},T}(\infty)}{C_X^{\lambda_{obs},T}(0)} \tag{10.76}$$

The last equation to define the system is derived from the expression of the rate at the photostationary state, $\left(dC_X^{\lambda_{isos},T}(t)/dt\right)_{pss} = 0$,

$$\left(a_2^{\lambda_{isos}} - a_1^{\lambda_{isos}}\right)\left(a_3^{\lambda_{isos}} a_4^{\lambda_{isos}} - a_1^{\lambda_{isos}} a_6^{\lambda_{isos}}\right) + \left(a_3^{\lambda_{isos}} - a_1^{\lambda_{isos}}\right)\left(a_2^{\lambda_{isos}} a_7^{\lambda_{isos}} - a_1^{\lambda_{isos}} a_8^{\lambda_{isos}}\right) - a_{20}^{\lambda_{isos}} a_1^{\lambda_{isos}} = 0 \tag{10.77}$$

Thus far, we have defined 11 linearly-independent equations (Eqs. 10.67, 10.68, 10.69, 10.70, 10.71, 10.72, 10.73, 10.74, 10.75, 10.76 and 10.77) to solve for the eight unknowns of our *NPY* kinetics (the thermal rate-constant $k_{Y_1 \rightarrow X}^T$ is obtained from the fitting of the thermal reaction trace, Y_2 is thermally stable, $k_{Y_2 \rightarrow Y_1}^T = 0$, and we assumed that $k_{Y_1 \rightarrow Y_2}^T = 0$). The excess over the quorum in terms of equations (equations/variable number, $11 > 8$) might be an advantage for the elucidation process. To improve the level of refinement of the final results and information about *NPY* kinetics, a number of extra equations are derived. These are not linearly-independent from the above set of 11 equations but must be verified by the kinetic solutions as extra constraints on the elucidation criteria.

One of these is obtained when considering $\left(dC_{Y_1}^{\lambda_{isos}}(t)/dt\right)_{pss}=0.$

$$a_4^{\lambda_{isos}}\left(a_2^{\lambda_{isos}}a_9^{\lambda_{isos}}-a_3^{\lambda_{isos}}a_8^{\lambda_{isos}}\right)-a_5^{\lambda_{isos}}\left(a_3^{\lambda_{isos}}a_4^{\lambda_{isos}}-a_1^{\lambda_{isos}}a_6^{\lambda_{isos}}\right)-a_6^{\lambda_{isos}}\left(a_2^{\lambda_{isos}}a_7^{\lambda_{isos}}-a_1^{\lambda_{isos}}a_8^{\lambda_{isos}}\right)=0 \quad (10.78)$$

Equation 10.77 can be worked out with a different set of coefficients (Eq. 10.79) without alteration of the theoretical meaning.

$$a_{13}^{\lambda_{isos}}a_{20}^{\lambda_{isos}}\left(a_2^{\lambda_{isos}}a_9^{\lambda_{isos}}-a_3^{\lambda_{isos}}a_8^{\lambda_{isos}}\right)C_X^{\lambda_{isos},\mathrm{T}}(0)+a_{12}^{\lambda_{isos}}a_{14}^{\lambda_{isos}}-a_{11}^{\lambda_{isos}}+a_{15}^{\lambda_{isos}}=0 \quad (10.79)$$

where the $a_{13}^{\lambda_{isos}}$–$a_{15}^{\lambda_{isos}}$ coefficients are not independent variables themselves and are defined by the following expressions.

$$a_{13}^{\lambda_{isos}}=a_1^{\lambda_{isos}}-\left(\frac{a_2^{\lambda_{isos}}}{\varepsilon_{Y_1}^{\lambda_{obs}}}-\frac{a_3^{\lambda_{isos}}}{\varepsilon_{Y_2}^{\lambda_{obs}}}\right)\frac{\varepsilon_{Y_1}^{\lambda_{obs}}\varepsilon_{Y_2}^{\lambda_{obs}}}{\varepsilon_{Y_2}^{\lambda_{obs}}-\varepsilon_{Y_1}^{\lambda_{obs}}} \quad (10.80)$$

$$a_{14}^{\lambda_{isos}}=\frac{a_3^{\lambda_{isos}}+a_{13}^{\lambda_{isos}}-a_1^{\lambda_{isos}}}{\varepsilon_{Y_2}^{\lambda_{obs}}} \quad (10.81)$$

$$a_{15}^{\lambda_{isos}}=\left(a_1^{\lambda_{isos}}-a_{13}^{\lambda_{isos}}\right)C_X^{\lambda_{isos},T}(0) \quad (10.82)$$

The above equations are all numerically defined, except Eqs. 10.69, 10.70, 10.71 and 10.72.

10.8.2 The Elucidation

Mechanisms involving less than three quantum yields are solved analytically but they are found unfit for *NPY* kinetics, as the values of the intrinsic reactions' parameters they yielded were inconsistent, a feature confirming a similar, previously obtained conclusion from the analysis undertaken in Sect. 10.7.2 that showed that the true photomechanism of *NPY* is surely a different sequence compared to $XY_2(2\Phi, 2k)$ [8].

A further elucidation of *NPY* kinetics using the 11 above equations and under the assumptions employed therein leads to a total of seven (S_8^{PT} and S_{14}^{PT}) discriminated kinetic solutions, including an $XY_2(6\Phi, 1k)$, three $XY_2(5\Phi, 1k)$, and three $XY_2(4\Phi, 1k)$ sequences (Table 10.7).

These results indicate that when no preconceived mechanism is imposed, the mathematical framework allowed a reduction of the number of the possible kinetic solutions but do not necessarily lead to the single "true" mechanism. It is obvious that resolving the distinguishability problem is not achievable. If we adhere to the discussion on the thermal mechanism provided above (only $k_{Y_1\rightarrow X}^T$ is different from zero, whereas $k^T_{Y_2 \to Y_1} = k^T_{Y_1 \to Y_2} = 0$), then the mechanistic options S^{PT}_{12} and S^{PT}_{13} have to be excluded because they do not allow the formation of Y_2. This observation might point to the weakness of the assumptions proposed for the thermal mechanism (e.g. $k^T_{Y_1 \to Y_2}$ might have a non-zero value).

A review of the data laid out in Table 10.7 shows that two photoreaction-steps are present in all the mechanistic options, namely, $X \to Y_1$ and $Y_2 \to X$. These are presumably present in the true *NPY* mechanism. Also, the quantum yield of the latter photochemical step ($\Phi^{\lambda_{isos}}_{Y_2 \to X}$) is invariably characterised by the highest value of the photochemical quantum yields, irrespective of the number of photochemical steps involved in the actual sequence. This particular reaction-step ($Y_2 \to X$) as well as the $Y_1 \to X$ step were not considered in a preconceived *NPY* mechanism involving three, $X \to Y_1$, $Y_1 \to Y_2$, and $Y_2 \to Y_1$, photochemical steps but without putting forward experimental evidence in support of that sequence [27]. However, another study [28] concluded that the excited-state of *NPY* leads to both Y_1 and Y_2 (excluding the exclusivity of the formation of Y_2 from its open-form counterpart, $Y_1 \to Y_2$) and the possibility of generating Y_2, Y_1, and *NPY* from the deactivation of the excited-state Y_2^* (i.e. the *TT* isomer, where *CT* was attributed to Y_1).

The data of Table 10.7 agree with most of the conclusions reached by the latter study, where an occurrence of step $Y_2 \to X$ should allow us to assume the occurrence of similar processes from Y_1, $Y_1 \to X$, and possibly, $Y_1 \to Y_2$. Such mechanistic considerations are in agreement with some results relative to several photochromic systems, e.g. spirooxazine, spiropyran, and benzopyran derivatives [22], where at least the $Y_1 \to X$ photochemical step is accepted to take place.

Table 10.7 The seven ($S^{PT}_8 - S^{PT}_{14}$) kinetic solutions predicted for *NPY* kinetics, producing a mono-exponential trace of the photocolouring reactions in toluene at 30 °C

	S^{PT}_8	S^{PT}_9	S^{PT}_{10}	S^{PT}_{11}	S^{PT}_{12}	S^{PT}_{13}	S^{PT}_{14}
$\Phi^{\lambda_{isos}}_{X \to Y_1}$	0.02.00	0.0229	0.0059	0.0122	0.0013	0.0017	0.0297
$\Phi^{\lambda_{isos}}_{Y_1 \to X}$	0.0148	0.0074	–	0.0309	0.0391	0.0392	–
$\Phi^{\lambda_{isos}}_{Y_1 \to Y_2}$	0.0085	–	0.0395	0.0049	0.0027	–	–
$\Phi^{\lambda_{isos}}_{Y_2 \to Y_1}$	0.0184	0.0419	0.0054	–	–	0.0019	0.0388
$\Phi^{\lambda_{isos}}_{X \to Y_2}$	0.0060	0.0047	0.0093	0.0012	–	–	0.0150
$\Phi^{\lambda_{isos}}_{Y_2 \to X}$	0.1218	0.106	0.1294	0.1404	0.1404	0.1459	0.1061
$\varepsilon^{423}_{Y_1}/M^{-1}\ cm^{-1}$	4.295 10^3	3.127 10^3	1.462 10^4	7.326 10^3	6.779 10^4	5.017 10^4	2.560 10^3
$\varepsilon^{423}_{Y_2}/M^{-1}\ cm^{-1}$	3.369 10^3	5.912 10^3	2.000 10^3	1.312 10^4	5.024 10^4	2.541 10^4	2.000 10^3
$C_X(pss)/M$	9.31 10^{-5}	8.96 10^{-5}	1.00 10^{-4}	1.03 10^{-4}	1.03 10^{-4}	1.11 10^{-4}	7.97 10^{-5}
$C_{Y_1}(pss)/M$	1.49 10^{-5}	2.06 10^{-5}	4.40 10^{-6}	8.78 10^{-6}	9.48 10^{-7}	1.28 10^{-6}	2.51 10^{-5}
$C_{Y_2}(pss)/M$	4.89 10^{-6}	2.79 10^{-6}	8.24 10^{-6}	1.22 10^{-6}	3.28 10^{-7}	6.48 10^{-7}	8.24 10^{-6}

Another confirmation of a dual radiationless decay of excited NPY^*, such as $Y_1 \leftarrow X \rightarrow Y_2$, is provided by the theoretical basis of the conical intersection mechanism [34, 35]. Computational studies have shown that ultra-fast photochemical reactions (with life spans of hundreds of *fs*) are most likely to take place via energy surfaces crossing through a nodal or intersection point that represents a structure of a common intermediate (e.g. between an excited- and a ground-state) that can lead to more than a one possible photoproduct. This general perspective can be anticipated for the photomechanism of *NPY*, which is neither fluorescent nor phosphorescent [28, 36], and no triplet state was proven to operate in its photoreaction [22, 37]. The absence of competition between the *NPY* singlet-state deactivation to Y_1 and Y_2 and photophysical events or intersystem crossing fulfils rather the requirements for a conical intersection mechanism than those for an avoided crossing. This is supported by a femtosecond investigation of *NPY* [37] that strongly suggested the completion of that molecule's photochemical processes within picoseconds after the initial excitation pulse, indicating a singlet-state pathway favouring the concomitant formation of both Y_1 and Y_2 (with the final spectrum of the reactive medium, measured after a 100 *ps*, being the same as the one obtained by continuous irradiation). The fast *NPY* photochemistry corroborates the information of Chap. 4.

According to the above discussion, the likelihood of the $Y_1 \rightarrow X$ and $Y_2 \rightarrow Y_1$ occurrence automatically discards the mechanistic options S_{10}^{PT}, S_{11}^{PT}, and S_{14}^{PT} from the series laid out in Table 10.7. We are hence left with only S_8^{PT} and S_9^{PT} as possible mechanisms to interpret *NPY* kinetics.

The overall analysis proves that *NPY* kinetics must, most probably, be more complex than predicted in the literature through preconceived mechanisms, and the information available on this dynamics is far less than sufficient to allow a complete and consistent analysis, ultimately leading to the true mechanism of *NPY*.

Although no literature offered an $XY_2(6\Phi, 1k)$ proposal of *NPY* mechanism, there are, nonetheless, examples of $XY_2(5\Phi)$ and $XY_2(4\Phi)$ that have been suggested for the photoisomerisations of di(α-naphthyl)ethylene [38] and dithienylrethene photochromes [39], respectively.

10.9 Experiment Design for Analytical Kinetic Solutions for $XY_2(3-5\Phi, 2k)$ Systems

The difficulty in reaching a kinetic solution for trimolecular XY_2 systems has prompted a research effort to propose alternative ways to reach a solution without considering a particular preconceived mechanism for the reaction at hand. In this context, the starting point for such kinetic problems is to adequately address a shortage in the number of the mostly non-linear equations describing the reactive system, in comparison to the number of the reaction's unknown parameters, a situation that generally results in unsolvable kinetics. For the specific case of *NPY* kinetics, it was only possible to develop a method capable of analytically solving the

$XY_2(3 - 5\Phi, 2k)$ kinetics (Sect. 10.8). Accordingly, an attempt to solve the imbalance of the ratio equations'/unknowns' numbers has been proposed on the basis of an experiment design [40]. Here, we will review a theoretical model that was established for $XY_2(3 - 5\Phi, 2k)$ reactions by putting to advantage the variation of the total absorbance measured at the *pss* when the temperature of the medium is changed. Such data usually serve the determination of the activation energy but do not find use for kinetic analytical purposes. The $XY_2(3 - 5\Phi, 2k)$ reactions are considered under isosbestic irradiation but which neither include thermally reversible reaction-steps nor have a mono-exponential trace for the photothermal reaction (as seen in the last section). It is relevant to emphasise that such XY_2 systems, including $XY_2(3 - 5\Phi, 2k)$ reactions, have neither been solved analytically in the literature nor by using numerical integration methods [2, 10, 41, 42]. As a consequence, the true mechanisms governing many photochromes (including but not limited to fulgides, naphthopyrans, and diarylethenes) have remained, thus far, incompletely characterised (because assumptions were usually introduced).

10.9.1 The Unknowns

The kinetics of $XY_2(3 - 5\Phi, 2k)$ photothermal reactions is characterised by up to nine unknowns (two rate-constants of consecutive thermal reaction-steps, two absorption coefficients of the photoproducts, and up to five quantum yields). The choice of two consecutive thermal reaction-steps is dictated by a practical consideration imposed by spectrophotometry, which is usually employed for these kinds of kinetic studies and allows us to unambiguously determine, at most, two rate-constant values from bi-exponential kinetic traces that correspond to two thermal relaxations. Whenever the mechanism of the thermal reaction encompasses more than two thermal reactions, the resulting exponential factors are no longer specific to individual reaction-steps (but rather represent complex formulations and combinations of all rate-constants, e.g. Eq. 9.9).

10.9.2 The Knowns

The present theoretical method requires that the reactant is thermally stable and photoactive (e.g. $\Phi_{X \to Y_1}^{\lambda_{isos}} \neq 0$) and that its initial reactant concentration, the incident radiation intensity, the full spectrum of the initial species, the irradiated volume and area of the sample, and the optical path lengths of irradiation and observation are all known. Also, we assume that both absorption coefficients and quantum yields involved in the laid-out expressions describing the reactions must be invariant over temperature changes (within the range selected for the investigation). The traces are acquired by spectrophotometry and look, overall, like the example shown in Fig. 10.4.

10.9.3 Fitting Equations

The total absorbance traces for the photothermal ($A_{tot}^{\lambda_{isos},T/\lambda_{obs}}(t)$) and purely thermal ($A_{tot}^{\lambda_{obs},T}(t)$) reactions are fitted by the general expression, Eqs. 8.60. 8.62, which is adapted to each reaction (e.g. $A_{tot}^{\lambda_{obs},T}(\infty) = 0$).

10.9.4 Simulated Patterns of $A_{tot}^{\lambda_{isos},T/\lambda_{obs}}(pss) = f(T)$

The variation of $A_{tot}^{\lambda_{isos},T/\lambda_{obs}}(pss)$ with temperature is theoretically expected to occur since both the denominator ($a_{20}^{\lambda_{isos}}$) and the numerator ($a_{21}^{\lambda_{isos}}$) of its formula, Eq. 10.58, are dependent on the thermal rate-constants (ρ_{+}^{T} and ρ_{-}^{T}), which are themselves dependent on temperature (Fig. 10.5).

In order to verify the variation of the *pss* absorbance with temperature, several reactions with different sets of fundamental parameters were considered over a specific range of temperature. Parabolic, sinusoidal, and apparently linear trends of $A_{tot}^{\lambda_{isos},T/\lambda_{obs}}(pss) = f(T)$ could be observed, with the latter characterised by high correlation coefficient values for both types of lines, with either negative or positive slopes [40]. Such a behaviour (inverse proportionality, as $A_{tot}^{\lambda_{isos},T/\lambda_{obs}}(pss) = \alpha T^{-1}$) was experimentally observed for photochromic naphthopyran derivatives [43].

10.9.5 A Set of New Equations

The effect of temperature on the *pss* total absorption is mainly evaluated by Eq. 10.58; the coefficients of this equation (Eqs. 8.33 and 8.34) can be expressed as a linear combination of rate-constants and their products. The inverse of Eq. 10.58 for the general case $XY_2(6\Phi, 2k)$, whose thermal reaction is $X \leftarrow Y_1 \leftarrow Y_2$, can be expressed as

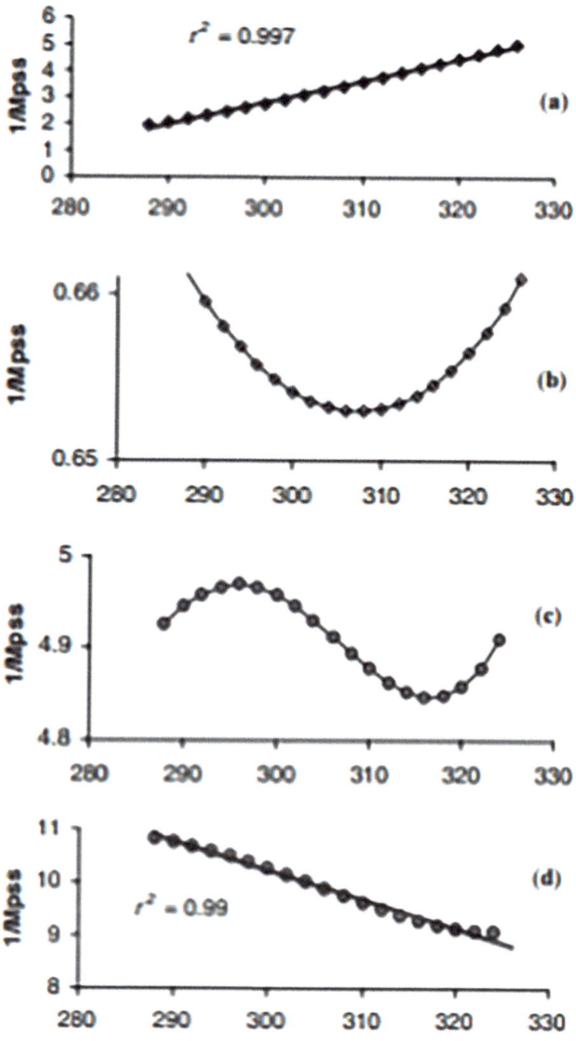

Fig. 10.5 Different patterns of the variation of the reverse total absorbance of coloured *NPY* isomers at pss with wavelength. (Reproduced with permission from [40], *International Journal of Chemical Kinetics*)

$$\frac{1}{A_{tot}^{\lambda_{isos},T/\lambda_{obs}}(pss)} = -\frac{a_{20}^{\lambda_{isos}}}{a_{21}^{\lambda_{isos}}} = \frac{\alpha_{20}\,k_{Y_1\to X}^{T} + \beta_{20}\,k_{Y_2\to Y_1}^{T} + \delta_{20}\,\gamma}{\alpha_{21}\,k_{Y_1\to X}^{T} + \beta_{21}\,k_{Y_2\to Y_1}^{T} + \delta_{21}\,\gamma} + \frac{k_{Y_1\to X}^{T}\,k_{Y_2\to Y_1}^{T}}{\alpha_{21}\,k_{Y_1\to X}^{T}\,\gamma + \beta_{21}\,k_{Y_2\to Y_1}^{T}\,\gamma + \delta_{21}\,\gamma^2} \quad (10.83)$$

with the thermal rate-constants, $-\rho_+^T = k_{Y_1\to X}^T$ and $-\rho_-^T = k_{Y_2\to Y_1}^T$, obtained from the fitting of the thermal reaction trace; the constant γ_1 is given by Eq. 8.13, and the numerical values of coefficients $a_{20}^{\lambda_{isos}}$ and $a_{21}^{\lambda_{isos}}$ can be calculated from the fitting parameters as given by Eqs. 10.84 and 10.85,

$$a_{20}^{\lambda_{isos}} = -\rho_{+}^{PT}\,\rho_{-}^{PT} \tag{10.84}$$

$$a_{21}^{\lambda_{isos}} = A_{tot}^{\lambda_{isos},T/\lambda_{obs}}(pss)\,\rho_{+}^{PT}\,\rho_{-}^{PT} \tag{10.85}$$

and the temperature-independent coefficients α_i, β_i, and δ_i have the following expressions.

$$\alpha_{20} = \Phi_{X\to Y_2}^{\lambda_{isos}} + \Phi_{Y_2\to X}^{\lambda_{isos}} + \Phi_{Y_2\to Y_1}^{\lambda_{isos}} \tag{10.86}$$

$$\beta_{20} = \Phi_{X\to Y_1}^{\lambda_{isos}} + \Phi_{Y_1\to X}^{\lambda_{isos}} + \Phi_{X\to Y_2}^{\lambda_{isos}} \tag{10.87}$$

$$\begin{aligned}\delta_{20} = {} & \Phi_{X\to Y_2}^{\lambda_{isos}}\,\Phi_{Y_1\to X}^{\lambda_{isos}} + \Phi_{Y_2\to X}^{\lambda_{isos}}\left(\Phi_{X\to Y_1}^{\lambda_{isos}} + \Phi_{Y_1\to X}^{\lambda_{isos}}\right)\\ & + \Phi_{Y_2\to Y_1}^{\lambda_{isos}}\left(\Phi_{X\to Y_1}^{\lambda_{isos}} + \Phi_{Y_1\to X}^{\lambda_{isos}} + \Phi_{X\to Y_2}^{\lambda_{isos}}\right)\\ & + \Phi_{Y_1\to Y_2}^{\lambda_{isos}}\left(\Phi_{X\to Y_1}^{\lambda_{isos}} + \Phi_{X\to Y_2}^{\lambda_{isos}} + \Phi_{Y_2\to X}^{\lambda_{isos}}\right)\end{aligned} \tag{10.88}$$

$$\alpha_{21} = \Phi_{X\to Y_2}^{\lambda_{isos}}\,\varepsilon_{Y_2}^{\lambda_{obs}}\,C_{X}^{\lambda_{isos}}(0) \tag{10.89}$$

$$\beta_{21} = \left(\Phi_{X\to Y_1}^{\lambda_{isos}} + \Phi_{X\to Y_2}^{\lambda_{isos}}\right)\varepsilon_{Y_1}^{\lambda_{obs}}\,C_{X}^{\lambda_{isos}}(0) \tag{10.90}$$

$$\begin{aligned}\delta_{21} = {} & \left(\Phi_{X\to Y_2}^{\lambda_{isos}}\,\Phi_{Y_1\to X}^{\lambda_{isos}} + \Phi_{Y_1\to Y_2}^{\lambda_{isos}}\left(\Phi_{X\to Y_1}^{\lambda_{isos}} + \Phi_{X\to Y_2}^{\lambda_{isos}}\right)\right)\varepsilon_{Y_2}^{\lambda_{obs}}\,C_{X}^{\lambda_{isos}}(0)\\ & + \left(\Phi_{X\to Y_1}^{\lambda_{isos}}\,\Phi_{Y_2\to X}^{\lambda_{isos}} + \Phi_{Y_2\to Y_1}^{\lambda_{isos}}\left(\Phi_{X\to Y_1}^{\lambda_{isos}} + \Phi_{X\to Y_2}^{\lambda_{isos}}\right)\right)\varepsilon_{Y_1}^{\lambda_{obs}}\,C_{X}^{\lambda_{isos}}(0)\end{aligned} \tag{10.91}$$

The model developed here is specific to the reaction indicated above (whose thermal reaction is a consecutive process starting at Y_2 and ending at the reactant), but it is possible to generate the adequate equations for any $XY_2(6\Phi, 2k)$ sequence, provided that its thermal trace has a bi-exponential character.

The above six equations (involving eight unknowns) might be useful for elucidation, but the process can be tedious to handle owing to the non-linearity of most of these equations. So in order to alleviate the mathematical treatment, a linearisation is proposed. This is achieved by using the a_{19} coefficients,

$$a_{19}^{T} = \rho_{+}^{T} + \rho_{-}^{T} = -k_{Y_1\to X}^{T} - k_{Y_2\to Y_1}^{T} \tag{10.92}$$

$$\begin{aligned}a_{19}^{\lambda_{isos}} = {} & \rho_{+}^{PT} + \rho_{-}^{PT}\\ = {} & -\left(\Phi_{X\to Y_1}^{\lambda_{isos}} + \Phi_{Y_1\to X}^{\lambda_{isos}} + \Phi_{Y_1\to Y_2}^{\lambda_{isos}} + \Phi_{Y_2\to Y_1}^{\lambda_{isos}} + \Phi_{X\to Y_2}^{\lambda_{isos}} + \Phi_{Y_2\to X}^{\lambda_{isos}}\right)\gamma\\ & - k_{Y_1\to X}^{T} - k_{Y_2\to Y_1}^{T}\end{aligned} \tag{10.93}$$

to define the numerical value of the sum of the reaction quantum yields as

$$\Delta_{a_{19}} = \frac{a_{19}^{T} - a_{19}^{\lambda_{isos}}}{\gamma_1} = \Phi_{X\to Y_1}^{\lambda_{isos}} + \Phi_{Y_1\to X}^{\lambda_{isos}} + \Phi_{Y_1\to Y_2}^{\lambda_{isos}} + \Phi_{Y_2\to Y_1}^{\lambda_{isos}} + \Phi_{X\to Y_2}^{\lambda_{isos}} + \Phi_{Y_2\to X}^{\lambda_{isos}} \tag{10.94}$$

Combining Eq. 10.94 with Eqs. 10.86, 10.87 and 10.88 leads to a series of six linear equations. The left-hand side terms of these equations are numerically defined from the available fitting and experimental data.

$$\Delta_{a_{19}} - \alpha_{20} = \Phi_{X\to Y_1}^{\lambda_{isos}} + \Phi_{Y_1\to X}^{\lambda_{isos}} + \Phi_{Y_1\to Y_2}^{\lambda_{isos}} \tag{10.95}$$

$$\Delta_{a_{19}} - \beta_{20} = \Phi_{Y_1\to Y_2}^{\lambda_{isos}} + \Phi_{Y_2\to Y_1}^{\lambda_{isos}} + \Phi_{Y_2\to X}^{\lambda_{isos}} \tag{10.96}$$

$$\Delta_{a_{19}} - \alpha_{20} - \beta_{20} = \Phi_{Y_1\to Y_2}^{\lambda_{isos}} + \Phi_{X\to Y_2}^{\lambda_{isos}} \tag{10.97}$$

$$\mu = \frac{\delta_{20} - \alpha_{20}\,\beta_{20}}{\Delta_{a_{19}} - \alpha_{20} - \beta_{20}} = \Phi_{X\to Y_1}^{\lambda_{isos}} + \Phi_{X\to Y_2}^{\lambda_{isos}} + \Phi_{Y_2\to X}^{\lambda_{isos}} \tag{10.98}$$

$$\mu - \alpha_{20} = \Phi_{X\to Y_1}^{\lambda_{isos}} + \Phi_{Y_2\to Y_1}^{\lambda_{isos}} \tag{10.99}$$

$$\mu - \beta_{20} = \Phi_{Y_1\to X}^{\lambda_{isos}} + \Phi_{Y_2\to X}^{\lambda_{isos}} \tag{10.100}$$

10.9.6 Extracting the Values of α_i, β_i, and δ_i Coefficients

For each temperature (T_i) of the range selected, the thermal relaxation and bi-exponential traces are fitted, and the respective rate-constant values, $k_{Y_1\to X}^{T_i}$ and $k_{Y_2\to Y_1}^{T_i}$, are determined. To ensure reliability and a high degree of accuracy of the results of the subsequent steps, it is recommended that the range considered encompasses as many temperature data points as possible.

The obtained table of rate-constant values is used to conduct separate multilinear correlations on either $a_{20}^{\lambda_{isos}}$ and $a_{21}^{\lambda_{isos}}$, given by Eqs. 10.101 and 10.102, or on Eq. 10.83. This treatment results in assigning a numerical value to each of the six α_i, β_i, and δ_i coefficients (hence defining Eqs. 10.95, 10.96, 10.97, 10.98, 10.99 and 10.100).

$$a_{20}^{\lambda_{isos}} = -\alpha_{20}\, k_{Y_1\to X}^{T}\,\gamma - \beta_{20}\, k_{Y_2\to Y_1}^{T}\,\gamma - \delta_{20}\,\gamma^2 - k_{Y_1\to X}^{T}\, k_{Y_2\to Y_1}^{T} \tag{10.101}$$

$$a_{21}^{\lambda_{isos}} = \alpha_{21}\, k_{Y_1\to X}^{T}\,\gamma + \beta_{21}\, k_{Y_2\to Y_1}^{T}\,\gamma + \delta_{21}\,\gamma^2 \tag{10.102}$$

We will proceed now to the kinetic elucidation. In the following, several solving procedures are presented individually for specific reactive systems, and when no analytical solution can be reached, alternative ways for those particular cases will be suggested.

10.9.7 The $XY_2(2\Phi, 2k)$ Systems

We have seen that the kinetics of $XY_2(2\Phi, 2k)$ can be analytically solved without the need for multiple measurements at different temperatures (Sect. 10.7). But in the present context, this kinetics can also be solved by using Eqs. 10.95, 10.96, 10.97, 10.98, 10.99 and 10.100. An $XY_2(2\Phi, 2k)$ mechanism means that the reactant photoreacts (i.e. $\Phi^{\lambda_{isos}}_{X\to Y_1} \neq 0$) and Y_2 is produced (i.e. either $\Phi^{\lambda_{isos}}_{X\to Y_2} \neq 0$ or $\Phi^{\lambda_{isos}}_{Y_1\to Y_2} \neq 0$). The four remaining quantum yields are equal to zero. This translates into two mechanistic options of the photochemical reaction, either $S^{PT}_{15}: Y_1 \leftarrow X \to Y_2$ or $S^{PT}_{16}: X \to Y_1 \to Y_2$, both overlapped with a unique thermal reaction $X \leftarrow Y_1 \leftarrow Y_2$. For S^{PT}_{15} ($\Phi^{\lambda_{isos}}_{Y_1\to X} = \Phi^{\lambda_{isos}}_{Y_2\to X} = \Phi^{\lambda_{isos}}_{Y_1\to Y_2} = \Phi^{\lambda_{isos}}_{Y_2\to Y_1} = 0$), Eqs. 10.97 and 10.99 deliver $\Phi^{\lambda_{isos}}_{X\to Y_2}$ and $\Phi^{\lambda_{isos}}_{X\to Y_1}$, respectively. In the case of S^{PT}_{16} ($\Phi^{\lambda_{isos}}_{Y_1\to X} = \Phi^{\lambda_{isos}}_{X\to Y_2} = \Phi^{\lambda_{isos}}_{Y_2\to X} = \Phi^{\lambda_{isos}}_{Y_2\to Y_1} = 0$), the photochemical quantum yields $\Phi^{\lambda_{isos}}_{X\to Y_1}$ and $\Phi^{\lambda_{isos}}_{Y_1\to Y_2}$ can be calculated by using the same equations, Eqs. 10.99 and 10.97, respectively.

10.9.8 Solving $XY_2(3\Phi, 2k)$ Reactions

These systems, involving three quantum yields and two absorptivities, can be fully solved using the series of equations available. To facilitate the elucidation, $XY_2(3\Phi, 2k)$ reactions will be divided into two groups of sub-mechanisms.

In the first group, the three missing quantum yields (which are equal to zero) are evenly distributed between Eqs. 10.97, 10.99, and 10.100. In this case, the values of the reaction (non-zero) quantum yields will easily be worked out from the latter equations. The absorption coefficients ($\varepsilon^{\lambda_{obs}}_{Y_1}$ and $\varepsilon^{\lambda_{obs}}_{Y_2}$) are then determined from any two equations among the Eqs. 10.89, 10.90 and 10.91 series.

The second group of mechanisms encompasses sequences where two of the missing quantum yields belong to one equation among the last series (either Eqs. 10.97, 10.99 or 10.100), and hence, that particular equation takes a zero value (because both its quantum yields have zero values). However, in accordance with our thermal reaction ($X \leftarrow Y_1 \leftarrow Y_2$, as postulated here), the equation equating zero is certainly not Eq. 10.97 because if $\Phi^{\lambda_{isos}}_{Y_1\to Y_2} = \Phi^{\lambda_{isos}}_{X\to Y_2} = 0$, Y_2 is not formed in solution (which contradicts the bi-exponential trend of the thermal reaction). Also, that equation might not be Eq. 10.97 if we consider that the reactant forms the *TT* species (i.e. $\Phi^{\lambda_{isos}}_{X\to Y_2} \neq 0$). In any case, from one of the two remaining equations, which must contain the third quantum yield, the first of the three unknown photochemical quantum yields of this reactive system is defined. The remaining equation of the series could well be a difference of two unknown quantum yields and, hence, will not be solvable by itself. In such a configuration, we make use of Eq. 10.94, which is equal to the sum of the two remaining unknown quantum yields (since four of its six terms are now known). Solving this pair of equations will lead to exact values of the final parameters. From here, $\varepsilon_{Y_1}^{\lambda_{obs}}$ and $\varepsilon_{Y_2}^{\lambda_{obs}}$ are easily worked out by using again the series of equations described above.

This overall procedure is clearly capable of solving any sequence derived from the $XY_2(3\Phi, 2k)$ reaction (whose thermal trace is bi-exponential).

10.9.9 Elucidation of $XY_2(4\Phi, 2k)$ Kinetics

For these reactions, we can distinguish three mechanistic groups. The first group includes the sequences for which the two zero-valued quantum yields fit in two different equations of the series, Eqs. 10.95, 10.99, and 10.100. These cases are solved in the same way as was described for the second group of $XY_2(3\Phi, 2k)$ mechanisms seen in the previous section.

The $XY_2(4\Phi, 2k)$ reactions of the second group are characterised by $\Phi_{Y_1 \to X}^{\lambda_{isos}} = \Phi_{Y_2 \to X}^{\lambda_{isos}} = 0$ or Eq. 10.100 equal to zero. The elucidation is then derived by using the two equations, Eqs. 10.97 and 10.99 (which are dependent on four unknowns); the two equations expressing α_{21} and β_{21} (Eqs. 10.89 and 10.90); and the initial reaction rate measured for the photochemical reaction, Eq. 10.103.

$$r_{0,A}^{\lambda_{isos},T/\lambda_{obs}} = \left(\Phi_{X \to Y_1}^{\lambda_{isos}} \varepsilon_{Y_1}^{\lambda_{obs}} + \Phi_{X \to Y_2}^{\lambda_{isos}} \varepsilon_{Y_2}^{\lambda_{obs}}\right) C_X^{\lambda_{isos},T}(0)\, l_{obs} \tag{10.103}$$

The numerical value of the ratio of the forward quantum yields (Eq. 10.104) is worked out from a combination of the latter equations.

$$\frac{\Phi_{X \to Y_1}^{\lambda_{isos}}}{\Phi_{X \to Y_2}^{\lambda_{isos}}} = \frac{r_{0,A}^{\lambda_{isos},T/\lambda_{obs}} - \alpha_{21}}{\beta_{21} - \left(r_{0,A}^{\lambda_{isos},T/\lambda_{obs}} - \alpha_{21}\right) C_X^{\lambda_{isos},T}(0)} \tag{10.104}$$

Introducing this equation in Eqs. 10.97 and 10.99 removes one unknown so that they are now expressed by only three unknowns. The third linear equation provided by Eq. 10.94 allows the complete elucidation of this second group of reaction.

The last group of $XY_2(4\Phi, 2k)$ reactions includes the sequences for which either Eq. 10.97 or Eq. 10.99 is equal to zero (i.e. the quantum yields of Eq. 10.100 are both non-zero). This means that either $\Phi_{X \to Y_1}^{\lambda_{isos}}$ or $\Phi_{X \to Y_2}^{\lambda_{isos}}$ is necessarily equal to zero, and hence, Eq. 104 cannot be used. Unfortunately, the elucidation of this type of kinetics cannot be achieved analytically, where its six unknowns can only be achieved through iterative calculations of eight linear and non-linear equations (Eqs. 10.89, 10.90, 10.91, 10.94, 10.97 or 10.98, 10.99, 10.100, and 10.103). In the case where $\Phi_{X \to Y_2}^{\lambda_{isos}} = 0$, the latter equation set reduces to six since $\alpha_{21} = 0$ and $\beta_{21} = r_{0,A}^{\lambda_{irr},T/\lambda_{obs}}$.

10.9.10 *Elucidation of* $XY_2(5\Phi, 2k)$ *Kinetics*

The numerical value of one quantum yield is obtained from one of the linear equations, Eqs. 10.97, 10.99, or 10.100, and the two remaining ones are expressed by four non-zero quantum yields. The mathematical situation is analogous to the two last groups of $XY_2(4\Phi, 2k)$ discussed in the previous section. Hence, for the sequences whose both forward quantum yields $\Phi_{X\to Y_1}^{\lambda_{isos}}$ and $\Phi_{X\to Y_2}^{\lambda_{isos}}$ are non-zero, an analytical solution can be devised. But if $\Phi_{X\to Y_1}^{\lambda_{isos}} \neq 0$ and $\Phi_{X\to Y_2}^{\lambda_{isos}}$ is non-zero, then a computational procedure is required on the basis of the eight-equations system described above, with either Eq. 10.95 or 10.96 added to the computed set of equations for the same reasons discussed above for $XY_2(4\Phi, 2k)$ kinetics.

10.9.11 *Elucidation of* $XY_2(6\Phi, 2k)$ *Kinetics*

It is obvious that for $XY_2(6\Phi, 2k)$ reactions, a full analytical elucidation is not achievable, even though there are eight equations dependent on eight unknowns. Numerical integration can be used, but it is important to underline that by applying such procedure, there is a realistic possibility of facing identifiability issues for any kinetic system found for $XY_2(6\Phi, 2k)$ and $XY_2(5\Phi, 2k)$ or $XY_2(4\Phi, 2k)$ kinetics.

10.10 Concluding Remarks

Analytical solutions were attempted for a majority of trimolecular reactions obeying $XY_2(6\Phi, 2k)$ sub-mechanisms. Some are straightforward, and others need some experimental design.

The approach adopted here aims to avoid the preconception of the mechanism. Rather, it displays a methodology by which to search for the mechanistic options that might fit a set of kinetic data acquired for a trimolecular photothermal reaction. The output might not be unique.

There are unsolvable cases in the present approach, which, perhaps, will be solved if another, more adequate experiment design is proposed. Also, if this approach solves the identifiability problems for the solvable cases, it might, overall, lead to identifying (but not solving) a distinguishability problem.

The advantage of the approach presented here is to give an example of kinetic analysis, and to underline that a good fit of the experimental (or simulated data), in addition to finding plausible values for the reaction parameters, are two fundamental (and necessary) but not sufficient conditions to unravel the true kinetics governing a photothermal reaction.

References

1. Maafi M, Maafi W (2015) Quantification of unimolecular photoreaction kinetics: determination of quantum yields and development of actinometers - the photodegradation case of cardiovascular drug nisoldipine. Int J Photoenergy 454895. https://doi.org/10.1155/2015/454895
2. Brown GH (1971) Techniques of chemistry, Vol. III, Photochromism. Wiley, New York. ISBN. 978-0471928942
3. Sakugari M, Aoki K, Tamaki T, Ichimura K (1990) The role of triplet state of nitrospiropyran in their photochromic reaction. Bull Chem Soc Japan 63:74–79. https://doi.org/10.1246/bcsj.63.74
4. Gorner H (2001) Photochromism of nitrospiropyrans: effects of structure, solvent and temperature. Phys Chem Chem Phys 3:416–423. https://doi.org/10.1039/B007708I
5. Deniel MH, Lavabre D, Micheau JC (1999) Topics in applied chemistry. In: Crano JC, Guglielmetti RJ (eds) Organic photochromic and thermochromic compounds, vol 1. Plenum Press, New York. ISBN. 978-0-306-45883-5
6. Rini M, Holm A-K, Nibbering ETJ, Fidder H (2003) Ultrafast vibrational relaxation processes induced by intramolecular excited state hydrogen transfer. J Am Chem Soc 125:3028–3034. https://doi.org/10.1016/S0009-2614(03)00650-X
7. Maafi M, Brown RG (2007) An analytical solution for the kinetics of $AB(1k, 1\Phi)$ systems and its application to a spirobenzopyran. Int J Chem Kinet 39:539–545. https://doi.org/10.1002/kin.20267
8. Maafi M, Brown RG (2006) A general procedure to uniquely identify the kinetic and spectroscopic parameters of $ABC(2\Phi, 2k)$ systems and application to naphthopyran reactions. Int J Chem Kinet 38:421–430. https://doi.org/10.1002/kin.20192
9. Appriou P, Guglielmetti R, Garnier F (1978) Étude du processus photochimique impliqué dans la réaction d'ouverture du cycle benzopyrannique des spiropyrannes photochromiques. J Photochem 8:145–165. https://doi.org/10.1016/0047-2670(78)80016-1
10. Bouas-Laurent H, Durr H (2001) Organic photochromism. Pure Appl Chem 73:639–665. https://doi.org/10.1351/pac200173040639
11. Van Gemert B (1999) Chapter 3: Benzo and naphthopyrans (chromenes). In: Crano JC, Guglielmetti RJ (eds) Organic photochromic and thermochromic compounds - topics in applied chemistry, vol 1. Plenum Press, New York, pp 111–140. ISBN. 978-0-306-45883-5 and 978-0-306-45882-8
12. Maafi M, Lee L-Y (2015) Actinometric and Φ-order photodegradation properties of anti-cancer Sunitinib. J Pharm Biomed Anal 110:34–41. https://doi.org/10.1016/j.jpba.2015.02.026
13. Maafi M, Maafi W (2014) Montelukast photodegradation: elucidation of K-order kinetics, determination of quantum yields and application to actinometry. Int J Pharm Sci 471:544–452. https://doi.org/10.1016/j.ijpharm.2014.05.016
14. Maafi M, Maafi W (2015) Quantitative assessment of photostability and photostabilisation of Fluvoxamine and its design for actinometry. Photochem Photobiol Sci 14:982–994. https://doi.org/10.1039/c5pp0022j
15. Maafi M, Al-Qarni MA (2019) Dependence of pinosylvin photochemical quantum yield on excitation wavelength. Trends Photochem Photobiol 18:45–58
16. Maafi M, Al-Qarni MA (2018) Φ-order spectrophotokinetic characterisation and quantification of trans-cis oxyresveratrol reactivity, photodegradation and actinometry. Spectrochim Acta A Mol Biomol Spectrosc 188:64–71. https://doi.org/10.1016/j.saa.2017.06.067
17. Naito A, Makino Y, Tasei Y, Kawamura I (2018) Photoirradiation and microwave irradiation NMR spectroscopy. In: The Nuclear Magnetic Resonance Society of Japan (ed) Experimental approaches of NMR spectroscopy. Springer, Singapore. https://doi.org/10.1007/978-981-10-5966-7_5
18. Cowen DO, Drisko RL (1976) Elements of organic photochemistry. Plenum Press, New York. https://doi.org/10.1007/978-1-4684-2130-9

19. Turro NJ (1978) Modern molecular photochemistry. The Benjamin/Cummings Publishing Company, San Francisco, CA. ISBN: 0-8053-9354-4
20. El'tsov AV (1990) Organic photochromes. Consultants Bureau, New York. ISBN: 978-1461585879
21. Durr H, Bouas-Laurent H (1990) Photochromism. Elsevier, Amsterdam. ISBN:0-444-87432-1
22. Crano JC, Guglielmetti RJ (eds) (1999) Organic photochromic and thermochromic compounds - topics in applied chemistry, vol 1 and 2. Plenum Press, New York. ISBN: 978-0-306-45883-5 and 978-0-306-45882-8
23. Borderie B, Lavabre D, Micheau JC, Laplante JP (1992) Nonlinear dynamics, multiple steady states, and oscillations in photochemistry. J Phys Chem 96:2953–2961. https://doi.org/10.1021/j100186a035
24. Gou Z, Tang Y, Zhang F, Zhao F, Song XJ (1997) Photokinetic study of a novel phenylfulgide: effects of solvents on its photochromic reaction. Photochem Photobiol A Chem 110:29–33. https://doi.org/10.1016/S1010-6030(97)00158-5
25. Maafi M, Brown RG (2005) Photophysics and kinetics of naphthopyran derivatives, part 2: analysis of Diarylnaphthopyran kinetics. Degeneracy of the kinetic solution. Int J Chem Kinet 37:717–727. https://doi.org/10.1002/kin.20130
26. Hobley J, Malatesta V, Hatanaka K, Kajimoto S, Williams SL, Fukumura H (2002) Picosecond and nanosecond photo-dynamics of a naphthopyran merocyanine. Phys Chem Chem Phys 4: 180–184. https://doi.org/10.1039/B107973P
27. Ottavi G, Favaro G, Malatesta V (1998) Spectrokinetic study of 2,2-diphenyl-5,6-benzo(2H) chromene: a thermoreversible and photoreversible photochromic system. J Photochem Photobiol A Chem 115:123–128. https://doi.org/10.1016/S1010-6030(98)00254-8
28. Gorner H, Chibisov AK (2002) Photoprocesses in 2,2-diphenyl-5,6-benzo(2H)chromene. J Photochem Photobiol A Chem 149:83–89. https://doi.org/10.1016/S1010-6030(02)00002-3
29. Frigoli M, Pimienta V, Moustrou C, Samat A, Guglielmetti R, Aubard J, Maurel F, Micheau JC (2003) Continuous irradiation and flash-photolysis studies of new[3*H*]naphtho[2,1-*b*]pyrans linked by covalent bonds to oligothiophene units. Effect of thiophene substituents on the photochromism. Photochem Photobiol Sci 2:888–892. https://doi.org/10.1039/b301999c
30. Maafi M, Brown RG (2006) Investigation of [3*H*]-naphthopyran kinetics considered as an ABC (2k, 6φ) system yielding a monoexponential trace. Int J Chem Kinet 38:431–438. https://doi.org/10.1002/kin.20123
31. Coelho PJ, Carvalho LM, Abrantes S, Oliviera MM, Oliviera-Campos AMF, Samat A, Gugglielmetti R (2002) Synthesis and spectrokinetic studies of spiro[thioxanthene-naphthopyrans]. Tetrahedron 58:9505–9511. https://doi.org/10.1016/S0040-4020(02)01278-4
32. Brun P, Gugglielmetti R, Anguille S (2002) Metallocenyl-[2*H*]naphtho[1,2-b]pyrans: metal effect on the photochromic behaviour. Appl Organometal Chem 16:271–276. https://doi.org/10.1002/aoc.295
33. Arnall-Culliford JC, Teral Y, Sgarabotto P, Campedron M, Gusti G (2003) New halogenated diphenyl-2H-benzo[h]chromene derivatives: synthesis and optical properties. J Photochem Photobiol A Chem 159:7–16. https://doi.org/10.1016/S1010-6030(03)00112-6
34. Bernardi F, Olivucci M, Robb MA (1996) Potential energy surface crossings in organic photochemistry. Chem Soc Rev 25:321–328. https://doi.org/10.1039/CS9962500321
35. Robb MA, Garavelli M, Olivucci M, Bernardi F (2006) Chapter 2: A computational strategy for organic photochemistry. In: Lipkowitz KB, Boyd DB (eds) Reviews in computational chemistry, vol 15. Wiley-VCH Wiley, New York. ISBN. 978-0-470-12592-2
36. Barachevsky VA (2000) Photofluorochromic spirocompounds and their application. J Fluoresc 10:185–191. https://doi.org/10.1023/A:1009403411765
37. Aubard J, Maurel F, Buntinx G, Poizat O, Levi G, Guglielmetti R, Samat A (2000) Femto/picosecond transient absorption spectroscopy of photochromic 3,3-diphenylnaphtho[2,1-b] pyran. Mol Cryst Liq Cryst 345:215–220. https://doi.org/10.1080/10587250008023921

38. Budyka MF, Laukhina OD, Razumov VF (1997) Evidence for an adiabatic reaction pathway for trans-to-cis photoisomerization of di(α-naphthyl)ethylene. Chem Phys Lett 279:327–332. https://doi.org/10.1016/S0009-2614(97)01062-2
39. Higashiguchi K, Matsuda K, Irie M (2003) Photochromic reaction of a fused dithienylethene: multicolor photochromism. Angew Chem Int Ed 42:3537–3540. https://doi.org/10.1002/anie.200351751
40. Maafi M, Brown RG (2007) An analytical model utilizing temperature effects to evaluate ABC (2k,3–5φ) kinetic parameters. Int J Chem Kinet 39:472–479. https://doi.org/10.1002/kin.20251
41. Mauser H, Gauglitz G, Compton RG, Hancock G (eds) (1998) Comprehensive chemical kinetics, photokinetics: theoretical fundamentals and applications, vol 36. Elsevier, Amesterdam. ISBN-13. 978-0444825360
42. Feringa BL (ed) (2001) Molecular switches. Wiley-VCH, Weinheim. ISBN: 3-527-29965-3
43. Kolc J, Becker RS (1970) The spectroscopy and photochemistry of naturally occurring and synthetic chromenes. Photochem Photobiol 12:383. https://doi.org/10.1111/j.1751-1097.1970.tb06069

Part IV
Analytical Description of Kinetics Under Non-isosbestic Irradiation

Chapter 11
Defining the Φ-Order Kinetics

11.1 Introduction

The description of photokinetics under isosbestic irradiation (at λ_{isos}), detailed in Part III, has advanced our understanding of the photothermal reaction's kinetic behaviour and characteristics but did not inform about the features of the reaction when the employed monochromatic light is non-isosbestic (at $\lambda_{n\ -\ isos}$). Our investigation of photokinetics under this type of irradiation light would naturally start from the basic photoreaction, the simplest that occurs in the realm of photochemical transformations, that is, the primary photoprocess, $X \rightarrow Y$. This reaction is also considered because despite its simplicity, its kinetics has not been described in the literature at the required level of detail (i.e. when the monochromatic irradiation light is non-isosbestic). It turns out that unravelling and understanding the properties of this primary photoprocess under $\lambda_{n\ -\ isos}$-irradiation shapes a template that will allow setting up the fundamental bases for a larger photokinetics endeavour (as shall be seen in Chaps. 14 and 19).

11.2 The Kinetics of the Primary Photoprocess in the Literature

The primary photoreaction is a light-driven phototransformation of a photoactive but thermally stable reactant (X) into a single, photochemically and thermally inert photoproduct (Y, Scheme 11.1). The reaction is characterised by a photochemical quantum yield, $\Phi_{X\rightarrow Y}$. This reaction is the simplest possible photochemical process when it is driven by a monochromatic collimated light.

The kinetic description of this reaction, as available in the literature, is unfortunately incomplete, mainly limited to its rate-law, as has been discussed in Chap. 7. Eq. 11.1, which is equivalent to Eq. 7.6, is written here for the particular case where

M. Maafi, *Photokinetics*, https://doi.org/10.1007/978-3-031-98985-8_11

$$X \; - h\nu \rightarrow \; Y$$

Scheme 11.1 The primary photoreaction

only the reactant absorbs once the reaction medium is exposed to a non-isosbestic irradiation at $\lambda_{n\ -\ isos}$ [1–3]. The limited description of the photokinetics of the primary photoreaction is confirmed by a remarkable absence of its integrated rate-law from photochemistry textbooks, and thus far, no known consensual equation has been adopted by the photochemistry community for this reaction.

$$\frac{dC_X^{\lambda_{n-isos},T}(t)}{dt} = -\Phi_{X\rightarrow Y}^{\lambda_{n-isos}} P_0^{\lambda_{n-isos}} \left(1 - 10^{-A_X^{\frac{\lambda_{n-isos},T}{\lambda_{n-isos}}}(t)}\right) \tag{11.1}$$

In the absence of an analytically derived integrated rate-law for this photoreaction, several alternative approaches were proposed to facilitate its kinetic analysis. We will limit our discussion here to the most recurrent in the literature.

The most consistent among those approaches employed numerical integration methods (*NIMs*) [4]. Because *NIMs* are based on the true rate-law of the reaction, the results they provide are consistent with experiments. The undeniable good performance of *NIMs* falls short, however, of delivering a mathematical formulation of the integrated rate-law for the reaction. Such an equation would capture the dynamics of the process, allow a better intuitive understanding of the kinetics, and offer a simple tool for experimentalists.

The latter reasons are probably what justified the need for the development of other approaches that focused on proposing mathematical formulations for the integrated rate-law of this reaction. This, however, came at a cost. Given that closed-form integration was not achieved, the suggested derivations relied on introducing approximations and/or simplifications that led to a modification of the reaction rate-law.

In this context, as a general observation, models of integrated rate-laws that take similar formulations to those corresponding to zeroth-, first-, or second-order thermal reaction kinetics are ubiquitous in the literature [5–8].

One approach [6], presumably considering that an analytical integration of the rate-law (Eq. 11.1) was not accessible, employed power expansion series (Sect. 3.8) in order to simplify the rate-law to an integrable form. The rate-law of the primary photoreaction is accordingly rewritten, for any monochromatic irradiation beam of wavelength λ, as

$$\frac{dC_X^{\lambda_{n-isos},T}(t)}{dt}=-\Phi_{X\to Y}^{\lambda_{n-isos}}P_0^{\lambda_{n-isos}}\left(1-10^{-A_X^{\lambda_{n-isos},T/\lambda_{n-isos}}(t)}\right)=-\Phi_{X\to Y}^{\lambda_{n-isos}}P_0^{\lambda_{n-isos}}$$
$$\times\left(1-\left[1+A_X^{\lambda_{n-isos},T/\lambda_{n-isos}}(t)+\frac{\left(A_X^{\lambda_{n-isos},T/\lambda_{n-isos}}(t)\right)^2}{2}+\frac{\left(A_X^{\lambda_{n-isos},T/\lambda_{n-isos}}(t)\right)^3}{6}\cdots\right]\right) \tag{11.2}$$

The latter equation is further simplified by imposing that the absorbance must be very low, $A_X^{\lambda_{n-isos},T/\lambda_{n-isos}}(t)<0.02$, a condition making second- and higher-order terms of the expansion series negligible and which can be removed from the equation. Hence, by introducing the expression of the explicit reactant absorbance formula, $A_X^{\lambda_{n-isos},T/\lambda_{n-isos}}(t)=\varepsilon_X^{P_0^{\lambda_{n-isos}}}\ l_{irr}\ C_X^{\lambda_{n-isos},T}(t)$, in the resulting simplified equation, it takes the form

$$\frac{dC_X^{\lambda_{n-isos},T}(t)}{dt}=-\Phi_{X\to Y}^{\lambda_{n-isos}}\ P_0^{\lambda_{n-isos}}\ A_X^{\lambda_{n-isos},T/\lambda_{n-isos}}(t)=$$
$$-\Phi_{X\to Y}^{\lambda_{n-isos}}\ P_0^{\lambda_{n-isos}}\ \varepsilon_X^{\lambda_{n-isos}}\ l_{irr}\ C_X^{\lambda_{n-isos},T}(t) \tag{11.3}$$

We notice that because the rate-law proposed by the authors involves a single absorption coefficient ($\varepsilon_X^{\lambda_{n-isos}}$), it must correspond to the situation where only the reactant X absorbs the excitation light. This specification should exclude the application of Eq. 11.3 to the case where X and Y concomitantly absorb the incident light (Chap. 12). Also, the optical path length (l_{irr}) occurring in this equation, must correspond to that of the irradiation light, which, here, is supposed equal to that of the spectrophotometer monitoring light used to measure the absorbance (since it makes part of the absorbance equation). Eq. 11.3 corresponds, therefore, to a case where $l_{obs}=l_{irr}$.

Interestingly, the above approach has been expanded further to encompass the situation of exposure of the sample to polychromatic light [9]. Because, now, $P_0^{\lambda_{n-isos}}$ and $\varepsilon_X^{\lambda_{n-isos}}$ will be varying with wavelength, the integration of the rate-law is performed over the relevant wavelength range $\lambda_a-\lambda_b$. Even though this approach does not provide any other details on the kinetic system, we are however forced to observe that Eq. 11.4 implicitly assumes that the polychromatic light is absorbed by only X (since only one absorptivity, ε_X^{λ}, is present). This represents a simplification that reduces considerably the field of application of this equation in practice.

$$\frac{dC_X^{\lambda_{n-isos},T}(t)}{dt}=-\Phi_{X\to Y}^{\lambda_{n-isos}}\ l_{irr}\ C_X^{\lambda_{n-isos},T}(t)\int_{\lambda_a}^{\lambda_b}\left(P_0^{\lambda_{n-isos}}\ \varepsilon_X^{\lambda_{n-isos}}\right)d\lambda \tag{11.4}$$

The overlap integral, $\int_{\lambda_a}^{\lambda_b}\left(P_0^{\lambda_{n-isos}}\ \varepsilon_X^{\lambda_{n-isos}}\right)d\lambda$, is considered constant for a given light source and absorbing substance. Hence, by grouping the constants in a factor α ($\alpha=\Phi_{X\to Y}^{\lambda_{n-isos}}\ l_{irr}\int_{\lambda_a}^{\lambda_b}\left(P_0^{\lambda_{n-isos}}\ \varepsilon_X^{\lambda_{n-isos}}\right)d\lambda$), we end up with a simplified rate-law

(Eq. 11.5). Notice that Eq. 11.5 has an equivalent mathematical form to that of Eq. 11.3 (for which the constant would be $\alpha = \Phi_{X \to Y}^{\lambda_n - isos} P_0^{\lambda_n - isos} \varepsilon_X^{\lambda_n - isos} l_{irr}$).

$$\frac{dC_X^{\lambda_n - isos, T}(t)}{dt} = -\alpha \, C_X^{\lambda_n - isos, T}(t) \tag{11.5}$$

The integration of the equation yields

$$ln \, C_X^{\lambda_n - isos, T}(t) = ln \, C_X^{\lambda_n - isos, T}(0) - \alpha \, t \tag{11.6}$$

According to this approach, the primary photoreaction, when only X absorbs the incident light, obeys first-order kinetics, irrespective of whether the irradiation is monochromatic or polychromatic.

The authors of the approach [9] mentioned a caveat relative to the application of Eq. 11.5 by different laboratories, in that the rate-constant derived from the experiment carried out in one laboratory will not be the same as that recorded in another due to the difficulty in reproducing exactly both the irradiated geometry of the sample and the features of the light source.

It was also stated that the condition on the absorbance magnitude could be extended to a 0.4 limit value of the absorbance, but they did not justify how this can be compatible with the necessary condition of low $A_X^{\lambda_n - isos/\lambda_n - isos, T}(t)$ (e.g. <0.02), imposed by a consistent use of the power expansion series (the latter condition allows us to consider only the first term of the expansion series and, therefore, leads the integration to a first-order kinetic model).

Incidentally, it is surprising that each of the application examples supplied for illustration of the application of the above method [9] involved a reactant and a photoproduct whose absorption spectra overlapped with their quasi-whole absorption wavelength ranges.

The previous approach can be held as a limiting case within another line of thought that considers the true total order of a bimolecular elementary photoreaction to always be two (Chap. 7). According to this approach, the primary photoreaction must be a second-order reaction because it is fundamentally bimolecular, involving two "reactants," namely, the photoactive species X and the photon, $C_{h\nu}^{\lambda_n - isos}$ [2].

$$\frac{dC_X^{\lambda_n - isos, T}(t)}{dt} = -k \, C_{h\nu}^{\lambda_n - isos} \, C_X^{\lambda_n - isos, T}(t) \tag{11.7}$$

This bimolecular reaction adopts the pseudo-first-order kinetic type if the "concentration" of the photons, $C_{h\nu}^{\lambda_n - isos}$, remains constant (i.e. irradiation with constant radiation intensity). That is, the apparent reaction order becomes one.

$$\frac{dC_X^{\lambda_n-isos,T}(t)}{dt} = -k\, C_{h\nu}^{\lambda_n-isos}\, C_X^{\lambda_n-isos,T}(t) = -k' C_X^{\lambda_n-isos,T}(t) \tag{11.8}$$

On the other hand, if in addition to $C_{h\nu}^{\lambda_n-isos}$ being constant, the variation of the absorbance (or the concentration of the reactant) is relatively small or if the initial concentration of X is very large, then the rate may appear approximately constant. Hence, the apparent order of the primary photoreaction becomes zero.

In essence, this approach states that the primary photoreaction must behave as one of the three classical reaction orders defined for thermal reactions. The true order of the reaction is second, but the kinetics of the reaction can obey either a first or a zeroth -order if a constant light intensity of the source or a high initial concentration of the reactant, is imposed. In each case, the corresponding integrated rate-law (Sect. 3.5) is used.

It is important to underline that the literature, over the past century, has adopted the latter point of view in a relatively large extent and has stretched this way of dealing with photokinetics to more complex reactions than the primary photoreaction. We have seen in Chap. 7 the reasons brought forward to evaluate and debunk the usefulness of the second-order kinetic proposal for the primary photoreaction (or any other photoreaction). The issue with these types of rate-laws has been very early on underlined by Toby and Shaw in the 1960s [5]. As a matter of fact, there is a quasi-consensus within the photokinetics community on the rate-law formulated in Eq. 11.1 [5, 10] for the primary photoreaction. However, the community remained far from reaching such a consensus for the integrated rate-law of Eq. 11.1.

In 2007, an assumption-free treatment of the rate-law (Eq. 11.1) proved that an analytical solution can mathematically be derived for this reaction [11]. The mathematical form of the analytically obtained integrated rate-law is significantly different from the approximated solutions proposed in the literature. Further details of this treatment will be discussed in Sect. 11.4.

Mathematically equivalent expressions to the latter were later reported elsewhere [12, 13] but without providing the underlying derivations.

11.3 Light Absorption and Irradiation Types: Is It a Vain, Thin Hair-Splitting Exercise or an Essential Part of Photokinetics?

The primary photoprocess, as defined in Sect. 11.2 and Scheme 11.1, is the simplest photoreaction that can exist in nature. As stated above, studying its dynamics is an obvious starting point in the exploration of photokinetics under non-isosbestic irradiation light.

If we start by considering the case of a purely unimolecular thermal reaction (the thermal equivalent of the photoreaction shown in Scheme 11.1), the reaction may obey either zeroth- or first-order kinetics (Sect. 3.5 and Chap. 7). This difference, totally independent of the reaction conditions or the experimentalist's choices, is mainly due to the specific features and reactivity of the reactant. Such reaction-order options underline a typical case of an intrinsic distinctability issue (Chap. 9). For both kinetic options, a speeding up of the reaction can be induced by increasing the temperature of the reactive medium (which, as a whole, has a homogenous distribution of that temperature in the reactor).

Photoreactions' kinetics is different from thermal kinetics in the sense that the reaction rate depends on the amount of light absorbed by the species (Eq. 11.1). The rate of a photoreaction may as well depend on the amount of light absorbed by the products of the reaction (see Sect. 2.16, Chap. 16). Such a concept cannot be mapped out for a thermal reaction. In an imaginary situation of a thermal reaction, this would mean that two impossible conditions must be realised: the overall rate of the thermal reaction must (a) depend on the species (both X and Y) present in the reactive medium, and (b) it varies when reactant and product are maintained at different temperatures.

This essential difference between thermal reactions and photoreactions imposes, to the photokinetic investigation procedure, to carefully define which of the species present in the reactive medium absorb the incident light. This should be held as a general concept to be applied irrespective of the actual mechanism undergone by the overall reaction (that might encompass several photo- and/or thermal reaction-steps).

A photokinetic investigation needs, therefore, to look at the irradiation condition under which the reaction at hand is performed. The knowledge of the irradiation wavelength and the absorption spectra of the involved species will allow determining the species absorbing the incident light.

The corollary to this concept is a mandatory knowledge of all the (appearing and disappearing) species during the reaction progress. This is hardly achieved if the mechanism undergone by the reactive system is unknown. Also, the identification of which of these species absorb the used irradiation light requires the knowledge of the individual absorption spectra of the species present in the medium during the reaction time.

Let us examine the options offered to the primary photoreaction (Scheme 11.1) when subjected to a monochromatic light.

It is obvious that for the primary photoreaction, there are two situations: (i) only the reactant absorbs the incident light (i.e. $\varepsilon_Y^{\lambda_{n-isos}} = 0$), in which case the wavelength of the non-isosbestic irradiation is labelled as $\lambda_{n-isos} = \lambda_X$, and (ii) both the reactant and the photoproduct absorb (labelled simply as λ_{n-isos}). The latter case can give rise to a further distinction: (ii.a) the wavelength of the excitation light corresponding to that of an isosbestic point (λ_{isos}), i.e. by performing an isosbestic irradiation where $\varepsilon_X^{\lambda_{isos}} = \varepsilon_Y^{\lambda_{isos}}$ (Chap. 8), and (ii.b) an irradiation occurring in a non-isosbestic region, λ_{n-isos}, i.e. a non-isosbestic irradiation of both X and Y (with $\varepsilon_X^{\lambda_{n-isos}} \neq \varepsilon_Y^{\lambda_{n-isos}}$).

In photokinetics, the irradiation conditions are defined relative to the wavelength at which the irradiation is performed, generically set as λ_{irr}, which can take three different values, $\lambda_{irr} = \lambda_X$, λ_{isos}, or λ_{n-isos} (Fig. 11.1).

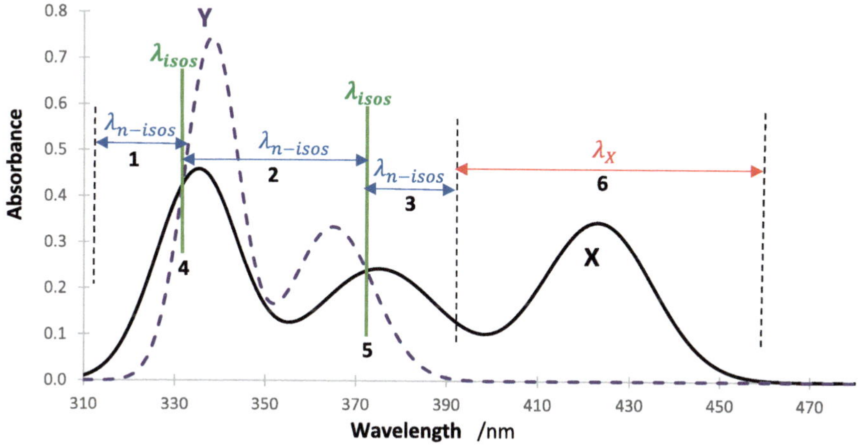

Fig. 11.1 An example of reactant X (plain) and photoproduct Y (dashed) absorption spectra and the wavelength sections where irradiation of the sample can be carried out by a monochromatic light. Possible monochromatic irradiation with λ_{n-isos} in sections 1, 2, and 3 (where $\varepsilon_X^{\lambda_{n-isos}} \neq \varepsilon_Y^{\lambda_{n-isos}}$); λ_{isos} in sections 4 and 5 ($\varepsilon_X^{\lambda_{isos}} = \varepsilon_Y^{\lambda_{isos}}$); and λ_X in section 6 ($\varepsilon_X^{\lambda_{n-isos}} \neq 0$ and $\varepsilon_Y^{\lambda_{n-isos}} = 0$)

The consequence of such irradiation differences is clearly evident in the subtle alterations of the rate-laws describing the three cases. Let the total absorbance be $A_{tot}^{\lambda_{irr},T}(t) = A_X^{\lambda_{irr},T}(t) + A_Y^{\lambda_{irr},T}(t)$; the reaction medium is continuously and homogenously stirred at a constant temperature; the optical path lengths are supposed to have the same value, $l_{irr} = l_{obs}$; the initial concentrations of the species belong to the linearity ranges of their respective calibration graphs; and the concentrations of excited-state species and intermediates are negligible.

The rate-laws for the different irradiation conditions can be written according to the general framework laid out in Sect. 3.6.

$$\frac{dC_X^{\lambda_X,T}(t)}{dt} = -\Phi_{X\to Y}^{\lambda_X} P_0^{\lambda_X}\left(1 - 10^{-A_X^{\lambda_X,T}(t)}\right) \tag{11.9}$$

$$\frac{dC_X^{\lambda_{isos},T}(t)}{dt} = -\Phi_{X\to Y}^{\lambda_{isos}} P_0^{\lambda_{isos}} A_X^{\lambda_{isos},T} \frac{1 - 10^{-A_{tot}^{\lambda_{isos},T}}}{A_{tot}^{\lambda_{isos},T}} \tag{11.10}$$

$$\frac{dC_X^{\lambda_{n-isos},T}(t)}{dt} = -\Phi_{X\to Y}^{\lambda_{n-isos}} P_0^{\lambda_{n-isos}} A_X^{\lambda_{n-isos},T}(t) \frac{1 - 10^{-A_{tot}^{\lambda_{n-isos},T}(t)}}{A_{tot}^{\lambda_{n-isos},T}(t)} \tag{11.11}$$

The equation characterising the primary photoreaction under λ_X-irradiation (Eq. 11.9) is different from the two others because it lacks a time-dependent absorbance occurring outside the term in brackets. The second equation, corresponding to an isosbestic irradiation (Eq. 11.10) is different from that of the non isosbestic irradiation equation (Eq. 11.11) by the time-invariance of its photokinetic factor (i.e. the fraction on its right-hand side term is constant with time because $A_{tot}^{\lambda_{isos},T}$ does not vary during the progress of the photoconversion).

According to the general mathematical rule (see Sect. 3.5), these different ordinary differential equations (*ODE*s) are expected to lead to specific solutions that are not necessarily equivalent (or the same).

Therefore, from a qualitative viewpoint, depending on irradiation, the primary photoreaction may possess up to three different behaviours (i.e. being a case of extrinsic distinctability (Sect. 6.5)). At this stage, we can already conclude that the primary photoreaction cannot be considered to have a kinetic description equivalent to that proposed for a similar purely thermal reaction. This also means that the second-order kinetics (Eq. 11.7) as well as the more restrictive option of the first-order kinetics that might be derived through the power series approximation (Eq. 11.5) are not reasonable models for the primary photoreaction because they cannot explicitly account for the effects induced by the irradiation conditions and cannot predict such variety of the reaction kinetic behaviours.

Since this has been established for the primary photoprocess, it is only sensible to consider that, in general, a careful consideration of the irradiation conditions is essential in photokinetics of any photomechanism. This would lead to deriving the correct rate-laws and, hence, to work out, whenever possible, the correct integrated rate-laws for both reactants and photoproducts (examples are provided in the following chapters).

11.4 Solving the Rate-Law of the Primary Photoreaction Under λ_X-Irradiation

The rate-law, Eq. 11.9, of a primary photoprocess driven by a monochromatic λ_X-excitation light can be presented with an explicit expression of the absorbance as

$$\frac{dC_X^{\lambda_X,T}(t)}{dt} = -\Phi_{X\to Y}^{\lambda_X} P_0^{\lambda_X}\left(1 - 10^{-\varepsilon_X^{\lambda_X} l_{irr} C_X^{\lambda_X,T}(t)}\right) \tag{11.12}$$

We apply here the universal procedure to solve this type of differential equations [11]. The first step in the process of solving this rate-law is to start with a separation of the variables $C_X^{\lambda_X,T}$ and t (Sect. 3.4).

$$\frac{dC_X^{\lambda_X,T}(t)}{1 - 10^{-\varepsilon_X^{\lambda_X} l_{irr} C_X^{\lambda_X,T}(t)}} = -\Phi_{X\to Y}^{\lambda_X} P_0^{\lambda_X}\, dt \tag{11.13}$$

This defines the two terms of the equation (Eq. 11.13) that will be integrated separately. The right-hand-side term integration (over dt) is straightforward.

$$-\int_0^t \Phi_{X\to Y}^{\lambda_X} P_0^{\lambda_X}\, dt = -\Phi_{X\to Y}^{\lambda_X} P_0^{\lambda_X}\,(t-0) = -\Phi_{X\to Y}^{\lambda_X} P_0^{\lambda_X}\, t \tag{11.14}$$

The integration of the left-hand-side term of Eq. 11.13 cannot be performed directly. It requires first transforming the integrand into a form that has a known antiderivative (Chap. 3). One way to achieve this is to change the integration variable from the actual $dC_X^{\lambda_X,T}$ to the more convenient $d\left(10^{\varepsilon_X^{\lambda_X} l_{irr} C_X^{\lambda_X,T}(t)}\right)$. This transformation leads ultimately to a simpler integrand (whose antiderivative is known).

So let us first consider the result of differentiating the power term, $10^{\varepsilon_X^{\lambda_X} l_{irr} C_X^{\lambda_X,T}(t)}$.

$$d\left(10^{\varepsilon_X^{\lambda_X} l_{irr} C_X^{\lambda_X,T}(t)}\right) = \ln 10\; \varepsilon_X^{\lambda_X}\, l_{irr}\, 10^{\varepsilon_X^{\lambda_X} l_{irr} C_X^{\lambda_X,T}(t)}\, dC_X^{\lambda_X,T}(t) \tag{11.15}$$

which can be rearranged into

$$dC_X^{\lambda_X,T}(t) = \frac{d\left(10^{\varepsilon_X^{\lambda_X} l_{irr} C_X^{\lambda_X,T}(t)}\right)}{\ln 10\; \varepsilon_X^{\lambda_X}\, l_{irr}\, 10^{\varepsilon_X^{\lambda_X} l_{irr} C_X^{\lambda_X,T}(t)}} \tag{11.16}$$

By replacing $dC_X^{\lambda_X,T}(t)$ by its expression (Eq. 11.16) on the left-hand-side term of Eq. 11.13, we obtain

$$\frac{dC_X^{\lambda_X,T}(t)}{1-10^{-\varepsilon_X^{\lambda_X} l_{irr} C_X^{\lambda_X,T}(t)}} = \frac{1}{1-10^{-\varepsilon_X^{\lambda_X} l_{irr} C_X^{\lambda_X,T}(t)}}\, \frac{d\left(10^{\varepsilon_X^{\lambda_X} l_{irr} C_X^{\lambda_X,T}(t)}\right)}{\ln 10\; \varepsilon_X^{\lambda_X}\, l_{irr}\, 10^{\varepsilon_X^{\lambda_X} l_{irr} C_X^{\lambda_X,T}(t)}} \tag{11.17}$$

The integration of this simplified form yields

$$\begin{aligned} &\frac{1}{\ln 10\; \varepsilon_X^{\lambda_X}\, l_{irr}} \int_{C_X^{\lambda_X,T}(0)}^{C_X^{\lambda_X,T}(t)} \frac{d\left(10^{\varepsilon_X^{\lambda_X} l_{irr} C_X^{\lambda_X,T}(t)}\right)}{10^{\varepsilon_X^{\lambda_X} l_{irr} C_X^{\lambda_X,T}(t)} - 1} \\ &= \frac{1}{\ln 10\; \varepsilon_X^{\lambda_X}\, l_{irr}} \ln \frac{10^{\varepsilon_X^{\lambda_X} l_{irr} C_X^{\lambda_X,T}(t)} - 1}{10^{\varepsilon_X^{\lambda_X} l_{irr} C_X^{\lambda_X,T}(0)} - 1} \end{aligned} \tag{11.18}$$

The complete expression of the integration of Eq. 11.13 is obtained by combining the right-hand-side terms of Eqs. 11.15 and 11.18 as

$$\frac{1}{\ln 10\; \varepsilon_X^{\lambda_X}\, l_{irr}} \ln \frac{10^{\varepsilon_X^{\lambda_X} l_{irr} C_X^{\lambda_X,T}(t)} - 1}{10^{\varepsilon_X^{\lambda_X} l_{irr} C_X^{\lambda_X,T}(0)} - 1} = -\Phi_{X\to Y}^{\lambda_X} P_0^{\lambda_X}\, t \tag{11.19}$$

The analytically derived integrated rate-law of the present primary photoprocess is worked out from Eq. 11.19 as

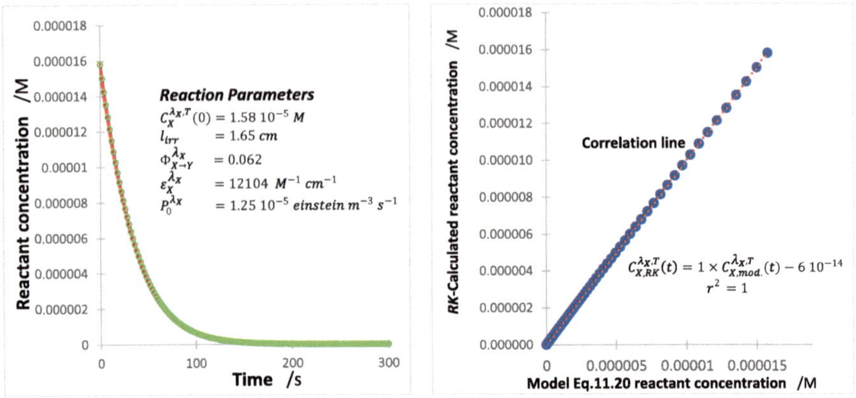

Fig. 11.2 *Left*: a good correspondence between the numerical calculated trace (green circles) and the data generated by the integrated rate-law, Eq. 11.20 (red line). The $XY(1\Phi)$ reaction parameters are indicated. *Right*: linear correlation of the *RK*- and model equation sets of the reactant concentration data (dots), as indicated by the equation of the line

$$C_X^{\lambda_X,T}(t) = \frac{1}{\varepsilon_X^{\lambda_X}\ l_{irr}} \log\left[1 + \left(10^{\varepsilon_X^{\lambda_X}\ l_{irr}\ C_X^{\lambda_X,T}(0)} - 1\right) e^{-\Phi_{X\to Y}^{\lambda_X}\ P_0^{\lambda_X}\ \varepsilon_X^{\lambda_X}\ l_{irr}\ \ln 10\ t}\right] \quad (11.20)$$

Notice that at the boundary conditions, $t = 0$ and $t = \infty$, the expected concentrations of X are equal to $C_X(0)$ and 0, respectively, as should be observed experimentally.

The argument of the exponential is, for coherence, necessarily dimensionless. Accordingly, the term ($\Phi_{X\to Y}^{\lambda_X}\ P_{\lambda_X}^0 \varepsilon_X^{\lambda_X}\ l_{irr}\ ln\ 10$) must be expressed in s^{-1} (and t in s). Also importantly, the former term is assimilated to a rate-constant for the unimolecular photoreaction.

$$k_r^{\lambda_X,T} = \Phi_{X\to Y}^{\lambda_X}\ P_0^{\lambda_X}\ \varepsilon_X^{\lambda_X}\ l_{irr}\ ln\ 10 \quad (11.21)$$

The dimension analysis of $k_r^{\lambda_X,T}$ confirms the unit predicted above. Using the dimensions of the different coefficient of Eq. 11.21 (as defined in Sect. 2.18) gives

$$\left[k_r^{\lambda_X,T}\right] = \left[\Phi_{X\to Y}^{\lambda_X}\right]\left[P_0^{\lambda_X}\right]\ \left[\varepsilon_X^{\lambda_X}\right]\ \left[l_{irr}\right] = 1 \times einstein\ dm^{-3}\ s^{-1}$$
$$\times\, mol^{-1}\ dm^3\ cm^{-1} \times cm = s^{-1} \quad (11.22)$$

Accordingly, the kinetic traces of the primary photoreaction under irradiation at λ_X will be faithfully modelled by the integrated rate-law, Eq. 11.20. This is shown by the good correspondence obtained between the data generated by numerical integration and the model equation, Eq. 11.20 (Fig. 11.2). The excellent agreement could also be shown for a great number of simulations using a large number of sensibly different reaction parameters. The model equation also fits experimental data well (Chaps. 17 and 18).

The analytical derivation of the solution for this reaction meant that there is neither a need for methods using approximations to solve the rate-equation, such as the ones described in Sect. 11.2, nor for the use of numerical integration methods.

The integrated rate-law is most useful in the form given by Eq. 11.20, a concentration function of time, as classically presented in kinetics. However, it can also be rearranged into another formulation.

Starting from Eq. 11.19, factorising the terms $10^{\varepsilon_X^{\lambda_X}\, l_{irr} C_X^{\lambda_X,T}(t\ or\ 0)}$ gives

$$\frac{1}{ln\,10\;\varepsilon_X^{\lambda_X}\;l_{irr}}\;ln\left(\frac{10^{\varepsilon_X^{\lambda_X}\,l_{irr}C_X^{\lambda_X,T}(t)}}{10^{\varepsilon_X^{\lambda_X}\,l_{irr}C_X^{\lambda_X,T}(0)}}\;\frac{1-10^{-\varepsilon_X^{\lambda_X}\,l_{irr}C_X^{\lambda_X,T}(t)}}{1-10^{-\varepsilon_X^{\lambda_X}\,l_{irr}C_X^{\lambda_X,T}(0)}}\right)=-\Phi_{X\to Y}^{\lambda_X}\;P_0^{\lambda_X}\;t \tag{11.23}$$

and by developing the logarithm of Eq. 11.23, we obtain

$$\frac{1}{ln\,10\;\varepsilon_X^{\lambda_X}\;l_{irr}}\;ln\left(\frac{10^{\varepsilon_X^{\lambda_X}\,l_{irr}C_X^{\lambda_X,T}(t)}}{10^{\varepsilon_X^{\lambda_X}\,l_{irr}C_X^{\lambda_X,T}(0)}}\right)+\frac{1}{ln\,10\;\varepsilon_X^{\lambda_X}\;l_{irr}}\;ln\left(\frac{1-10^{-\varepsilon_X^{\lambda_X}\,l_{irr}C_X^{\lambda_X,T}(t)}}{1-10^{-\varepsilon_X^{\lambda_X}\,l_{irr}C_X^{\lambda_X,T}(0)}}\right)=-\Phi_{X\to Y}^{\lambda_X}\;P_0^{\lambda_X}\;t \tag{11.24}$$

or expanding the first logarithm, we get

$$\left(C_X^{\lambda_X,T}(t)-C_X^{\lambda_X,T}(0)\right)+\frac{1}{ln\,10\;\varepsilon_X^{\lambda_X}\;l_{irr}}\;ln\left(\frac{1-10^{-\varepsilon_X^{\lambda_X}\,l_{irr}C_X^{\lambda_X,T}(t)}}{1-10^{-\varepsilon_X^{\lambda_X}\,l_{irr}C_X^{\lambda_X,T}(0)}}\right)=-\Phi_{X\to Y}^{\lambda_X}\;P_0^{\lambda_X}\;t \tag{11.25}$$

Finally, the latter equation can also be rearranged into Eq. 11.26.

$$\left(A_X^{\lambda_X,T}(t)-A_X^{\lambda_X,T}(0)\right)+Log\left(\frac{1-10^{-A_X^{\lambda_X,T}(t)}}{1-10^{-A_X^{\lambda_X,T}(0)}}\right)=-\Phi_{X\to Y}^{\lambda_X}\;P_0^{\lambda_X}\;\varepsilon_X^{\lambda_X}\;l_{irr}\;t \tag{11.26}$$

Equation 11.26 can also be derived by applying a standard antiderivative (#11 in Table 3.1) to the right-hand-side term of Eq. 11.13 when this equation is multiplied by $\varepsilon_X^{\lambda_X}\,l_{irr}$ (Eq. 11.27), such that the integration variables are $dA_X^{\lambda_X,T}(t)$ and dt [12, 14].

$$\frac{dA_X^{\lambda_X,T}(t)}{1-10^{-A_X^{\lambda_X,T}(t)}}=-\Phi_{X\to Y}^{\lambda_X}\;P_0^{\lambda_X}\;\varepsilon_X^{\lambda_X}\;l_{irr}\;dt \tag{11.27}$$

In addition, the integrated rate-law (Eq. 11.20) can be formulated as a three-parameter curve (Eq. 11.28), where $C_X(0)$ and the product $(k_a k_b)$ are considered scaling parameters, whereas $(k_b\ C_X(0))$ is postulated to be a shape parameter [15].

$$C_X^{\lambda_X,T}(t) = \frac{1}{k_b} Log\left[1 + \left(e^{k_b\ C_X^{\lambda_X,T}(0)} - 1\right) e^{-k_a\ k_b\ t}\right] \tag{11.28}$$

with

$$k_a = \Phi_{X\to Y}^{\lambda_X}\ P_0^{\lambda_X} \tag{11.29}$$

and

$$k_b = \varepsilon_X^{\lambda_X}\ l_{irr}\ \ln 10 \tag{11.30}$$

It is important to keep in mind that Eqs. 11.20, 11.25, 11.26, and 11.28 are formulations of the same equation, i.e. all carry exactly the same kinetic information for a unimolecular photoreaction under a λ_X-irradiation. We might say that Eqs. 11.20 and 11.28 represent a curved function of the concentration with reaction time, whereas Eqs. 11.25 and 11.26 can be thought of as linear relationships between their respective left-hand-side terms (as one function making up the ordinate) and reaction time (as the abscissa).

11.5 Defining Φ-Order Kinetics

It has already been established (Chap. 7) that the three most prominent kinetic orders of thermal reactions, i.e. zeroth-, first-, and second-order kinetics, describe different reaction behaviours because they obey independent integrated rate-laws (Table 7.1). The four integrated rate-laws shown therein cannot be worked out from each other. The differences in kinetic behaviour or reaction-order are primarily due to mathematical differences in the reactions' respective differential equations, that is, different differential equations will necessarily yield different solutions (Sect. 3.5).

The above mathematical principle should also apply to the unimolecular photoreaction's differential equation, Eq. 11.9. Indeed, since the latter is fundamentally different from the *ODE*s describing thermal reactions, its integrated rate-law should be different from those already established for the thermal reactions, as demonstrated by Eq. 11.20. It is important to underline here that Eq. 11.20 is not possibly derived from Eq. 8.55 (or vice versa). Also, as far as we are aware, the formula of the integrated rate-law of the primary photoprocess when only the reactant absorbs (exposed to λ_X-irradiation, Eq. 11.20) is a unique expression in kinetics. Its typical *log-exponential* combination does not, for instance, resemble the integrated rate-laws derived for thermal reactions.

Some of the inherent properties of the integrated rate-law, Eq. 11.20, are as follows:

(a) Its formulation involves a logarithmic function whose argument carries an exponential time function.
(b) It cannot be worked out from any of the known integrated rate-laws in kinetics.
(c) It is applicable only within the spectrophotometrically defined linearity range of the reactive species' calibration graph (where the absorbance, $A_X^{\lambda_X,T}(t)$, is consistently defined and linear with $C_X(t)$, according to the Beer–Lambert law, $A_X^{\lambda_X,T}(t) = \varepsilon_X^{\lambda_X} l_{irr} C_X^{\lambda_X,T}(t)$, Sect. 2.10).
(d) The variation of the concentration of the reactant X, $C_X^{\lambda_X,T}(t)$, during the reaction is dependent on the reactant attributes ($\Phi_{X\to Y}^{\lambda_X}$ and $\varepsilon_X^{\lambda_X}$) and the reaction's experimental conditions (l_{irr} and $P_0^{\lambda_X}$), including its own initial concentration $C_X(0)$. Notice that all these parameters are defined at λ_X (Sect. 2.10).
(e) The argument of the exponential function includes a constant defined by an explicit equation, Eq. 11.21 ($k_r^{\lambda_X,T} = \Phi_{X\to Y}^{\lambda_X} \varepsilon_X^{\lambda_X} P_0^{\lambda_X} l_{\lambda_{irr}} ln(10)$), that determines the relative depletion velocity of X (depletion is faster for higher values of $k_r^{\lambda_X,T}$).
(f) The dimensional analysis of $k_r^{\lambda_X,T}$ shows that it has s^{-1} unit. This unit gives $k_r^{\lambda_X,T}$ a physical sense as a reaction rate-constant. In this sense, it could be viewed as the overall rate-constant of the primary photoreaction under λ_X. Technically, $k_r^{\lambda_X,T}$ does not correspond to the usual ratio of the rate to a function of the concentration of the reactive species, as defined for thermal reactions (Eqs. 1.1, 1.2 and 1.3).
(g) $k_r^{\lambda_X,T}$ is irradiation wavelength (λ_X)–dependent because of the variation of $\varepsilon_X^{\lambda_X}$, $P_0^{\lambda_X}$, and, a priori (see Chaps. 4 and 18), $\Phi_{X\to Y}^{\lambda_X}$ with wavelength.
(h) $k_r^{\lambda_X,T}$ is proportional not only to the light intensity but also to the absorptivity, quantum yield of the reaction at λ_X, the irradiation optical path length, and the sample area and volume irradiated (implicitly part of the formula of $P_0^{\lambda_X}$, Sect. 2.17).
(i) $k_r^{\lambda_X,T}$ is constant for a given experiment, i.e. for a set of well-defined experimental conditions, including a constant flux of the λ_X-monochromatic incident light (similarly, a thermal reaction rate-constant is only constant at a particular temperature).
(j) The primary photoreaction considered here has a concentration-dependent rate (present within $A_X^{\lambda_X,T}(t)$, Eq. 11.9 or Eq. 11.12) and a concentration-independent rate-constant (Eq. 11.21).
(k) The reaction half-life time ($t_{\frac{1}{2}}$) has a logarithmic increasing trend with $C_X(0)$. $t_{\frac{1}{2}}$ is defined for the reactive medium exposed to a given λ_X-irradiation as

$$t_{\frac{1}{2}}^{\lambda_X} = \frac{-1}{k_r^{\lambda_X,T}} Ln\left(\frac{\sqrt{10^{\varepsilon_X^{\lambda_X} l_{irr} C_X^{\lambda_X,T}(0)}} - 1}{10^{\varepsilon_X^{\lambda_X} l_{irr} C_X^{\lambda_X,T}(0)} - 1}\right) = \frac{\ln\left(\sqrt{10^{\varepsilon_X^{\lambda_X} l_{irr} C_X^{\lambda_X,T}(0)}} + 1\right)}{k_r^{\lambda_X,T}} \tag{11.31}$$

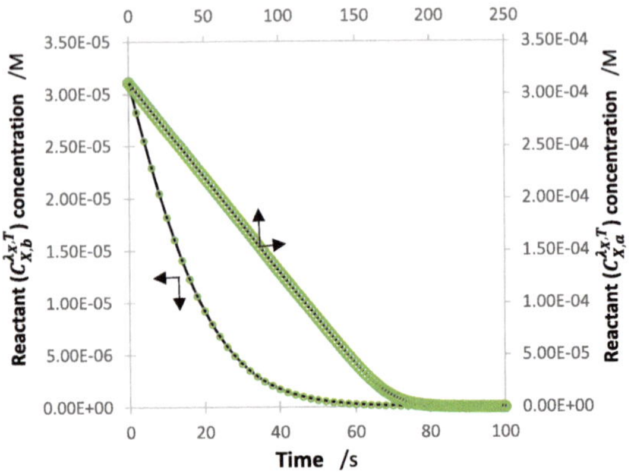

Fig. 11.3 An example of the shape changes on the trace when the concentration varies. For high initial concentration ($C_{X,a}^{\lambda_X,T}(0) = 3.11\ 10^{-4}\ M$), a quasi-linear pattern is recorded, compared to the smooth curve observed when the concentration is tenfold lower ($C_{X,b}^{\lambda_X,T}(0) = 3.11\ 10^{-5}\ M$). The circles were generated by *RK* calculation, and the lines correspond to the fittings with Eq. 11.20. The reaction features are $l_{irr} = 2.37\ cm$, $P_0^{\lambda_X} = 1.25\ 10^{-5}\ einstein\ dm^{-3}\ s^{-1}$, $\Phi_{X \to Y}^{\lambda_X} = 0.145$, and $\varepsilon_X^{\lambda_X} = 9504\ M^{-1}\ cm^{-1}$. The initial reaction-rates and half-life times are $r_{X,a}^{\lambda_X,T}(0) = -1.81\ 10^{-6}\ M\ s^{-1}$, $t_{\frac{1}{2},a}^{\lambda_X} = 86\ s$, and $r_{X,b}^{\lambda_X,T}(0) = -1.45\ 10^{-6}\ M\ s^{-1}$, $t_{\frac{1}{2},b}^{\lambda_X} = 12.6\ s$ for the reaction's initial concentrations $C_{X,a}^{\lambda_X,T}(0)$ and $C_{X,b}^{\lambda_X,T}(0)$, respectively. The rate-constants (Eq. 11.21) for the latter reactions are equal, $k_{r,a}^{\lambda_X,T} = k_{r,b}^{\lambda_X,T} = 0.094\ s^{-1}$ (i.e. they are initial concentration–invariant)

(l) The rate (Eq. 11.9) and the initial velocity of the reaction ($r_X^{\lambda_X,T}(0)$, Eq. 11.32) are proportional to the factor $\left(1 - 10^{-\varepsilon_X^{\lambda_X} l_{irr} C_X^{\lambda_X,T}(0)}\right)$. $r_X^{\lambda_X,T}(0)$ increases with increasing initial concentration.

$$r_X^{\lambda_X,T}(0) = -k_r^{\lambda_X,T}\ \frac{\left(1 - 10^{-\varepsilon_X^{\lambda_X}\ l_{irr}\ C_X^{\lambda_X,T}(0)}\right)}{\varepsilon_X^{\lambda_X}\ l_{irr}\ Ln(10)} = -0.4343\ k_r^{\lambda_X,T}\ C_X^{\lambda_X,T}(0)\ PKF^{\lambda_X}(0) \qquad (11.32)$$

The behaviour of the reaction tends to a linear variation of the concentration with reaction time when the initial concentration tends to have higher values (Eq. 11.28, Fig. 11.3). An equivalent overall behaviour can be observed by varying the irradiation optical path length, l_{irr}, values.

A possible confusion could arise from points (*k*) and (*l*) of the above list, as an increase of both the half-life time ($t_{\frac{1}{2}}^{\lambda_X}$) and the initial velocity ($-r_X^{\lambda_X,T}(0)$) are predicted with the increase of $C_X(0)$. The apparent contradiction comes from the fact that the reaction seems to both speed up with an increase of $-r_X^{\lambda_X,T}(0)$ and slow down with an increase of $t_{\frac{1}{2}}^{\lambda_X}$, when $C_X(0)$ increases. In fact, for our primary photoreaction under λ_X-irradiation, when $C_X(0)$ increases, the reaction speeds up, but because $C_X(0)$ has a higher value, it reaches $t_{\frac{1}{2}}^{\lambda_X}$ later in time compared to the same reaction performed with a smaller value of $C_X(0)$. It is all due to the fact that $\left(1-10^{-\varepsilon_X^{\lambda_X}\, l_{irr} C_X^{\lambda_X,T}(0)}\right)$ increases faster than $ln\left(\sqrt{10^{\varepsilon_X^{\lambda_X}\, l_{irr} C_X^{\lambda_X,T}(0)}}+1\right)$ in Eqs. 11.31 and 11.32, respectively.

For illustration, small increments (0.1) of the initial absorbance, $A_X^{\lambda_X,T}(0)=\varepsilon_X^{\lambda_X}\, l_{irr} C_X^{\lambda_X,T}(0)$ $(1 \geq A_X^{\lambda_X,T}(0) \geq 0.1)$, induce a fairly constant $t_{\frac{1}{2}}^{\lambda_X}$, which varies between 1.064 and 1.083 (1.78%), whereas, the initial velocity increases by 74.34% $(1.029 \geq -r_X^{\lambda_X,T}(0) \geq 1.794)$ for the same increments. If we take only two values of $A_X^{\lambda_X,T}(0)$ as 1 and 0.1, the ratios $-r_X^{\lambda_X,T}(0)/t_{\frac{1}{2}}^{\lambda_X}$ for these two examples are 4.37 and 1.89, respectively (both >1). This indicates that $-r_X^{\lambda_X,T}(0)$ increases faster than $t_{1/2}$ when $A_X^{\lambda_X,T}(0)$ increases, and this is more pronounced when the difference between the considered $A_X^{\lambda_X,T}(0)$ values is bigger.

Accordingly, the integrated rate-law of the primary photoreaction under λ_X-irradiation possesses particular kinetic features (*Points* ***a***–***l***) that cannot be mapped out by any other kinetic equation known today. Its specific behaviour (Eq. 11.20) is also unique in kinetics.

We are, therefore, in the presence of a new reaction order. This particular reaction order and the kinetics reflected by it are dubbed the "Φ-*order kinetics.*" As far as we know, Φ-order reactions are necessarily photoreactions. This terminology will be employed in subsequent chapters.

The Φ-order is then the true kinetic order of the primary photoprocess performing under monochromatic irradiation at λ_X (with $\varepsilon_Y^{\lambda_X}=0$).

We shall discuss the cases of photo- and photothermal reactions driven by mono- and polychromatic irradiation, in relation to Φ-order kinetics in Parts V and VII.

11.6 Approximating Φ-Order to Thermal Kinetic Models

We can review two cases where the Φ-order kinetics can seemingly be estimated by thermal kinetic formulae.

The first case is based on a limit property of the power numbers. It is known that if the argument of the power number is small, we can use the first-order Maclaurin expansion of the exponential (Sect. 3.8), where the difference $1-e^{-x} \cong 1-1-(-x) \cong x$, which can be extended to the power 10 expression, as $1-10^{-x}=1-e^{-x\, ln\, 10} \cong 1-1-(-x\, ln\, 10) \cong x\, ln\, 10$ or $(1-10^{-x})/\, ln\, 10 \cong x$.

To a certain extent, if the absorbance is also very small ($\varepsilon_X^{\lambda_X}\, l_{irr} C_X^{\lambda_X,T} \ll 1$), then the Φ-order kinetics of the primary photoreaction under λ_X-irradiation, expressed by Eq. 11.12, becomes

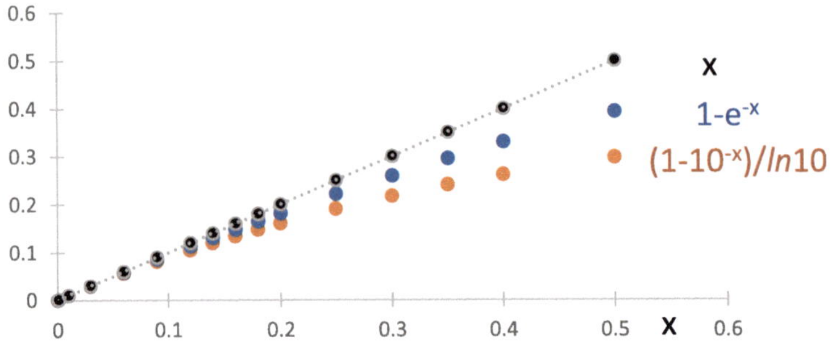

Fig. 11.4 A comparison of the value of x to its approximation with both $(1-10^{-x})/\ln 10$ and $1-e^{-x}$

$$\frac{dC_X^{\lambda_X,T}(t)}{dt}=-\Phi_{X\to Y}^{\lambda_X}\,P_0^{\lambda_X}\left(1-10^{-\varepsilon_X^{\lambda_X}\,l_{irr}\,C_X^{\lambda_X,T}(t)}\right)\frac{\ln10}{\ln10}=$$
$$-\left(\Phi_{X\to Y}^{\lambda_X}\,P_{\lambda_X}^{0}\,\varepsilon_X^{\lambda_X}\,l_{irr}\,\ln10\right)C_X^{\lambda_X,T}(t) \qquad (11.34)$$

In these conditions, the resulting rate is linearly dependent on the concentration, and therefore, the reaction obeys a first-order kinetics.

However, the approximation is only valid for very small values of the absorbances. Figure 11.4 indicates that this only holds for absorbance values much smaller than 0.1. Hence, for the Φ-order kinetics to be approximated to a first-order kinetics, the required absorbance values that are allowed to be used (<0.1) are not those commonly employed in practice for reasons relating to accuracy of absorbance measurements and/or to real-life samples.

In the second case, the approximation considers the vanishing of the power number when its argument is high and negative, $10^{-x}\cong 0$ when $x\gg 1$. When applied to Eq. 11.12, the values of the concentration (and/or the optical path length, l_{irr}) must be high enough so that the absorbance is high. In these conditions, we have

$$\frac{dC_X^{\lambda_X,T}(t)}{dt}=-\Phi_{X\to Y}^{\lambda_X}\,P_0^{\lambda_X}\left(1-10^{-\varepsilon_X^{\lambda_X}\,l_{irr}\,C_X^{\lambda_X,T}(t)}\right)=-\Phi_{X\to Y}^{\lambda_X}\,P_0^{\lambda_X} \qquad (11.35)$$

Here, the rate is constant, and the reaction kinetics has a 0th-order character.

This approximation is, in principle, not valid at all owing to the mandatory condition imposed to the absorbance, that is, the condition relative to a standing Beer–Lambert law (the absorbance must belong to the linearity range of the calibration graph), which is most likely not satisfied, as the calibration graphs of organic molecules are seldom linear beyond $A_X^{\lambda_X,T}(0)>1.5$. Owing to this discrepancy, the reaction rate-law, as written in Eq. 11.12, is not valid anymore when $A_X^{\lambda_X}(0)\gg 1$. And because of the invalidity of the rate-law, the approximation becomes void (see Sect. 2.13).

11.7 The Quantum Yield Expression

A similar equation to that set out for the classical differential quantum yield (Eq. 1.5) can be worked out from the rate-law of the unimolecular reaction considered under λ_X-irradition, Eq. 11.12.

$$\Phi_{X\to Y}^{\lambda_X} = -\frac{\frac{dC_X^{\lambda_X,T}(t)}{dt}}{P_0^{\lambda_X}\left(1-10^{-\varepsilon_X^{\lambda_X} l_{irr} C_X^{\lambda_X,T}(t)}\right)} \tag{11.36}$$

A second explicit expression of the quantum yield can also readily be obtained from the analytically derived integration of the above rate-law, Eq. 11.20.

$$\Phi_{X\to Y}^{\lambda_X} = -\frac{1}{P_0^{\lambda_X}\varepsilon_X^{\lambda_X} l_{irr}\ln 10}\ln\frac{10^{\varepsilon_X^{\lambda_X} l_{irr} C_X^{\lambda_X,T}(t)}-1}{10^{\varepsilon_X^{\lambda_X} l_{irr} C_X^{\lambda_X,T}(0)}-1} \tag{11.37}$$

Despite their differences, these two equations relay the same physical meaning. However, the latter (Eq. 11.37) substantially differs from the classical integral quantum yield equation (Eq. 1.4). The same assessment is reached for the formula of the quantum yield worked out from Eq. 11.25.

$$\Phi_{X\to Y}^{\lambda_X} = \left(\frac{C_X^{\lambda_X,T}(0) - C_X^{\lambda_X,T}(t)}{P_0^{\lambda_X}\, t}\right) + \frac{1}{P_0^{\lambda_X}\, t}\frac{1}{\ln 10\,\varepsilon_X^{\lambda_X} l_{irr}}\ln\frac{1-10^{-\varepsilon_X^{\lambda_X} l_{irr} C_X^{\lambda_X,T}(0)}}{1-10^{-\varepsilon_X^{\lambda_X} l_{irr} C_X^{\lambda_X,T}(t)}} \tag{11.38}$$

The first fraction (in brackets) on the right-hand-side term of Eq. 11.38 might be considered to equate the equation of the classical integral quantum yield of the reaction (Eq. 1.4). This might be justified by a ubiquitous practice assimilating the number of photons delivered by the incident light to the reactor over an interval of time to the number of quanta absorbed (Eq. 5.1), i.e. $n_{tot.abs.h\nu} = P_0^{\lambda_X}\, t$. In this perspective, the reactant supposedly absorbs all the photons entering the reactor. Even though this condition was not relevant or necessary for the derivation of Eq. 11.25, let us assume that it is valid. In this case, the second term of the sum on the right-hand side of Eq. 11.38 cannot be equal to zero if the reaction progresses over the time interval t (i.e. $C_X^{\lambda_X,T}(0) \neq C_X^{\lambda_X,T}(t)$ in the last ratio of Eq. 11.38).

This discussion proves that the classical integral quantum yield formula does not fulfil the requirements of the actual physical system. However, both Eqs. 11.37 and 11.38 are valid for any time interval, value of $P_0^{\lambda_X}$, and/or initial concentration value (as long as $A_X^{\lambda_X,T}(0)$ belongs to the linearity range of the reactant spectrophotometric calibration graph).

To expand this point further, let us look into the reaction under an isosbestic irradiation. Here, the integrated rate-law is obtained from Eq. 8.54 that is adapted to the unimolecular reaction, i.e. where $a_5^{\lambda_{isos}} = a_9^{\lambda_{isos}} = C_Y(0) = 0$ and $\rho_{\pm} = a_{19}^{\lambda_{isos}} = a_1^{\lambda_{isos}} = -a_4^{\lambda_{isos}} = -\Phi_{X\to Y}^{\lambda_{isos}}\,\gamma_1$.

$$C_X^{\lambda_{isos},T}(t) = C_X^{\lambda_{isos},T}(0)\; e^{-\Phi_{X\to Y}^{\lambda_{isos}}\,\gamma_1\, t} \qquad (11.39)$$

which is rearranged to yield a formula for the quantum yield as

$$\Phi_{X\to Y}^{\lambda_{isos}} = \frac{1}{\varepsilon_X^{\lambda_{isos}}\; l_{irr}\; F^{\lambda_{isos}}}\,\frac{1}{P_0^{\lambda_X}\; t}\,\ln\frac{C_X^{\lambda_{isos},T}(0)}{C_X^{\lambda_{isos},T}(t)} \qquad (11.40)$$

Again, this equation cannot be rearranged to give the classical integral quantum yield formula (Eq. 1.4).

11.8 A Change of the Photoreaction Kinetic Order

It is interesting to observe that one of the inherent and very special kinetic properties of a photoreaction is a true change of its kinetic order. Indeed, every photoreaction can have at least two distinct kinetic behaviours, whose respective integrated rate-law equations cannot be worked out from one another, a situation that occurs in the instance of a primary photoreaction when the wavelength of the monochromatic irradiation beam is either equal to that of an isosbestic point (λ_{isos}) or different from it ($\lambda_X \neq \lambda_{isos}$).

We have seen that the isosbestic irradiation imposes a pure first-order kinetics to the primary photoreaction (Chap. 8) if assuming the situation shown in Fig. 11.1. Here, the kinetic trace of the reaction is described by a mono-exponential function, Eq. 8.54 [16, 17]. However, under λ_X-excitation, the trace follows Eq. 11.20, indicating a Φ-order kinetics.

Therefore, a genuine change of the kinetic order obeyed by the unimolecular reaction is attributable to the particular choice of the irradiation wavelength (λ_X or λ_{isos}). It is also a compelling case of distinctability (the kinetic behaviour of the reaction cannot be deduced from its mechanism).

Such a shift of the reaction order is not specific to primary photoreactions. It is expected to occur for photoreactions in general.

It is relevant to stress that such a true change of reaction-order is explicitly and uniquely attributable to a particular choice of the irradiation wavelength (λ_X or λ_{isos}) and not to any change in the magnitude of the initial concentration of the reactant, $C_X(0)$. In this context, the supposed change of the primary photoreaction order that was ascribed, in the literature (Chap. 7), to a variation of $C_X(0)$ is a mere approximation. In the approaches described therein, no concern has been directed towards the irradiation type, where the excitation light was assumed to have an undifferentiated wavelength, λ_{irr}. However, since the argument of the exponential factor involves a single term ($\varepsilon^{\lambda_{irr}} l_{irr} C_X(t)$), it theoretically corresponds to the primary photoreaction under non-isosbestic λ_X-irradiation. The argument for a change of reaction order when low and high initial concentrations of X are used, follows from some of the derivations presented in Sects. 11.2 and 11.6. That is, when $C_X(0)$ is very low, the absorbance is low, $\varepsilon_X^{\lambda_{irr}} l_{irr} C_X(t) < < 1$, and the exponential factor in Eq. 11.12 can be developed into an expansion series (Eq. 11.2), resulting in a rate-law similar to that of first-order kinetics (Eq. 11.3 or Eq. 11.34). On the other hand, if the absorbance is high, $\varepsilon_X^{\lambda_{irr}} l_{irr} C_X(t) > > 1$, as in the case where $C_X(0)$ is high, then the exponential factor in Eq. 11.12 can be assumed negligible (so that $1 - 10^{-\varepsilon_X^{\lambda_{irr}} l_{irr} C_X(t)} \approx 1$), and the rate-law is approximated to that of a zeroth-order reaction (Eq. 11.35). Hence, this approach concluded that the kinetic order of the primary photoreaction changes from the first order (low $C_X(0)$) to the zeroth-order (high $C_X(0)$), depending on the concentration. These types of approximations were, however, proven to be limiting cases for the general Φ-order kinetics (Sect. 11.6).

Incidentally, the latter approach did not hint or explore the increase of $P_0^{\lambda_{irr}}$. As a matter of fact, Fig. 11.5 clearly indicates that by changing the light intensity, the shape of the kinetic curve is invariant ($k_r^{\lambda_{irr}}$ or the product ($k_a\, k_b$) in Eq. 11.28 is only a scaling factor). However, the higher the incident radiation intensity, $P_0^{\lambda_{irr}}$, the higher the $r_X^{\lambda_X,T}(0)$ and $k_r^{\lambda_{irr}}$ and the shorter the $t_{\frac{1}{2}}^{\lambda_X}$. In general, these behaviours described in this chapter are specific to the unimolecular reaction whose product is transparent to the incident light but do not automatically extend to other photo- or photothermal reactions (as will be evidenced in subsequent chapters).

11.9 Unsuitability of the First-Order Model for Φ-Order Kinetic Traces

The limiting cases where Φ- and first-order kinetics tend to have a similar behaviour, following a mono-exponential template, have been reviewed in Sect. 11.6. However, it is understood that this is not, in general, true. A proof of principle is brought by testing the mono-exponential function on $RK4$-calculated or Eq. 11.20 kinetic traces.

For illustration, we consider the trace of a unimolecular reaction exposed to a non-isosbestic irradiation, performed at a constant temperature and characterised by $C_X(0) = 1.47 \times 10^{-5} M$, $l_{irr} = 2.3\ cm$, $P_0^{\lambda_{irr}} = 9 \times 10^{-5}\ einstein\ dm^{-3}\ s^{-1}$, $\Phi_{X\to Y}^{\lambda_X} = 0.063$, $\varepsilon_X^{\lambda_X} = 29403\ M^{-1}\ cm^{-1}$, $\varepsilon_Y^{\lambda_X} = 0$, $\varepsilon_X^{\lambda_{obs}} = 5811\ M^{-1}\ cm^{-1}$, and $\varepsilon_Y^{\lambda_{obs}} = 14287\ M^{-1}\ cm^{-1}$. A $A_{tot}^{\lambda_{obs},T}(t)$ trace is obtained by feeding an $RK4$ integration of Eq. 11.9 with the latter set of data. The calculation assumes that $A_X^{\lambda_X,T}(0)$ lies within the linearity range of the calibration graph that is constructed for the reactant X by measuring its absorbance at the wavelength λ_X.

For the reaction with the above features (Fig. 11.6), as for many other examples, the Φ-order equation (Eq. 11.20) is obeyed by numerically calculated traces of

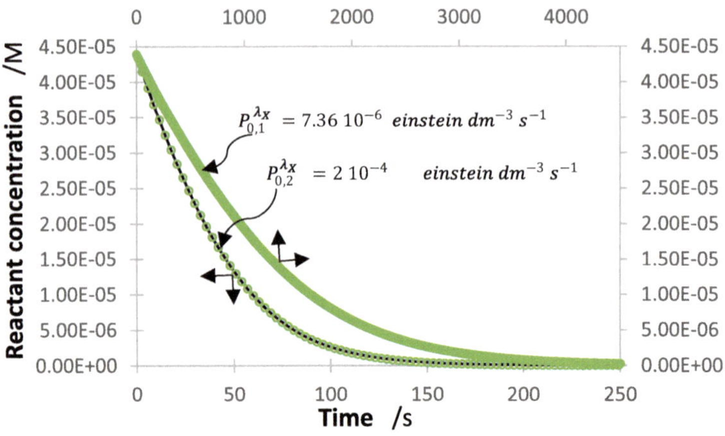

Fig. 11.5 Increasing $P_0^{\lambda_{irr}}$ ($P_{0,2}^{\lambda_{irr}} > P_{0,1}^{\lambda_{irr}}$) translates into a faster reaction ($k_{r,2}^{\lambda_{irr}} = 0.037\ s^{-1} > k_{r,1}^{\lambda_{irr}} = 0.00135\ s^{-1}$) but does not affect the shape of the trace. The traces generated by *RK*-calculation (circles) were fitted with Eq. 11.20 for the reaction $C_X^{\lambda_X}(0) = 4.39\ 10^{-5} M$, $l_{irr} = 1.41\ cm$, $\Phi_{X \to Y}^{\lambda_X} = 0.0051$, and $\varepsilon_X^{\lambda_X} = 11094\ M^{-1}\ cm^{-1}$. The initial reaction-rates and half-life times evolve in opposite directions, $r_{X,1}^{\lambda_X,T}(0) = -2.98\ 10^{-8}\ M\ s^{-1}$, $t_{\frac{1}{2},1}^{\lambda_X} = 861\ s$, and $r_{X,2}^{\lambda_X,T}(0) = -1.45\ 10^{-6}\ M\ s^{-1}$, $t_{\frac{1}{2},2}^{\lambda_X} = 104.5\ s$, for the reaction's incident light intensities, $P_{0,1}^{\lambda_{irr}}$ and $P_{0,2}^{\lambda_{irr}}$, respectively

unimolecular reactions driven by λ_X-excitation (Fig. 11.6(a$_1$)). The same is also shown for the *RK*4 total absorbance trace that is fitted to the formula $A_{tot}^{\lambda_{obs},T}(t) = \varepsilon_X^{\lambda_{obs}}\, l_{irr} C_X(t) + \varepsilon_Y^{\lambda_{obs}}\, l_{irr}(C_X(0) - C_X(t))$, the *RK*4 data for the reactant concentration, and the absorbance at λ_{obs} (Fig. 11.6(b$_1$)). This might perhaps be considered as an additional confirmation of the model equation, even though, in principle, analytically derived equations need no confirmation by alternative methods.

Conversely, both reactant concentration (Fig. 11.6(a$_2$)) and total absorbance traces (Fig. 11.6(b$_2$)) are very poorly fitted to a mono-exponential model [11]. The latter being a property of first-order kinetics, the poor fitting stands as undisputable evidence for the disparity that exists between first- and Φ-order kinetics.

11.10 A Matter for Reflection

A twofold argument can perhaps be put forward to question the relevance of precisely determining the order of a photoreaction for practical quantification of that process.

On the one hand, one might argue that the important features determined by the kinetic study are things like the proportion of reaction progress in an interval of time, such as the 10% degradation or the half-life of the reaction. This type of information

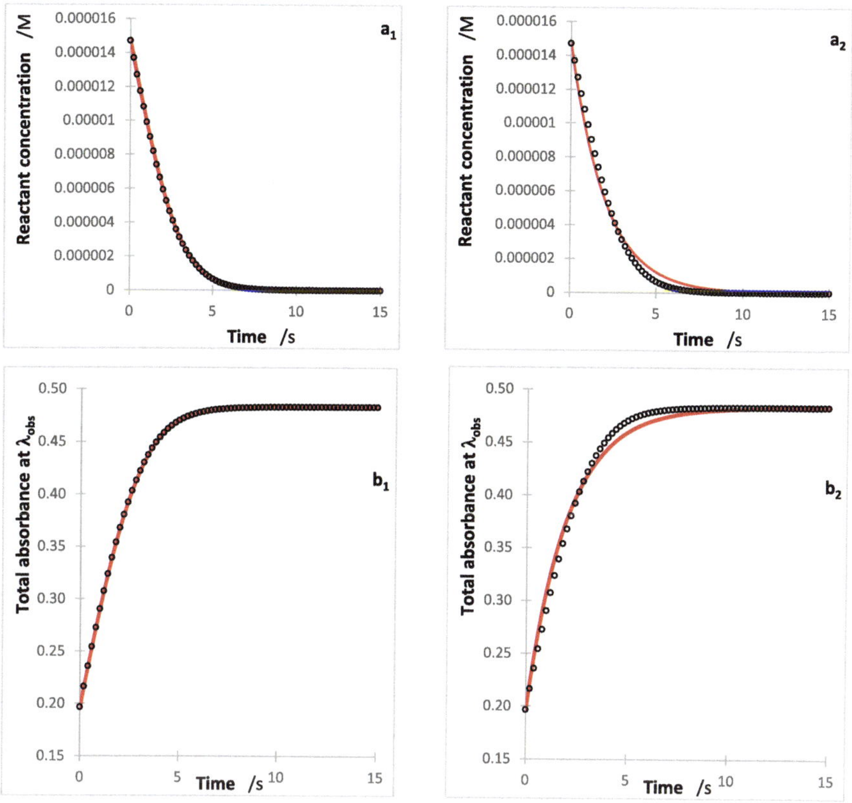

Fig. 11.6 Trace of an $XY(1\Phi)$ reaction under a λ_X-irradiation, calculated by $RK4$ integration of Eq. 11.12 (black circles), which has been fed with the parameter values provided in the text. The disappearance of the reactant species in the reaction medium is shown as a time-variation of either the concentration of the reactant ((a_1) and (a_2)) or the total absorbance at λ_{obs} ((b_1) and (b_2)). The kinetic trace of X is incompatible with a mono-exponential model, Eq. 8.54 (red line in (a_2) and (b_2)), but are well reproduced by the Φ-order theoretical model (red line in (a_1) and (b_1))

will not considerably change if the mathematical model employed for their determination does not exactly correspond to the true order of the reaction.

On the other hand, the essential property of the photoreaction, i.e. the photochemical quantum yield, is obtained by a photokinetic-independent method based on actinometric measurements.

Therefore, as long as a mathematical model, for instance, that of the first-order kinetics, is fitting relatively well with the experimental kinetic trace of the reaction, it is sufficient to obtain some practical features of that reaction.

Ultimately, one can argue that even if the order of the reaction is not known and even if a kinetic study is not conducted at all, the quantum yield of the reaction can still be determined (by actinometry). This makes the photoreaction kinetic order determination look like the earthen pot facing the brass kettle, represented by the efficiency of actinometry for the determination of the quantum yield. From this point of view, detailed kinetic studies of photoreactions seem less relevant. This might also be supported by the fact that experimental photochemistry has continuously developed during the past century despite the lack of precise knowledge of photoreactions' orders, including that of the simplest photoreaction.

Such a strong argument cannot, for instance, be addressed without a detailed analysis of the strategy employed for the determination of the quantum yield. We have seen in Chap. 4 a plethora of ways that have been adopted for quantum yield determination. The lack of consensus around this matter is somewhat deepened by the discussion carried out in Chap. 5, relative to the most popular integral and differential quantum yield formulae. Also, the use of thermal kinetic models in photochemistry is not suitable (Chap. 7). This leaves us with an obvious conclusion, that is, a standardisation of photokinetics has become an urgent requirement.

For the moment, let us agree that the knowledge of the ϕ-order kinetic established for the unimolecular reaction ($\varepsilon_Y^{\lambda_{irr}} = 0$) certainly achieves what *NIM*s do but also brings more advantages, among which are the following:

a-. The natural way in kinetics of thermal reactions is to determine the integrated rate-law and then unleash a kinetic study. So, in principle, it should be the same in photokinetics. It is a kind of a harmonisation of the general subject of kinetics along the same rules.
b-. The order of a reaction might be deduced from the *ODE* of the reaction, but it is more easily known from the integrated rate-law. The latter is a handy equation to use in experimental photochemistry, facilitating the quantification of the reaction attributes.
c-. We shall see that the integrated rate-law is very useful for the rational determination of the quantum yield (of each photoreaction-step of the mechanism).
d-. It also allows a predictive power for the kinetics used. For instance, how would the rate of the reaction (of a given kinetic order) vary with initial concentration? The few examples set out in this chapter are a useful illustration.
e-. The integrated rate-law and the mathematical analysis of the reaction equations are necessary to reach a rational interpretation of the origin of distinctability, distinguishability, and identifiability issues.

The above points are not accessible to *NIM*s.

11.11 Conclusion

Because the differential equations (e.g., Eq. 3.26) written for thermal reactions (i) do not accommodate absorptivity terms and irradiation path lengths, (ii) lead to inconsistent units for photochemical reactions rate-constants, and (iii) are incapable of explaining the effects of absorbing spectator species (light absorption competitors), they are deemed inadequate to describe photoreaction behaviours and should not be used in photokinetics.

The photokinetics of the unimolecular photoreaction ($\varepsilon_{product} = 0$) has been proven to obey a different behaviour from that of the usual thermal kinetic reactions. This behaviour has been mathematically characterised, and the integrated rate-law of the reaction has been derived analytically.

From the mathematical formulation of that unimolecular reaction kinetics, it is undoubtedly established that the reaction obeys a distinctive order and kinetics, namely, *the Φ-order kinetics*, whose characteristic formulation is given by Eq. 11.20, or as a general format, Eq. 11.41 (with ϕ_i being constants).

$$C_X(t) = \phi_1 \; log\left(1 + \phi_2 \, e^{-\phi_3 \, t}\right) \tag{11.41}$$

The Φ-order kinetics, however, consistently rationalises all behavioural aspects of the unimolecular photoreaction ($\varepsilon_{product} = 0$) and predicts its experimentally observed trends.

For future research, there might be two questions of interest that must be addressed in order to sustain progress in this underdeveloped area of physical chemistry: (i) whether closed-form integration methods will be developed to solve photoreactions' differential equations and (ii) whether more than one (*Φ*-) order type exist in photoreaction kinetics.

The conjecture for the latter question would be to assume the existence of such variety, as differential equations are different (and have not yet been set/solved) for each different photoreaction.

Any progress on these questions would certainly have impact, not only academically but also in a wide range of research disciplines, as photokinetics has found its way into many modern technologies.

References

1. Logan RS (1997) Does a photochemical reaction have a reaction order? J Chem Edu 74:1303. https://doi.org/10.1021/ed074p1303
2. Hippler M (2003) Photochemical kinetics: reaction orders and analogies with molecular beam scattering and cavity ring-down experiments. J Chem Edu 80:1074–1077. https://doi.org/10.1021/ed080p1074
3. Toby S (2005) Does a photochemical reaction have a reaction order? J Chem Edu 82:37–38. https://doi.org/10.1021/ed082p37.2
4. Nassar N, Trivella A, Mokh S, Al-Iskandarani M, Budzinski H, Mazellier P (2017) Photodegradation of sulfamethazine, sulfamethoxypiridazine, amitriptyline, and clomipramine drugs in aqueous media. J Photochem Photobiol A Chem 336:176–182. https://doi.org/10.1016/j.jphotochem.2016.12.008
5. Shaw H, Toby S (1966) Light absorption in photochemistry. J Chem Edu 43:408–410. https://doi.org/10.1021/ed043p408
6. Jorfi S, Barzegar G, Ahmadi M, Soltani RDC, Haghighifard NAJ, Takdastan A, Saeedi R, Abtahi M (2016) Enhanced coagulation-photocatalytic treatment of acid red 73 dye and real

textile wastewater using UVA/synthesized MgO nanoparticles. J Env Manag 177:111–118. https://doi.org/10.1016/j.jenvman.2016.04.005
7. Uher G, Pillans JJ, Hatton AD, Upstill-Goddard R (2017) Photochemical oxidation of dimethylsulphide to dimethylsulphoxide in estuarine and coastal waters. Chemosphere 186: 805–816. https://doi.org/10.1016/j.chemosphere.2017.08.050
8. Hu C, Xu M, Zhang J, Hu B, Yu G (2019) Recyclable MoO_3 nanobelts for photocatalytic degradation of Rhodamine B by near infrared irradiation. Int J Chem Kinet 51:3–13. https://doi.org/10.1002/kin.21223
9. Tonnesen HH (ed) (2004) Photostability of drugs and drug formulations. CRC Press, Boca Raton. ISBN. 0-415-30323—0
10. Mauser H, Gauglitz G (1998) Comprehensive chemical kinetics. In: Compton RG, Hancock G (eds) Photokinetics: theoretical fundamentals and applications, vol 36. Elsevier, Amesterdam. ISBN. 978-0444825360
11. Maafi M, Brown RG (2007) The kinetic model for AB(1Φ) systems. A closed-form integration of the differential equation with a variable photokinetic factor. J Photochem Photobiol A Chem 187:319–324. https://doi.org/10.1016/j.jphotochem.2006.10.030
12. Klan P, Wirz J (2009) Photochemistry of organic compounds: from concepts to practice. Wiley-Blackwell, Chicester, UK. https://doi.org/10.1002/9781444300017
13. Sumi T, Takagi Y, Yagi A, Morimoto M, Irie M (2014) Photoirradiation wavelength dependence of cycloreversion quantum yields of diarylethenes. Chem Commun 50:3928–3930. https://doi.org/10.1039/c4cc00396a
14. Bolton JR (2020) A master equation for photochemical rates. Photochem Photobiol 96:1355–1357. https://doi.org/10.1111/php.13325
15. Lente G (2015) Deterministic kinetics in chemistry and systems biology. The dynamics of complex reaction networks. Springer, Cham/Heidelberg/New York/Dordrecht/London. https://doi.org/10.1007/978-3-319-15482-4
16. Brown RG, Maafi M (2005) General analytical solutions for the kinetics of AB(k, Φ) and ABC (k, Φ) systems. Int J Chem Kinet 37:162–174. https://doi.org/10.1002/kin.20061
17. Maafi M, Maafi W (2015) Quantitative assessment of the photostability and photostabilisation of Fluvoxamine and its design for actinomtry. Photochem Photobiol Sci 14:982–994. https://doi.org/10.1039/c5pp00022j

Chapter 12
The Hurdle

12.1 Introduction

The analytical solution achieved for the simplest photoreaction and the discovery of Φ-order kinetics, as described in the previous chapter, raise interest towards studying more photokinetic systems under non-isosbestic irradiation. The central question here is whether closed-form integrations can be achieved for other photo- and photothermal reactions. One can observe that, unlike what is found in thermal kinetics, answers to the latter question do not seem to have been explored in any detail within the literature of photochemistry. A way forwards to shed light on this aspect is to look at the integration of the rate-laws of reactions having a slight incrementation in complexity compared to that of the unimolecular reaction whose product is transparent.

12.2 The Unimolecular Reaction with an Absorbing Product

The next reaction on the ladder of complexity after the unimolecular reaction $XY(1\Phi)$, whose product is transparent ($\varepsilon_Y^{\lambda_X} = 0$, and $\lambda_{n-isos} = \lambda_X$), is the same reaction but with a photoproduct absorbing the monochromatic/non-isosbestic excitation light ($\varepsilon_Y^{\lambda_{n-isos}} \neq 0$) in a different proportion compared to the absorption of the reactant ($\varepsilon_X^{\lambda_{n-isos}} \neq \varepsilon_Y^{\lambda_{n-isos}}$).

The standard rate-law of the latter reaction is given by Eq. 11.11 but can be rearranged into a more appropriate form for integration. For that purpose, let us consider the formulation of the total absorbance, where, for simplicity, we assume $l_{obs} = l_{irr}$.

M. Maafi, *Photokinetics*, https://doi.org/10.1007/978-3-031-98985-8_12

$$A_{tot}^{\lambda_{n-isos},T/\lambda_{n-isos}} = A_X^{\lambda_{n-isos},T/\lambda_{n-isos}} + A_Y^{\lambda_{n-isos},T/\lambda_{n-isos}}$$
$$= \left(\varepsilon_X^{\lambda_{n-isos}} - \varepsilon_Y^{\lambda_{n-isos}}\right) l_{irr}\, C_X^{\lambda_{n-isos}}(t) + \varepsilon_Y^{\lambda_{n-isos}}\, l_{irr}\, C_X(0) = \delta_1\, C_X^{\lambda_{n-isos}}(t) + \delta_2 \quad (12.1)$$

When introducing Eq. 12.1 in Eq. 11.11, we have

$$\frac{dC_X^{\lambda_{n-isos},T}(t)}{dt} = \frac{\delta_3\, C_X^{\lambda_{n-isos},T}(t)\left(1 - 10^{-\delta_1\, C_X^{\lambda_{n-isos},T}(t) - \delta_2}\right)}{\delta_1\, C_X^{\lambda_{n-isos},T}(t) + \delta_2} \quad (12.2)$$

with

$$\delta_3 = -\Phi_{X\to Y}^{\lambda_{n-isos}}\, P_0^{\lambda_{n-isos}}\, \varepsilon_X^{\lambda_{n-isos}}\, l_{irr} \quad (12.3)$$

And by separating the variable of Eq. 12.2, we have

$$\frac{\delta_1\, C_X^{\lambda_{n-isos},T}(t) + \delta_2}{C_X^{\lambda_{n-isos},T}(t)}\; \frac{dC_X^{\lambda_{n-isos},T}(t)}{1 - 10^{-\delta_1\, C_X^{\lambda_{n-isos},T}(t) - \delta_2}} = \delta_3\, dt \quad (12.4)$$

The integration of Eq. 12.4 can be presented in the following form.

$$\underbrace{\delta_1 \int_{C_X(0)}^{C_X(t)} \frac{dC_X^{\lambda_{n-isos},T}(t)}{1 - 10^{-\delta_1\, C_X^{\lambda_{n-isos},T}(t) - \delta_2}}}_{Int_a}$$
$$+ \underbrace{\delta_2 \int_{C_X(0)}^{C_X(t)} \frac{1}{C_X^{\lambda_{n-isos},T}(t)}\; \frac{dC_X^{\lambda_{n-isos},T}(t)}{1 - 10^{-\delta_1\, C_X^{\lambda_{n-isos},T}(t) - \delta_2}}}_{Int_b} = \delta_3 \underbrace{\int_0^t dt}_{Int_c} \quad (12.5)$$

Hence, solving the rate-law of the photoreaction (Eq. 12.2) is equivalent to performing three separate integrations (of the three integrals Int_a, Int_b and Int_c in Eq. 12.5).

The integrations of Int_a and Int_c are straightforward, following the procedures previously adopted in Sect. 11.4. However, the closed-form integration of Int_b is impossible as its integrand has no known antiderivative. As a consequence, the analytical integrated rate-law for this reaction is not accessible. For now, the full photokinetic characterisation of the relatively simple $XY(1\Phi)$ with $\varepsilon_Y^{\lambda_{n-isos}} \neq 0$ under non-isobestic irradiation at $\lambda_{n\,-\,isos}$ is unreachable until the development of a new mathematical method capable of integrating Int_b in a closed form.

Incidentally, it is interesting to show that even if the integrated rate-law of the reaction is not known, the quantum yield of the reaction can still be defined by measurable quantities. Indeed, using the rate-law (Eq. 11.11) at time $t = 0$, results in the equation of the initial reaction-rate are

$$\left(\frac{dC_X^{\lambda_n-isos,T}(t)}{dt}\right)_{t=0} = -\Phi_{X\to Y}^{\lambda_n-isos} P_0^{\lambda_n-isos} A_X^{\lambda_n-isos,T/\lambda_n-isos}(0)\,\frac{1-10^{-A_{tot}^{\lambda_n-isos,T/\lambda_n-isos}(0)}}{A_{tot}^{\lambda_n-isos,T/\lambda_n-isos}(0)} \quad (12.6)$$

where, at $t = 0$, $A_X^{\lambda_n-isos,T/\lambda_n-isos}(0) = A_{tot}^{\lambda_n-isos}(0) = \varepsilon_X^{\lambda_n-isos}\, l_{irr} C_X^{\lambda_n-isos,T}(0)$.

Rearranging Eq. 12.6 leads to defining the formula of the quantum yield as

$$\Phi_{X\to Y}^{\lambda_n-isos} = \frac{-\left(\frac{dC_X^{\lambda_n-isos,T}(t)}{dt}\right)_{t=0}}{P_0^{\lambda_n-isos}\left(1-10^{-\varepsilon_X^{\lambda_n-isos}\, l_{irr} C_X^{\lambda_n-isos,T}(0)}\right)} \quad (12.7)$$

If, now, the initial concentration of the reactant is low so that $A_X^{\lambda_n-isos,T/\lambda_n-isos}(0) < 0.1$, the rate of the reaction can be simplified by considering $10^{-A_{tot}^{\lambda_n-isos,T/\lambda_n-isos}(t)} \approx 0$. The expected shape for the $C_X^{\lambda_n-isos,T}(t)$ trace should have a mono-exponential trend. The same conclusion has been found above for the unimolecular reaction whose product is transparent to the incident light (Sect. 11.6). In these conditions, the $C_X^{\lambda_n-isos,T}(t)$ trace can be fitted to a mono-exponential formula and its initial rate value easily determined. The latter numerical values stand for $\left(dC_X^{\lambda_n-isos,T}(t)/dt\right)_{t=0}$, the numerator of Eq. 12.7. The remaining factors in the denominator of Eq. 12.7 are experimentally accessible. This strategy gives reliable experimental results for $\Phi_{X\to Y}^{\lambda_n-isos}$ [1].

12.3 The Photoreversible Reaction

Photoreversibility, *XY*(2*Φ*), imposes an absorption of both species at the non-isosbestic wavelength. The rate-law of the reactive system depends on both the timely concentrations of the reactant ($C_X(t)$) and the product ($C_Y(t)$), and the quantum yields of the two photoreaction-steps.

$$\frac{dC_X^{\lambda_n-isos,T}(t)}{dt} = P_0^{\lambda_n-isos}\, l_{irr}\left(\Phi_{Y\to X}^{n-\lambda_{isos}}\, \varepsilon_Y^{\lambda_n-isos}\, C_Y^{\lambda_n-isos,T}(t) - \Phi_{X\to Y}^{n-\lambda_{isos}}\, \varepsilon_X^{\lambda_n-isos}\, C_X^{\lambda_n-isos,T}(t)\right) \times \frac{1-10^{-A_{tot}^{\lambda_n-isos,T/\lambda_n-isos}(t)}}{A_{tot}^{\lambda_n-isos,T/\lambda_n-isos}(t)} \quad (12.8)$$

The total absorbance in Eq. 12.8 has the form given by Eq. 12.1. The rate-law for the *XY*(2*Φ*) reaction can be rearranged with variable separation in mind as

$$\frac{\delta_1\, C_X^{\lambda_n-isos,T}(t)+\delta_2}{\delta_4\, C_X^{\lambda_n-isos}(t)+\delta_5}\,\frac{dC_X^{\lambda_n-isos,T}(t)}{1-10^{-\delta_1\, C_X^{\lambda_n-isos,T}(t)-\delta_2}}=\delta_6\, dt \tag{12.9}$$

where δ_1 to δ_3 coefficients have the same definitions set out in Eqs.12.1 and 12.3. The remaining δ_4, δ_5 and δ_6 have the following expressions.

$$\delta_4=-\left(\Phi_{X\to Y}^{\lambda_{isos}}\,\varepsilon_X^{\lambda_n-isos}+\Phi_{Y\to X}^{\lambda_{isos}}\,\varepsilon_Y^{\lambda_n-isos}\right)\,l_{irr} \tag{12.10}$$

$$\delta_5=\Phi_{Y\to X}^{\lambda_{isos}}\,\varepsilon_Y^{\lambda_n-isos}\,C_X^{\lambda_n-isos,T}(0) \tag{12.11}$$

$$\delta_6=-P_0^{\lambda_n-isos}\,l_{irr} \tag{12.12}$$

Converting the latter equation Eq. 12.9 into its integration form gives

$$\underbrace{\int_{C_X(0)}^{C_X(t)}\frac{dC_X^{\lambda_n-isos,T}(t)}{1-10^{-\delta_1\, C_X^{\lambda_n-isos,T}(t)-\delta_2}}}_{Int_a}+\left(\frac{\delta_2}{\delta_1}\delta_4-\delta_5\right)\underbrace{\int_{C_X(0)}^{C_X(t)}\frac{dC_X^{\lambda_n-isos,T}(t)}{\left(\delta_4\, C_X^{\lambda_n-isos,T}(t)+\delta_5\right)\left(1-10^{-\delta_1\, C_X^{\lambda_n-isos,T}(t)-\delta_2}\right)}}_{Int_d}=\frac{\delta_4\,\delta_6}{\delta_1}\underbrace{\int_0^t dt}_{Int_c} \tag{12.13}$$

This equation has the overall mathematical form of the differential equation previously derived for the primary photoprocess (Eq. 12.5), where the second integral, Int_d, does not have an elementary antiderivative, making it impossible to be analytically integrated in a closed form. Therefore, the true integrated rate-law of the photoreversible reaction remains unknown. Eq. 12.13 does not support the analysis, claiming that the rate-law can be solved by applying the integration formula #11 in Table 3.1, in the same way it is applicable to the kinetics of $XY(1\Phi)$ when $\varepsilon_Y^{\lambda_n-isos}=0$, as might have been stated in the literature [2].

12.4 The Basic Photothermal Reversible Reaction

The minimum number of reaction-steps that can constitute a photothermal process is two, one photo and one thermal reaction-step. The basic photothermal reaction has one of two mechanisms, either an $XY(1\Phi, 1k)$ or an $XY_2(1\Phi, 1k)$. For the purpose of simplicity, we will consider the reaction involving the minimum number of reactive molecules, i.e. $XY(1\Phi, 1k)$ reaction. Two mechanisms can be proposed for the latter reaction where $\Phi^{\lambda_{irr}}_{X\to Y}$ is either concurrent (with $k^T_{X\to Y}$) or opposed to $k^T_{Y\to X}$.

The rate-laws of both latter reactions have the same form and can be set in a single expression.

$$\frac{dC_X^{\lambda_{n-isos},T}(t)}{dt} = -P_0^{\lambda_{n-isos}}\, l_{irr}\Phi^{n-\lambda_{isos}}_{X\to Y}\, \varepsilon_X^{\lambda_{n-isos}}\, C_X^{\lambda_{n-isos},T}(t) \times \frac{1-10^{-A_{tot}^{\lambda_{n-isos},T/\lambda_{n-isos}}(t)}}{A_{tot}^{\lambda_{n-isos},T/\lambda_{n-isos}}(t)} \pm k^T\, C_X^{\lambda_{n-isos},T}(t) \tag{12.14}$$

where the plus and minus signs before the thermal rate-constant indicate the reversed ($k^T_{Y\to X}$) and concurrent reactions, respectively.

The variable separation on the rate-law equation leads to two integrals. The integral (Int_e) on the left-hand side of Eq. 12.15 does not have a known antiderivative.

$$\underbrace{\int_{C_X(0)}^{C_X(t)} \frac{dC_X^{\lambda_{n-isos},T}(t)}{\gamma_3\, C_X^{\lambda_{n-isos},T}(t)\, \frac{1-10^{-\gamma_1 C_X^{\lambda_{n-isos},T}(t)-\gamma_2}}{\gamma_1\, C_X^{\lambda_{n-isos},T}(t)+\gamma_2} \pm k^T\, C_X^{\lambda_{n-isos},T}(t)}}_{Int_e} = \underbrace{\int_0^t dt}_{Int_c} \tag{12.15}$$

As for the previous reactions, here as well, the integrated rate-law of the simplest $XY(1\Phi, 1k)$ photothermal reaction is not accessible when the monochromatic light of a non-isosbestic wavelength is absorbed by both the reactant and the product.

12.5 Conclusion

The recurrent failure to reach the integrated rate-laws by closed-form integration for the set of the simplest reactions ($XY(1\Phi)$, $XY(2\Phi)$, $XY(1\Phi, 1k)$ or $XY_2(1\Phi, 1k)$) is undeniably an indication of the general situation in photokinetics. In fact, such mathematical hurdles should be expected for photo- and photothermal reactions subjected to a non-isosbestic/monochromatic light that is absorbed by more than one species in the reactive medium (i.e. the reactant and at least one of the products). Therefore, we can draw a conclusion to summarise the situation: *the integrated rate-laws in photokinetics are, in general, unachievable analytically for reactions exposed to non-isosbestic/monochromatic light*. This also means that the order of these reactions will remain unknown.

The exception to that rule is the unimolecular reaction (with $\varepsilon_Y^{\lambda_{n-isos}}=0$) and equivalent reactions, i.e. photo- and photothermal reactions, where the reactant is the only absorbing species in the medium at λ_{n-isos} (here, the reactant-integrated rate-law will have a similar general form of Eq. 11.20, and the reactant obeys Φ-order kinetics).

References

1. Stadler E, Eibel A, Fast D, Freißmuth H, Holly C, Wiech M, Moszner N, Gescheidt G (2018) A versatile method for the determination of photochemical quantum yields via online UV-vis spectroscopy. Photochem Photobiol Sci 17:660–669. https://doi.org/10.1039/C7PP00401J
2. Klan P, Wirz J (2009) Photochemistry of organic compounds: from concepts to practice. Wiley. https://doi.org/10.1002/9781444300017

Part V
Modelling Approaches for the Unsolvable Kinetics Under Monochromatic/Non-isosbestic Irradiation

Chapter 13
The Power Series Expansion Approach

13.1 Introduction

In Chap. 12, three types of unsolvable integrals were identified for the most basic photo- and photothermal reactions. The insolvability of those integrals made the solutions of their respective rate-laws not possibly achievable by analytical means. We will attempt here to bring approximated solutions to these integrals by using the Maclaurin expansion series [1]. The approximation using first-order expansion to the primary photoreaction (with $\varepsilon_Y^{\lambda_n-isos}=0$) has previously been reported in Sect. 11.6. Such a first-order expansion treatment is the only level of approximation reported in the literature [2]. The approximation in the present chapter will be performed with second-order expansion series. It will be applied to the series of unsolvable integrands laid out in the previous chapter. The second-order expansion is employed here because treatments based on third and higher expansion orders are not practically useful since they will ultimately lead to complicated integrands that do not have known antiderivatives.

13.2 Second-Order Maclaurin Expansion Series Applied to the Kinetics of the Primary Photoprocess

Our review of the photokinetic description of the primary photoreaction ($XY(1\Phi)$ with $\varepsilon_Y^{\lambda_n-isos}\neq 0$) has identified one unsolvable integral (Int_b) in the rate-law equation (Eq. 12.5) that impeded the derivation of an integrated rate-law for that reaction.

$$Int_b=\delta_2\int_{C_X(0)}^{C_X(t)}\frac{1}{C_X^{\lambda_n-isos,T}(t)}\frac{dC_X^{\lambda_n-isos,T}(t)}{1-10^{-\delta_1\,C_X^{\lambda_n-isos,T}(t)-\delta_2}}\tag{13.1}$$

M. Maafi, *Photokinetics*, https://doi.org/10.1007/978-3-031-98985-8_13

In a first approximation, let us apply a second-order Maclaurin expansion series (Eq. 3.51) to Int_b and, more precisely, to the term in the denominator of its second fraction under the integral (with $A_{tot}^{\lambda_{n-isos},T/\lambda_{n-isos}}(t) = \delta_1\ C_X(t) + \delta_2$, Eq. 12.1).

$$1 - 10^{-A_{tot}^{\lambda_{n-isos},T/\lambda_{n-isos}}(t)} = 2.3\ A_{tot}^{\lambda_{n-isos},T/\lambda_{n-isos}}(t) - \frac{(2.3)^2}{2}\left(A_{tot}^{\lambda_{n-isos},T/\lambda_{n-isos}}(t)\right)^2$$
$$= -2.645\ A_{tot}^{\lambda_{n-isos},T/\lambda_{n-isos}}(\mathrm{t})\left(A_{tot}^{\lambda_{n-isos},T/\lambda_{n-isos}}(\mathrm{t}) - 0.8696\right) \tag{13.2}$$

In order to proceed to a change of variable from concentration as the variable of Int_b to total absorbance, let us work out the expression of the concentration from that of the total absorbance and perform a differentiation.

$$C_X^{\lambda_{n-isos},T}(t) = \frac{A_{tot}^{\lambda_{n-isos},T/\lambda_{n-isos}}(\mathrm{t}) - \delta_2}{\delta_1} \tag{13.3}$$

with

$$dC_X^{\lambda_{n-isos},,T}(t) = \frac{1}{\delta_1}\ dA_{tot}^{\lambda_{n-isos},T/\lambda_{n-isos}}(\mathrm{t}) \tag{13.4}$$

Introducing Eqs. 13.3 and 13.4 in Eq. 13.1 gives

$$Int_b = \frac{-\delta_2}{2.645}\int_{C_X(t)}^{C_X(t)} \frac{dA_{tot}^{\lambda_{n-isos},T/\lambda_{n-isos}}(\mathrm{t})}{\left(A_{tot}^{\lambda_{n-isos},T/\lambda_{n-isos}}(\mathrm{t}) - \delta_2\right) A_{tot}^{\lambda_{n-isos},T/\lambda_{n-isos}}(\mathrm{t})\left(A_{tot}^{\lambda_{n-isos},T/\lambda_{n-isos}}(t) - 0.8696\right)} \tag{13.5}$$

Decomposing the improper fraction of the ratio under the integral in Eq. 13.5 into partial fractions, as described by Eq. 3.19 (Sect. 3.4), the integrand of the latter equation can take the form of a sum of simpler fractions of known antiderivatives.

$$Int_b = \int_{C_X(t)}^{C_X(t)} \left(\delta_{13a} \frac{dA_{tot}^{\lambda_{n-isos},T/\lambda_{n-isos}}(\mathrm{t})}{A_{tot}^{\lambda_{n-isos},T/\lambda_{n-isos}}(\mathrm{t})} + \delta_{13b} \frac{dA_{tot}^{\lambda_{n-isos},T/\lambda_{n-isos}}(\mathrm{t})}{\left(A_{tot}^{\lambda_{n-isos},T/\lambda_{n-isos}}(\mathrm{t}) - \delta_2\right)} \right.$$
$$\left. - \delta_{13c} \frac{dA_{tot}^{\lambda_{n-isos3},T/\lambda_{n-isos}}(\mathrm{t})}{\left(A_{tot}^{\lambda_{n-isos},T/\lambda_{n-isos}}(t) - 0.8696\right)} \right) \tag{13.6}$$

with

$$\delta_{13a} = \frac{-\delta_2}{2.645}\ \frac{1}{0.8696\ \delta_2} \tag{13.7}$$

$$\delta_{13b} = \frac{-\delta_2}{2.645} \frac{1}{0.8696\,(0.8696 + \delta_2)} \tag{13.8}$$

$$\delta_{13c} = \frac{-\delta_2}{2.645} (\delta_{13a} + \delta_{13b}) \tag{13.9}$$

The integration of Int_b yields,

$$\begin{aligned} Int_b = \delta_{13a}\ ln \frac{A_{tot}^{\lambda_{n-isos},T/\lambda_{n-isos}}(\mathrm{t})}{A_{tot}^{\lambda_{n-isos},T/\lambda_{n-isos}}(0)} + \delta_{13b}\ ln \frac{A_{tot}^{\lambda_{n-isos},T/\lambda_{n-isos}}(\mathrm{t}) - \delta_2}{A_{tot}^{\lambda_{n-isos},T/\lambda_{n-isos}}(0) - \delta_2} \\ - \delta_{13c}\ ln \frac{A_{tot}^{\lambda_{n-isos},T/\lambda_{n-isos}}(t) - 0.8696}{A_{tot}^{\lambda_{n-isos},T/\lambda_{n-isos}}(0) - 0.8696} \end{aligned} \tag{13.10}$$

Now, to complete the approximated integrated rate-law for the reaction, we proceed with the two remaining integrals (Int_a and Int_c) of Eq. 12.5.

The solution for Int_c is

$$Int_c = \delta_3 \int_0^t dt = \delta_3\ t \tag{13.11}$$

and the integration of Int_a is derived by the procedure used in Sect. 11.4 to yield

$$Int_a = \int_{A_{tot}^{\lambda_{n-isos}}(0)}^{A_{tot}^{\lambda_{n-isos}}(\mathrm{t})} \frac{dA_{tot}^{\lambda_{n-isos},T/\lambda_{n-isos}}(\mathrm{t})}{1 - 10^{-A_{tot}^{\lambda_{n-isos},T/\lambda_{n-isos}}(\mathrm{t})}} = \frac{1}{ln\ 10}\ ln \frac{10^{A_{tot}^{\lambda_{n-isos},T/\lambda_{n-isos}}(t)} - 1}{10^{A_{tot}^{\lambda_{n-isos},T/\lambda_{n-isos}}(0)} - 1} \tag{13.12}$$

The complete approximated form of the integrated rate-law of Eq. 12.5 that describes the $XY(1\Phi)$ reaction (with $\varepsilon_Y^{\lambda_{n-isos}} \neq 0$) is obtained by combining the three integral components.

$$\begin{aligned} \frac{1}{ln\ 10}\ ln \frac{10^{A_{tot}^{\lambda_{n-isos},T/\lambda_{n-isos}}(t)} - 1}{10^{A_{tot}^{\lambda_{n-isos},T/\lambda_{n-isos}}(0)} - 1} + \delta_{13a}\ ln \frac{A_{tot}^{\lambda_{n-isos},T/\lambda_{n-isos}}(\mathrm{t})}{A_{tot}^{\lambda_{n-isos},T/\lambda_{n-isos}}(0)} \\ + \delta_{13b}\ ln \frac{A_{tot}^{\lambda_{n-isos},T/\lambda_{n-isos}}(\mathrm{t}) - \delta_2}{A_{tot}^{\lambda_{n-isos},T/\lambda_{n-isos}}(0) - \delta_2} - \delta_{13c}\ ln \frac{A_{tot}^{\lambda_{n-isos},T/\lambda_{n-isos}}(t) - 0.8696}{A_{tot}^{\lambda_{n-isos},T/\lambda_{n-isos}}(0) - 0.8696} \\ = -\Phi_{X \to Y_j}^{n-\lambda_{isos}}\ P_{\lambda_{n-isos}}^0\ \varepsilon_X^{\lambda_{n-isos}}\ l_{irr}\ t \end{aligned} \tag{13.13}$$

13.3 The Photoreversible Reaction

The second unsolved integral (Int_d, Eq. 13.14) was part of the rate-law of the photoreversible reaction (Eq. 12.13). It is characterised by the same overall mathe matical formulation as that of Int_b (Eq. 13.1).

$$Int_d = \left(\frac{\delta_2}{\delta_1}\delta_4 - \delta_5\right) \int_{C_X(0)}^{C_X(t)} \frac{dC_X^{\lambda_{n-isos},T}(t)}{\left(\delta_4\, C_X^{\lambda_{n-isos},T}(t) + \delta_5\right)\left(1 - 10^{-\delta_1\, C_X^{\lambda_{n-isos},T}(t) - \delta_2}\right)} \tag{13.14}$$

Hence, by using the expansion of the power number in the denominator given in Eq. 13.12, as well as the definition and the differentiation of concentration laid out in Eq. 13.3 and Eq. 13.4, respectively, we obtain

$$Int_d = -\frac{\delta_2\,\delta_4 - \delta_1\,\delta_5}{2.645} \times \int_{C_X(0)}^{C_X(t)} \frac{dA_{tot}^{\lambda_{n-isos},T/\lambda_{n-isos}}(\mathrm{t})}{\left(\delta_4\, A_{tot}^{\lambda_{n-isos},T/\lambda_{n-isos}}(\mathrm{t}) - (\delta_2\,\delta_4 - \delta_1\,\delta_5)\right) A_{tot}^{\lambda_{n-isos},T/\lambda_{n-isos}}(\mathrm{t}) \left(A_{tot}^{\lambda_{n-isos},T/\lambda_{n-isos}}(\mathrm{t}) - 0.8696\right)} \tag{13.15}$$

The integrand is converted into a sum of fractions as

$$Int_d = \int_{C_X(t)}^{C_X(t)} \left(\delta_{13d} \frac{dA_{tot}^{\lambda_{n-isos},T/\lambda_{n-isos}}(\mathrm{t})}{A_{tot}^{\lambda_{n-isos},T/\lambda_{n-isos}}(\mathrm{t})} + \delta_{13e} \frac{dA_{tot}^{\lambda_{n-isos},T/\lambda_{n-isos}}(\mathrm{t})}{\left(A_{tot}^{\lambda_{n-isos},T/\lambda_{n-isos}}(t) - 0.8696\right)} + \delta_{13f} \frac{dA_{tot}^{\lambda_{n-isos},T/\lambda_{n-isos}}(\mathrm{t})}{\delta_4\, A_{tot}^{\lambda_{n-isos},T/\lambda_{n-isos}}(\mathrm{t}) - (\delta_2\,\delta_4 - \delta_1\,\delta_5)} \right) \tag{13.16}$$

with

$$\delta_{13d} = -\frac{1}{2.645 \times 0.8696} \tag{13.17}$$

$$\delta_{13e} = \frac{0.8696\,(\delta_2\,\delta_4 - \delta_1\,\delta_5)^2}{0.8696\,\delta_4 - (\delta_2\,\delta_4 - \delta_1\,\delta_5)}\,\delta_{13d} \tag{13.18}$$

$$\delta_{13f} = -\frac{\delta_2\,\delta_4 - \delta_1\,\delta_5}{2.645}\,(\delta_{13e} + 1)\,\delta_4 \tag{13.19}$$

and after integration of Int_d, we have

$$Int_d = \delta_{13d}\ ln\,\frac{A_{tot}^{\lambda_n-isos,T/\lambda_n-isos}(\mathrm{t})}{A_{tot}^{\lambda_n-isos,T/\lambda_n-isos}(0)} + \delta_{13e}\ ln\,\frac{A_{tot}^{\lambda_n-isos,T/\lambda_n-isos}(t) - 0.8696}{A_{tot}^{\lambda_n-isos,T/\lambda_n-isos}(0) - 0.8696}$$
$$+\,\delta_{13f}\ ln\,\frac{\delta_4\,A_{tot}^{\lambda_n-isos,T/\lambda_n-isos}(\mathrm{t}) - (\delta_2\,\delta_4 - \delta_1\,\delta_5)}{\delta_4\,A_{tot}^{\lambda_n-isos,T/\lambda_n-isos}(\mathrm{t}) - (\delta_2\,\delta_4 - \delta_1\,\delta_5)} \tag{13.20}$$

The approximated integrated rate-law of Eq. 12.13 for $XY(2\Phi)$ reaction, is a combination of the latter equation and those for Int_a (Eq. 13.12) and Int_c (Eq. 13.11) as

$$\frac{1}{ln\,10}\,ln\,\frac{10^{A_{tot}^{\lambda_n-isos,T/\lambda_n-isos}(t)} - 1}{10^{A_{tot}^{\lambda_n-isos,T/\lambda_n-isos}(0)} - 1} + \delta_{13d}\,ln\,\frac{A_{tot}^{\lambda_n-isos,T/\lambda_n-isos}(\mathrm{t})}{A_{tot}^{\lambda_n-isos,T/\lambda_n-isos}(0)}$$
$$+\,\delta_{13e}\,ln\,\frac{A_{tot}^{\lambda_n-isos,T/\lambda_n-isos}(t) - 0.8696}{A_{tot}^{\lambda_n-isos,T/\lambda_n-isos}(0) - 0.8696} + \delta_{13f}\,ln\,\frac{\delta_4\,A_{tot}^{\lambda_n-isos,T/\lambda_n-isos}(\mathrm{t}) - (\delta_2\,\delta_4 - \delta_1\,\delta_5)}{\delta_4\,A_{tot}^{\lambda_n-isos,T/\lambda_n-isos}(\mathrm{t}) - (\delta_2\,\delta_4 - \delta_1\,\delta_5)}$$
$$= -\Phi_{X\to Y_j}^{n-\lambda_{isos}}\,P_{\lambda_n-isos}^{0}\,\varepsilon_X^{\lambda_n-isos}\,l_{irr}\,t \tag{13.21}$$

13.4 The Case of a Photoreaction Reversed by a Thermal Reaction

In the integrand of the unsolvable integral (Int_e) for this reaction (Eq. 12.15), the concentration can be factorised.

$$Int_e = \int_{C_X(0)}^{C_X(t)} \frac{1}{C_X^{\lambda_n-isos,T}(t)}\,\frac{dC_X^{\lambda_n-isos,T}(t)}{\delta_3\,\frac{1-10^{-\delta_1\,C_X^{\lambda_n-isos,T}(t)-\delta_2}}{\delta_1\,C_X^{\lambda_n-isos,T}(t)+\delta_2} \pm k^T} \tag{13.22}$$

Following the procedure employed above, i.e. by introducing the approximation (Eq. 13.2) and the concentration equations (Eq. 13.3 and Eq. 13.4), the integrand takes the form of a single fraction,

$$Int_e = \frac{1}{-2.645\,\delta_3} \int_{C_X(0)}^{C_X(t)} \frac{dA_{tot}^{\lambda_{n-isos},T/\lambda_{n-isos}}(\mathrm{t})}{\left(A_{tot}^{\lambda_{n-isos},T/\lambda_{n-isos}}(\mathrm{t}) - \delta_2\right)\left(A_{tot}^{\lambda_{n-isos},T/\lambda_{n-isos}}(\mathrm{t}) - 0.8696 \pm \frac{-k^T}{2.645\,\delta_3}\right)} \tag{13.23}$$

that is easily transformed into the sum of two fractions.

$$Int_e = \delta_{13g}\left(\int_{C_X(0)}^{C_X(t)} \frac{dA_{tot}^{\lambda_{n-isos},T/\lambda_{n-isos}}(\mathrm{t})}{\left(A_{tot}^{\lambda_{n-isos},T/\lambda_{n-isos}}(\mathrm{t}) - \delta_2\right)} - \int_{C_X(0)}^{C_X(t)} \frac{dA_{tot}^{\lambda_{n-isos},T/\lambda_{n-isos}}(\mathrm{t})}{\left(A_{tot}^{\lambda_{n-isos},T/\lambda_{n-isos}}(\mathrm{t}) - 0.8696 \pm \frac{-k^T}{2.645\delta_3}\right)}\right) \tag{13.24}$$

where

$$\delta_{13g} = \frac{-1}{2.645\,\delta_3\left(0.8696 \pm \frac{k^T}{2.645\,\delta_3} + \delta_2\right)} \tag{13.25}$$

The integration of Eq. 13.24 is straightforward, yielding

$$Int_e = \delta_{13g}\ ln\ \frac{A_{tot}^{\lambda_{n-isos},T/\lambda_{n-isos}}(\mathrm{t}) - \delta_2}{A_{tot}^{\lambda_{n-isos},T/\lambda_{n-isos}}(0) - \delta_2} - \delta_{13g}\ ln\ \frac{A_{tot}^{\lambda_{n-isos},T/\lambda_{n-isos}}(\mathrm{t}) - 0.8696 \pm \frac{-k^T}{2.645\,\delta_3}}{A_{tot}^{\lambda_{n-isos},T/\lambda_{n-isos}}(0) - 0.8696 \pm \frac{-k^T}{2.645\,\delta_3}} \tag{13.26}$$

The global integrated rate-law of the reaction, incorporating Int_c (Eq. 13.11), is given as

$$\begin{aligned} &\delta_{13g}\ ln\ \frac{A_{tot}^{\lambda_{n-isos},T/\lambda_{n-isos}}(\mathrm{t}) - \delta_2}{A_{tot}^{\lambda_{n-isos},T/\lambda_{n-isos}}(0) - \delta_2} - \delta_{13g}\ ln\ \frac{A_{tot}^{\lambda_{n-isos},T/\lambda_{n-isos}}(\mathrm{t}) - 0.8696 \pm \frac{-k^T}{2.645\,\delta_3}}{A_{tot}^{\lambda_{n-isos},T/\lambda_{n-isos}}(0) - 0.8696 \pm \frac{-k^T}{2.645\,\delta_3}} \\ &= -\Phi_{X \to Y_j}^{n-\lambda_{isos}}\ P_{\lambda_{n-isos}}^0\ \varepsilon_X^{\lambda_{n-isos}}\ l_{irr}\ t \end{aligned} \tag{13.27}$$

For the reactions investigated here, the three final forms of the integrated rate-laws (Eq. 13.13, 13.21 and 13.27) are not possibly expressed by the absorbance as a function of time (i.e. $A_{tot}^{\lambda_{n-isos}/\lambda_{n-isos}}(t) = f(t)$), as usually the integrated rate-laws are presented in kinetics. However, we could depict them in a simpler form, as given by Eq. 13.28 (in an analogous manner stated for one of the forms, Eq. 11.26, projected for the analytically derived integrated rate-law of the unimolecular reaction whose $\varepsilon_Y^{\lambda_{n-isos}} = 0$). The left-hand side term of the equation (Eq. 13.27) is the dependent variable of a linear pattern, expressed as a time-dependent function through the variation of $A_{tot}^{\lambda_{n-isos}/\lambda_{n-isos}}(\mathrm{t})$.

$$F\left(A_{tot}^{\lambda_{n-isos},T/\lambda_{n-isos}}(\mathrm{t})\right) = \delta_3\, t \qquad (13.28)$$

13.5 Conclusion

The common use of power expansion series in the literature of photokinetics has been limited to the first order of the expansion. It has been shown in this chapter that the approximation can as well be achieved on the basis of the second order of the expansion. Even though extending the expansion to the second order might represent an improvement, it is nonetheless important to remember that the limit on the acceptable magnitude of the absorbance is still very low (Chap. 3). Unless the expansion includes an infinite number of terms, these kinds of approximations are not expected to deliver useful integrated rate-laws for photokinetics. Alternative approaches must be considered. We will describe one of these, the semi-empirical method, in the following chapter.

The insolvable integrals found here for the three simplest reactive systems are, in fact, of a wider occurrence in photokinetics. The rate-laws of any photo- or photothermal reaction under any type of irradiation (other than a monochromatic/isosbestic irradiation) belong, with very few exceptions, to this category of analytically insolvable systems of differential equations.

This situation may explain, at least in part, the poor development of mathematical descriptions of photokinetic processes.

References

1. Abrmowitz M, Stegun IA (eds) (1970) Handbook of mathematical functions with formulas, graphs and mathematical tables. Dover Publications, New York. Available at https://personal.math.ubc.ca/~cbm/aands/abramowitz_and_stegun.pdf
2. Tonnesen HH (ed) (2004) Photostability of drugs and drug formulations. CRC Press, Boca Raton. ISBN. 0-415-30323—0

Chapter 14
A Semi-Empirical Method

14.1 Introduction

To palliate both the restricted absorbance range associated with using Maclaurin power series expansion and the lack of typical formulations of the integrated rate-law for the unsolved integrals, a new method based on a semi-empirical approach [1–3] is considered.

14.2 A Retrospective on Establishing Integrated Rate-Laws

The usually employed approaches found in the literature favour simplifications of the rate-law in order to turn it into an easily integrable form. Most often, an exponential form is ultimately found for the approximated integrated rate-law equation. These equations are applied for the treatment of the kinetic data, irrespective of whether the reaction is driven by monochromatic (isosbestic or, most frequently, non-isosbestic) or polychromatic irradiation. Our discussion in the previous chapters showed that such a uniformity causes many contradictions with the theoretical framework of photokinetics. For instance, this type of approach seems to implicitly consider that the photoreaction is driven by isosbestic irradiation since the proposed exponential formulae are mathematically similar to the ones (of first-order character) derived for photoreactions performed under isosbestic irradiation (Chap. 8). The analyses undertaken till now tend to show that the Φ-order character is inherent to photokinetics performed under non-isosbestic light irradiation, which makes postulating a purely mono-exponential model, not a strictly valid option. Another strong indication of the inconsistency of such amalgamation of irradiation conditions is also provided by the general mathematical rule (Sect. 3.5) stating that "*differences in the rate-laws inevitably lead to differences in the integrated rate-laws.*" Thus far, the literature does not propose model equations that

M. Maafi, *Photokinetics*, https://doi.org/10.1007/978-3-031-98985-8_14

could stand for reasonable approximations of the integration rate-laws of many kinetic systems, including those discussed in Chap. 12.

One way to achieve good approximations for these situations was proposed by introducing power series in the integrands to facilitate the integrations (Chap. 13). Despite the possible practical usefulness granted to such equations, the result of the integration obtained by this type of approximation cannot provide a simple representation of the concentration (or the absorbance) as a function of time ($C_X^{\lambda_n-isos}(t) = f(t)$), i.e. the format in which experimental kinetic data are recorded.

The kinetic treatment of unsolvable kinetic cases can arguably be performed by using numerical integration methods (*NIM*s). However, an approach involving *NIM*s might have two limitations for the modelling and elucidation of photokinetics. On the one hand, it requires the prior knowledge of all values of the photoreaction parameters (to solve the indentifiability issue), except that of the photochemical quantum yield. This might not practically be achievable for all cases and reactions. On the other hand, *NIM*s are not capable of generating any kind of mathematical formula for the integrated rate-law, hence limiting a direct description of the data obtained experimentally.

This situation implies that it would be beneficial for the experimentalist if there was an approach capable of providing integrated rate-laws in the form $C_X^{\lambda_n-isos,T}(t) = f(t)$ (or equivalent) for the analytically unsolvable rate-laws. Obviously, such integrated rate-law equations are necessarily approximated forms of the unsolvable kinetic cases. This is why an important condition must always be imposed to any of such approaches, namely, the proposed equations must be tested and proven accurate. It is then mandatory to satisfy this condition on the basis of tangible evidence.

This retrospective indicates clearly that there is need to advance research in this field for a better description of photokinetics and a higher accuracy of the results. In this context, the semi-empirical approach discussed below might contribute to supporting this research theme.

14.3 The Basic Principle of the Semi-Empirical Method

The semi-empirical method is based on both a large set of data generated by a numerical integration method that is applied to the true rate-law of the selected reaction, and an empirical equation.

Generally speaking, writing down the true rate-law system infers that we must have a prior knowledge of the detailed overall reaction mechanism, a condition that must be satisfied either by conducting preliminary investigation or by using the mechanism already established in the literature.

The empirical data obtained by *NIM* calculations are considered as an exact representation of the reaction kinetic evolution if a highly performing *NIM* is used (which ensures reasonably high levels of data precision) [4]. The data hence obtained are no different from those generated by an integrated rate-law equation (as testified by the very good agreement between such data and those of known integrated rate-laws (Chaps. 8 and 11)). It is also important to stress that there is no need to introduce fluctuations and variations in the *NIM* calculations for the purpose of simulating the experimental errors. Indeed, the final explicit mathematical equations reached by any method must deliver, within the limits of their application domains, the same results provided by the corresponding integrated rate-law equations (if the latter were analytically obtained for the reactions studied). For instance, the exponential model equation derived for a photoreaction subjected to an isosbestic irradiation generates exactly the same data for $C_X(t)$ trace (e.g. of a mono- or bi-exponential trend) as those obtained from an *NIM* calculation on the rate-law of the reaction. This correspondence between these two types of data is necessary and sufficient to confirm the validity of the proposed explicit equation. Hence, for the purpose of establishing the validity of a proposed equation, there is no objective requirement for the introduction of small variation on the data to simulate possible experimental fluctuations. The analytically derived integrated rate-laws are pure mathematical objects that do not consider fluctuations, yet they are used to treat experimental kinetic data.

The advantage offered by the empirical simulation of data, as generated by *NIM*, is the absence of any imposed restrictions on the values of the parameters, as long as these values have a physical meaning (e.g. $0 < \Phi_{X \to Y_j}^{\lambda_{n-isos}} < 1$). Therefore, it is possible to simulate a greater number of different reactions and kinetic situations by feeding the *NIM*s with a substantial number of different but plausible sets of reaction parameters (selected within the range allowed for each parameter). It is evident that as the number of the reaction's parameters increases, the number of combinations increases. It is very important to carefully select useful sets for the parameter values and avoid redundancies.

NIM calculations also have the advantage of providing the data for all the individual traces of the species that are involved in the selected reaction mechanism. This important feature, together with the large number of traces, allow better testing of the reliability of the proposed equation (or equations). Such a procedure is not usually considered for the approximated formulae proposed in the literature, where the test is only performed on the experimental data at hand (i.e. most often, of a single reaction).

Once a substantial amount of *NIM*-simulated data is generated, we can proceed to the next main step of the semi-empirical method, where an equation is proposed. Since our purpose is to find an integrated rate-law that corresponds to the type of data generated experimentally, we expect it to be in the form $C_X(t) = f(t)$. The simplest equation of this type that we know with certainty for photokinetics under non-isosbestic irradiation is the Φ-order model (i.e. the log-exponential function), obtained analytically for the primary photoreaction (Eq. 11.20). Therefore, this becomes our basic model. However, the exact formulation of the log-exponential model, as it stands in that equation (Eq. 11.20), might not necessarily be a template for all reactions. Therefore, we can suppose that our proposed final model equation for the integrated rate-law equation of a given reaction will have the same general form laid out for Eq. 11.41, with its parameters being, most likely, different for each specific reaction. As a consequence, it is expected that limits of applicability of the proposed integrated rate-law model will have to be imposed. This condition guarantees the reliability of the data that are extracted from such a proposed explicit model equation.

Finally, the procedure should include validity tests, whereby the proposed integrated rate-law's reliability is confirmed. The validity tests would concern a good fitting of the *NIM*-generated ($C(t)$ and $A(t)$) traces by the model equation (as a first and necessary condition). It must also prove that the initial reaction-velocities and the exponential arguments (which are calculated from their respective expressions worked out from the proposed model equation and the parameters that originally fed the *NIM* calculations), should imperatively linearly correlate with their values obtained from the *NIM* traces (e.g. by fitting). An agreement between all these features must be attained for each simulated reaction. Disagreement between the three sets of data (*NIM*s, fitting, and proposed model equation) would determine the limits of applicability for the proposed explicit model equation. The magnitude of the error between these sets of data can be defined by the user. For the following studies, an error of no more than 10% was selected.

The method is set here for strictly monochromatic/non-isosbestic light, and the *NIM* used is either a fourth- or fifth-order Runge–Kutta (*RK*4 or *RK*5) numerical integration method (Sect. 3.9).

In summary, the robustness of this method comes from the fact that (i) it uses the true rate-laws of the reactive system to (ii) generate, by a performant *NIM*, a great and very reliable number of simulated reactions' kinetic traces, (iii) whose properties and features agree with those obtained from the proposed explicit model equations, (iv) according to tightly defined criteria. Methods with such a level of testing and conditions on reliability are lacking in photokinetics.

The step-by-step procedure of the general semi-empirical method is presented hereafter. It is set here for the cases of the primary photoreaction but can be extended to virtually any reaction.

The Reaction and Its Differential Equations.

(a) Establish the differential equation or the system of differential equations for the reaction selected. This means that the photochemical mechanism governing the selected reactive system is known.

Simulated Data.

(b) Select values for both initial intrinsic (e.g. $\varepsilon_{X\ or\ Y_j}^{\lambda_{n-isos}}$, $\Phi_{X\to Y_j}^{\lambda_{n-isos}}$, $\Phi_{Y_j\to X}^{\lambda_{n-isos}}$, $k_{X\to Y_j}^{T}$, and $k_{Y_j\to X}^{T}$) and experimental extrinsic (e.g. $C_X(0)$, $C_{Y_j}(0)$, l_{irr}, and $P_0^{\lambda_{n-isos}}$) parameters of the reaction to feed the numerical calculation.

(c) Use a highly performant *NIM* to generate the kinetic traces and temporal variations of the concentrations and absorbances of both the reactant and the products.
(d) Record the traces, making sure that the end of the reaction (i.e. a plateau region on the traces or a zero value after a long reaction time) is reached. At the final stages of the reaction, the concentrations of the reactant and products do not vary (or vary in negligible proportions).
(e) Change the values of the initial parameters, and perform a new numerical integration calculation (according to the same conditions set out in b–d).
(f) Repeat steps b–e for different combinations of parameter values (at least 100 different sets). Realistic values could be considered in the following ranges: $0 \le \varepsilon \le 10^5$, $0 \le \Phi \le 1$, $0 \le C(0) \le 10^{-4}$, $0 \le l_{irr} \le 10$, and $0 \le P_0^{\lambda_{n-isos}} \le 10^6$. The process of *NIM* calculations considers, de facto, that the species concentrations fall within the linearity ranges of their respective calibration graph (Sect. 2.12) (even for concentrations of the order of 10^{-4}).

Model Equation.

(g) Fit the recorded traces corresponding to the reactant concentration to the model equation of the type given by Eq. 11.41.

$$C_X(t) = \partial_1 + \partial_2\, Log\left[1 + \left(10^{\partial_3} - 1\right) e^{-\partial_4\, t}\right] \tag{14.1}$$

(h) Determine the limits of application of Eq. 14.1. This is performed by finding which of the simulated traces are not well-fitted to the model equation, Specially looking at the limit of the total absorbance value ($A_{lim}^{\lambda_{irr}}$) for which a lower level of goodness of fit of the kinetic traces is recorded. It is clear that the success of this procedure is subordinate to the number of tested kinetic traces. This underlines the importance of generating a large number of traces by significantly varying the sets of the parameter values (where a 100 examplars should be a minimum).
(i) Propose similar explicit model equations to Eq. 14.1 for the products (C_{Y_j}) of the selected reaction.
(j) Perform the procedure seen in point h, and optimise the limit value of the total absorbance ($A_{lim}^{\lambda_{irr}}$) based on the information obtained from the fitting of the products' concentration traces. The total absorbance is preferred here because it depends on three important quantities, ε, C, and l_{irr}.
(k) Establish the optimal value of $A_{lim}^{\lambda_{irr}}$, i.e. all the traces characterised by a value of the total absorbance equal to or lower than $A_{lim}^{\lambda_{irr}}$ ($A_{tot}^{\lambda_{irr}} \le A_{lim}^{\lambda_{irr}}$) are well-fitted by Eq. 14.1.
(l) Record the values of fitting ∂ coefficients (∂_{1-4}^{fit}) for each validated trace ($A_{tot}^{\lambda_{irr}} \le A_{lim}^{\lambda_{irr}}$).

Identifying the Analytical Expressions of the ∂_i Coefficients for C_X.

(m) Propose formulae for the coefficients ∂_i of Eq. 14.1. These formulae must only use the parameters of the reaction (e.g. $\varepsilon_{X\ or\ Y_j}^{\lambda_{n-isos}}$, $\Phi_{X \to Y_j}^{\lambda_{n-isos}}$, $\Phi_{Y_j \to X}^{\lambda_{n-isos}}$, $C_X(0)$, $C_{Y_j}(0)$, l_{irr}, and/or $P_0^{\lambda_{n-isos}}$).

(n) Calculate the values of coefficients using the proposed model formulae ($\partial_{1-4}^{mod.}$) and the initial values of the reaction parameters that fed the *NIM* calculation (for each individual trace).

(o) Compare ∂_{1-4}^{fit} and $\partial_{1-4}^{mod.}$ values.

(p) If the numerical differences between any ∂_{1-4}^{fit} and its counterpart ∂_{1-4}^{mod} is higher than 10%, the corresponding formula is discarded.

(q) Propose new improved explicit formulae to replace the discarded ones.

(r) Perform points n, o, and p until satisfactory results, in point p, are obtained.

Identifying the ∂_i-Algebraic Expressions for the Products ($C_{Y_j}(t)$)

(s) Proceed in a similar way for each $C_{Y_j}(t)$ obtained in point i to determine their respective individual ∂_i coefficients.

Validation of the Semi-empirical Integrated Rate-Laws

(t) Draw graphs of each ∂_{1-4}^{fit} coefficient against its corresponding ∂_{1-4}^{mod} coefficient (calculated from their formulae) for all the traces that fitted the proposed explicit equations defined above for the reactant and the products well. Each graph should be a straight line characterised by an intercept close to zero and a correlation coefficient close to unity.

Application of the Validated Model Equation

(u) Accordingly, the equations proposed for $C_{Y_j}(t)$ and the coefficients ∂_{1-4} are valid to be applied to characterise the experimental kinetics of the reaction at hand, provided that $A_{tot}^{\lambda_{irr}} \leq A_{lim}^{\lambda_{irr}}$ at λ_{irr}.

14.4 Conclusion

The recurrent failure to reach the integrated rate-laws by closed-form integration, opened an avenue for proposing methods that are able to generate approximated formulae that stand for integrated rate-laws. The way, generally adopted in the literature, is developed using simplifications to convert the originally insolvable integrands into forms having antiderivatives. So far, in photokinetics, *NIM* data were, most often, used to fit one experimental set of traces of a specific reaction.

These approaches lack a testing process to prove their reliability, limits of applicability, and generality for the reactive systems concerned.

The semi-empirical method is a new way to develop approximated but reliable explicit formulae that play the role of integrated rate-laws. It assembles *NIM*-precise data, a template model equation for photokinetics (i.e. the Φ-order formula), and a rigorous testing process.

The methodology of the semi-empirical method is applied, in the next chapter, to some reactions, to demonstrate its usefulness in photokinetics.

References

1. Maafi W, Maafi M (2013) Modelling nifedipine photodegradation, photostability and actinometry properties. Int J Pharm 456:153–164. https://doi.org/10.1016/j.ijpharm.2013.07.075
2. Maafi M, Maafi W (2014) F-order kinetics of photoreversible-drug reactions. Int J Pharm 471: 536–543. https://doi.org/10.1016/j.ijpharm.2014.05.017
3. Maafi M, Maafi W (2016) Modeling and elucidation of the kinetics of multiple consecutive photoreactions AB4(4Φ) with F-order kinetics. Application to the photodegradation of riboflavin. J Pharm Sci 105:3537–3548. https://doi.org/10.1016/j.xphs.2016.06.030
4. Mauser H, Gauglitz G, Compton RG, Hancock G (1998) Comprehensive chemical kinetics, photokinetics: theoretical fundamentals and applications, vol 36. Elsevier, Amesterdam. ISBN-13. 978-0444825360

Chapter 15
Established Semi-Emperical Model Equations (SEM)

15.1 Introduction

The semi-empirical method can, in principle, apply to any photo- or photothermal reaction. So far, it has produced equations for the kinetics of five photoreactions, namely, the unimolecular reaction, $XY(1\Phi)$ with $\varepsilon_Y^{\lambda_n - isos} \neq 0$; the reversible $XY(2\Phi)$, $XY(1\Phi, 1k)$, and $XY(2\Phi, 1k)$ reactions (where the thermal reaction-step is $Y \rightarrow X$); and the multi-consecutive reaction involving four photoproducts and four photochemical reaction-steps, $XY_4(4\Phi)$. The order in which these reactions are listed follows, in fact, the chronology that should be adopted for such studies, adhering to a progressive incrementation of the complexity of the mechanism. This strategy is justified by the novelty of *SEM*, which requires that each new reaction is to be individually checked, the details of which are presented hereafter.

15.2 SEM Equations for $XY(2\Phi, 1k)$ Reactions

The findings relative to the four reactions involving two species (X and Y) show that they obey the same equations obeyed by the $XY(2\Phi, 1k)$ system [1, 2]. Hence, we will review, in the following, the equations for the latter reaction.

The application of the semi-empirical method described in the previous section is performed by considering the general case where the initial concentrations are $C_X^{\lambda_n - isos, T}(\mathrm{t}) \neq 0$ and $C_Y^{\lambda_n - isos, T}(0) = 0$; the temperature ($T$) of the reactive medium is constant; the wavelength of irradiation is non-isosbestic, $\lambda_{n\ -\ isos}$, crossing the exposed solution along l_{irr}; and the reactive system may be observed by spectrophotometry at a different wavelength (λ_{obs}, corresponding to measurements across the observation path length, l_{obs}), a provision that frees the experimentalist from being restricted to measuring the traces at the specific $\lambda_{n\ -\ isos}$. It also extends the field of validation to more criteria for the semi-empirical integrated rate-law (see

M. Maafi, *Photokinetics*, https://doi.org/10.1007/978-3-031-98985-8_15

Table 15.1 Examples of parameter sets for various reactions that fed Runge–Kutta calculations to generate the corresponding *XY*(2*Φ*, 1*k*)-simulated kinetic traces

Sets	Parameters							
	$C_X(0)$ /M	$P^0_{\lambda_{n-irr}}$ /M^{-1} s^{-1}	l_{irr} /cm	$\varepsilon_X^{\lambda_{n-irr}}$ /M^{-1} cm^{-1}	$\varepsilon_Y^{\lambda_{n-irr}}$ /M^{-1} cm^{-1}	$\Phi_{Y\to X}^{\lambda_{n-isos}}$	$\Phi_{X\to Y}^{\lambda_{n-isos}}$	$k^T_{Y\to X}$ s^{-1}
Set_1	2.0 10^{-5}	1.0 10^{-6}	2	6748	32,159	0.50	0.20	0
Set_2	2.0 10^{-5}	9.3 10^{-7}	1	8587	30,269	0.85	0.17	0
Set_3	2.0 10^{-5}	3.4 10^{-6}	1	8587	30,269	0.17	0.17	0
Set_4	7.5 10^{-5}	1.5 10^{-6}	1	55,984	10,837	0.60	0.20	0
Set_5	1.0 10^{-5}	2.0 10^{-6}	1.7	35,984	10,837	0.70	0.40	0
Set_6	6.0 10^{-5}	7.0 10^{-6}	1	5231	30,273	0.10	0.50	0
Set_7	1.0 10^{-5}	3.0 10^{-7}	2	22,231	71,273	0.10	0.30	0
Set_8	8.0 10^{-7}	1.35 10^{-6}	1	45,000	5000	0.90	0.30	4 10^{-3}
Set_9	8.0 10^{-6}	1.35 10^{-6}	1	18,000	22,500	0.90	0.10	9 10^{-4}
Set_{10}	6.0 10^{-6}	1.35 10^{-6}	1	31,000	37,500	0.50	0.50	7 10^{-3}
Set_{11}	2.0 10^{-6}	1.35 10^{-6}	1	4000	40,000	0.30	0.60	0.040
Set_{12}	1.0 10^{-5}	1.35 10^{-6}	1	21,000	35,000	0.70	0.50	5 10^{-3}
Set_{13}	8.0 10^{-7}	1.35 10^{-6}	1	40,000	30,500	0.80	0.15	0.030
Set_{14}	4.0 10^{-6}	1.35 10^{-6}	1	30,000	10,000	0.10	0.50	0.010

point (*t*) of the procedure in Sect. 14.3). The absorbances are therefore defined by this dual wavelength ($\lambda_{n-isos}/\lambda_{obs}$).

The equations are given below for the temporal variation of concentration and the total absorbance of *XY*(2*Φ*, 1*k*) reaction with time that mirror the *RK*-simulated data collected for this reaction. The simulated data of the traces have been generated by a fourth-order Runge–Kutta numerical integration method.

For the purpose of the *SEM*, more than 700 *RK*-simulated traces were employed to investigate the four reactions. The traces correspond to reactive systems differing by their mechanism, intrinsic features ($\varepsilon_X^{\lambda_{n-isos}}$, $\varepsilon_Y^{\lambda_{n-isos}}$, $\Phi_{X\to Y}^{\lambda_{n-isos}}$, $\Phi_{Y\to X}^{\lambda_{n-isos}}$, and $k^T_{Y\to X}$), and experimental conditions ($C_X(0)$, $P_0^{\lambda_{n-isos}}$, and l_{irr}). A large variety of sets of the parameters' values, in diverse combinations covering a wide number of possible reaction situations, including high and low values of the parameters, in addition to relative values of $\varepsilon_X^{\lambda_{n-isos}} > \varepsilon_Y^{\lambda_{n-isos}}$, $\varepsilon_X^{\lambda_{n-isos}} < \varepsilon_Y^{\lambda_{n-isos}}$, $\Phi_{X\to Y}^{\lambda_{n-isos}} > \Phi_{Y\to X}^{\lambda_{n-isos}}$, $\Phi_{X\to Y}^{\lambda_{n-isos}} < \Phi_{Y\to X}^{\lambda_{n-isos}}$, $\Phi_{X\to Y}^{\lambda_{n-isos}} = \Phi_{Y\to X}^{\lambda_{n-isos}}$, etc. Examples of such sets are given in Table 15.1.

The semi-empirical integrated rate-law for the *XY*(2*Φ*, 1*k*) reaction has the form of the general expression given by Eq. 14.1 but written for the total absorbance.

$$A_{tot}^{\lambda_{n-isos},T/\lambda_{obs}}(t) = \partial_1 + \partial_2\, Log\left(1 + \left(10^{\partial_3} - 1\right)\, e^{-\partial_4\, t}\right) \quad (15.1)$$

The coefficients ∂_{1-4} of Eq. 15.1, determined for this reaction, take the following formulae.

$$\partial_1 = A_{tot}^{\lambda_{n-isos},T/\lambda_{obs}}(\infty) = \varepsilon_X^{\lambda_{obs}}\, l_{obs}\, C_X^{\lambda_{n-isos},T}(\infty) + \varepsilon_Y^{\lambda_{obs}}\, l_{obs} C_Y^{\lambda_{n-isos},T}(\infty) \tag{15.2}$$

$$\partial_2 = \frac{A_{tot}^{\lambda_{n-isos},T/\lambda_{obs}}(0) - A_{tot}^{\lambda_{n-isos},T/\lambda_{obs}}(\infty)}{A_{tot}^{\lambda_{n-isos},T/\lambda_{n-isos}}(0) - A_{tot}^{\lambda_{n-isos},T/\lambda_{n-isos}}(\infty)} \frac{l_{obs}}{l_{irr}} \tag{15.3}$$

$$\partial_3 = \left(A_{tot}^{\lambda_{n-isos},T/\lambda_{n-isos}}(0) - A_{tot}^{\lambda_{n-isos},T/\lambda_{n-isos}}(\infty)\right) \frac{l_{irr}}{l_{obs}} \tag{15.4}$$

$$\partial_4 = \left(\Phi_{X\to Y}^{\lambda_{n-isos}}\, \varepsilon_X^{\lambda_{n-isos}} + \Phi_{Y\to X}^{\lambda_{n-isos}}\, \varepsilon_Y^{\lambda_{n-isos}}\right) P_0^{\lambda_{n-isos}}\, l_{irr} F^{\lambda_{n-isos}}(\infty) + k_{Y\to X}^T \tag{15.5}$$

with the photokinetic factor, $F_{\lambda_{n-isos}}(\infty)$, expressed as

$$F_{\lambda_{n-isos}}(\infty) = \frac{1 - 10^{-A_{tot}^{\lambda_{n-isos},T/\lambda_{obs}}(\infty)\,(l_{irr}/l_{obs})}}{A_{tot}^{\lambda_{n-isos},T/\lambda_{obs}}(\infty)\,(l_{irr}/l_{obs})} \tag{15.6}$$

The values of the total absorbances appearing in Eqs. 15.2, 15.3, 15.4, 15.5 and 15.6 ($A_{tot}^{\lambda_{n-isos},T/\lambda_{obs}}$) correspond to those directly measured by spectrophotometry (across the observation optical path length, l_{obs}). Because the total absorbances $A_{tot}^{\lambda_{n-isos},T/\lambda_{n-isos}}$ might not be accessible from the experiment, in the above equations, the latter total absorbances are calculated, whenever required, by multiplying the measured $A_{tot}^{\lambda_{n-isos},T/\lambda_{obs}}$ by the factor l_{irr}/l_{obs}. Also, in the case where the total absorbance at the end of the reaction is zero, the value of the photokinetic factor tends to $ln(10)$.

In the same perspective, we complete the set of equations characterising the traces of the photothermal reaction by the individual concentration expressions. They have the overall mathematical form of Eq. 14.1.

$$C_X^{\lambda_{n-isos},T}(\mathrm{t}) = C_X^{\lambda_{n-isos},T}(\infty) + \frac{1}{\left(\varepsilon_X^{\lambda_{n-isos}} - \varepsilon_Y^{\lambda_{n-isos}}\right) l_{irr}} \times Log\left(1 + \left(10^{\left(\varepsilon_X^{\lambda_{n-isos}} - \varepsilon_Y^{\lambda_{n-isos}}\right)\left(C_X^{\lambda_{n-isos},T}(0) - C_X^{\lambda_{n-isos},T}(\infty)\right) l_{irr}} - 1\right) e^{-k_r^{\lambda_{n-isos},T} t}\right) \tag{15.7}$$

and

$$C_Y^{\lambda_{n-isos},T}(\mathrm{t}) = C_X^{\lambda_{n-isos},T}(0) - C_X^{\lambda_{n-isos},T}(\mathrm{t}) = C_Y^{\lambda_{n-isos},T}(\infty) - \frac{1}{\left(\varepsilon_X^{\lambda_{n-isos}} - \varepsilon_Y^{\lambda_{n-isos}}\right) l_{irr}} \times Log\left(1 + \left(10^{\left(\varepsilon_X^{\lambda_{n-isos}} - \varepsilon_Y^{\lambda_{n-isos}}\right)\left(C_X^{\lambda_{n-isos},T}(0) - C_X^{\lambda_{n-isos},T}(\infty)\right) l_{irr}} - 1\right) e^{-k_r^{\lambda_{n-isos},T} t}\right) \tag{15.8}$$

The formula of the reaction rate-constant, $k_r^{\lambda_{n-isos},T}$, is the same as that of ∂_4 (Eq. 15.5). It is expressed in units of s^{-1}, as the dimension analysis of this parameter shows

$$\left[k_r^{\lambda_{n-isos},T}\right]=\left[\Phi_{X\to Y}^{\lambda_{n-isos}}\,or\,\Phi_{Y\to X}^{\lambda_{n-isos}}\right]\left[\varepsilon_X^{\lambda_{n-isos}}\,or\,\varepsilon_Y^{\lambda_{n-isos}}\right]\left[P_0^{\lambda_{n-isos}}\right]\left[l_{irr}\right]\left[F^{\lambda_{n-isos}}(\infty)\right]$$
$$+\left[k_{Y\to X}^T\right]=1\times mol^{-1}\,dm^3\,cm^{-1}\times einstein\,dm^{-3}\,s^{-1}\times cm\times 1+s^{-1}=s^{-1} \tag{15.9}$$

The condition of applicability of this equation imposes that the total absorbance recorded at the end of the reaction ($t = \infty$, which corresponds to the *pss* for reversible reactions) must not exceed a value of 0.6 and that the value of $F_{\lambda_{n-isos}}(\infty)$ is higher than 1.2. The traces that did not fit those criteria are discarded and not counted for the development of the *SEM* equations. From an experimental point of view, a final total absorbance of 0.6 or less is more than comfortable to investigate these types of reactions.

The efficiency of the *SEM* equations of the integrated rate-laws (Eqs. 15.1, 15.7, and 15.8) is approved by their excellent fit of the 700 tested *RK*-simulated traces. An example is depicted in Fig. 15.1.

It is important to underline the advantage provided by the explicit expressions of the coefficients $\partial_1 - \partial_4$ in the formula of the total absorbance (Eq. 15.1) and the fully explicit equations of the concentrations (Eqs. 15.7 and 15.8). Such explicit formulations leave $k_r^{\lambda_{n-isos},T}$ (or ∂_4) as the single fitting parameter of the traces, where

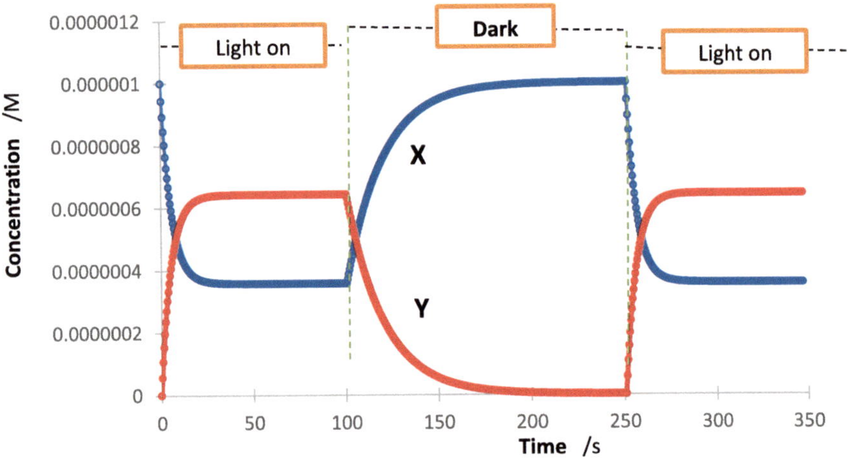

Fig. 15.1 Example of the good fit of the *RK*-generated traces (circles) with *SEM* rate-law Eqs. 15.7, 15.8, and 15.15 (lines). The kinetic data of the *XY*(2Φ, 1*k*) reaction (*X* in blue and *Y* in red) are recorded for phases where the excitation monochromatic/non-isosbestic light is switched on and off. The *RK*-simulation was fed by the following parameters: $C_X^{\lambda_{n-isos},T}(0)=10^{-6}\,M$, $\varepsilon_X^{\lambda_{n-isos}}=4.4\;10^4\,M^{-1}\,cm^{-1}$, $\varepsilon_Y^{\lambda_{n-isos}}=3.5\;10^4\,M^{-1}\,cm^{-1}$, $l_{irr}=1\;cm$, $\Phi_{X\to Y}^{\lambda_{n-isos}}=0.9$, $\Phi_{Y\to X}^{\lambda_{n-isos}}=0.15$, $k_{Y\to X}^{\lambda_{n-isos}}=0.05$, and $P_0^{\lambda_{n-isos}}=1.35\;10^{-6}\,einstein\,dm^{-3}\,s^{-1}$

the remaining parameters of the equations are all available experimentally. The advantage here is the disappearance of the identifiability issues in the fitting process of each trace (i.e. a unique value for $k_r^{\lambda_{n-isos},T}$ will be determined for each trace).

The next validation step of the integrated rate-laws, according to the procedure set out in Chap. 14, involves the initial velocity of the reaction, whose values are determined by two independent ways, namely, from (i) the fitting parameters, and (ii) the theoretical equation calculated by using the values of the parameters that originally fed the *RK*-simulation. The theoretical initial rate-equation for the concentration trace of the reactant (and photoproduct) is

$$Theo: r_{0,X}^{\lambda_{n-isos},T} = -Theo: r_{0,Y}^{\lambda_{n-isos},T} = \left(\frac{dC_X^{\lambda_{n-isos},T}(t)}{dt}\right)_{t=0}$$
$$= -\Phi_{X\to Y}^{\lambda_{n-isos}}\,\varepsilon_X^{\lambda_{n-isos}}\,P_0^{\lambda_{n-isos}}\,l_{irr}\,F^{\lambda_{n-isos}}(0)\,C_X^{\lambda_{n-isos},T}(0) \tag{15.10}$$

The initial rate for the total absorbance trace,

$$Theo: r_{0,A}^{\lambda_{n-isos},T/\lambda_{obs}} = Theo: r_{0,X}^{\lambda_{n-isos},T/\lambda_{obs}}\,\varepsilon_X^{\lambda_{obs}}\,l_{obs}$$
$$+ Theo: r_{0,Y}^{\lambda_{n-isos},T/\lambda_{obs}}\,\varepsilon_Y^{\lambda_{obs}}\,l_{obs\,t=0} = \left(\frac{dA_{tot}^{\lambda_{n-isos},T/\lambda_{obs}}(t)}{dt}\right)_{t=0}$$
$$= \left(\varepsilon_Y^{\lambda_{obs}} - \varepsilon_X^{\lambda_{obs}}\right)l_{obs}\,\Phi_{X\to Y}^{\lambda_{n-isos}}\,\varepsilon_X^{\lambda_{n-isos}}\,P_0^{\lambda_{n-isos}}\,l_{irr}\,F^{\lambda_{n-isos}}(0)\,C_X^{\lambda_{n-isos},T}(0) \tag{15.11}$$

with $F^{\lambda_{n-isos}}(0)$ is calculated by Eq. 15.6, where the total absorbance the symbol "∞" is replaced by "0".

The $Fit: r_0^{\lambda_{n-isos},T/\lambda_{obs}}$ values obtained from the fitting data of the traces are calculated by the following general equation, derived by differentiation of either Eq. 14.1 or 15.1.

$$Fit: r_0^{\lambda_{n-isos},T} = -\frac{\partial_2\,\partial_4}{ln(10)}\left(1 - 10^{-\partial_3}\right) \tag{15.12}$$

Or when substituting the ∂ factors of Eq. 15.12 by their respective expressions (Eqs. 15.2, 15.3, 15.4 and 15.5), the initial rate of the total absorbance trace takes the form

$$Fit:r_{0,A}^{\lambda_{n-isos},T/\lambda_{obs}} = -\frac{k_r^{\lambda_{n-isos},T}}{ln(10)}\frac{A_{tot}^{\lambda_{n-isos},T/\lambda_{obs}}(0) - A_{tot}^{\lambda_{n-isos},T/\lambda_{obs}}(\infty)}{A_{tot}^{\lambda_{n-isos},T/\lambda_{n-isos}}(0) - A_{tot}^{\lambda_{n-isos},T/\lambda_{n-isos}}(\infty)}\frac{l_{obs}}{l_{irr}} \times \left(1 - 10^{-\left(A_{tot}^{\lambda_{n-isos},T/\lambda_{n-isos}}(0) - A_{tot}^{\lambda_{n-isos},T/\lambda_{n-isos}}(\infty)\right)\frac{l_{irr}}{l_{obs}}}\right) \tag{15.13}$$

The sign of $Theo:r_{0,A}^{\lambda_{n-isos},T/\lambda_{obs}}$ and $Fit:r_{0,A}^{\lambda_{n-isos},T/\lambda_{obs}}$ will depend on the relative values of the absorptivies of the reactive species, i.e. they are positive if $\varepsilon_Y^{\lambda_{obs}} > \varepsilon_X^{\lambda_{obs}}$ and negative otherwise.

The differentiation of Eq. 15.7 leads to the initial rate-equation for the reactant (concentration trace).

$$Fit:r_{0,X}^{\lambda_{n-isos},T} = -\frac{k_r^{\lambda_{n-isos},T}}{ln(10)}\frac{1 - 10^{\left(\varepsilon_Y^{\lambda_{n-isos}} - \varepsilon_X^{\lambda_{n-isos}}\right)\left(C_X^{\lambda_{n-isos},T}(0) - C_X^{\lambda_{n-isos},T}(\infty)\right) l_{irr}}}{\left(\varepsilon_X^{\lambda_{n-isos}} - \varepsilon_Y^{\lambda_{n-isos}}\right) l_{irr}} \tag{15.14}$$

For the tested *RK*-simulated traces, $Theo:r_0^{\lambda_{n-isos},T}$ is linearly correlated to $Fit:r_0^{\lambda_{n-isos},T}$. The good agreement between the two quantities is relayed by high correlation coefficients of the lines (≥0.99) and very small intercepts ($<10^{-4}$), with slopes practically equal to unity. Similar conclusions are reached for the plots of the reactions rate-constants, $Theo:k_r^{\lambda_{n-isos},T}$ and $Fit:k_r^{\lambda_{n-isos},T}$ (Fig. 15.2).

Finally, the thermal traces are fit by the general, mono-exponential equation (Eqs. 8.56 and 8.58),

$$C_X^{T/\lambda_{obs}}(t) = \partial_{1T} + \partial_{2T}\, e^{-\partial_{4T}\, t} \tag{15.15}$$

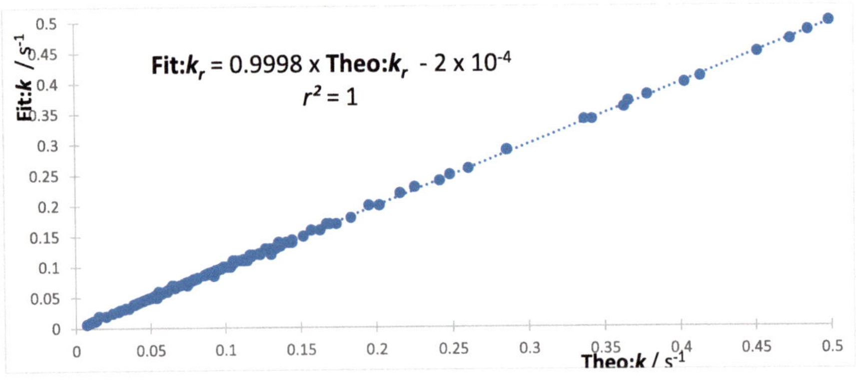

Fig. 15.2 Linear correlation between reaction rate-constants, $Fit:k_r^{\lambda_{n-isos},T}$ vs. $Theo:k_r^{\lambda_{n-isos},T}$, for $XY(2\Phi, 1k)$ photothermal reactions

The set of formulae laid out above are also useful for any of the sub-mechanisms of the photothermal $XY(2\Phi, 1k)$ reaction considered here. For instance, they reduce to those applicable to the photothermal reversible $XY(1\Phi, 1k)$ reaction by setting $\Phi_{Y\rightarrow X}^{\lambda_{n-isos}}$ equal to zero, the photoreversible $XY(2\Phi)$ reaction when $k_{Y\rightarrow X}^{T}=0$, and the basic primary $XY(1\Phi)$ photoreaction when $k_{Y\rightarrow X}^{T}=\Phi_{Y\rightarrow X}^{\lambda_{n-isos}}=0$ but where $\varepsilon_{Y}^{\lambda_{n-isos}}\neq 0$. It is also remarkable that the analytically derived equation (Eq. 11.20) of the unimolecular $XY(1\Phi)$ with its transparent photoproduct is recovered from the above equations if we set $k_{Y\rightarrow X}^{T}=\Phi_{Y\rightarrow X}^{\lambda_{n-isos}}=\varepsilon_{Y}^{\lambda_{n-isos}}=0$ (with $F^{\lambda_{n-isos}}(\infty)$ in Eq. 15.5, being equal to $ln(10)$ as $C_{X}^{\lambda_{n-isos},T}(\infty)$ tends to zero).

The model equation (Eq. 15.1) faithfully provides the exact values as well as the analytical expressions of the four characteristic markers of the reaction kinetic trace of the photo- and photothermal reactions involving a reactant and its photoproduct (*XY* systems): $A_{tot}^{\lambda_{n-isos},T/\lambda_{n-isos}}(0)$, $A_{tot}^{\lambda_{n-isos},T/\lambda_{n-isos}}(pss)$, $k_{r}^{\lambda_{n-isos},T}$, and $r_{0,A}^{\lambda_{n-isos},T/\lambda_{n-isos}}$. It can be considered as an unprecendented achievement in the field as the explicit model equation not only fits well the *RK*-simulated traces but also due to the fact that the method leading to the refinement of Eq. 15.1 is assumption-free (i.e. using the theoretical rate-laws). The limit of applicability ($A_{tot}^{\lambda_{n-isos},T/\lambda_{obs}}(\mathrm{t})<0.6$) is a direct consequence of the analysis of the data by assessing how good the above four markers of the kinetics are rendered by the equations (i.e. as a consequence of the *SEM* procedure and not preconceived). These criteria make the model equation a reliable tool to describe the kinetics of the reactions detailed above.

The ease afforded by the application of the semi-empirical model equation, Eq. 15.1, and the potential it allows to determine the kinetic parameters from experimental data probably confer to this formula a higher performance than that achieved by the expression derived by using the Maclaurin power expansion series (Chap. 13), along with offering a larger range for the total absorbance values (< 0.6).

Finally, it is important to underline that the above equations are valid for any radiation intensity. Figure 15.3 shows the good fit achieved by Eq. 15.1 (where *C*(t) replaces *A*(t)) to the reactant concentration traces obtained at a range of monochromatic/non-isosbestic radiation intensities impinging on an $XY(2\Phi, 1k)$ photothermal reversible reaction, $P_{0}^{\lambda_{n-isos}}=1.35\ 10^{-6}-2.5\ 10^{-8}\ einstein\ dm^{-3}\ s^{-1}$.

It might also be interesting to define the thermodynamic constant for reversible $XY(2\Phi, 1k)$ reactions, which takes the form

$$K_{X}^{\lambda_{n-isos},T}=\frac{C_{Y}^{\lambda_{n-isos},T}(pss)}{C_{X}^{\lambda_{n-isos},T}(pss)}=\frac{\Phi_{Y\rightarrow X}^{\lambda_{n-isos}}\varepsilon_{Y}^{\lambda_{n-isos}}+k_{Y\rightarrow X}^{T}}{\Phi_{X\rightarrow Y}^{\lambda_{n-isos}}\varepsilon_{X}^{\lambda_{n-isos}}} \tag{15.16}$$

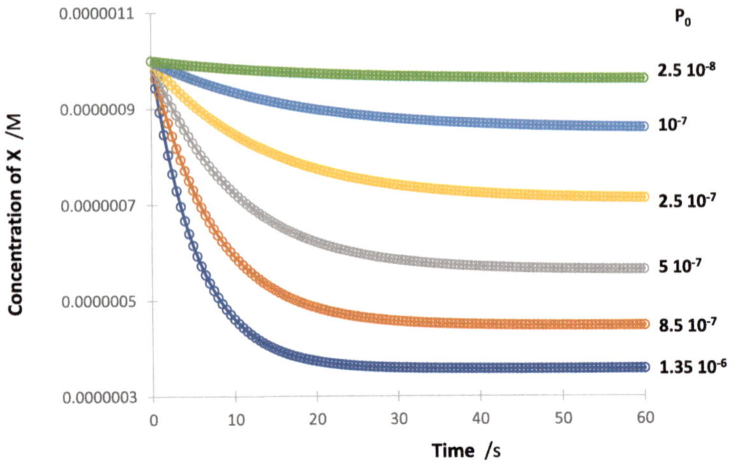

Fig. 15.3 Validity of Eq. 15.1, irrespective of the values of $P_0^{\lambda_n-isos}$ (in *einstein dm*$^{-3}$ s^{-1}). The photothermal $XY(2\Phi, 1k)$ reaction was an *RK*-simulation using the reaction parameters indicated in Fig. 15.1, and $P_0^{\lambda_n-isos}$ values are indicated next to the traces

$$X \longrightarrow Y_1 \longrightarrow Y_2 \longrightarrow Y_3 \longrightarrow Y_4$$

Scheme 15.1 The multi-consecutive $XY_4(4\Phi)$ photoreaction

15.3 The $XY_4(4\Phi)$ Reaction Modelling with *SEM*

The family of consecutive photoreactions is one of the largest among the mechanistic sequences encountered within reactive photochemical systems. The literature reports on the kinetic description of the consecutive, two-reaction-steps $XY_2(2\Phi)$ mechanism according to the classical treatments used for thermal reactions [3, 4]. Using thermal kinetic models and treatments, it corresponds effectively to the situation where the reactive system is being subjected to an isosbestic irradiation (Chap. 8) (which does not correspond to the irradiation conditions (λ_{n-isos}) employed in the present experiment).

Mathematical derivations for longer consecutive reactions under isosbestic irradiation are very rare. The situation is even more acute for non-isosbestic irradiation conditions where, untill now, the challenge of solving this type of reaction kinetics has not been met.

We consider here the case of a photo-transformation occurring in four consecutive photoreaction-steps, $XY_4(4\Phi)$, starting from a single reactant, X (Scheme 15.1). Each of the reaction-steps is characterised by an independent quantum yield ($\Phi_{Y_j \to Y_j'}$), and all four products ($Y_{1\ to\ 4}$), in addition to X, absorb the non-isosbestic irradiation light ($\varepsilon_{Y_j}^{\lambda_n-isos}$, which are all different and non-zero).

The rate-laws of the reaction are composed by a system of differential equations that describe the temporal variation of the species concentrations.

$$\frac{dC_X^{\lambda_n-isos,T}(\mathrm{t})}{dt} = -\Phi_{X\rightarrow Y_1}^{\lambda_n-isos}\,\varepsilon_X^{\lambda_n-isos}\,P_0^{\lambda_n-isos}\,l_{irr}\,F^{\lambda_n-isos}(t)\,C_X^{\lambda_n-isos,T}(t) \tag{15.17}$$

$$\frac{dC_{Y_1}^{\lambda_n-isos,T}(\mathrm{t})}{dt} = \left(\Phi_{X\rightarrow Y_1}^{\lambda_n-isos}\,\varepsilon_X^{\lambda_n-isos}\,C_X^{\lambda_n-isos,T}(t) - \Phi_{Y_1\rightarrow Y_2}^{\lambda_n-isos}\,\varepsilon_{Y_1}^{\lambda_n-isos}\,C_{Y_1}^{\lambda_n-isos,T}(t)\right) \times P_0^{\lambda_n-isos}\,l_{irr}\,F^{\lambda_n-isos}(t) \tag{15.18}$$

$$\frac{dC_{Y_2}^{\lambda_n-isos,T}(\mathrm{t})}{dt} = \left(\Phi_{Y_1\rightarrow Y_2}^{\lambda_n-isos}\,\varepsilon_{Y_1}^{\lambda_n-isos}\,C_{Y_1}^{\lambda_n-isos,T}(t) - \Phi_{Y_2\rightarrow Y_3}^{\lambda_n-isos}\,\varepsilon_{Y_2}^{\lambda_n-isos}\,C_{Y_2}^{\lambda_n-isos,T}(t)\right) \times P_0^{\lambda_n-isos}\,l_{irr}\,F^{\lambda_n-isos}(t) \tag{15.19}$$

$$\frac{dC_{Y_3}^{\lambda_n-isos,T}(\mathrm{t})}{dt} = \left(\Phi_{Y_2\rightarrow Y_3}^{\lambda_n-isos}\,\varepsilon_{Y_2}^{\lambda_n-isos}\,C_{Y_2}^{\lambda_n-isos,T}(t) - \Phi_{Y_3\rightarrow Y_4}^{\lambda_n-isos}\,\varepsilon_{Y_3}^{\lambda_n-isos}\,C_{Y_3}^{\lambda_n-isos,T}(t)\right) \times P_0^{\lambda_n-isos}\,l_{irr}\,F^{\lambda_n-isos}(t) \tag{15.20}$$

$$\frac{dC_{Y_4}^{\lambda_n-isos,T}(\mathrm{t})}{dt} = \Phi_{Y_3\rightarrow Y_4}^{\lambda_n-isos}\,\varepsilon_{Y_3}^{\lambda_n-isos}\,C_{Y_3}^{\lambda_n-isos,T}(t)\,P_0^{\lambda_n-isos}\,l_{irr}\,F^{\lambda_n-isos}(t) \tag{15.21}$$

with the photokinetic factor expressed by

$$FKP^{\lambda_n-isos}(t) = \frac{1-10^{-A_{tot}^{\lambda_n-isos,T/\lambda_n-isos}(\mathrm{t})\,\frac{l_{irr}}{l_{obs}}}}{A_{tot}^{\lambda_n-isos,T/\lambda_n-isos}(\mathrm{t})\,\frac{l_{irr}}{l_{obs}}} \tag{15.22}$$

The total absorbance corresponds here to the cumulative absorbances of the five species, and the factor l_{irr}/l_{obs} is introduced to correct the expression of $PKF^{\lambda_n-isos}(t)$, where the experimentally measured absorbance ($A_{tot}^{\lambda_n-isos,T/\lambda_n-isos}(t)$) differs from the one corresponding to the absorption of the excitation light, when irradiation and observation path lengths are different (see the previous section).

$$A_{tot}^{\lambda_n-isos,T/\lambda_n-isos}(t) = \sum_{j=0}^{4} \varepsilon_{Y_j}^{\lambda_n-isos}\,C_{Y_j}^{\lambda_n-isos,T}(t)\,l_{irr} \tag{15.23}$$

There is no known mathematical method capable of analytically solving this system of differential equation (Eqs. 15.17, 15.18, 15.19, 15.20 and 15.21), which includes a non-linear, time-dependent photokinetic factor (Eq. 15.22). Of course, the experimental data of $XY_4(4\Phi)$ reactions can be fitted using numerical integration methods (*NIM*s), as in the case of the trimolecular consecutive $XY_2(2\Phi)$ reaction [5], but no explicit integrated rate-law equations have, thus far, been proposed in the literaturc.

The application of the semi-empirical approach to the present kinetics is carried out by the analysis of 150 *RK*-simulation reaction-cases. The treatment of the consecutive $XY_4(4\Phi)$ yields the following integrated rate-laws for the individual concentrations of both the reactant and the photoproducts.

$$C_X^{\lambda_n-isos,T} = \rho_{15a}\, L\left(k_X^{\lambda_n-isos}\right) \tag{15.24}$$

$$C_{Y_1}^{\lambda_n-isos,T}(t) = k_{X\to Y_1}^{\lambda_n-isos}\, \frac{\rho_{15a}}{\rho_{15c}}\, \left(L\left(k_{X\to Y_1}^{\lambda_n-isos}\right) - L\left(k_{Y_1\to Y_2}^{\lambda_n-isos}\right)\right) \tag{15.25}$$

$$C_{Y_2}^{\lambda_n-isos,T}(t) = \rho_{15a}\, k_{X\to Y_1}^{\lambda_n-isos}\, k_{Y_1\to Y_2}^{\lambda_n-isos} \left(\frac{L\left(k_{X\to Y_1}^{\lambda_n-isos}\right)}{\rho_{15c}\,\rho_{15d}} - \frac{L\left(k_{Y_1\to Y_2}^{\lambda_n-isos}\right)}{\rho_{15c}\,\rho_{15e}} + \frac{L\left(k_{Y_2\to Y_3}^{\lambda_n-isos}\right)}{\rho_{15d}\,\rho_{15e}}\right) \tag{15.26}$$

$$\begin{aligned} C_{Y_3}^{\lambda_n-isos,T}(t) \\ = \rho_{15a}\, k_{X\to Y_1}^{\lambda_n-isos}\, k_{Y_1\to Y_2}^{\lambda_n-isos}\, k_{Y_2\to Y_3}^{\lambda_n-isos} & \left(\frac{L\left(k_{X\to Y_1}^{\lambda_n-isos}\right)}{\rho_{15c}\,\rho_{15d}\,\rho_{15f}} - \frac{L\left(k_{Y_1\to Y_2}^{\lambda_n-isos}\right)}{\rho_{15c}\,\rho_{15e}\,\rho_{15g}} + \frac{L\left(k_{Y_2\to Y_3}^{\lambda_n-isos}\right)}{\rho_{15d}\,\rho_{15e}\,\rho_{15h}}\right. \\ \times + & \left.\frac{L\left(k_{Y_3\to Y_4}^{\lambda_n-isos}\right)}{\rho_{15c}\,\rho_{15e}\,\rho_{15h}} - \frac{L\left(k_{Y_3\to Y_4}^{\lambda_n-isos}\right)}{\rho_{15c}\,\rho_{15d}\,\rho_{15f}} - \frac{L\left(k_{Y_3\to Y_4}^{\lambda_n-isos}\right)}{\rho_{15d}\,\rho_{15e}\,\rho_{15h}}\right) \end{aligned} \tag{15.27}$$

$$\begin{aligned} C_{Y_4}^{\lambda_n-isos,T}(t) \\ = \rho_{15a}\, k_{X\to Y_1}^{\lambda_n-isos}\, k_{Y_1\to Y_2}^{\lambda_n-isos}\, k_{Y_2\to Y_3}^{\lambda_n-isos} & \left(\frac{1}{\rho_{15c}\,\rho_{15d}\,\rho_{15f}} - \frac{1}{\rho_{15c}\,\rho_{15e}\,\rho_{15g}} + \frac{1}{\rho_{15d}\,\rho_{15e}\,\rho_{15h}}\right) \\ \times \left(L\left(k_{Y_3\to Y_4}^{\lambda_n-isos}\right) - \left|\varepsilon_X^{\lambda_n-isos} - \varepsilon_{Y_1}^{\lambda_n-isos}\right| l_{irr} C_X(0)\right) \\ + \rho_{15a}\, k_{Y_1\to Y_2}^{\lambda_n-isos}\, k_{Y_2\to Y_3}^{\lambda_n-isos}\, k_{Y_3\to Y_4}^{\lambda_n-isos} & \left(\frac{L\left(k_{X\to Y_1}^{\lambda_n-isos}\right) - \left|\varepsilon_X^{\lambda_n-isos} - \varepsilon_{Y_1}^{\lambda_n-isos}\right| l_{irr} C_X(0)}{\rho_{15c}\,\rho_{15d}\,\rho_{15g}}\right) \\ + \rho_{15a}\, k_{X\to Y_1}^{\lambda_n-isos}\, k_{Y_2\to Y_3}^{\lambda_n-isos}\, k_{Y_3\to Y_4}^{\lambda_n-isos} & \left(\frac{L\left(k_{Y_1\to Y_2}^{\lambda_n-isos}\right) - \left|\varepsilon_X^{\lambda_n-isos} - \varepsilon_{Y_1}^{\lambda_n-isos}\right| l_{irr} C_X(0)}{\rho_{15c}\,\rho_{15e}\,\rho_{15g}}\right) \\ + \rho_{15a}\, k_{X\to Y_1}^{\lambda_n-isos}\, k_{Y_1\to Y_2}^{\lambda_n-isos}\, k_{Y_3\to Y_4}^{\lambda_n-isos} & \left(\frac{L\left(k_{Y_2\to Y_3}^{\lambda_n-isos}\right) - \left|\varepsilon_X^{\lambda_n-isos} - \varepsilon_{Y_1}^{\lambda_n-isos}\right| l_{irr} C_X(0)}{\rho_{15d}\,\rho_{15e}\,\rho_{15h}}\right) \end{aligned} \tag{15.28}$$

where the factors $L\left(k_{i\to j}^{\lambda_n-isos}\right)$ are defined as

$$L\left(k_{Y_j \to Y_{j'}}^{\lambda_n - isos}\right) = Log\left(1 + \rho_{15b}\, e^{-k_{Y_j \to Y_{j\prime}}^{\lambda_n - isos}\, t}\right) \tag{15.29}$$

The rate-constants of the individual photochemical reaction-steps ($k_{Y_j \to Y_{j'}}^{\lambda_n - isos}$) have a dimension of s^{-1}, consistent with quantities in the argument of the exponential of Eq. 15.29. The index "*j*" corresponds to the first species (reactant or photoproduct) starting a given reaction-step, and "*j'*" is the generated photoproduct from that species *j*. The rate-constants formulae are

$$k_{Y_j \to Y_{j'}}^{\lambda_n - isos} = \Phi_{Y_j \to Y_{j'}}\, \varepsilon_{Y_j}^{\lambda_n - isos}\, l_{irr}\, P_0^{\lambda_n - isos}\, \frac{1 - 10^{-\varepsilon_{Y_j}^{\lambda_n - isos}\, l_{irr}\, C_X^{\lambda_n - isos, T}(0)}}{\varepsilon_{Y_j}^{\lambda_n - isos}\, l_{irr}\, C_X^{\lambda_n - isos, T}(0)} \tag{15.30}$$

The coefficients ρ in Eqs. 15.22, 15.23, 15.24, 15.25, 15.26 and 15.27 have the following expressions.

$$\rho_{15a} = \frac{1}{\left|\varepsilon_X^{\lambda_n - isos} - \varepsilon_{Y_1}^{\lambda_n - isos}\right| l_{irr}} \tag{15.31}$$

$$\rho_{15a} = e^{\left|\varepsilon_X^{\lambda_n - isos} - \varepsilon_{Y_1}^{\lambda_n - isos}\right| l_{irr}\, C_X^{\lambda_n - isos, T}(0)} - 1 \tag{15.32}$$

$$\rho_{15c} = \left(k_{Y_1 \to Y_2}^{\lambda_n - isos} - k_{X \to Y_1}^{\lambda_n - isos}\right) \tag{15.33}$$

$$\rho_{15d} = \left(k_{Y_2 \to Y_3}^{\lambda_n - isos} - k_{X \to Y_1}^{\lambda_n - isos}\right) \tag{15.34}$$

$$\rho_{15e} = \left(k_{Y_2 \to Y_3}^{\lambda_n - isos} - k_{Y_1 \to Y_2}^{\lambda_n - isos}\right) \tag{15.35}$$

$$\rho_{15f} = \left(k_{Y_3 \to Y_4}^{\lambda_n - isos} - k_{X \to Y_1}^{\lambda_n - isos}\right) \tag{15.36}$$

$$\rho_{15g} = \left(k_{Y_3 \to Y_4}^{\lambda_n - isos} - k_{Y_1 \to Y_2}^{\lambda_n - isos}\right) \tag{15.37}$$

$$\rho_{15h} = \left(k_{Y_3 \to Y_4}^{\lambda_n - isos} - k_{Y_2 \to Y_3}^{\lambda_n - isos}\right) \tag{15.38}$$

The above equations are valid for any of the sub-mechanisms of the pentamolecular system. In case the reaction involves only one photochemical step (only $k_{X \to Y_1}^{\lambda_n - isos} \neq 0$), the above equations reduce down to the characteristic equation of the primary photoreaction (Eq. 11.20).

We complete the set of equations that serve the validation by the initial reaction-velocity expression. The theoretical initial velocity of the reactant is worked out from its rate-law (Eq. 15.17), considered at $t = 0$, as

$$Theo : r_{0,X}^{\lambda_n-isos,T} = -Theo : r_{0,Y_1}^{\lambda_n-isos,T}$$

$$= -\Phi_{X\to Y_1}^{\lambda_n-isos}\,\varepsilon_X^{\lambda_n-isos}\,P_0^{\lambda_n-isos}\,l_{irr}\,F^{\lambda_n-isos}(0)\,C_X^{\lambda_n-isos,T}(0) \quad (15.39)$$

and its value is calculated from the differentiation of Eq. 15.24 at $t = 0$ and using the fitting data of the reactant concentration trace.

$$Fit : r_{0,X}^{\lambda_n-isos,T} = \rho_{15a}\,\frac{dL\left(k_X^{\lambda_n-isos}\right)}{dt} = -\frac{\rho_{15a}\,\rho_{15b}}{ln(10)}\,\frac{k_{X\to Y_1}^{\lambda_n-isos}}{1+\rho_{15b}} \quad (15.40)$$

It is also interesting to define the initial rate for the reaction by considering a differentiation of the equation of the total absorbance (Eq. 15.23) at $t = 0$ and introducing the theoretical initial reactant rate-equation (Eq. 15.39),

$$Theo : r_{0,A}^{\lambda_n-isos,T/\lambda_{obs}} = -\left(\varepsilon_{Y_1}^{\lambda_n-isos} - \varepsilon_X^{\lambda_n-isos}\right) l_{obs}\; Theo : r_{0,X}^{\lambda_n-isos,T} \quad (15.41)$$

The numerical values obtained for each pair of equations, i.e. Eqs. 15.39 and 15.40 and Eqs. 15.41 and 15.42, should be the same.

$$Fit : r_{0,A}^{\lambda_n-isos,T/\lambda_{obs}} = \frac{l_{obs}}{l_{irr}}\,\frac{k_{X\to Y_1}^{\lambda_n-isos}}{ln(10)}\left(1 - 10^{\left|\varepsilon_X^{\lambda_n-isos} - \varepsilon_{Y_1}^{\lambda_n-isos}\right|\,l_{irr}\,C_X^{\lambda_n-isos,T}(0)}\right) \quad (15.42)$$

The testing of the integrated rate-laws (Eqs. 15.24, 15.25, 15.26, 15.27 and 15.28) has been performed on more than 150 kinetic traces (Table 15.2) that were generated by a fifth-order Runge–Kutta method. In order to simulate the widest array of reaction eventualities, the parameter sets included carefully selected scenarios of parameter values.

For all 150 cases, an excellent fit was obtained for each of the individual five species' kinetic traces ($C_X^{\lambda_n-isos,T}(t)$ or $C_{Y_j}^{\lambda_n-isos,T}(t)$) by their respective semi-empirical model equations (Figs. 15.4(left)).

Similarly, a good fit was obtained for the reaction traces ($A_{tot}^{\lambda_n-isos}(t)$, Fig. 15.4 (right)). Cases of sub-mechanisms involving two or three photoproducts were also successfully tested.

As has been previously mentioned, a good fit is far from providing a definite proof of the validity of an equation. For example, *Trace #1* of Fig. 15.4(right) could well be fitted to the primary photoreaction equation (Eq. 11.20), but this good fit does not amount to a determining argument for the true mechanism of the reaction.

The general trend of the kinetic behaviour for this kind of consecutive photoreactions includes a monotonical depletion of the reactant until its disappearance from the solution and an increase of the concentration of the photoproducts Y_1, Y_2, and Y_3 to a specific maximum value of a Gaussian-type curve followed by a decrease to zero, whereas the trace of the last photoproduct (Y_4) has a sigmoid kind of shape for its appearance in the medium, starting by a slow regime followed by a relatively

Table 15.2 Examples of parameter sets for pentamolecular consecutive photoreactions used to generate *RK*-simulated kinetic traces using fifth-order Runge–Kutta calculations [6]

	Traces case						
Parameters	1	2	3	4	5	6	7
$C_X(0) \times 10^6$/M	10	8	4	8	9	7	4
$P^0_{\lambda_n-irr} \times 10^6$/M^{-1} s^{-1}	5	3	5	6	8	6	7
l_{irr}/cm	1	2.5	5	1.4	1	2	2
$\varepsilon_X^{\lambda_n-irr}$/M^{-1} cm^{-1}	3500	5500	16,000	2500	8550	2200	3750
$\varepsilon_{Y_1}^{\lambda_n-irr}$/M^{-1} cm^{-1}	5800	3200	14,000	12,400	6500	6500	4800
$\varepsilon_{Y_2}^{\lambda_n-irr}$/M^{-1} cm^{-1}	5900	1500	12,000	1100	4200	8700	6700
$\varepsilon_{Y_3}^{\lambda_n-irr}$/M^{-1} cm^{-1}	6300	3666	10,000	5000	3100	5000	3200
$\varepsilon_{Y_4}^{\lambda_n-irr}$/M^{-1} cm^{-1}	7000	5700	8000	4000	2249	1050	2100
$\Phi_{X\to Y_1}^{\lambda_n-isos}$	0.1	0.02	0.03	0.1	0.01	0.05	0.02
$\Phi_{Y_1\to Y_2}^{\lambda_n-isos}$	0.2	0.1	0.001	0.1	0.2	0.03	0.08
$\Phi_{Y_2\to Y_3}^{\lambda_n-isos}$	0.3	0.1	0.02	0.1	0.2	0.01	0.1
$\Phi_{Y_3\to Y_4}^{\lambda_n-isos}$	0.1	0.02	0.03	0.1	0.01	0.05	0.02
$k_{X\to Y_1}^{\lambda_n-isos}$/s^{-1}	0.00387	0.00175	0.01955	0.00490	0.00155	0.00290	0.00234
$k_{Y_1\to Y_2}^{\lambda_n-isos}$/s^{-1}	0.01250	0.00495	0.00059	0.02300	0.02270	0.00520	0.01185
$k_{Y_2\to Y_3}^{\lambda_n-isos}$/s^{-1}	0.01905	0.00230	0.01061	0.00230	0.01510	0.00240	0.02032
$k_{Y_3\to Y_4}^{\lambda_n-isos}$/s^{-1}	0.00675	0.00320	0.02768	0.00900	0.00480	0.00255	0.00501

much faster increase; then finally, the curve flattens asymptotically towards its final value that corresponds to the initial concentration of the reactant ($C_{Y_4}^{\lambda_n-isos,T}(\infty) = C_X^{\lambda_n-isos,T}(0)$). An example for the traces of a pentamolecular system is shown in Fig. 15.4 (left).

The variation of the total absorbance of the reactive medium with time cannot, however, be mapped out by a particular trend, as its overall shape may adopt a variety of different profiles (Fig. 15.4(right)). Some of the traces may, in appearance, look like mono- or bi-exponential patterns, while others adopt a smooth Gaussian curve or present peaks and troughs. Such a variability makes it difficult to work out the total number of species involved in the reaction from the sole analysis of the total absorbance trace of the reaction.

To strengthen the validation of the rate-law equations, the overall rate-constants obtained as fitting parameters ($Fit : k_{Y_j\to Y_{j'}}^{\lambda_n-isos}$, where the data of Table 15.2 are supposed to be not known) were compared to their counter parts ($RK : k_{Y_j\to Y_{j'}}^{\lambda_n-isos}$) that were calculated from Eq. 15.30 by using the original input set of parameters feeding the Runge–Kutta simulations (as those laid out in Table 15.2). Each trace was characterised by up to four overall rate-constants. A good linear relationship linked the two types ($Fit : k_{Y_j\to Y_{j'}}^{\lambda_n-isos}$ and $RK : k_{Y_j\to Y_{j'}}^{\lambda_n-isos}$) of values varying

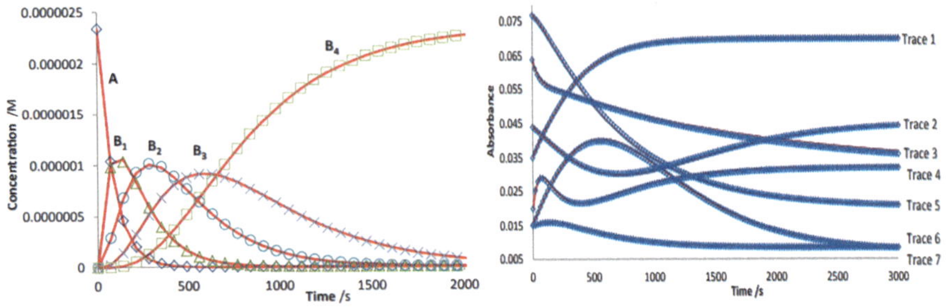

Fig. 15.4 *Left*: the temporal evolution of the individual species concentrations for the consecutive $XY_4(4\Phi)$ reaction (*case 6* in Table 15.2) undergone by X and generating four products. The species X and Y_i are labelled here as A and B_i. *Right*: some of the possible shapes of the absorbance traces obtained for pentamolecular consecutive photoreactions. The gemeric shapes correspond to the variation with time of the fifth-order Runge–Kutta-simulated total absorbances of the medium and the red lines to the corresponding model equation (Eq. 15.23, where the concentrations are replaced by their expressions Eqs. 15.24, 15.25, 15.26, 15.27 and 15.28). (Reproduced with permission from [6], *Journal of Pharmaceutical Sciences*)

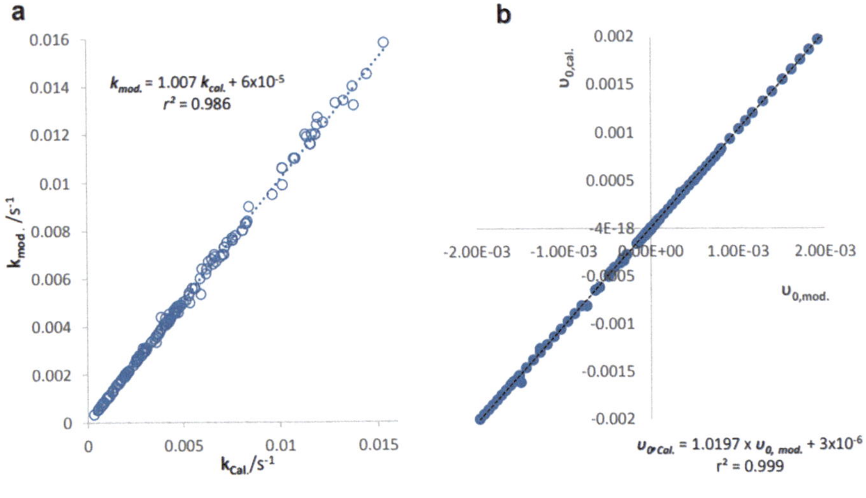

Fig. 15.5 Excellent linear correlations between (**a**) the rate-constants, $Fit : k_{Y_j \to Y_{j'}}^{\lambda_{n-isos}}$ and $RK : k_{Y_j \to Y_{j'}}^{\lambda_{n-isos}}$, and (**b**) the theoretical initial velocity ($Theo : r_{0,A}^{\lambda_{n-isos},T/\lambda_{obs}}$ and $Fit : r_{0,A}^{\lambda_{n-isos},T/\lambda_{obs}}$) values calculated theoretically using Eqs. 15.37 and 15.39, respectively, and those determined from the fitting of the traces, Eqs. 15.38 and 15.40, respectively. Over 150 *RK*-simulated traces are represented in the plots. (Reproduced with permission from [6], Journal of Pharmaceutical Sciences)

between 10^{-4} and $10^{-2}\,s^{-1}$ (Fig. 15.5a). The correlation coefficient and slope of the line approached unity, and its intercept is practically zero.

The last validation criterion concerns the correspondence of two values of the initial reaction-velocity. The first one ($Theo : r_{0,X\ or\ A}^{\lambda_{n-isos},T}$) is the theroretical rate-law of the pentamolecular photoreaction at $t = 0$ (Eqs. 15.39 and 15.41), using the specific parameter values that fed the Runge–Kutta (e.g. Table 15.2). The second set corresponding to the numerical values of $Fit : r_{0,X\ or\ A}^{\lambda_{n-isos},T}$ (Eqs. 15.40 and 15.41), supposed to be equal to the previous ones, use the fitting data of the corresponding traces. The values delivered by these equations do not vary by more than 10% for the 150 reaction-cases and correlate well in a linear trend (Fig. 15.5b).

The validation tests accredit the integrity of the semi-empirical equations to faithfully describe the kinetics of the pentamolecular consecutive reaction. These equations have been found to be reliable (with errors below 10%) when the final absorption is not exceeding a value of 0.22 (i.e. $A_{tot}^{\lambda_{n-isos},T/\lambda_{obs}}(\infty) < 0.22$, and $FKP^{\lambda_{n-isos}}(\infty) \geq 1.8$). This represents the validity limit for the semi-empirical model for $XY_4(4\Phi)$. The low absorption limit might be a constraint, but experiments can be conducted reliably by reducing either the initial concentration of the reactant (as we shall see in *Part VI*) or the optical path length of irradiation (l_{irr}).

15.4 Conclusion

The semi-empirical method has proven to work rather well in the description of several photoreactions. The explicit *SEM*-model equations for the temporal variation of both concentration and total absorbance reproduced, with high precision, the corresponding *RK*-generated traces. In addition to an excellent fitting, the *SEM*-equations for the initial rates and reaction rate-constants showed good agreement with these quantities calculated by the theoretical equations using the original values of their parameters (these parameters fed the *RK*-calculation). These criteria, all together, validate the *SEM*-model formulae of the integrated rate-law.

Despite the number of compounds that undergo one of the reactions encompassed within the general $XY(2\Phi, 1k)$ and $XY_4(4\Phi)$ sequences being large, it remains effectively a relatively small group when compared to the litany of photo- or photothermal reactions obeying other mechanisms.

The examples presented here can represent a template for the application of the semi-empirical method to other reactions as the *SEM* is expected to apply to virtually any reaction, it is a viable way to circumvent the impossibility in analytically solving the rate-laws.

However, even if the *SEM* works well, it is a time-consuming approach and would greatly benefit from the introduction of a computational element for the treatment and optimisation of the procedure.

References

1. Maafi W, Maafi M (2013) Modelling nifedipine photodegradation, photostability and actinometry properties. Int J Pharm 456:153–164. https://doi.org/10.1016/j.ijpharm.2013.07.075

2. Maafi M, Maafi W (2014) Φ-order kinetics of photoreversible-drug reactions. Int J Pharm 471: 536–543. https://doi.org/10.1016/j.ijpharm.2014.05.017
3. Gilbert A, Bagott J (1991) Essentials of molecular photochemistry. Blackwell Science, Oxford, MA. ISBN. 0-632-02428-3
4. Albini A, Fasani E (1998) Drugs photochemistry and photostability. The Royal Society of Chemistry, Cambridge. ISBN. 0-85404-743-3
5. Mauser H, Gauglitz G, Compton RG, Hancock G (eds) (1998) Comprehensive chemical kinetics, photokinetics: theoretical fundamentals and applications, vol 36. Elsevier, Amesterdam. ISBN-13. 978-0444825360
6. Maafi M, Maafi W (2016) Modeling and elucidation of the kinetics of multiple consecutive photoreactions $AB_4(4\Phi)$ with F-order kinetics. Application to the Photodegradation of Riboflavin. J Pharm Sci 105:3537–3548. https://doi.org/10.1016/j.xphs.2016.06.030

Chapter 16
Elucidation Methods Using SEM Equations

16.1 Introduction

The full set of information required for a complete kinetic analysis should include an elucidation method, as was reviewed for the kinetics performed under isosbestic irradiation (Chap. 10). The analysis of the reaction data by the semi-empirical method (*SEM*)-model equations is followed by detailed elucidation methods for each individual reaction so that the cases where the identifiability issues can be solved will be identified and separated from those where they cannot. This process is discussed in this chapter.

16.2 The Experimental Conditions

The photoreactions are exposed to collimated, monochromatic, and non-isosbestic continuous irradiation at constant temperature. The reactive medium solution is permanently and vigorously stirred. The observation and irradiation wavelengths and optical path lengths are λ_{obs} and λ_{n-isos} and l_{obs} and l_{irr}, respectively. The incident radiation is characterised by a known and constant $P_0^{\lambda_{n-isos}}$.

Each reaction is defined by a set of known and unknown parameters. For instance, the photoreversible system is fully defined by a set of ten parameters. Six of them relate to the experimental conditions in which the system is performing, and six characterise the reactive species and their photochemical reactivities. The experimental conditions are described by the extrinsic parameters, namely, the initial concentration ($C_X(0)$), the incident light intensity ($P^0_{\lambda_{irr}}$), the optical path lengths of irradiation and observation (l_{irr} and l_{obs}, respectively), and the selected wavelengths at which the system will be irradiated and observed or monitored (λ_{irr} and λ_{obs}, respectively). The four other features that are the intrinsic parameters, proper to the reactive system, include the absorption coefficients of the species at both irradiation

M. Maafi, *Photokinetics*, https://doi.org/10.1007/978-3-031-98985-8_16

($\varepsilon_Y^{\lambda_{irr}}$ and $\varepsilon_Y^{\lambda_{obs}}$, respectively) and the photochemical quantum yields ($\Phi_{X\rightarrow Y}^{\lambda_{irr}}$ and $\Phi_{Y\rightarrow X}^{\lambda_{irr}}$) of each individual reaction-step.

Most of these quantities are easily accessible to the experimentalist. The parameters corresponding to the experimental conditions can all be precisely determined, with the incident light intensity at the irradiation wavelength being obtained from actinometric measurements (see Chaps. 17 and 23). The absorptivities of the reactant at irradiation and observation wavelengths are measured by the calibration graph of this species.

Therefore, the kinetic elucidation method for this reaction has the precise purpose of determining the true values of four, not directly accessible, intrinsic parameters, namely, the two absorptivities of the product, $\varepsilon_Y^{\lambda_{irr}}$ and $\varepsilon_Y^{\lambda_{obs}}$, and the two quantum yields, $\Phi_{X\rightarrow Y}^{\lambda_{irr}}$ and $\Phi_{Y\rightarrow X}^{\lambda_{irr}}$. These make up the set of the unknown parameters of the kinetic system.

16.3 The Simplest *XY*(1*Φ*) Reaction

The quantum yield of the photoconversion with the absorptivity of the photoproduct are the two unknowns of the primary *XY*(1*Φ*) photoreaction whose photoproduct is transparent. The elucidation of the photokinetics of this reaction is simple not only because it records the least number of (two) unknowns but also because the reaction is described by an analytically derived integrated rate-law (Eq. 11.20).

The absorption coefficient of the reactant, $\varepsilon_X^{\lambda_{obs}}$, is determined from its electronic absorption spectrum in the medium at $t = 0$ and the initial concentration, $C_X^{\lambda_{n-isos},T}(0)$, whereas, the absorption coefficient of the photoproduct, $\varepsilon_Y^{\lambda_{n-isos}}$, is worked out from the spectrophotometric spectrum of the medium at the end of the reaction (at $t = \infty$, when no variation is observed in the electronic spectrum, i.e. the reactant has completely been depleted). The absorption spectrum recorded at $t = \infty$ corresponds to that of Y with its concentration equal to $C_Y^{\lambda_{n-isos},T}(\infty) = C_X^{\lambda_{n-isos},T}(0)$.

The spectrum, $\varepsilon_Y^{\lambda_{obs}}$, can also be calculated by

$$\varepsilon_Y^{\lambda_{obs}} = \frac{A_{tot}^{\lambda_{n-isos}/\lambda_{obs},T}(\infty)}{l_{obs}\, C_X^{\lambda_{n-isos},T}(0)} \tag{16.1}$$

For a monochromatic and non-isosbestic irradiation ($\lambda_{n-isos} = \lambda_X$), exclusively absorbed by the reactant X, the quantum yield of the process is derived from the Φ-order kinetic equation (Eq. 11.20) and defined by experimentally available quantities at any reaction time t.

$$\Phi_{X\rightarrow Y}^{\lambda_X} = -\frac{ln\left(\dfrac{10^{C_X^{\lambda_{n-isos},T}(t)\,\varepsilon_X^{\lambda_X}\,l_{irr}} - 1}{10^{\varepsilon_X^{\lambda_X}\,l_{irr}\,C_X^{\lambda_{n-isos},T}(0)} - 1}\right)}{P_0^{\lambda_X}\,\varepsilon_X^{\lambda_X}\,l_{irr}\,ln\,10\,t} \tag{16.2}$$

16.4 The $XY(1\Phi)$ Reaction with $\varepsilon_X^{\lambda_X} \neq 0$

The integrated rate-law useful for the reactant concentration is given by Eq. 15.7. The determination of the absorptivity of Y follows the same procedure applied above for the simplest photoreaction (Eq. 16.1). The total absorbance trace of the reaction is fit to Eq. 15.1, with the only fitting parameter being the reaction rate-constant (labelled ∂_4 in that equation). The quantum yield of the photochemical reaction-step is then obtained from the equation (Eq. 15.5, where $\Phi_{Y \to X}^{\lambda_{n-isos}} = k_{Y \to X}^{T} = 0$ and $A_{tot}^{\lambda_{n-isos},T/\lambda_{obs}}(\infty) = \varepsilon_Y^{\lambda_{n-isos}}\, l_{irr}\, C_X^{\lambda_{n-isos},T}(0)$) and the value of $\partial_4 = k_r^{\lambda_{n-isos},T}$ as

$$\Phi_{X \to Y}^{\lambda_{n-isos}} = \frac{k_r^{\lambda_{n-isos},T}}{\varepsilon_X^{\lambda_{n-isos}}\, P_0^{\lambda_{n-isos}}\, l_{irr}\, F^{\lambda_{n-isos}}(\infty)} \tag{16.3}$$

16.5 The Reversible $XY(1\Phi, 1k)$ Photokinetics

The kinetics of the reversible $XY(1\Phi, 1k)$ reaction is solved when the solution is exposed to monochromatic/isosbestic irradiation (Sect. 10.5). The knowledge of the full absorptivity spectrum of Y could be used for a subsequent solution of this kinetics, when considered under non-isosbestic irradiation. It means that two separate irradiations are carried out, i.e. one at an isosbestic wavelength to recover the $\varepsilon_Y^{\lambda_{obs}}$ spectrum (according to the method laid out in Sect. 10.5) and, later, one at a non-isosbestic wavelength to determine $\Phi_{X \to Y}^{\lambda_{n-isos}}$, a procedure that is a valid way to unravel the kinetic parameters of the reaction. However, it is not necessary, in practice, to employ such a double irradiation unless the quantum yield of the isosbestic wavelength is specifically required.

The full elucidation of the kinetics can also be achieved from the use of the data obtained by a non-isosbestic irradiation. The procedure comprises three steps. In the first instance, the total absorbance trace of the thermal reaction is fitted by mono-exponential equation, equivalent to Eq. 15.15 (set out for the total absorbance) and whose fitting parameters (∂_{1T} and ∂_{4T}) are not yet explicitly defined. The starting point of the trace is the time when the light is switched off; this could well be measured after the *pss* is reached, i.e. starting from $A_{tot}^{T,\lambda_{obs}}(0) = A_{tot}^{\lambda_{n-isos},T/\lambda_{obs}}(pss)$, as stated in Eq. 16.4 (and $A_{tot}^{T,\lambda_{obs}}(\infty) = A_X^{T,\lambda_{obs}}(0)$, i.e. a full regeneration of X). Hence, there is a single fitting parameter, $Fit : k_r^T = Fit : k_{Y \to X}^T$, in that equation.

$$A_{tot}^{T,\lambda_{obs}}(t) = A_{tot}^{T,\lambda_{obs}}(\infty) + \left(A_{tot}^{\lambda_{n-isos},T/\lambda_{obs}} - A_{tot}^{T,\lambda_{obs}}(\infty)\right) e^{-k_{Y \to X}^{T}\, t} \tag{16.4}$$

The equation that will be used to fit the total absorbance trace of the photothermal reaction under $\lambda_{n\ -\ isos}$-irradiation, Eq. 16.5 (which is Eq. 15.1), also has a single fitting parameter, $k_r^{\lambda_{n-isos},T}$. The full trace is considered, i.e. when it has reached the plateau at *pss* ($A_{tot}^{\lambda_{n-isos},T/\lambda_{obs}}(\infty) = A_{tot}^{\lambda_{n-isos},T/\lambda_{obs}}(pss)$).

$$A_{tot}^{\lambda_{n-isos},T/\lambda_{obs}}(t) = A_{tot}^{\lambda_{n-isos},T/\lambda_{obs}}(pss) + \frac{A_{tot}^{\lambda_{n-isos},T/\lambda_{obs}}(0) - A_{tot}^{\lambda_{n-isos},T/\lambda_{obs}}(pss)}{A_{tot}^{\lambda_{n-isos},T/\lambda_{n-isos}}(0) - A_{tot}^{\lambda_{n-isos},T/\lambda_{n-isos}}(pss)} \frac{l_{obs}}{l_{irr}} \times Log\left(1 + \left(10^{\left(A_{tot}^{\lambda_{n-isos},T/\lambda_{obs}}(0) - A_{tot}^{\lambda_{n-isos},T/\lambda_{obs}}(pss)\right)\frac{l_{irr}}{l_{obs}}} - 1\right) e^{-k_r^{\lambda_{n-isos},T}\, t}\right) \quad (16.5)$$

The value of quantum yield is calculated from Eq. 15.5 and the fitting value of $k_r^{\lambda_{n-isos},T}$ as

$$\Phi_{X\to Y}^{\lambda_{n-isos}} = \frac{Fit : k_r^{\lambda_{n-isos},T} - Fit : k_{Y\to X}^{T}}{\varepsilon_X^{\lambda_{n-isos}} P_0^{\lambda_{n-isos}} l_{irr} F^{\lambda_{n-isos}}(pss)} \quad (16.6)$$

16.6 The *XY*(2*Φ*) photoreaction

The kinetics of the reversible *XY*(2*Φ*) photoreaction has been studied under isosbestic light irradiation in Sect. 10.6. When the irradiation is non-isosbestic, several methods were proposed in the literature to solve such a kinetics. We will first review the most interesting of these methods; then, we will lay down the procedure facilitated by the *SEM*-equations.

16.6.1 *Solving* XY(2Φ) *Kinetics in the Literature*

As stated in Sect. 16.2, an elucidation method is required to unambiguously yield the true values of all the unknowns of a kinetics. The elucidation of the photoreversible reaction kinetics is complete when the true values of the four intrinsic parameters, $\varepsilon_Y^{\lambda_{n-isos}}$, $\varepsilon_Y^{\lambda_{obs}}$, $\Phi_{X\to Y}^{\lambda_{n-isos}}$, and $\Phi_{Y\to X}^{\lambda_{n-isos}}$, are known ($\varepsilon_X^{\lambda_{obs}}$ is directly accessible from spectrophotometry).

A rich documentation was dedicated to solving photoreversible reactions' kinetics. The earliest papers date back to the late 1930s [1–4], and one of the latest studies proposing an elucidation method for these kinds of reactions was published in recent years [5, 6]. This very particular interest stems from the ubiquity of these types of photosystems in chemistry, biology, and technology. In this regard, the control of the reaction kinetics is primordial to promote specific designs, increase efficiency, and enlarge the range of applications of such systems.

As we have briefly mentioned earlier, the endeavour of the research community to solve this problem has resulted in a multiplication of method propositions. Such a multitude is, in itself, a sufficient indication of the absence of a solid consensus, a situation that most often permits the end user to select an elucidation method of their own choice, but in general, the end users do not justify their choice of a particular method in comparison to the other available elucidation options. In this context, it seems like, for each study, the particular elucidation method employed is presumably a true and valid method. Yet this cannot be a satisfactory argument since each published elucidation method implicitly claims its difference from the other published ones.

The literature has accommodated this variety well. Nonetheless, the community is yet to accommodate a debate about the lack of consensus on this particularly old problem in regard to its undisputable importance in research, technology, and academia.

In this context, a review of the most prominent elucidation methods proposed for photo- and photothermal reversible photoreactions seems a pertinent contribution to such an eventual debate. It would not only shed light on the fundamental differences that exist between the proposed methods but, more importantly, might clarify their relative individual performance. This is attempted hereafter.

Let us first lay out the general setting in which these kinetic elucidation methods were developed. The reversible photothermal systems can involve either of both photochemical and thermal reactions (with the condition that the reactant is only photoreactive). This generality is not detrimental to the purely photochemical reversible systems since any kinetic method that addresses a complex mechanism should, in theory, also apply to all its sub-mechanisms. The light beam used for irradiation is (supposed) monochromatic and collimated (however, non-collimated, polychromatic light is seldom effectively used). The analytical technique of choice is spectrophotometry. The electronic absorption spectra of the reactive species X and Y totally overlap most often (in some cases, however, they might overlap only in the *UV* region of the electromagnetic spectrum, as it is notorious for photochromes). In the *UV* region, the reactant, X, is neither irradiated nor observed (by spectrophotometry) separately from its photoproduct. Finally, the rate-laws (or the systems of differential equations) describing these reactions were not solved by closed-form integrations (their integrated rate-laws are unknown except for the isosbestic irradiation (Chap. 12) and the *SEM*-equations (Chap. 15)).

The lack of analytically derived integrated rate-laws has been stressed by many authors [7–17], who concluded to the non-triviality of the mathematical process that is supposed to achieve an analytical kinetic solution for the reactive photosystem, i.e. the mathematics that would allow unravelling the true kinetic solution of the reactive system was deemed beyond reach. Such a conclusion might make up the starting point of the debate, suggested above, on the kinetic solutions of the photoreversible reaction only because, if it is to be considered true, it might automatically undermine any credit of validity for many of the elucidation options proposed in the literature. It certainly calls for a scrutiny of the foundations of these options before granting merit to any.

Anyway, the consequence of the unreachable analytical elucidation procedure has forced many of those proposed methods to introduce at least one simplifying assumption in order to achieve the kinetic solution (the treatment of the latter being based on purely photometric kinetic data) [7, 18–36].

Zimmermann and co-workers [7] were perhaps the first to attempt a full solution of the kinetics of photoreversible azobenzene system. The method was laid out, assuming the irradiation light to be monochromatic, the stirring of the solution to be complete, the Beer–Lambert law to apply, and the quantum yields of both forward and reverse photoreactions to be independent of both $P_0^{\lambda_{n-isos}}$ and $C_X^{\lambda_{n-isos},T}(0)$ (where some high initial reactant concentrations, $10^{-5}-10^{-3}$ M, were used). The full rate-law was stated, but its photokinetic factor was approximated by a first-order power series, assuming values of the total absorbance higher than 5, and the term involving the rate-constant of the thermal reaction-step was neglected ($k_{Y\to X}^T \cong 0$).

Fischer's method [25, 25, 37] advocates using a sufficiently high incident radiation intensity with the aim to make the photochemical processes the dominant ones in kinetics, hence allowing to neglect the contribution of the thermal reaction-step to kinetics [14]. The operation transforms the $XY(2\Phi, 1k)$ kinetic system into an $XY(2\Phi)$ reaction. Furthermore, this treatment requires that the ratio of forward/reverse quantum yields is independent of the irradiation wavelength. Finally, this method imposes that a double irradiation must be carried out at two different (non-isosbestic) wavelengths, where $\varepsilon_X^{\lambda_{n-isos}} \neq \varepsilon_Y^{\lambda_{n-isos}}$. Accordingly, the method can only be employed if all these conditions are fully satisfied (notably, that the quantum yield ratio is constant with wavelength).

A technique that uses millisecond flash photolysis [20, 21, 23] allows the determination of the product $\left(\varepsilon_Y^{\lambda_{n-isos}}\, \Phi_{X\to Y}^{\lambda_{n-isos}}\right)$. Such a product was deemed useful for the measurement of the colourability, but the approach does not allow the determination of the values for each individual parameter in that product, unless it is assumed that the reaction undergoes a nearly complete conversion of X into Y [38].

Micheau and co-workers [11, 13, 14, 26–28, 38] developed a software programme built on both a numerical integration method and a non-linear minimisation algorithm that aims at optimising the simultaneous fitting of the theoretical differential equations describing the investigated system to a number of experimental total absorbance traces (between four [11, 14] and nine [28, 29]) acquired by exposing the reactive system to two or three non-isosbestic irradiations. The kinetic solution for the $XY(2\Phi, 1k)$ system is then attained by assuming that the two reaction quantum yields are wavelength-independent. The computation did not include a mathematical analysis to show how the identifiability issues could be solved. This is a crucial concern, as in some instances, the treatment generated a multitude of different sets of values for the reaction parameters, with each set leading to satisfactory fitting of the experimental traces [11, 38]. In one instance, a quantum yield value exceeding unity was reported for a spironaphthoxazine photoproduct, $\Phi_{Y\to A}^{313} = 1.4$, when the quantum yields were considered wavelength-dependent. This particular finding has been attributed by the authors to either or both experimental errors and defects in the model [38].

The approach proposed by Atabekyan and co-workers [29] evaluated the kinetic and spectroscopic parameters of $XY(2\Phi, 1k)$ kinetics by using a couple of wavelengths, $\lambda_{n-isos}/\lambda_{obs}$, for the total absorbance trace, together with an optimisation based on Newton's method, while Sun and co-workers [31] employed a semi-empirical approach that considers a single total absorbance trace but assuming $\Phi_{Y \to X}^{\lambda_{n-isos}} = 0$ (i.e. deliberatly converting $XY(2\Phi, 1k)$ into $XY(1\Phi)$). They could graphically evaluate the integral of the photokinetic factor by fitting the kinetic data to a mono-exponential function (i.e. assuming that the reaction obeys first-order kinetics while $\lambda_{n-isos} \neq \lambda_{isos}$).

A few more methods have also been reported for the elucidation of *XY* systems' kinetics [30, 32–36, 39]. Jeong and his team treated the kinetic data by a mono-exponential model [33, 34], and the quantum yields of the reaction were determined by comparison to reference compounds [34]. A more recent study [6] equally used the first-order kinetic model equation to fit the total absorbance of a $XY(2\Phi, 1k)$ reaction exposed to an *LED*-excitation light for the determination of the quantum yields.

Favaro and her colleagues proposed the determination of the reaction unknowns by combining two differential equations, one for the colour-forming under *UV* irradiation and one for the photobleaching by visible light, and assuming a wavelength-invariant quantum yield [36]. In another method developed by the authors for $XY(1\Phi, 1k)$, the unknown $\Phi_{X \to Y}^{\lambda_{n-isos}}$ and $\varepsilon_{Y}^{\lambda_{n-isos}}$ values were obtained by plotting the reciprocal of the total absorbance of *Y* (the coloured isomer) at the photostationary state (*pss*) against the rate-constant of the thermal process [39].

The one-photon excitation absorption method [35] relies on prior knowledge of the spectra of both photoisomers (this method, was in fact, originally developed for the particular unimolecular photoreaction of 5,7-dimethylcoumarin [30] and requires a high reactant concentration, $C_{X}^{\lambda_{n-isos}}(0) \geq 10^{-4}\ M$.)

This literature review indicates that even though several methods to solve the $XY(2\Phi, 1k)$ or $XY(2\Phi)$ kinetics have been proposed, almost all these approaches lack the mechanisms to prove, identify, properly address, and/or solve the identifiability problems inherently associated with such kinetic elucidations (Chaps. 6 and 9). Also, beyond the fact that all introduced at least one important assumption, most of the proposed elucidation methods have not themselves been tested by any means, i.e. their validity seems to be uniquely based on the plausibility of the assumptions and perhaps on the mathematical treatment that followed.

16.6.2 Elucidation of XY(2Φ) Kinetics

We have previously reached the conclusion that the elucidation of $XY(2\Phi)$ kinetics under isosbestic irradiation cannot be achieved by uniquely using total absorbance traces (Sect. 10.6). This conclusion is also valid when the irradiation is performed by a monochromatic/non-isosbestic light. Therefore, extra information is required to solve this kinetics. The same overall strategy implemented for the elucidation of the $XY(2\Phi)$ kinetics under isosbestic irradiation (Sect. 10.6) is applied for that reaction under non-isosbestic irradiation. One elucidation is presented hereafter, but more options could otherwise be implemented.

Let us also consider here (as was earlier adopted in Sect. 10.6) that $C_X^{\lambda_{n-isos},T}(pss)$ and $C_Y^{\lambda_{n-isos},T}(pss)$ can be experimentally determined, e.g. by *high performance liquid chromatography (HPLC)*. The absorptivity spectrum of the product ($\varepsilon_Y^{\lambda_{obs}}$) can then be constructed from a similar equation to Eq. 10.46, where λ_{isos} is replaced by λ_{n-isos}.

The total absorbance trace is fit with Eq. 15.1 and the value of its parameter, $Fit : k_r^{\lambda_{n-isos},T}$, determined. The value of the initial rate of the reaction, $Fit : r_{0,A}^{\lambda_{n-isos},T/\lambda_{obs}}$, is calculated from the fitting data and Eq. 15.12. The forward quantum yield is obtained from the latter equation and a rearranged Eq. 15.11 as

$$\Phi_{X\to Y}^{\lambda_{n-isos}} = \frac{Fit : r_{0,A}^{\lambda_{n-isos},T/\lambda_{obs}}}{\left(\varepsilon_Y^{\lambda_{obs}} - \varepsilon_X^{\lambda_{obs}}\right) l_{obs}\, \varepsilon_X^{\lambda_{n-isos}}\, P_0^{\lambda_{n-isos}}\, l_{irr}\, F^{\lambda_{n-isos}}(0)\, C_X^{\lambda_{n-isos},T}(0)} \tag{16.7}$$

The last kinetic parameter is determined from the formula of the rate-constant, $k_r^{\lambda_{n-isos},T}$, Eq. 15.5.

$$\Phi_{Y\to X}^{\lambda_{n-isos}} = \frac{Fit : k_r^{\lambda_{n-isos},T}}{\varepsilon_Y^{\lambda_{n-isos}}\, P_0^{\lambda_{n-isos}}\, l_{irr} F^{\lambda_{n-isos}}(pss)} - \Phi_{X\to Y}^{\lambda_{n-isos}} \frac{\varepsilon_X^{\lambda_{n-isos}}}{\varepsilon_Y^{\lambda_{n-isos}}} \tag{16.8}$$

The kinetic analyses presented so far in this chapter, either constructed on *SEM*- or analytically derived equations, have been tested on *RK*-simulated data, where only the quantities available experimentally for each elucidation procedure were considered to be known. This successful testing is evidence of the validity of the elucidation procedures. This elucidation method for *XY*(2Φ) turns out to be the only one obtaining the true kinetic solution (i.e. solves the identifiability problem) without introducing a single assumption.

Clearly, the availability of handy integrated rate-laws greatly alleviates the mathematical developments, the kinetic analyses, and the elucidation procedures and solves the identifiability issues.

16.7 The Elucidation of *XY*(2Φ, 1*k*) Photokinetics

The kinetic elucidation of the *XY*(2Φ, 1*k*) reaction (with the thermal reaction-step being $k_{Y\to X}^T$) is hampered by the thermal instability of the product. Indeed, the determination of the *pss* concentration of the reactant is compromised if *HPLC* is used. However, the elucidation is possible whenever $C_X^{\lambda_{n-isos},T}(pss)$ is available, e.g. experimentally determined by any other analytical means. Here, the total absorbance of the thermal reaction performed in the dark after the reaction has attained *pss* is fitted by Eq. 16.4 and the value of its parameter, $k_{Y\to X}^T$, determined. The procedure follows in the same order adopted for the elucidation of *XY*(2Φ)

kinetics (Sect. 16.6) by employing the equivalent equations, Eqs. 10.46, 15.1, 16.7, and 16.8. The latter has the form

$$\Phi_{Y\to X}^{\lambda_n-isos} = \frac{Fit : k_r^{\lambda_n-isos,T} - k_{Y\to X}^T}{\varepsilon_Y^{\lambda_n-isos} P_0^{\lambda_n-isos} l_{irr} F^{\lambda_n-isos}(pss)} - \Phi_{X\to Y}^{\lambda_n-isos} \frac{\varepsilon_X^{\lambda_n-isos}}{\varepsilon_Y^{\lambda_n-isos}} \quad (16.8)$$

We have to underline the fact that if the *pss* concentration is not available, this kinetics is not solvable.

Also, in the case where the thermal reaction-step is $X \to Y$ ($XY(1k, 2\Phi)$), the kinetic solution follows similar conditions and constraints as stated above for the sister $XY(2\Phi, 1k)$ reaction but where now the spectrum of Y is known but not that of X (by letting the reaction progress to completion in the dark, the medium will contain only Y), and the photoreaction is initiated by the irradiation of Y.

16.8 The Elucidation Options for Consecutive $XY_4(4\Phi)$ Sequences

The basic set of unknowns for the $XY_4(4\Phi)$ is constituted by four quantum yields and four absorption coefficients ($\varepsilon_X^{\lambda_n-isos}$ being known). The monochromatic/non-isosbestic irradiation should simultaneously be absorbed by the reactant and all the photoproducts (a test in the procedure, developed below, allows a confirmation of this aspect). The total absorbance should not exceed a value of 0.22 (Sect. 15.3) at any reaction time to ensure that the SEM-equations and the Beer–Lambert law apply (otherwise the value of the initial reactant concentration must be reduced).

The elucidation of both $XY_4(4\Phi)$ and $XY_3(3\Phi)$ kinetics is not achievable by using the total absorbance trace only. The concentration traces of the individual species are required for these sequences. However, the consecutive $XY_2(2\Phi)$ kinetics is solvable by using only the absorbance trace (Sect. 16.8.6). The unimolecular reaction $XY(1\Phi)$ kinetics is not considered here, as it has already been treated in Sect. 16.3. The general elucidation of all $XY_4(4\Phi)$ reactions is presented, hereafter, in a stepwise procedure.

16.8.1 Fitting the Traces

The concentration traces will be fitted in succession, i.e. in the order of appearance of the species in the reaction mechanism (first fitting that of X, then that of Y_1, etc). Each subsequent fitting equation includes the terms obtained for the previous species' fitting equations, as indicated by Eqs. 15.24, 15.25, 15.26, 15.27 and 15.28. The values of the factors ρ and $k_{Y_j\to Y_{j'}}^{\lambda_n-isos}$ are optimised towards achieving the fitting of all species' concentration traces with a unique set of ρ-factors.

The obtained fitting equations for the individual species are combined into Eq. 15.23, and the resulting equation is applied to fit the total absorbance trace of the reaction. An excellent fit of the latter trace indicates the validity of the individual fitting equations of the species' concentration traces.

16.8.2 Determining the Reactant Quantum Yield and the Absorptivity of the Last Species

The absorption coefficient of the last species of the reaction mechanism (e.g. Y_4) is calculated from the absorbance reached at the end of the reaction ($t = \infty$), where only this last species remains in the medium, with its concentration being equal to that of the reactant at the beginning of the reaction.

$$\varepsilon_{Y_4}^{\lambda_{n-isos}} = \frac{A_{tot}^{\lambda_{n-isos},T/\lambda_{obs}}(\infty)}{C_X^{\lambda_{n-isos},T}(0)\, l_{obs}} = \frac{A_{Y_4}^{\lambda_{n-isos},T/\lambda_{obs}}(\infty)}{C_X^{\lambda_{n-isos},T}(0)\, l_{obs}} \tag{16.9}$$

The quantum yield of the first reaction-step is deduced from the rate-constant of the reactant (Eq. 15.30) as

$$\Phi_{X\to Y_1}^{\lambda_{n-isos}} = -\frac{Fit : k_{X\to Y_1}^{\lambda_{n-isos}}}{\varepsilon_X^{\lambda_{n-isos}}\, P_0^{\lambda_{n-isos}}\, l_{irr}\, F^{\lambda_{n-isos}}(0)} \tag{16.10}$$

16.8.3 Absorptivity and Quantum Yield of the First Photoproduct

The absorptivity of Y_1 can be worked out from the value (Eq. 15.42) and equation (Eq. 15.41) of the initial rate of the total absorbance trace.

$$\varepsilon_{Y_1}^{\lambda_{n-isos}} = -\frac{Fit : r_{0,A}^{\lambda_{n-isos},T/\lambda_{obs}}}{l_{obs}\, Fit : r_{0,X}^{\lambda_{n-isos},T/\lambda_{obs}}} + \varepsilon_X^{\lambda_{n-isos}} \tag{16.11}$$

The quantum yield of the second reaction-step is then obtained from the fitting value and equation of $k_{Y_1\to Y_2}^{\lambda_{n-isos}}$ (Eq. 15.30) as

$$\Phi_{Y_1\to Y_2}^{\lambda_{n-isos}} = \frac{Fit : k_{Y_1\to Y_2}^{\lambda_{n-isos}}\, C_X^{\lambda_{n-isos},T}(0)}{1 - 10^{-\varepsilon_{Y_1}^{\lambda_{n-isos}}\, l_{irr}\, C_X^{\lambda_{n-isos},T}(0)}} \tag{16.12}$$

16.8.4 The Last Unknown Absorptivities, $\varepsilon_{Y_2}^{\lambda_{n-isos}}$ and $\varepsilon_{Y_3}^{\lambda_{n-isos}}$

The fitting equation of the total absorbance trace is used to determine the values of the absorbance at two time intervals (t_1 and t_2), preferably selected in the middle of the reaction time. At the same two time intervals, the values of the five species concentrations are also obtained from their respective fitting equations. We then have two values for the total absorbance and ten values of concentrations. These values are introduced into two Eq. 15.23. The system of equations is

$$C_{Y_2}^{\lambda_{n-isos},T}(t_1)\,\varepsilon_{Y_2}^{\lambda_{n-isos}} + C_{Y_3}^{\lambda_{n-isos},T}(t_1)\,\varepsilon_{Y_3}^{\lambda_{n-isos}} = \frac{A_{tot}^{\lambda_{n-isos},T/\lambda_{obs}}(t_1)}{l_{obs}} - C_{X}^{\lambda_{n-isos},T}(t_1)\,\varepsilon_{X}^{\lambda_{n-isos}} + C_{Y_1}^{\lambda_{n-isos},T}(t_1)\,\varepsilon_{Y_1}^{\lambda_{n-isos}} + C_{Y_4}^{\lambda_{n-isos},T}(t_1)\,\varepsilon_{Y_4}^{\lambda_{n-isos}} \tag{16.13}$$

and

$$C_{Y_2}^{\lambda_{n-isos},T}(t_2)\,\varepsilon_{Y_2}^{\lambda_{n-isos}} + C_{Y_3}^{\lambda_{n-isos},T}(t_2)\,\varepsilon_{Y_3}^{\lambda_{n-isos}} = \frac{A_{tot}^{\lambda_{n-isos},T/\lambda_{obs}}(t_2)}{l_{obs}} - C_{X}^{\lambda_{n-isos},T}(t_2)\,\varepsilon_{X}^{\lambda_{n-isos}} + C_{Y_1}^{\lambda_{n-isos},T}(t_2)\,\varepsilon_{Y_1}^{\lambda_{n-isos}} + C_{Y_4}^{\lambda_{n-isos},T}(t_2)\,\varepsilon_{Y_4}^{\lambda_{n-isos}} \tag{16.14}$$

The solution of the system of these two equations will deliver the values of $\varepsilon_{Y_2}^{\lambda_{n-isos}}$ and $\varepsilon_{Y_3}^{\lambda_{n-isos}}$ (which are the only unknowns of this system of linear equations, as $\varepsilon_{X}^{\lambda_{n-isos}}$, $\varepsilon_{Y_1}^{\lambda_{n-isos}}$, and $\varepsilon_{Y_4}^{\lambda_{n-isos}}$ are already known). The absorbance spectra of the individual species should be reconstituted by the individual Beer–Lambert equation, $C_{Y_j}^{\lambda_{n-isos},T}\ \varepsilon_{Y_j}^{\lambda_{n-isos}}\ l_{obs}$, a confirmation that the sum of the latter individual absorbance traces gives the experimentally recorded one (the total absorbance trace), must be carried out. This will not only prove the correctness of the procedure, but also that the incident light at λ_{n-isos} is absorbed by all species, otherwise the procedure must be repeated with another, more convenient irradiation wavelength.

16.8.5 The Remaining Quantum Yields

The last two unknowns to be determined by this elucidation method are $\Phi_{Y_2 \to Y_3}^{\lambda_{n-isos}}$ and $\Phi_{Y_3 \to Y_4}^{\lambda_{n-isos}}$. Their values are obtained from the respective rate-constants obtained from their concentration traces in a similar manner as that used for the determination of $\Phi_{Y_1 \to Y_2}^{\lambda_{n-isos}}$ (Eq. 16.12).

$$\Phi_{Y_2 \to Y_3}^{\lambda_{n-isos}} = \frac{Fit : k_{Y_2 \to Y_3}^{\lambda_{n-isos}}\ C_{X}^{\lambda_{n-isos},T}(0)}{1 - 10^{-\varepsilon_{Y_2}^{\lambda_{n-isos}}\ l_{irr}\ C_{X}^{\lambda_{n-isos},T}(0)}} \tag{16.15}$$

and

$$\Phi_{Y_3 \to Y_4}^{\lambda_n - isos} = \frac{Fit : k_{Y_3 \to Y_4}^{\lambda_n - isos} \, C_X^{\lambda_n - isos, T}(0)}{1 - 10^{-\varepsilon_{Y_3}^{\lambda_n - isos} \, l_{irr} \, C_X^{\lambda_n - isos, T}(0)}} \tag{16.16}$$

16.8.6 The Case of $XY_2(2\Phi)$ Consecutive Reaction

The kinetics of this reaction is defined by four parameters. The values of $\varepsilon_{Y_2}^{\lambda_n - isos}$ and $\Phi_{X \to Y_1}^{\lambda_n - isos}$ are derived by Eqs. 16.9 and 16.10. The initial rate of the reactant value (Eq. 15.39) can then be calculated and used to determine the value of the second absorption coefficient, $\varepsilon_{Y_1}^{\lambda_n - isos}$, using Eq. 16.11. Finally, the value of the quantum yield of the second reaction-step, $\Phi_{Y_1 \to Y_2}^{\lambda_n - isos}$, is obtained by Eq. 16.12. Hence, the complete elucidation of $XY_2(2\Phi)$ can exceptionally be achieved by using a single total absorbance trace (the rest of $XY_2(n\Phi)$ cannot be solved by only using such a trace).

16.9 Conclusion

A cornerstone of kinetic analysis has been set out in this chapter for the reactions described in Chap. 15 by the *SEM*-method. It has been demonstrated that most of these kinetics can readily be solved and their unique (true) set of parameters determined. The performance is unprecedented in the literature for most of these reaction systems, and the achievement is attributable to the usefulness of the *SEM*-equations. We will see the implementation of these elucidation methods for various experimental situations in the following part, Part VI.

References

1. Hartley GS (1938) The *cis*-form of azobenzene and the velocity of the thermal *cis→trans*-conversion of azobenzene and some derivatives. J Chem Soc:633–642. https://doi.org/10.1039/JR9380000633
2. Hartley GS, LeFevre RJW (1939) The dipole moments of *cis*- and *trans*-azobenzenes and of some related compounds. J Chem Soc:531–535. https://doi.org/10.1039/JR9390000531
3. Hausser I (1949) Die Quantenbilanz der cis ⇆ trans-Umwandlung von Azobenzol. Naturwissenschaften 36:315–317. https://doi.org/10.1007/BF00590336
4. Birnbaum PP, Style DWG (1954) The photo-isomerization of some azobenzene derivatives. Trans Faraday Soc 50:1192–1196. https://doi.org/10.1039/TF9545001192
5. Sharma A, Bekir M, Lomadze N, Santer S (2024) Photo-isomerization kinetics of azobenzene containing surfactant conjugated with polyelectrolyte. Molecules 26:19. https://doi.org/10.3390/molecules26010019
6. Solanki UB, Vikova M, Vik M (2024) Spectrokinetic investigation of the photochromic system under continuous UV irradiance using reflectance vs. time curves. Fibr Textil 31:35–41. https://doi.org/10.15240/tul/008/2024-2-005

7. Zimmerman G, Chow L-Y, Paik U-J (1958) The photochemical isomerization of azobenzene. J Am Chem Soc 80:3528–3531. https://doi.org/10.1021/ja01547a010
8. Hanazawa M, Sumiya R, Horikawa Y, Irie M (1992) Thermally irreversible photochromic systems. Reversible photocyclization of 1,2-bis (2-methylbenzo[b]thiophen-3-yl) perfluorocyclo- coalkene derivatives. J Chem Soc Chem Commun:206–207. https://doi.org/10.1039/C39920000206
9. Gauglitz G, Scheerer E (1993) Method for the determination of absorption coefficients, reaction rate constants and thermodynamic data in the system *AB*(2*Φ*, 1*k*). J Photochem Photobiol A Chem 71:205–212. https://doi.org/10.1016/1010-6030(93)85001-O
10. Favaro G, Malatesta V, Matzzucato U, Ottavi G, Romani A (1995) Thermally reversible photoconversion of spiroindoline-naphtho-oxazines to photomerocyanines: a photochemical and kinetic study. J Photochem Photobiol A Chem 87:235–241. https://doi.org/10.1016/1010-6030(94)03991-3
11. Pimienta V, Lavabre D, Levy G, Samat A, Guglielmetti R, Micheau JC (1996) Kinetic analysis of photochromic systems under continuous irradiation. Application to spiropyrans. J Phys Chem 100:4485–4490. https://doi.org/10.1021/jp9531117
12. Gade R, Porada T (1997) Determination of quantum yields and product spectra by studying reversible photoisomerizations in solution. J Photochem Photobiol A Chem 107:27–34. https://doi.org/10.1016/S1010-6030(96)04594-7
13. Deniel MH, Texier J, Lavabre D, Micheau JC, Durr H (1997) Kinetic modelling of the photochromism of dihydroindolizines. Mol Cryst Liq Cryst 298:129–135. https://doi.org/10.1080/10587259708036152
14. Crano C, Guglielmetti RJ (eds) (1999) Organic photochromic and thermochromic compounds, vol 1 and 2. Plenum Press, New York. ISBN: 0-3064-5882-9
15. Feringa BL (ed) (2001) Molecular switches. Wiley-VCH, Verlag. ISBN. 9783527299652
16. Durr H, Bouas-Laurent H (eds) (2003) Photochromism: molecules and systems. Elsevier, Amsterdam, Boston, London, New York, Oxford, Paris, San Diego, San Francisco, Singapore, Sydney, Tokyo. ISBN: 9780444513229
17. Maafi, M. (2008). Kinetic analysis and elucidation options for *AB*(1*k*, 2*Φ*) systems. New spectrokinetic methods for photochromes. Photochem Photobiol Sci 7, 1360–1372 https://doi.org/10.1039/b807556e
18. Fischer, E. (1967). Calculation of photostationary states in systems $A \leftrightharpoons B$ when only *A* is known. J Phys Chem 71, 3704–3706. https://doi.org/10.1021/j100870a063
19. Brown GH (ed) (1971) Photochromism, vol 3. Wiley, New York, London, Sydney, Toronto. ISBN: 10.0471928941
20. Appriou P, Gugglielmetti R, Garnier F (1978) Étude du processus photochimique impliqué dans la réaction d'ouverture du cycle benzopyrannique des spiropyrannes photochromiques. J Photochem 8:145–165. https://doi.org/10.1016/0047-2670(78)80016-1
21. Pottier E, Dubest R, Guglielmetti R, Tardieu P, Kellman A, Tfibel F, Levoir P, Aubard J (1990) Effets de substituant, d'hétéroatome et de solvant sur les cinétiques de décoloration thermique et les spectres d'absorption de photomérocyanines en série spiro[indoline-oxazine]. Helv Chim Acta 73:303–315. https://doi.org/10.1002/hlca.19900730210
22. Rau H, Greiner G (1991) Determination of quantum yields of the X/Y isomerization system from spectroscopic data. EPA Newslett 41:40–55
23. Kellmann A, Tfibel F, Gugglielmetti R (1995) Effect of substituents on the photochromism of a spiro[indoline-napthoxazine] under laser excitation. J Photochem Photobiol A Chem 91:131–136. https://doi.org/10.1016/1010-6030(95)04110-2
24. Mauser H, Gauglitz G, Compton RG, Hancock G (eds) (1998) Comprehensive chemical kinetics, photokinetics: theoretical fundamentals and applications, vol 36. Elsevier, Amesterdam. ISBN-13. 978-0444825360
25. Wong-Wah-Chung P, Maillot G, Bolte M (2001) 4,4′-Diaminostilbene-2,2′-disulfonate (DSD) behaviour: under irradiation in water: decrease of its activity as a fluorescent whitening agent. J Photochem Photobiol A Chem 138:275–280. https://doi.org/10.1016/S1010-6030(00)00400-7

26. Metelitsa AV, Micheau JC, Voloshin NA, Voloshina EN, Minkin VI (2001) Kinetic and thermodynamic investigations of the photochromism and solvatochromism of semipermanent merocyanines. J Phys Chem 105:8417–8422. https://doi.org/10.1021/jp010419r
27. Metelitsa AV, Lokshin V, Micheau JC, Samat A, Guglielmetti R, Minkin VI (2002) Photochromism and solvatochromism of push–pull or pull–push spiroindolinenaphthoxazines. Phys Chem Chem Phys 4:4340–4345. https://doi.org/10.1039/B204603B
28. Metelitsa AV, Micheau JC, Besugli SO, Gaeva EB, Voloshin NA, Voloshina EN, Samat A, Minkin VI (2004) Photochromic properties of six 5-O-n-alkyl, 6′-CN substituted spironaphthoxazines. Int J Photoenergy 6:199–204. https://doi.org/10.1155/S1110662X0400025X
29. Atabekyan LS, Zakharova GV, Ogienko VN, Chibisov AK (2002) Photochromic transformations of spiro compounds: 2. The kinetics of photocoloration under continuous irradiation. High Energy Chem 36:322–325. https://doi.org/10.1023/A:1020294500261
30. Belfield KD, Bondar MV, Liu Y, Przhonska OV (2003) Photophysical and photochemical properties of 5,7-dimethoxycoumarin under one- and two-photon excitation. J Phys Org Chem 16:69–78. https://doi.org/10.1002/poc.576
31. Sun F, Zhang F, Zhao F, Zhou X, Pu S (2004) Preparation and photochromic properties of diarylethene nanoparticles. Chem Phys Lett 380:206–212. https://doi.org/10.1016/j.cplett.2003.08.119
32. Matsuda K, Irie M (2004) Diarylethene as a photoswitching unit. J Photochem Photobiol C 2: 169–182. https://doi.org/10.1016/j.jphotochemrev.2004.07.003
33. Jeong Y-C, Yang SI, Kim E, Ahn K-K (2006) Development of highly fluorescent photochromic material with high fatigue resistance. Tetrahedron 62:5855–5861. https://doi.org/10.1016/j.tet.2006.04.029
34. Pu S, Yang T, Li G, Xu J, Chen B (2006) Substituent position effect on the optoelectronic properties of photochromic diarylethenes. Tetrahedron Lett 47:3167–3171. https://doi.org/10.1016/j.tetlet.2006.02.124
35. Corredor CC, Belfield KD, Bondar MV, Przhonska OV, Hernandez FE, Kachkovsky OD (2006) One- and two-photon photochromism of 3,4-bis-(2,4,5-trimethyl-thiophen-3-yl)furan-2,5-dione. J Photochem Photobiol A Chem 184:177–183. https://doi.org/10.1016/j.jphotochem.2006.04.012
36. Favaro G, Matzzucato U, Ortica F, Smimmo P (2007) Thermal reversibility and bistability in photochromic diarylethenes. Inorg Chim Acta 360:995–999. https://doi.org/10.1016/j.ica.2006.07.007
37. Fischer E, Frankel M, Wolovsky R (1955) Wavelength dependence of photoisomerization equilibria in azocompounds. J Chem Phys 23:1367. https://doi.org/10.1063/1.1742302
38. Pimienta V, Froute C, Deniel MH, Lavabre D, Gugglielmetti R, Micheau JC (1999) Kinetic modeling of the photochromism and photodegradation of a spiro[indoline-naphthoxazine]. J Photochem Photobiol A Chem 122:199–204. https://doi.org/10.1016/S1010-6030(99)00023-4
39. Di Nunzio MR, Gentili PL, Romani A, Favaro G (2008) Photochromic, thermochromic, and fluorescent spirooxazines and naphthopyrans: a spectrokinetic and thermodynamic study. ChemPhysChem 9:768–775. https://doi.org/10.1002/cphc.200700789

Part VI
Quantification by SEM Equations

Chapter 17
Reaction Photostabilisation and Quantum Yield Variability with Wavelength

17.1 Introduction

The photokinetic investigation also expands to quantification of various effects induced on photo- and photothermal reactions by changes occurring on extrinsic experimental parameters and reaction conditions. The tools used for that purpose are available in the *SEM*-equations (semi-empirical method equations) and the elucidation methods developed in Chaps. 15 and 16. These photokinetic tools are applied here to the experimental data of reactive systems. The quantification, in this chapter, aims at a dual objective: assessing extrinsic and intrinsic parameters, namely, on the one hand, the extrinsic factors causing photostabilisation and, on the other hand, the wavelength dependence of the intrinsic quantum yield parameter.

17.2 Autophotostabilisation: The $C_X^{\lambda_n-isos,T}(0)$ Effect

Examining the theoretical rate-law of the reactant of any photothermal reaction indicates that the photokinetics must be dependent on the initial reactant concentration, $C_X^{\lambda_n-isos,T}(0)$. More specifically, either the reactant concentration is explicitly present in that rate-law, or it is implicitly part of the photokinetic factor.

Theoretically, when the selected $C_X^{\lambda_n-isos,T}(0)$ values belong to the linearity range of the reactant calibration graph (assuming that all the remaining extrinsic factors are the same), the Φ-order *SEM*-equations apply, including the expressions for the reaction rate-constants and initial reaction-rates. The considered reaction is exposed to a monochromatic and non-isosbestic irradiation, supplied at a constant flux, $P_0^{\lambda_n-isos}$.

The phenomenon of the impact of $C_X^{\lambda_n-isos,T}(0)$ on reactivity has qualitatively been reported in the literature, though without providing a definitive fundamental interpretation for it [1–9].

M. Maafi, *Photokinetics*, https://doi.org/10.1007/978-3-031-98985-8_17

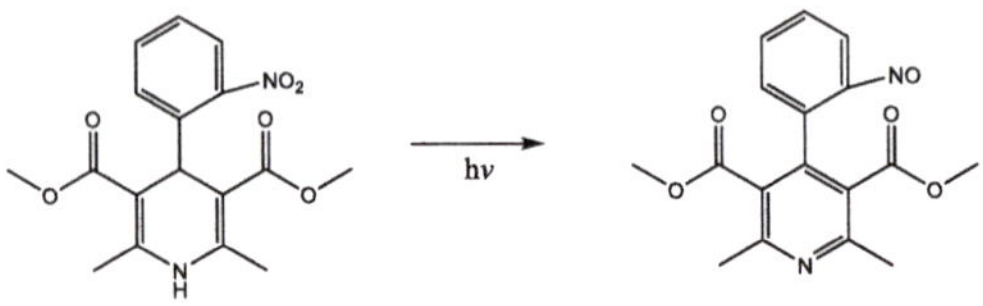

Scheme 17.1 Photodegradation reaction of the cardiovascular drug nifedipine

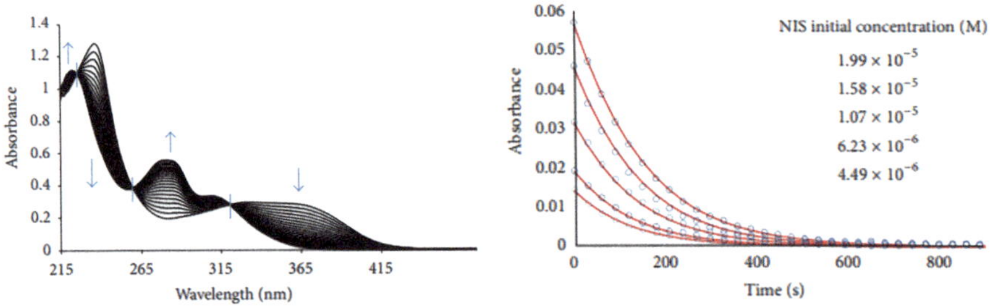

Fig. 17.1 *Left*: progressive evolution of the electronic absorption spectra of a 5.88 10^{-5} M nisoldipine (NIS) in ethanol that is subjected to a continuous monochromatic/non-isosbestic irradiation at 390 nm (total irradiation time 2000 s at a radiant power of $P_0^{390} = 7.76\ 10^{-7}$ $einstein\ dm^{-3}\ s^{-1}$.) The arrows indicate the direction of the evolution of the peaks during the photoreaction, and the vertical lines indicate the isosbestic points at 232, 259, and 320 nm. *Right*: total absorbance traces of *NIS* in ethanolic solutions of various concentrations (4.49 10^{-6} – 1.99 10^{-5} M). Irradiation performed at 350 nm and observation at 390 nm. The incident radiation intensities for the different experiments were in the range of $1.749\ 10^{-6} \leq P_0^{350} \leq 1.841\ 10^{-6}$ $einstein\ dm^{-3}\ s^{-1}$. Optical path lengths, $l_{obs} = 1\ cm$ and $l_{irr} = 2\ cm$. (Reproduced with permission from [11], *International Journal of Photoenergy*)

An example of the experimental data for the photokinetic reactivity variation with $C_X^{\lambda_{n-isos},T}(0)$ relates to photodegradations of nifedipine (*NIF*) [10] and nisoldipine (*NIS*) [11] (see Scheme 17.1). The two drugs obey a unimolecular reaction (*X*→*Y*). Both reaction species absorb the incident monochromatic/non-isosbestic light, except when $\lambda_{n\ -\ isos} > 370\ nm$, where only the reactant absorbs. The observation wavelength for the drugs'experiments was $\lambda_{obs} = 390\ nm$. The calibration graphs of the drugs have been constructed using both high performance liquid chromatography (*HPLC*) and spectrophotometry, and their linearity ranges were determined. The concentrations of the drugs used in the study fall within their respective linearity ranges (for spectrophotometric calibration graphs).

During irradiation, the spectrum of the medium variably evolves, with some peaks undergoing an absorbance reduction and others having their intensities increase, whereas new transition bands might appear (Fig. 17.1). In all cases, the overlapped medium spectra over time show distinct isosbestic points (as a confirmation for the smoothness of the reactions that progress without the emergence of by-products in the medium during the exposure time).

The experimental total absorbance traces are well-fitted to Eq. 15.1, irrespective of the value of the initial reactant concentration (Fig. 17.1).

The initial reaction-rate of the reactant for the reactions described by the *SEM*-equations (Chap. 15) is proportional to the term $\left(1-10^{-A_{tot}^{\lambda_n-isos,T/\lambda_{obs}}(0)}\right)$ that represents the only concentration-dependent factor in the expression of $r_{0,X}^{\lambda_n-isos,T}$ (Eqs. 15.10 and 15.37). The value of this term increases with the value of the initial total absorbance and, hence, with an increasing value of the initial concentration (since here, $A_{tot}^{\lambda_n-isos,T/\lambda_{obs}}(0)=A_X^{\lambda_n-isos,T/\lambda_{obs}}(0)$). Therefore, the absolute value of initial reactant-rate ($|-r_{0,X}^{\lambda_n-isos,T}|=r_{0,X}^{\lambda_n-isos,T}$) increases, i.e. $r_{0,X}^{\lambda_n-isos,T}\rightarrow-\infty$, which practically corresponds to a slowdown of the depletion of the reactant (e.g. the $r_{0,X}^{\lambda_n-isos,T}$ values, $-0.1>-1$). This means that the reactant is consumed faster when its initial concentration has smaller values. This purely photokinetic phenomenon can be viewed as an autophotostabilisation of the reactant (by higher values of $C_X^{\lambda_n-isos,T}(0)$). Experimental data collected on the photoreactions of the above drugs confirm this behaviour (Table 17.1).

The observed effect of varying $C_X^{\lambda_n-isos,T}(0)$ on the kinetics of the reactant is also passed onto the photoproducts. Hence, the overall reaction reactivity also shows a reduction when the initial reactant concentration increases. This is relayed by the reaction rate-constant, $k_r^{\lambda_n-isos}$, whose variation occurs in an opposite direction to that of the total absorbance (Table 17.1). This is caused by a reduction of the photokinetic factor (when $A_{tot}^{\lambda_n-isos,T/\lambda_{obs}}(\infty)$ or $A_{tot}^{\lambda_n-isos,T/\lambda_{obs}}(0)$ increases), in accordance with the expressions of this quantity (Eqs. 15.5 and 15.28). A linear relationship is expected between $k_r^{\lambda_n-isos}=f\left(PKF^{\lambda_n-isos}(\infty)\right)$ when $C_X^{\lambda_n-isos,T}(0)$ changes within the limits of linearity range of the calibration graph of the species (Fig. 17.2).

Up to a 58% reduction in the value of $k_r^{\lambda_n-isos,T}$ is recorded for up to a 6.2-fold increase of the initial concentration of the above drugs [10, 11].

Photostabilisation by incrementation of $C_X^{\lambda_n-isos,T}(0)$ might explain some experimental observations reported in the literature [4, 12]. For instance, the change of reaction-order postulated for *NIS* photokinetics, which attributed an apparent first-order kinetics for $C_X^{\lambda_n-isos,T}(0)\leq 3\ 10^{-5}\ M$ and a zeroth-order kinetics for $C_X^{\lambda_n-isos,T}(0)>2\ 10^{-4}\ M$ [13], might well be interpreted as a slowdown of *NIS*-photoreaction due to increasing initial concentrations. The kinetic trace for lower initial concentrations would have been faster (following a more curved trend in a given time interval) than for higher concentrations (having a more linear trend in the same time interval) and, therefore, might have justified the fitting of the experimental trace of the former to the exponential model (first-order reaction) and the latter to the linear model (zeroth-order reaction). Strongly concentrated solutions, resulting in slower photodegradation rates (investigated by first-order kinetics). were also attributed to a higher absorption at the edges or the top of a sample, which results in relatively fewer photons reaching the bulk of the solution, as in the case of the photodegradation of riboflavin in a 12-cm-long liquid-core waveguide cell [8, 9]. Such possible confusion confirms the need for using adequate kinetic methods (e.g. *SEM* equations) since good fittings can be obtained with classical thermal

Table 17.1 Experimental data of the effect of $C_X^{\lambda_n-isos,T}(0)$ on photoreactivity parameters of *NIF* and *NIS*

	Parameters					
Molecules	λ_{n-irr}/nm	$A_{tot}^{\lambda_n-isos/\lambda_{obs}}$	$P_0^{\lambda_n-isos} \times 10^6/M^{-1}\ s^{-1}$	$r_{0,X}^{\lambda_n-isos} \times 10^8/s^{-1}$	$C_X^{\lambda_n-isos}(0) \times 10^6/M$	$k_r^{\lambda_n-isos} \times 10^3/s^{-1}$
Nifedipine	354	0.027	2.105	- 7.821	8.40	4.3
		0.069	2.276	- 19.68	24.1	4.14
		0.144	2.215	- 34.11	52.1	3.89
Nisoldipine	350	0.014	1.749	- 3.682	4.49	8.2
		0.019	1.764	- 4.984	6.23	8.0
		0.032	1.784	- 8.239	10.7	7.7
		0.046	1.841	- 11.85	15.8	7.5
		0.057	1.841	- 14.33	19.9	7.2

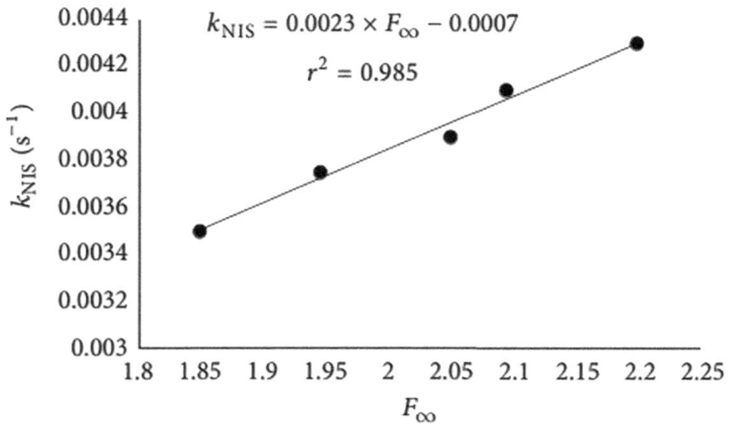

Fig. 17.2 A linear plot of the overall reaction rate-constant ($k_r^{\lambda_n - isos, T}$ labelled here k_{NIS}) of the photodegradation of nisoldipine and the photokinetic factor ($PKF^{\lambda_n - isos}(\infty)$ labelled here as F_∞), when ethanolic *NIS* solutions of concentrations ranging between 0.39 and 2.06 10^{-5} M are exposed to a monochromatic irradiation at 320 nm and observed at 390 nm. (Reproduced with permission from [11], *International Journal of Photoenergy*)

kinetic treatments, but these are not informative enough to reach valid conclusions on the true reactivity of photosystems.

17.3 Presence of Spectator Molecules in the Solution

The solution of the sample may contain one or more molecules that are both different from the species of the photothermal reactive system investigated and thermally and photochemically inert. Such molecules were termed "spectator molecules" (*SPMs*) since they do not chemically or photochemically interfere with the reaction studied (Sect. 2.16). Nonetheless, the question remains on whether the presence of *SPM*s in the sample solution affects the kinetics of photo- and photothermal reactions.

In thermal kinetics, the presence of *SPM*s is not expected to influence reactivity patterns. Hence, the pure thermal reaction-steps of photothermal reactions should not be concerned by *SPM*s' presence in the reactive medium. From a photochemical viewpoint, the effect of *SPM*s must be related to the extent of their absorption of the incident light. If so, the *SPM*s would compete for light absorption with the photoreactive species of the system investigated. This becomes the condition, i.e. the effect of *SPM*s is proportional to their capacity of absorbing the incident radiation.

It is known that *SPM*s have been proposed for photostabilisation of photolabile drugs [4, 12, 14–16]. Also, red-coloured iron oxides are currently employed as *SPM*s for many solid (tablet) drug formulations [12, 17]. The observed photodegradation reaction slowdown of drugs was interpreted, for instance, when increasing the concentrations of indigotin and azorubine dyes [5], as resulting from a decrease in the drug's photodegradation quantum yield in solution. The dyes were believed to perform as quenchers of the excited triplet-state of *NIS*, as was supported by Stern–Volmer constructs. The rate-constant of *NIS* photodegradation in ethanol was also found to decrease with an increase of β-carotene concentrations [18]. In these studies, the kinetic treatments were based on first-order kinetics, and the conclusions were reached by comparing the values of first-order rate-constants.

Technically, the absorption of the incident light by *SPM*s is implemented specifically within the total absorbance of the medium, which then includes the absorbances of both the "n_{sp+1}" species of the reaction and the "n_{SPM}" *SPM*s that absorb at λ_{n-isos} (at either $t = 0$ or $t = \infty$, according to the *SEM*-equations laid out in Chap. 15).

$$A_{tot}^{\lambda_{n-isos},T/\lambda_{obs}}(t) = \sum_{j=0}^{n_{sp}} A_{Y_j}^{\lambda_{n-isos},T/\lambda_{obs}}(t) + \sum_{i=1}^{n_{SPM}} A_{SPM,i}^{\lambda_{n-isos},T/\lambda_{obs}} \tag{17.1}$$

Amending the reactive medium total absorbance by *SPM* absorption (Eq. 17.1) directly affects the value of the rate-constant of the reaction (e.g. Eq. 15.5). The total absorbance is expected to reach higher values compared to those recorded in the absence of *SPM*s (reaction species alone); the value of the photokinetic factor will certainly decrease, hence, inducing a decrease of the rate-constant of the reaction. This kinetic interpretation does not require the involvement of either a variation of the reaction quantum yields or the occurrence of a quenching of molecular excited-states.

The rationalisation of this analysis relative to the effect of *SPM* concentration on reactivity is assessed by testing Eqs. 15.1 and 15.5 on *RK*-synthetic traces that describe the reactive systems studied in Chap. 15. The reactive medium is irradiated at a wavelength where both initial species and *SPM* absorb (the latter must have a constant concentration during a given experiment). The generated *RK* data were well-fitted, in all cases, to the model equations. Furthermore, linear relationships are found between $k_r^{\lambda_{n-isos},T}$ and $PKF^{\lambda_{n-isos}}(\infty)$. Generally, the *SPM* absorbances do not exceed a value of ca. 2, which corresponds to $PKF^{\lambda_{n-isos}}(\infty) > 0.4$.

Experimental quantification of this phenomenon was achieved for, on the one hand, *NIF* and *NIS* (both obeying an $X \rightarrow Y$ reaction) and was undertaken by three *SPM*s used individually, namely, tartrazine (*TRZ*), sun set yellow (*SSY*), and quinoline yellow (*QY*) [10, 11] and, on the other hand, for fluvoxamine (*FLU*) (undergoing a photoreversible reaction) by using *TRZ* [19]. The *SPM* dyes are experimentally found to be photo- and thermally stable when tested in the same experimental conditions in which the photodegradation of drugs is performed.

The electronic absorption spectra of these *SPM* dyes are characterised by two main broad absorption bands situated in the 200–320 and 320–550 nm spectral regions. The *SPM*-spectra overlap with those of the three drugs, hence, offering a potential photoprotection from UVA and visible radiations [16].

Table 17.2 *SPM* absorbances; reaction rate-constants of *NIF*, *NIS*, and *FLU*; and photoreactivity reduction percentages with increasing *SPM* concentration

	Parameters				
Molecules	*SPM*	$A_{SPM}^{\lambda_{n-isos}/\lambda_{obs}}$ Ɨ	$k_r^{\lambda_{n-isos}} \times 10^3$ $/s^{-1}$	$\frac{k_r^{\lambda_{n-isos}}\left(A_{SPM}^{\lambda_{n-isos}/\lambda_{obs}}=0\right)}{k_r^{\lambda_{n-isos}}\left(A_{SPM}^{\lambda_{n-isos}/\lambda_{obs}}\neq 0\right)}$	*% reduction*
Nifedipine ╠		0.000	6.7	1	0
	TRZ	0.253	4.3	1.558	33.48
		0.518	3.15	2.127	53.98
		0.978	2.1	3.191	69.22
		1.368	1.64	4.085	74.86
		1.948	1.3	5.154	75.29
	QY	0.319	4.3	1.558	26.86
		0.741	3.2	2.094	38.80
		1.025	2.6	2.577	47.76
		1.940	1.8	3.722	64.18
	SSY	0.160	3.3	2.030	22.98
		0.324	2.6	2.566	33.58
		0.490	2.2	2.979	38.80
		0.742	1.5	4.351	49.22
Nisoldipine		0	8.0	1	0
	TRZ	0.379	5.2	1.54	35.00
		0.762	3.9	2.05	51.25
		1.147	3.2	2.50	60.00
		1.415	2.2	3.64	72.50
	QY	0.261	6.7	1.19	16.25
		0.472	5.5	1.45	31.35
		0.792	4.0	2.00	50.00
		1.137	3.2	2.50	60.00
Fluvoxamine		0	0.87	1	0
	TRZ	0.314	0.51	1.71	41.40
		0.406	0.44	1.98	49.40
		0.483	0.37	2.32	36.90
		0.933	0.22	3.87	74.10

Ɨ: an observation wavelength of 390 nm for *NIF* and *NIS* and 280 nm for *FLU*
╠: the constant concentrations of *NIF*, *NIS*, and *FLU* are 4.65 10^{-5}, 1.56 10^{-5}, and 2.95 10^{-6} M, respectively, and the ranges of radiant power used for their respective experiments are 9.5 10^{-7}, 1.74 10^{-6}–1.78 10^{-6}, and 5.06 10^{-6}–5.18 10^{-6} einsten dm^{-3} s^{-1}

Increasing concentrations of a given dye in each drug solution induce gradual reduction in the photodegradation reaction. The data are reported on Table 17.2. The example shown in Fig. 17.3 depicts the total absorbance traces of *NIF* in the presence of *SSY* at different concentrations, ranging between 1.14 10^{-5} M and 5.35 10^{-5} M. The recorded improvement of *NIF* photostability is typical of the behaviours undergone by *NIS* and *FLU*. These results confirm similar findings, usually observed for binary systems, i.e. where a single amount of the dye was tested on a drug. A series of spectator molecules was employed, including curcumin, fast yellow,

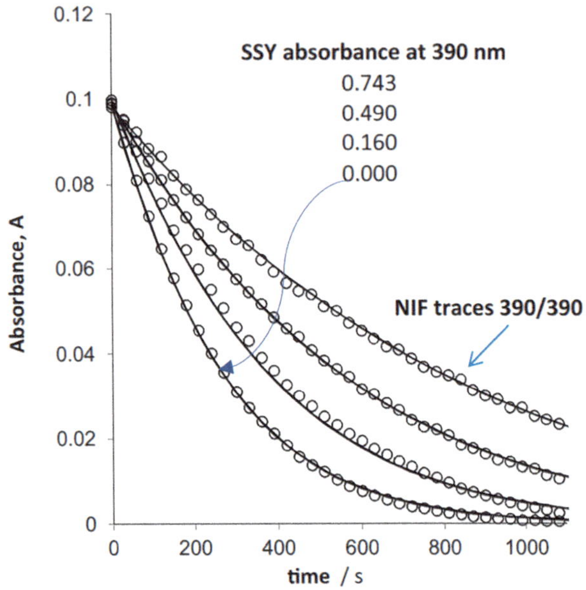

Fig. 17.3 Progressive reduction of reactivity of *NIF* with increasing concentration of the spectator *SSY* (1.14 – 5.35 10^{-5} M). The total absorbance traces of *NIF* photodegradation (circles) were obtained for ethanolic 3.52 10^{-5} M solutions. The solid lines correspond to the *SEM* fitting equations. (Reproduced with permission from [10], *International Journal Pharmaceutics*)

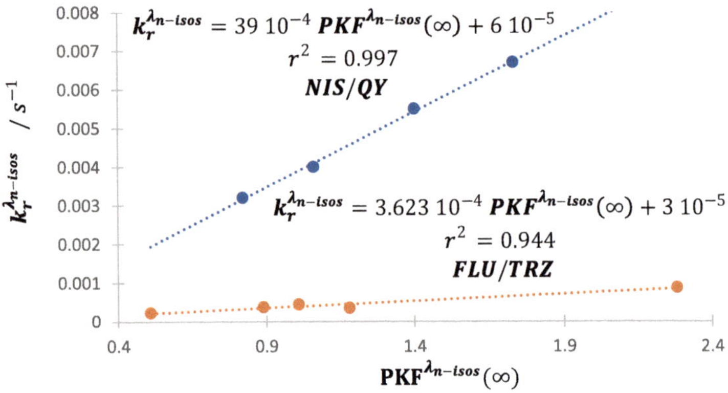

Fig. 17.4 Linear relationship between the photokinetic factors ($PKF^{\lambda_{n-isos}}(\infty)$) and the rate-constants ($k_r^{\lambda_{n-isos},T}$) of *FLU* and *NIS* photodegradations in the presence of increasing concentrations of *TRZ* and *QY*, respectively

chrysoine, apocarotenal, and cochineal in solutions [14] and curcumin [4] and tartrazine [20] in film-coating materials.

The experimental total absorbance traces of the three drugs were fitted to the *SEM*-equation, Eq. 15.1, and the value of the reaction rate-constant (Eq. 15.5) for each absorption competitor concentration was obtained as the fitting parameter. The *SPM*s caused retardation of the drugs' photodegradation in ethanol and was characterised by linearly correlated rate-constants with the corresponding photokinetic factor at $t = \infty$ (Fig. 17.4). The model equations, Eqs. 15.1 and 15.5, are reliably applied beyond the dynamic ranges of the organic *SPM*s (since, generally, linearity ranges of calibration graphs for organic molecules, including *TRZ*, *SSY*, and *QY*, do not reach a value of 2 in the absorbance of the compounds, Table 17.2). In practice, there is no requirement to comply with the Beer–Lambert law or impose it on *SPM*s. The experimental results confirm that the model equations are suitable to describe the kind of kinetics deployed in Chap. 15 in the presence *SPM*s (that are photo- and thermally stable).

The photostabilisation of the drugs reaches up to 75% in the presence of *SPM* compared to their reactivities in the absence of a spectator (Table 17.2). The sensitivity of the photodegradation of a given drug seems to be invariably affected by the nature of the *SPM*. The gradients of the linear plots for *NIF* in the presence of each of the three *SPM*s are very similar, being 0.00225 s^{-1} for *QY*, 0.0023 s^{-1} for *TRZ*, and 0.00208 s^{-1} for *SSY*. However, a given *SPM* affects distinct molecules differently (Table 17.2). Conceptually, a photoreaction could be virtually fully stopped if a high enough concentration of one or several *SPM*s is used.

17.4 Determination of the Quantum Yield Variation with Excitation Wavelength

From a general viewpoint, the models provided by the *SEM*-equations clearly indicate that the reaction rate-constant (Eq. 15.5) depends simultaneously on several reaction parameters, e.g. the quantum yield(s), the monochromatic radiation intensity, the absorptivies, the volume and area of irradiated sample, the optical path length of the irradiation, and the final absorbance at the irradiation wavelength. Therefore, the reaction rate-constant values are not directly indicative of the magnitude of the reaction quantum yield. Yet the quantum yield remains one of the essential physical quantities, objectively informing on photoreactivity, and ensures safely undertaking comparison between systems as well as between the effects of monochromatic lights impinging on the same molecule.

The *SEM* equations allow the determination of the absolute quantum yields of the photoreaction-steps as part of the set of the reaction parameters obtained by one of the straightforward elucidation procedures proposed in Chap. 16.

We will keep in mind that the phenomenon of the wavelength-dependent quantum yield has been reviewed in a general presentation proposed in Chap. 4. One of the limitations underlined therein related to the variation of the methods by which the values of the quantum yield were calculated (where the accuracy of most of these methods was met with limited success).

17.4.1 The Simplest Unimolecular Reaction

The best example of the unimolecular reaction, $X \rightarrow Y$, whose photoproduct is transparent, is represented by the ring-opening reaction of several photochromes. The reactant, in this case, is the coloured species (absorbing in the visible spectral region) that phototransforms into its closed-form isomer that absorbs in the *UV* but not in the visible. Generally, the reactant is synthesised in solution, in a previous step, by irradiating the open-form isomer by *UV* light.

The effect of the excitation light wavelength on the quantum yield of these types of reactions was measured for the ring-opening reaction-step (Scheme 17.3) of the photochromic diarylethene derivative (1,2-bis[2-methylbenzo[b]-thiophen-3,3,4,4,5,5-hexafluoro-1-cyclopentene) in ethanol at a constant temperature of 22 °C.

The initial concentrations of the diarylethene closed-form isomer were monitored by *HPLC*. The quantum yield was obtained spectrophotokinetically, by firstly, fitting the absorbance trace to Eq. 15.1, and subsequently, its absolute values were calculated using Eq. 16.2. The wavelengths of the monochromatic lights irradiating the sample (410, 420, 430, 450, 460, 480, 520, 535, and 555 nm) were selected to cover the whole absorption spectrum of the closed-form isomer at intervals less than 40 nm [21]. The wavelength of observation corresponded to the maximum of the long-wavelength absorption band ($\lambda_{obs} = 520$ nm) of the isomer. The absorption spectrum of closed-form isomer has two transitions situated between 400–450 and 450–600 nm, respectively (Fig. 17.5).

These findings indicate that the quantum yield values of ring opening are transition-dependent, higher, and decreasing between 400 and 450 nm and low and relatively constant in the region between 450 and 560 nm. The quantum yield values are relatively high (ranging between 0.86 and 0.12) in ethanol, compared to those reported in *n*-hexane, 0.35 and 0.28, measured in the excitation wavelength range between 450 and 580 nm [22].

A decrease of up to 75% of the values of the quantum yield were recorded with the irradiation wavelength (Fig. 17.6), a result that shows agreement with the reported wavelength-dependent quantum yields of diarylethene derivatives in the literature. For an ethynyl-phenyl diarylethene derivative, the quantum yield values, measured in acetonitrile, ranged between 0.016 and 0.0015 (~94% decrease) and correlated linearly with wavelength in the 404–609 nm interval [23]. The data of diarylethene derivative depicted in Fig. 17.6 are well-fitted to a sigmoid pattern represented by Eq. 17.2. Such a behaviour is corroborated for quantum yield dependence on irradiation wavelength for various molecules, as we shall see in further sections. The wavelength dependence of the quantum yield was either attributed to the existence of an energy barrier in the molecule's excited-state or to a reaction short timescale of the order of molecular vibrations [22].

$$\Phi_{X \rightarrow Y}^{\lambda_{irr}} = 0.12 + \frac{1.66}{1 + 1.2 \times e^{0.047\,(\lambda_{irr} - 410)}} \tag{17.2}$$

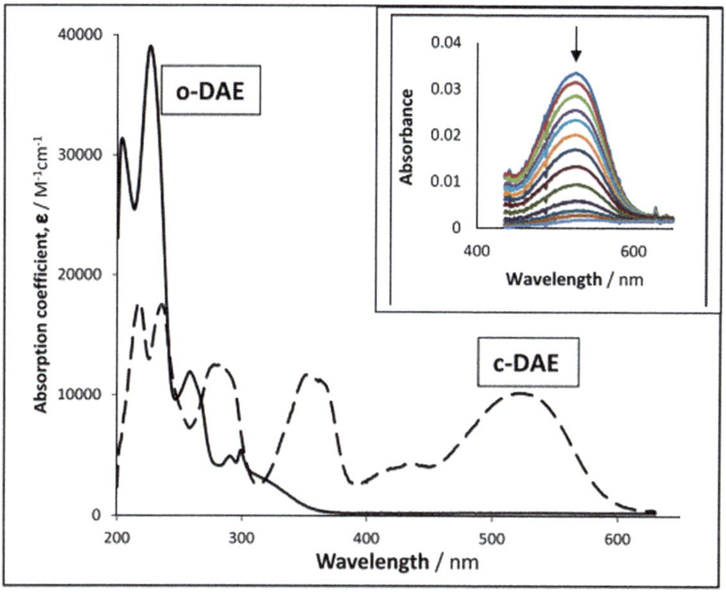

Fig. 17.5 Electronic spectra of closed- (*c-DAE*, dashed line) and open- (*o-DAE*, plain line) form isomers' spectra of *DAE*. The initial concentration of *o-DAE* was 1.39 10^{-5} M. Inset: photobleaching of *c-DAE* spectra under a 520 nm monochromatic light (3.42 10^{-6} M, 22 °C). (Reproduced with permission from [21], *Dyes and pigments)*

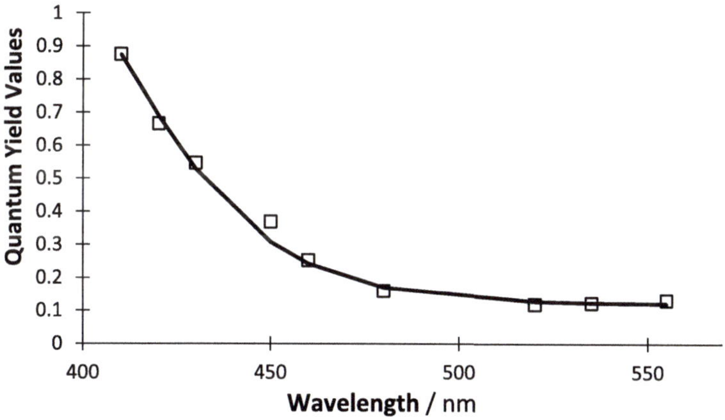

Fig. 17.6 Experimental quantum yield ($\Phi_{X\to Y}^{\lambda_{irr}}$) values of the closed-form diarylethene isomer (squares) for the selected irradiation wavelengths. Sigmoid pattern (solid line) is given by Eq. 17.2. (Reproduced with permission from [21], *Dyes and pigments)*

17.4.2 The XY(1Φ) Sequence with $\varepsilon_Y^{\lambda_{irr}} \neq 0$

Systematic irradiations (in steps of 10–20 nm intervals) were conducted on this type of sequence, operating the photodegradations of anticancer dacarbazine (*DBZ*, Scheme 17.4) and cardiovascular *NIF* and *NIS* drugs (Scheme 17.1).

The λ-effect for these molecules shows a reverse trend compared to that observed for diarylethene (Fig. 17.6). Here, monochromatic irradiations of the short-wavelength bands (corresponding to an S_2 state) translate into much smaller quantum yield values than those recorded for the irradiation of the long-wavelength absorption bands (S_1-state) [10, 11, 14, 25]. The variation of $\Phi_{X\rightarrow Y}^{\lambda_{irr}}$ values for these compounds with excitation wavelength showed a monotonical increase. The quantum yield values of the phototransformations (Fig. 17.7) fitted a full sigmoid function for *NIF* and *NIS* and a partial one for *DBZ* (Eqs. 17.3 and 17.4).

$$\Phi_{X\rightarrow Y}^{\lambda_{irr}}(NIF) = \frac{0.3317}{1 + 57.3684\, e^{-0.0547\,(\lambda_{irr}-223)}} \tag{17.3}$$

$$\Phi_{X\rightarrow Y}^{\lambda_{irr}}(DBZ) = \frac{0.263}{1 + 399\, e^{-0.0265\,(\lambda_{irr}-180)}} \tag{17.4}$$

The experiments were measured at a low initial reactant concentration (~5 10^{-6} M) and under strictly monochromatic/non-isosbestic irradiation over wavelength regions where both $\varepsilon_X^{\lambda_{irr}} \neq 0$ and $\varepsilon_Y^{\lambda_{irr}} \neq 0$. Eq. 15.1 fitted the total absorbance traces, and the quantum yields were measured by using the value of the reaction rate-

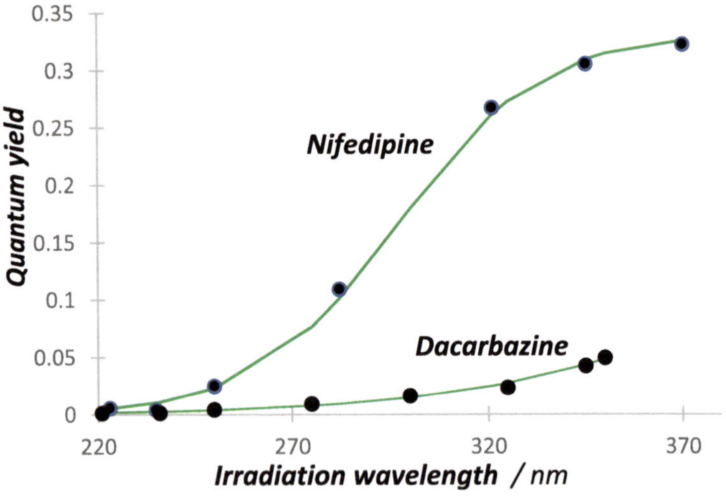

Fig. 17.7 λ-effect on the quantum yield ($\Phi_{X\rightarrow Y}^{\lambda_{irr}}$) values of *NIF* and *DBZ* (experimental values, plain circles). Sigmoid patterns (solid line) correspond to Eqs. 17.3 and 17.4. (Reproduced with permission from [26], *Photochem*)

constant (i.e. the fitting parameter) and a rearranged Eq. 15.5. The quantum yield values span the ranges of 0.004–0.35 for *NIF* and *NIS* and 0.001–0.049 for *DBZ*. The structural variation between *NIF* and *NIS* do not seem to affect photoreactivity in any significant manner.

The photoreaction quantum yields of these molecules indicate that *UVA-Vis*, unlike *UVB* radiations, are mostly responsible for photodegradation (i.e. the causative wavelength range situated between 300 and 390 nm). The significant differences between the two irradiation regions agree with the qualitative conclusions reached by comparing photoreaction first-order rate-constants [14]. The *NIF* results for long-wavelength irradiations agree with those reported in the literature, such as 0.24 in methanol at $\lambda_{irr} = 254\ nm$ [27]; 0.3 and 0.33 in 2-propanol and methanol, respectively, at $\lambda_{irr} = 313\ nm$ [28]; 0.27 in acetonitrile at λ_{irr}= 366 nm [27]; and 0.32 in water at $\lambda_{irr} = 313$ nm [28]. The methods of quantum yield determination used in these studies employed irradiation beams of a 35-nm bandwidth centred at 366 and 254 nm using HPLC data corresponding to only 25% depletion of *NIF* [27] or based on aberchrome actinometry [28].

Reactivity differences between *UVA-Vis* and *UVB* regions correlate well with the differences of the electronic absorption spectra of nitrobenzene and hydropyridine moieties [27]. If this hypothesis is adopted, then our results for *NIF* and *NIS* may emphasise the essential role of the excited-state of the hydropyridine group in the *H*-transfer mechanism operation in the phototransformation of *NIF*. The broad long-wavelength absorption band of *NIF* ($\lambda_{irr} > 300$ nm) was attributed to the absorption of the dihydropyridine chromophore [29] that undergoes aromatisation during the photoreaction (Scheme 17.3). The photoconversion of *NIF* takes place in the nanosecond time span [28], is not sensitive to the presence of molecular oxygen in the solution [28, 30], and proceeds without involving either a triplet-state pathway or a radical course [30].

17.4.3 The Photoreversible Reaction

Systematic monochromatic irradiations were conducted on a variety of molecules undergoing *E*/*Z* photoisomerisations (Scheme 17.5). The list includes the antioxidant stilbenoid oxyresveratrol (*OXY* [31]), the infection and stress protector phytoalexin pinosylvin (*PN* [32]), the *cis-SUT* of the angiogenesis inhibitor and anticancer sunitinib [33], the asthma drug montelukast (*MONTE* [34]), and the serotonin neuronal inhibitor fluvoxamine (*FLU* [19], Scheme 17.2). All the molecules but one (*FLU* involving $C = N$ double bond) have stilbene-like ($C = C$) configurations.

The selected monochromatic/non-isosbestic irradiations spanned the whole electronic absorption spectra of the molecules dissolved in ethanol. The quantum yields were determined at individual irradiation wavelengths for both *E*- and *Z*-isomers. Most often, the electronic absorption spectra of the isomers of a given molecule overlapped with large, if not all, sections of their wavelength absorption domains.

Scheme 17.2 Chemical structures of the drug fluvoxamine and the spectators tartrazine (*TRZ*), sunset yellow (*SSY*), and quinoline yellow (*QY*)

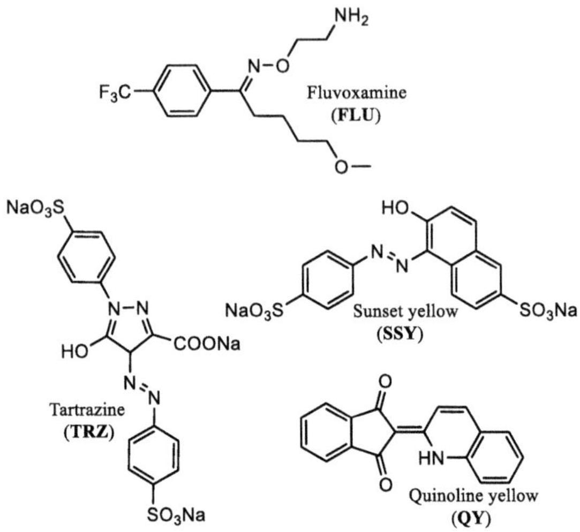

Scheme 17.3 The bleaching photoreaction of a closed-form diarylethene derivative under visible light

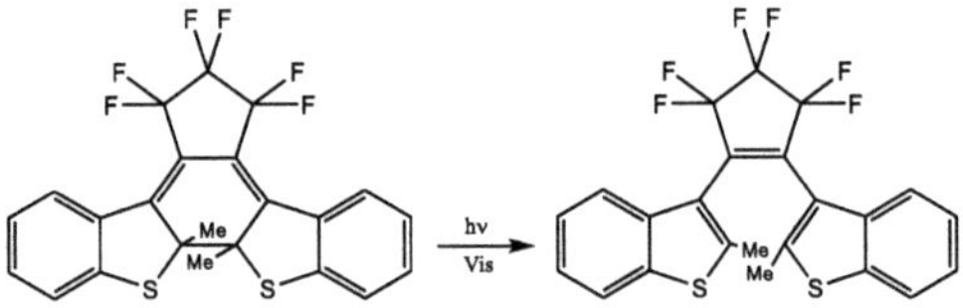

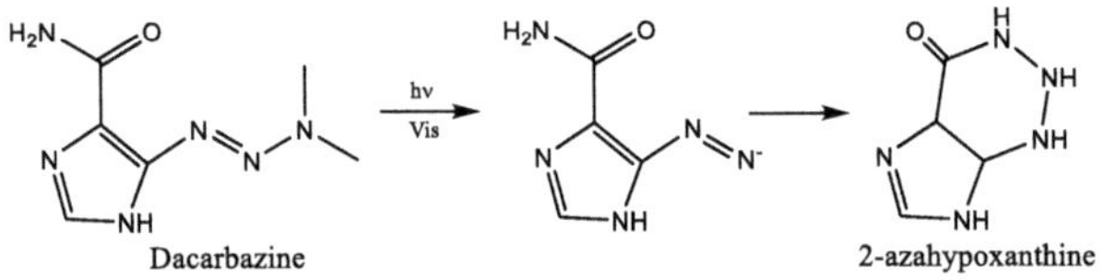

Scheme 17.4 Dacarbazine decomposition under light irradiation [24]

Calculation of the individual quantum yields followed the elucidation procedure described in Sects. 16.8.6 and 10.6 or a combination of both. The experimental traces of the temporal variation of the total absorbance were fitted to the model equation given by Eq. 15.1. The concentration traces of the isomers were obtained by *HPLC* analyses and were fitted to either Eqs. 15.7 and 15.8 (when irradiation is non-isosbestic) or Eqs. 10.28 and 10.29 (for isosbestic irradiation). Figure 17.8 relays an example of the concentration evolution with time for an isosbestic irradiation of a *SUT* solution. Typically, the photoproduct is formed, while the reactant is depleted, until they both reach their respective plateau regions (indicating the reaction has reached *pss*). It is important to monitor the reaction until its *pss* to fulfil the requirements of the elucidation methods.

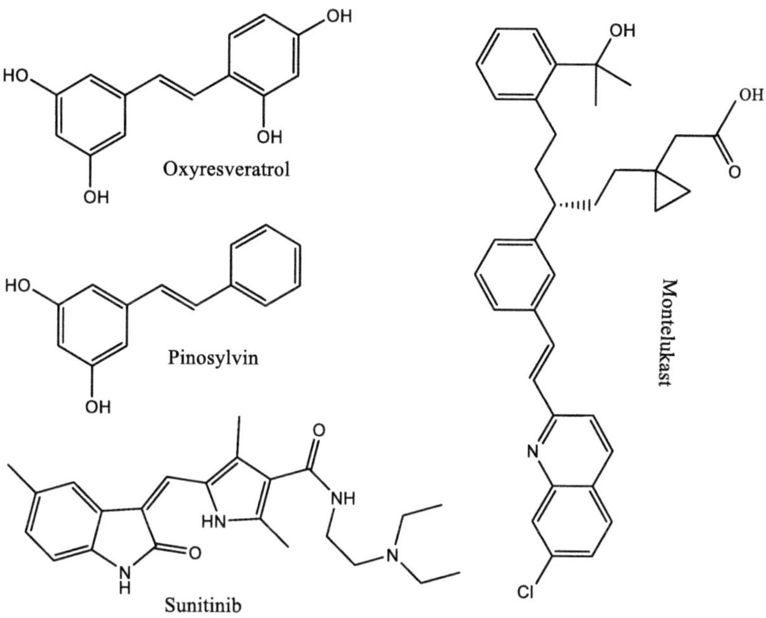

Scheme 17.5 Photoreversible phyto- (oxyresveratrol and pinosylvin) and drug (sunitinib and montelukast) molecules

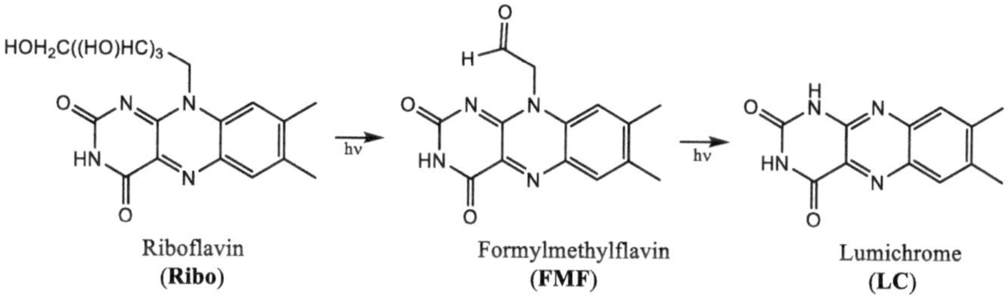

Scheme 17.6 Mechanism of riboflavin photoreaction in two consecutive steps

Among the patterns observed for the variation of the quantum yields of both *E*- and *Z*-isomers of these molecules (Table 17.3), only the *Z*→*E* photoisomerisation of sunitinib (*SUT*, Scheme 17.5) showed a flat, linear trend, indicating little or no wavelength dependence of the quantum yield on excitation wavelength [33]. The remaining quantum yields for both isomers showed a clear variation with irradiation wavelength (Table 17.4). Several patterns and shapes characterised the excitation wavelength dependence of the quantum yields, including a flat line (*Z- SUT*), a line with a positive slop (*Z-MONTE*), a sigmoid function (*E-SUT*, *Z-FLU*, and both isomers of *PN* and *OXY*), a curved shape (*E FLU*), and a triangular profile in the form of a reversed *V*-shape (*E-MONTE*). Thus far, both the outline and magnitude of

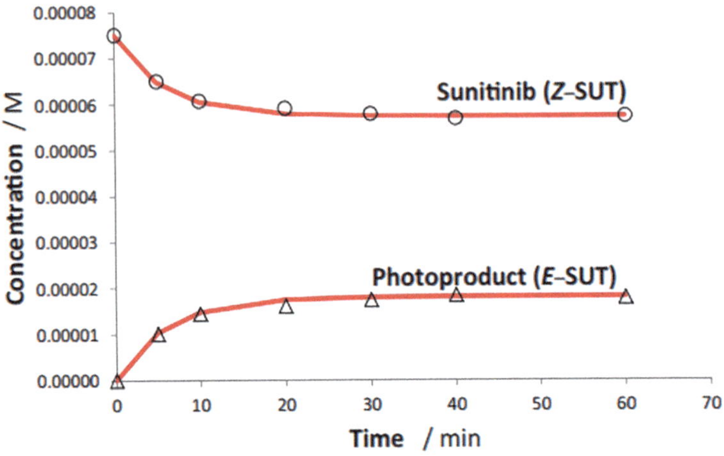

Fig. 17.8 Concentration traces of *Z*- and *E*- *SUT* isomers in ethanol. The evolution of the concentrations with time were monitored by HPLC when 2 mL of the sample was irradiated continuously over a 1 cm^2 area with a monochromatic/isosbestic beam at 345 nm, whose radiation intensity was 2.19 10^{-6} *einstein* dm^{-3} s^{-1}. The constant temperature of the medium was 22 °C. The experimental data are shown as circles or triangles and the fitting equations (Eqs. 10.28 and 10.29) as solid lines. Recorded initial and *pss* reactant concentrations were $C_X^{345}(0) = 7.53\ 10^{-5}\ M$ and $C_X^{345}(pss) = 5.73\ 10^{-5}\ M$, respectively, and the *pss* concentration of the formed *E*-isomer photoproduct was $C_Y^{345}(pss) = 1.8\ 10^{-5}\ M$. (Reproduced with permission from [33], *Journal of Pharmaceutical and Biomedical Analysis*)

the variation of the quantum yield with irradiation are not predictable by any known means and remain exclusive experimental findings. The wavelength sensitivity of the quantum yields of these molecules is clearly more prominent for the electronic transitions belonging to S_1-states than those of S_2-states, where the former are characterised by the longest-wavelength absorption bands. A similar pattern was also observed for azobenzene [1, 35], *NIF*, *NIS*, and *DBZ* (Sect. 17.4.2). Nonetheless, the pattern observed for these molecules clearly shows that their photoreactivity is electronic-transition-dependent.

It is interesting to underline the quantum yield variation with wavelength within the spectral range of a given electronic transition, e.g. the S_1-state band, as observed for the molecules described here. Such Φ-value changes for a single electronic state strongly suggest the involvement of vibrational states (e.g. $S_{1,v}$) of the considered excited electronic state in the photochemical transformation, a correlation that attests for a vibrational-mode dependence of the photochemistry of these molecules (this also holds for the closed-form diarylethene derivative reviewed in Sect. 17.4.1). Vibrational-mode-dependent photochemistry is an interpretation agreeing with the occurrence of a competition between vibrational relaxation (*VR*) and photochemistry [36–38], as shown by the studies discussed in Sect. 4.11. It is perhaps a better explanation than some other options put forward for that purpose [39–45]. Qualitatively, the Φ-variations with irradiation wavelength spanning a unique absorption

Table 17.3 Sigmoid, linear, and triangular equations characterising the variation with the excitation wavelength of the quantum yields of specific reaction-steps of the listed reactants

Reactant	Parameters					
	α	β	δ	γ	θ	σ
	Sigmoid: $\Phi^{\lambda_{irr}} = \alpha + \frac{\beta}{\delta + \gamma \times e^{-\theta\,(\lambda_{irr} - \sigma)}}$					
E-PN	0.1	0.5	1	210	0.0921	278
Z-PN	0.1	1.8	1	270	0.1060	280
E-FLU	0	1	0.07	400	0.13	250
E-OXY	0.068	0.417	1	440	0.0826	254
Z-OXY	0.025	0.28	1	135	0.0925	270
E-SUT	0.064	0.73	1	1300	0.03	300
RIBO	0.004965	0.0006	0.22	0.155	0.1	439.9
FMF	0.00019	0.0002	0.18	0.13	0.11	439.9
c-DAE	0.138	0.068	1	1.7	0.061	410
NIF	0	0.3317	1	57.37	0.0547	223
NIS	0	0.26	0.76	50	0.07	230
DBZ	0	0.263	1	399	0.0265	180
	Linear: $\Phi^{\lambda_{irr}} = \alpha + \beta\,\lambda_{irr}$					
E-MONTE	−0.2525	0.0012				
Z-SUT	0.02	0				
	Triangular (V-shaped): $\Phi^{\lambda_{irr}} = \alpha + \beta\,\lambda_{irr} + \delta + \gamma\,\lambda_{irr}$					
Z-MONTE	0.00005	−0.1143	−0.0013	0.471		
	Curved shape (of undefined equation)					
Z-FLU						

band form a sufficient argument to dismiss the application of Kasha's rule for such photochemistry.

For a majority of molecules, the absolute quantum yield values of $\Phi^{\lambda_{irr}}_{Z\to E}$ are higher than those of $\Phi^{\lambda_{irr}}_{E\to Z}$, but the cases of *PN* and *SUT* present a reversed behaviour ($\Phi^{\lambda_{irr}}_{E\to Z} > \Phi^{\lambda_{irr}}_{Z\to E}$).

The forward (*E*→*Z*) quantum yield of *OXY* increased 6.6-fold in the range 260–360 nm, a feature that might be due to the extra electron-donor hydroxyl-groups in *ortho*- and *para*- positions of *E-OXY* second benzenic moiety compared to *PN* structure, an interpretation corroborated by the red shift (ca. 30 nm) of the long-wavelength absorption band maximum of *E-OXY* ($\lambda^{max}_{E-OXY} = 328\ nm$) compared to the spectrum of either *E*-stilbene or *E-PN*. Comparatively, $\Phi^{\lambda_{irr}}_{Z\to E}(OXY)$ values ($0.069 - 0.454$) were higher than those obtained for $\Phi^{\lambda_{irr}}_{E\to Z}(OXY)$ (ranging between 0.023 and 0.26 in the same spectral interval). It is, however, noticeable that the efficiency of the reverse reaction is 3.3 times higher for *Z-PN* (0.87 at 330 nm) than for *Z-OXY* (0.26 at 350 nm). Since no other (e.g. cyclisation) products were identified in the medium, the difference in reactivity between *Z-PN* and *Z-OXY* might be due to the existence of strong intra-*H*-bond interactions between the spatially close *OH* groups occurring at *o*- and *m′*- positions of the *Z-OXY* ground-state conformation (but which is not possible in *Z-PN*). Even if, according to this

Table 17.4 Wavelength variations of forward and reverse quantum yield values, *pss*-absorbances, and photoreversible rate-constants for montelukast (*Monte*), fluvoxamine (*FLU*), oxyresveratrol (*OXY*), sunitinib (*SUT*), and pinosylvin (*PN*)

	Parameters					
Molecules	λ_{n-isos} /nm	$P_0^{\lambda_{n-isos}} \times 10^6$ /einst dm^{-3} s^{-1}	$A_{tot}^{\lambda_{obs}^{\lambda_{n-isos}}}(pss)^{I}$	$k_r^{\lambda_{n-isos}} \times 10^3$ /s^{-1}	$\Phi_{X\to Y}^{\lambda_{n-isos}}$	$\Phi_{Y\to X}^{\lambda_{n-isos}}$
Montelukast						
	220	0.800	0.088	2.1	0.012	0.0033
	258	0.636	0.044	3.9	0.065	0.024
	284	0.693	0.044	9.2	0.072	0.039
	328	1.187	0.033	28	0.14	0.061
	345	1.428	0.038	42	0.160	0.060
	360	1.332	0.026	35	0.180	0.021
Fluvoxamine						
	226	0.609	0.042	0.197	0.0038	0.0016
	245*	0.586	0.028	0.268	0.0061	0.0038
	260	0.460	0.019	0.380	0.0128	0.0085
	270	0.521	0.011	0.670	0.0361	0.0250
	280	0.551	0.003	0.875	0.0931	0.0535
	285	0.470	0.003	0.480	0.2167	0.0844
	290	0.256	0.028	1.520	0.4349	0.0573
Oxyresveratrol						
	260	1.95	0.097	0.28	0.069	0.023
	295	2.25	0.227	1.60	0.105	0.043
	310	2.29	0.232	2.60	0.154	0.091
	320	4.68	0.252	9.00	0.213	0.139
	328	2.01	0.239	5.00	0.233	0.219
	340	4.02	0.148	15.0	0.350	0.263
	350	4.55	0.096	15.0	0.438	0.262
	360	7.65	0.031	15.0	0.454	0.069
Sunitinib						
	240	1.22	0.215	0.57	0.003	0.006
	260	1.16	0.214	0.64	0.004	0.007
	340	1.98	0.221	3.00	0.020	0.065
	345	2.04	0.209	3.50	0.020	0.064
	360	2.67	0.226	9.40	0.014	0.067
	380	2.97	0.243	13.0	0.017	0.074
	400	3.48	0.239	21.0	0.022	0.089
	420	3.93	0.233	23.0	0.0165	0.087
	440	4.07	0.211	22.5	0.015	0.097
	460	4.30	0.208	19.5	0.020	0.124
	480	4.85	0.198	15.5	0.026	0.175
Pinosylvin						
	280	0.340	0.287	0.0023	0.010	0.125
	290	0.416	0.229	0.0038	0.113	0.097

(continued)

Table 17.4 (continued)

Molecules	Parameters					
	λ_{n-isos} /nm	$P_0^{\lambda_{n-isos}} \times 10^6$ /einst dm^{-3} s^{-1}	$A_{tot}^{\lambda_{obs}} \frac{\lambda_{n-isos}}{}$ $(pss)^f$	$k_r^{\lambda_{n-isos}} \times 10^3$ /s^{-1}	$\Phi_{X\to Y}^{\lambda_{n-isos}}$	$\Phi_{Y\to X}^{\lambda_{n-isos}}$
	300	0.545	0.179	0.0069	0.124	0.149
	310	0.629	0.089	0.0105	0.142	0.259
	320	0.609	0.043	0.0112	0.197	0.455
	330	0.699	0.044	0.011	0.277	0.880

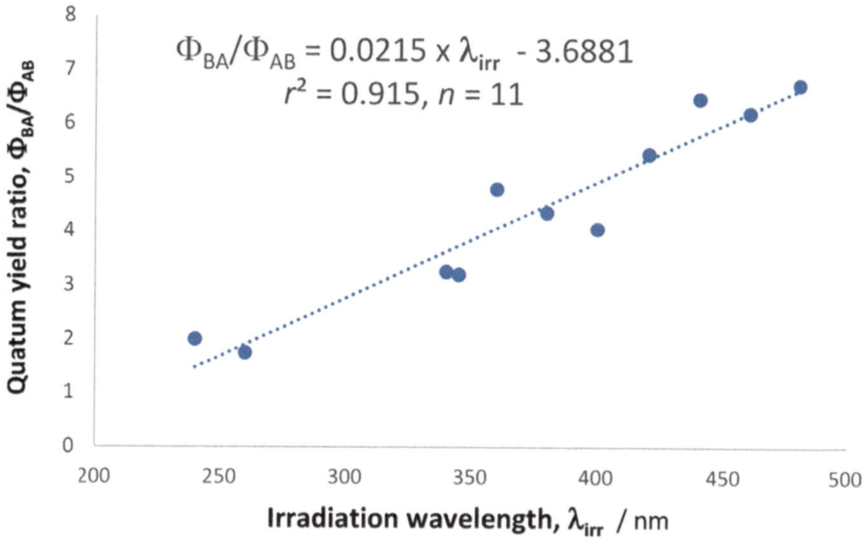

Fig. 17.9 Example of a linear correlation of *SUT* quantum yields' ratios with excitation wavelength (labelling: *A* stands for *Z- SUT* and *B* for *E- SUT*). (Reproduced with permission from [33], *Journal of Pharmaceutical and Biomedical Analysis*)

hypothesis, *Z-OXY* should have a more out-of-plane configuration than *Z-PN*, the *H*-bond is expected to confer a greater stability to this conformation (in comparison to the intuitive one where the *o*- and m'- are positioned at a distance). The additional energy barrier and the enhanced stabilisation due to *H*-bonding might concurrently contribute to hampering the smooth progress of *Z-OXY* photoreaction, hence, causing its quantum yield to have lower values compared to those of *Z-PN*. This case might be considered as an example of structure/quantum yield relationship.

For all the studied molecules, the ratio $\Phi_{Z\to E}^{\lambda_{irr}}/\Phi_{E\to Z}^{\lambda_{irr}}$ was either higher or lower than 1 depending on the molecule structure and its geometry. The data, hence, indicate that the *E* and *Z* isomers behave as if they were completely independent species during the photoisomerisation process. This interpretation is supported by the variation of the quantum yield ratios, i.e. $\Phi_{E\to Z}^{\lambda_{irr}}/\Phi_{Z\to E}^{\lambda_{irr}}$, with irradiation wavelength observed for all the molecules displayed in Scheme 17.5. Linear correlations of the type $\Phi_{E\to Z}^{\lambda_{irr}}/\Phi_{Z\to E}^{\lambda_{irr}} = f(\lambda_{irr})$ were recorded for *SUT* (Fig. 17.9), with its ratio

$\Phi^{\lambda_{irr}}_{E\to Z}/\Phi^{\lambda_{irr}}_{Z\to E}$ varying between 1.75 and 6.75, and $1.84 \leq \Phi^{\lambda_{irr}}_{E\to Z}/\Phi^{\lambda_{irr}}_{Z\to E} \leq 8.57$ for *E-MONTE* isomer [34], whose quantum yield is linear with excitation wavelength ($\Phi^{\lambda_{irr}}_{E\to Z}(MONTE) = 0.0012 \times \lambda_{irr} - 0.2525$). Such variations do not corroborate the occurrence of a common excited-state (intermediate) species between the isomers. Furthermore, postulating an intermediate species that emerges from the light-excitation of either *Z*- or *E*-isomers, can only be based on a supposed applicability of Kasha's rule to photochemistry. If the latter hypothesis was correct, then the prediction would favour a wavelength-independent quantum yield ratio. The very fact that the experimental ratios are wavelength-dependent ($\Phi^{\lambda_{irr}}_{E\to Z}/\Phi^{\lambda_{irr}}_{Z\to E} = f(\lambda_{irr})$) for these molecules indisputably represents strong evidence against an automatic extension of Kasha's rule to photochemistry. The conclusion reached here is confirmed by the previous results obtained for $XY(1\Phi)$ systems, those we shall see for $XY_2(2\Phi)$ reaction (see next Sect. 17.4.4), or those reported in Chap. 4, which all plead for a high probability to observe the sensitivity of the quantum yield (or end products' distribution) with excitation, and therefore, claim the invalidity of the concept stating that Kasha's rule extends to photochemistry.

The explicit formula of $k_r^{\lambda_{n-isos},T}$ (Eq. 15.5) for the photoreversible system stipulates that the reaction rate-constant does not exclusively depend on the absorption property of the reactant (at λ_{n-isos}), as might have been supposed, i.e. the reaction is expected to be faster whenever the absorption is higher, or simply put, *the reaction-rate depends only on absorption*. In fact, the reaction rate-constant of the photoreversible isomerisation is a much more inclusive function as it effectively includes all (intrinsic and extrinsic) parameters of the reaction at λ_{n-isos} (e.g. the irradiated volume (V_{irr}) and area (S_{irr}) of the reactive medium, both present in the formula of $P_0^{\lambda_{n-isos}}$ (Chap. 2)). This means that $k_r^{\lambda_{n-isos},T}$ is a relative quantity that cannot be, in any case, related to a specific parameter (such as $\varepsilon_X^{\lambda_{n-isos}}$), nor its value could be used to evaluate the quantum yield magnitude (of either forward or reverse reactions).

The thermodynamic constant of the reaction, $K_{rev}^{\lambda_{n-isos},T}$ (Eq. 15.16), expressed as the ratio of the isomers' concentrations at *pss* (here equating $\Phi_{Y\to X}^{\lambda_{n-isos}}\varepsilon_Y^{\lambda_{n-isos}}/\Phi_{X\to Y}^{\lambda_{n-isos}}\varepsilon_X^{\lambda_{n-isos}}$ for the photoreversible reaction) might vary as a result of changing one or more of the reaction parameters (λ_{n-isos}, for instance). Incidentally, the variation of $K_{rev}^{\lambda_{n-isos},T}$ with irradiation also indicates that the relative magnitude of the absorption coefficients is only one of several determinant factors influencing the photo-efficiency of photoreversible systems. The $K_{rev}^{\lambda_{n-isos},T}$ variation may explain many experimental observations of variable *E*/*Z* concentration ratios.

The most commonly reported quantum yield value in the stilbenoid series is that of the *E*-stilbene, approaching 0.5 (0.5 in acetonitrile and ethanol/methanol, 0.52 in ethanol, and 0.48 in methanol [3, 46, 47]), which was considered to be wavelength-invariant. Comparatively, it is relatively close to the maximal value obtained in ethanol for *E-OXY* (0.45 at 360 nm) but higher than that of *E-PN* (0.27 at 330 nm), Table 17.4. In general, the quantum yields of stilbenoids' *E*/*Z* isomerisation were smaller than the unsubstituted stilbene value (i.e. 0.5). In the mono-substituted derivatives, a slight reduction of the Φ-value was observed for *E-p*-hydroxy-stilbene (0.42 in cyclohexane) and *m*-hydroxystilbene (0.32 in cyclohexane) [48], whereas almost equal values of only 0.09 and 0.08 characterised *E*-*m*-aminostilbene [49] and *E*-*m*-dimethylaminostilbene [46], respectively, in methylcyclohexane.

E-disubstituted *p*-*p*′-dimethylaminostilbene [46] has a similar quantum yield value (0.1 in dioxane) as its mono-amino or dimethylamino-substituted analogues.

The quantum yield values of the *Z*-isomer of stilbene ($\Phi = 0.35$ [50]) and mono-substituted stilbene derivatives [3] were found to be smaller than those reported for their *E*-isomers (e.g. 0.50 for *E*-stilbene), a trend coherent with the value of *OXY* (0.27) but not with *PN (0.87)*, both in ethanol (Table 17.4). However, despite the scarcity of quantum yield values of *Z*-stilbenes in the literature, the case of *PN* is compatible with a number of examples of reported values for disubstituted stilbene derivatives, which have lower Φ-values for *E*-isomers than their *Z*-form counterparts, such as *p*-*p*′-dinitro- in *DMF*, *p*-*p*′-dimethylamino- in toluene, *m*-*p*′-dimethylamino- in methanol, *m*-*p*′-nitro-amino- in benzene, and *p*-*p*′-nitrodimethylamino- and *m*-*p*′-nitrodimethylamino- in cyclohexane, whose $\Phi_{Z\to E}/\Phi_{E\to Z}$ ratios varied between 1 and 12 [2, 3, 51, 52]. The above literature also showed the difficulty in deriving a general trend for the variation of the quantum yield with the substituent position on the structure of poly-substituted stilbene derivatives. Likewise, even with the limited literature data on stilbenoids in relation to irradiation wavelength effects, it still provides compelling evidence on the quantum yield sensitivity to excitation wavelength [2, 3, 37, 38, 44, 45, 51, 52]. The λ-effect origin and/or the validity of the quantum yield variation were not fully discussed/elucidated in those publications. One of the earliest papers on azobenzene [1] showed that the quantum yield of *E*-azobenzene was dependent on irradiation wavelength, with a twofold increase in the wavelength of 257–550 nm range and the photochemical efficiency being transition-dependent as $\Phi_{Z\to E} > \Phi_{E\to Z}$. The trend observed for the latter azobenzene features was recently confirmed [35].

The sum of forward and reverse quantum yields of the five molecules (Scheme 17.5) never added up to unity, i.e. $\Phi^{\lambda_{irr}}_{E\to Z} + \Phi^{\lambda_{irr}}_{Z\to E} \neq 1$ (Table 17.4). In some cases, $\Phi^{\lambda_{irr}}_{E\to Z} + \Phi^{\lambda_{irr}}_{Z\to E} > 1$. The sum of quantum yields at individual wavelengths of the range studied generally increased. For instance, $\Phi^{\lambda_{irr}}_{E\to Z} + \Phi^{\lambda_{irr}}_{Z\to E}$ for *PN* and *OXY* increases from 0.21 to 1.09 and from 0.09 to 0.7, respectively, when $\Phi^{\lambda_{irr}}_{E\to Z} + \Phi^{\lambda_{irr}}_{Z\to E} > 1$ can be seen as a confirmation for the independence of the isomers' reactions. The case where $\Phi^{\lambda_{irr}}_{E\to Z} + \Phi^{\lambda_{irr}}_{Z\to E} < 1$ might indicate the availability of other decay pathways to the excited molecule, other than photoreaction or reversing to the original ground-state, with the alternative pathways being equally or even more efficient than photoisomerisation. A similar interpretation was adopted for various mono- and di-aminostlibenes [48–50, 53]. For these compounds, it was suggested that the sum of the quantum yields of fluorescence (Φ_{fl}) and twice that of the isomerisation of the *E*-isomer, $\Phi_{fl} + 2\,\Phi_{E\to Z}$, was close to 1 except for *m*, *p*′-diaminostlibene (where the sum was 0.3). It is important to emphasise here that the analysis using $\Phi_{fl} + 2\Phi_{EZ} = 1$ is not rigorous since this equation implicitly supposes that the quantum yields of *E*- and *Z*-isomers are both wavelength-invariant and equal, $\Phi_{E\to Z} = \Phi_{Z\to E}$. This hypothesis has not been experimentally proven, even though it complies, as discussed above, with the observation for stilbenoids, in general, $\Phi_{E \to Z} + \Phi_{Z \to E} < 1$. More importantly, invoking an equality of forward and reverse quantum yields is not necessarily valid since the formula of $K_{rev}^{\lambda_{n-isos},T}$, $K_{rev}^{\lambda_{n-isos},T} = \varepsilon_Z \Phi_{Z \to E} / \varepsilon_E \Phi_{E \to Z}$, taken in the hypothesis $\Phi_{E \to Z} = \Phi_{Z \to E}$, reduces to the ratio of absorption coefficients, which is not corroborated by experiment. However, these results still prove the possible existence of deactivation pathways of the isomers' excited-states, other than photoisomerisation reaction and fluorescence. In the cases of the five reactions considered here (Scheme 17.5), it could be suggested that a higher competition between isomerisation, internal conversion, and/or other pathways occurs in the higher vibrational states of the excited electronic state. Photoreaction from lower vibrational states is, however, favoured, as the highest values of the quantum yields for almost all molecules are recorded for the long-wavelength bands, an observation that promotes $S_{1,0}$ to being the most efficient for photoisomerisation (which somewhat agrees with the spirit of Kasha's rule).

The inspection of a comprehensive collection of literature [3] on experimental photochemical quantum yield values of stilbenes and its derivatives allows us to draw five important observations:

(i) For almost every studied stilbene derivative, the quantum yield values for the *E*→*Z* and *Z*→*E* were different (in given experimental conditions).
(ii) When data are available, the absolute values of the quantum yields of a given stilbenoid, measured by independent laboratories, often show differences. For instance, five different values for *E*-stilbene photoconversion in *n*-hexane were reported as $0.27 < \Phi_{E \to Z} < 0.67$.
(iii) The sum $\Phi_{E \to Z}^{\lambda_{irr}} + \Phi_{Z \to E}^{\lambda_{irr}}$ of a given stilbene derivative scarcely equals unity (i.e. usually, it is less than 1).
(iv) The light used for irradiation was, in general, not monochromatic.
(v) Except for a few cases, $\Phi_{E \to Z}$ was found to be higher than $\Phi_{Z \to E}$.

The first two points ((i) and (ii)) suggest that the determination of Φ-values might be sensitive to either or both the technique/elucidation method adopted for the treatment of the kinetic data or/and the nature (mono- or polychromaticity) of the excitation beam irradiating the sample. The observed sum $\Phi_{E \to Z} + \Phi_{Z \to E} < 1$ (point (iii)) suggests the existence of underlying (radiative and/or non-radiative) processes. In any case, these features do not support an interpretation, assuming the existence of a unique common excited intermediate (the 90°-twisted isomer) emerging from the excitation of either the *E* or the *Z* form. The latter hypothesis also assumes that the so-called intermediate decays to the ground-state of *E*- or *Z*-isomers, either in equal or proportional amounts, as rendered by the equation, $\Phi_{E \to Z}^{\lambda_{irr}} = 1 - \Phi_{Z \to E}^{\lambda_{irr}}$ (assuming monochromatic λ_{irr}) [54, 55]. Further to the fact that the above experimental features (points (i)–(iii)) do not support the latter hypotheses, it is important to observe (point (iv)) that the concerned studies based their respective kinetic experiments on rather polychromatic/filtered light than monochromatic irradiations (which does not make them reliable to draw general or definitive conclusions).

17.4.4 The $XY_2(2\Phi)$ Consecutive Reaction

The published quantum yield values correspond, generally, to thermally stable reactants but not for every photochemical reaction-step of the reaction mechanism. We have seen in the previous section data relative to the isomers of photoreversible reactions, but it is not common to determine the quantum yields of each species involved in other sequences leading to a photostationary state. Likewise, the literature provides no data on quantum yields of photoproducts of consecutive reactions. The case of riboflavin (*RBF*) is perhaps the first example of a full characterisation of a consecutive photoreaction [56]. *RBF* undergoes a phototransformation into an intermediate formylmethylflavin (*FMF*), which itself photoreacts to produce lumichrome (*LC*) [57–61]. The two reaction-steps of this consecutive mechanism are irreversible (Scheme 17.6).

The intermediate aldehyde, *FMF*, is formed by intra-molecular photochemical reduction of the isoalloxazine ring system, followed by an oxidation of the ribityl side chain. The *RBF*→*FMF* reaction-step is thought to predominantly occur from the triplet-state of the photoexcited reactant *RBF* molecule [62, 63]. In this pathway, the ribityl chain, rich in hydroxyl oxygen atoms, was believed to promote the intersystem crossing (*S*→*T* transition) and stabilising it by steric protection [62]. The reported an intersystem crossing (*ISC*) quantum yield for *RBF* was relatively high, *ca.*0.6 in methanol [64] and 0.7 in ethanol [57]. Triplet-state lifetimes of 3.7 and 19 ms were also reported for *RBF* in aqueous and methanolic solutions, respectively [65, 66]. The second photochemical reaction-step (*FMF*→*LC*) in *RBF* sequence was also reported [59] to proceed from the lowest triplet excited-state of the photoexcited *FMF*, which undergoes a redox reaction to generate *LC*.

The electronic absorption spectrum of *RBF* in ethanol shows four distinct absorption bands in both *UV* and *Vis* spectral regions. The two short-wavelength peaks, situated at 220 and 265 nm, are more intense than the two broad absorption bands of longer wavelengths (with maxima at 360 and 445 nm and a shoulder at *ca.* 460 nm). The long-wavelength absorbance bands were mainly attributed to $\pi\rightarrow\pi^*$ transitions [59, 62, 65, 67].

RBF ethanolic solutions of initial concentrations *ca.* 10^{-6} *M* were exposed to collimated, monochromatic, and non-isosbestic light beams of ($\lambda_{n\ -\ isos}$) wavelengths 400, 420, 445, 460, and 480 nm. Under irradiation, the solution undergoes bleaching, monitored by the disappearance of the long-wavelength absorption peaks. This bleaching occurs concomitantly with a decrease of the *UV* region peaks' absorbances and the emergence of a shoulder at around 390 nm. The smoothness of the photoreaction was indicated by six isosbestic points, situated at λ_{isos} = 233, 260, 303, 320, 383, and 405 nm.

The total absorbance traces were monitored at the observation wavelength λ_{obs} = 445 nm, which is the maximum of the longest-wavelength absorption band. The traces were fitted to Eq. 15.23 that included the relevant concentration expressions given by Eqs. 15.24, 15.25 and 15.26. The elucidation method developed in Sect. 16.8 was applied for the determination of the four unknown parameters of the

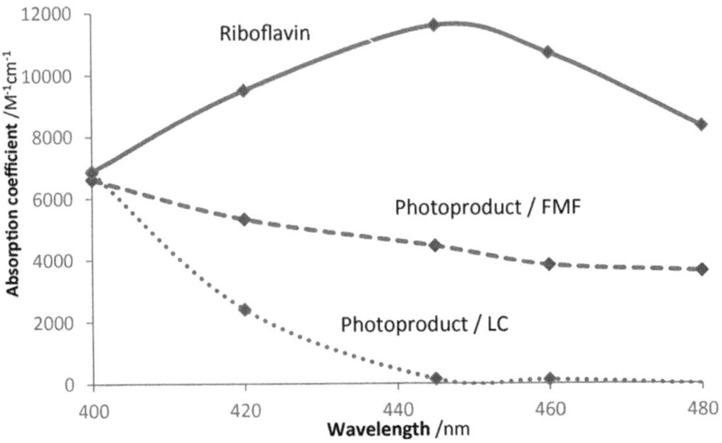

Fig. 17.10 Reconstructed absorption spectra of *RBF* photoproducts (*FMF* and *LC*), expressed as absorption coefficient units. (Reproduced with permission from [56], *Journal of Pharmaceutical Sciences*)

reaction, i.e. the products' absorption coefficients, $\varepsilon_{Y_1}^{\lambda_{rr}-isos}$ and $\varepsilon_{Y_2}^{\lambda_{rr}-isos}$, and the reaction-steps' quantum yields, $\Phi_{X\to Y_1}^{\lambda_{rr}-isos}$ and $\Phi_{Y_1\to Y_2}^{\lambda_{rr}-isos}$. The complete elucidation of the kinetics can be carried out for the consecutive $XY_2(2\Phi)$ photoreaction using the absorbance traces only with no need for the individual species concentration traces (Sect. 16.8).

The individual overall rate-constants, obtained by fitting the total absorbance traces for the consecutive photoreaction-steps, $k_{X\to Y_1}^{\lambda_{rr}-isos}$ and $k_{Y_1\to Y_2}^{\lambda_{rr}-isos}$, were relatively low ($0.1\ 10^{-4} - 5\ 10^{-4}\ s^{-1}$), with $k_{X\to Y_1}^{\lambda_{rr}-isos}$ being 10–40-times higher than $k_{Y_1\to Y_2}^{\lambda_{rr}-isos}$ in the 400–480 nm irradiation range. Overall, both rate-constants were found to increase with increasing irradiation wavelength, reaching a maximum at 460 nm and subsequently decreasing.

The reconstructed absorption spectra of the photoproducts show that the three reaction species are characterised by relatively low absorptivities ($\varepsilon^{\lambda_{rr}-isos} < 12 \times 10^3$ M^{-1} cm^{-1} for $\lambda > 400$ nm) and that *LC* (Y_2) does not absorb beyond 430 nm (Fig. 17.10). These features corroborate those obtained in the individual species, *RBF* and *LC*, in ethanolic solutions [68].

The individual concentration profiles of the three species are also easily retrieved by the substitution of the values of $\varepsilon_{Y_1}^{\lambda_{rr}-isos}$, $\varepsilon_{Y_2}^{\lambda_{rr}-isos}$, $k_{X\to Y_1}^{\lambda_{rr}-isos}$, and $k_{Y_1\to Y_2}^{\lambda_{rr}-isos}$ into the respective concentration equations (Eqs. 15.24, 15.25 and 15.26). These traces reveal that *RBF* undergoes a relatively fast photodepletion under visible irradiation at 445 nm, at which depletion is achieved in the early stages of the reaction (~3 hours), in comparison with the reactivity of its photoproducts, which require more than 20 hours to reach completion (Fig. 17.11).

The individual quantum yield values of both reaction-steps increased with irradiation wavelength, spanning the visible region of the spectrum. The magnitude of the variation was more pronounced for *FMF* photoreaction, displaying a fivefold

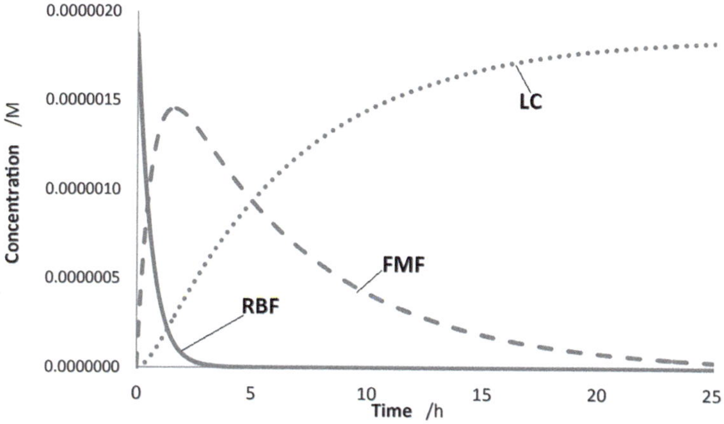

Fig. 17.11 Reconstructed concentration traces of the three reaction species using Eqs. 15.22, 15.23 and 15.24. Ethanol solution of *RBF*, with initial concentration 1.8 × 10^{-6} M, was continuously irradiated at 445 nm with a monochromatic beam of a constant flux, $P_0^{445} = 1.26 \times 10^{-6}$ *einstein* $dm^{-3}\ s^{-1}$. (Reproduced with permission from [56], *Journal of Pharmaceutical Sciences*)

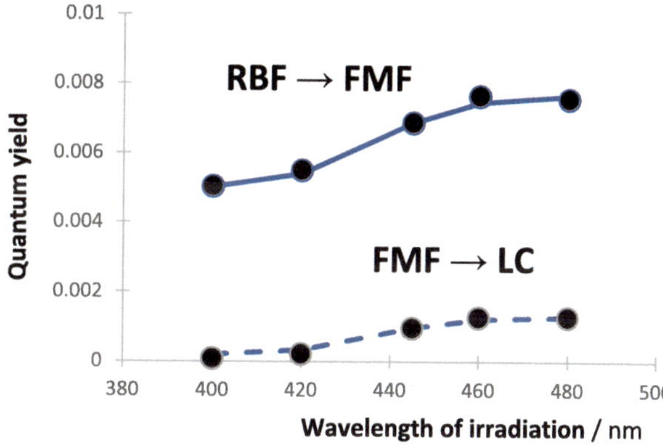

Fig. 17.12 Sigmoid patterns (Eqs. 17.5 and 17.6) of the experimental photochemical quantum yields of the two reaction-steps ($X{\rightarrow}Y_1$ and $Y_1{\rightarrow}Y_2$) involved in *RBF* photodegradation. Measurements performed in ethanol at 22 °C. (Reproduced with permission from [56], *Journal of Pharmaceutical Sciences*)

increment between the lowest and highest values of its quantum yields, compared to only a 1.5-fold observed for the quantum yield changes of the initial photoreaction-step (*RBF→FMF*). The calculated absolute quantum yield values of $\Phi_{X\rightarrow Y_1}^{\lambda_{n-isos}}$ were between 6 and 20 times higher than those obtained for the subsequent $\Phi_{Y_1\rightarrow Y_2}^{\lambda_{n-isos}}$. The variations of both sets of quantum yield values (Fig. 17.12) were well-described by sigmoid functions (Eqs. 17.5 and 17.6). The used excitation wavelength range (400–480 nm) corresponds to two overlapping absorption bands, situated at roughly 380–450 nm and 450–490 nm, respectively. Hence, the excitation operates on different electronic states, most likely, S_2 and S_1, respectively. Also, variation of the quantum yield within the wavelengths associated with each state indicate an implication of vibronic photochemistry.

$$\Phi_{X \to Y_1}^{\lambda_{irr}}(RBF) = 0.004965 + \frac{0.0006}{0.22 + 0.155\, e^{-0.1\,(\lambda_{irr} - 439.9)}} \tag{17.5}$$

$$\Phi_{Y_1 \to Y_2}^{\lambda_{irr}}(FMF) = 0.00019 + \frac{0.0002}{0.18 + 0.13\, e^{-0.11\,(\lambda_{irr} - 439.9)}} \tag{17.6}$$

Incidentally, the variation of quantum yield with excitation wavelength is a kind of distinctability issue since the phenomenon cannot be foreseen before conducting the experiment. The photostabilisation (Sect. 16.3) might, however, not be considered as a distinctability issue as it can be predicted from the rate-law once it is established for the reaction at hand (i.e. as soon as the reaction mechanism is known but before experiment).

17.5 Conclusion

The usefulness of the developed *SEM*-equations, as well as the elucidation methods for the various reaction mechanisms (Chaps. 15 and 16), has been established in the present chapter. These tools show excellent conformity with the experimental data and the behaviours of the reaction when changing either the initial concentration or the presence of spectator molecules. Photostabilisation has been experimentally observed for both latter reaction conditions, as predicted by the *SEM*-equations. The acute evolution of the measured quantum yield values clearly indicates that monochromatic light is the best tool for the characterisation of molecular photoreactivity. Such a combination of monochromaticity of the light and the kinetic mathematical tools offered by the semi-empirical method reaches a performance that cannot be achieved using the classical thermal kinetic models for photoreaction data. According to the experimental data laid out in the chapter, the quantum yield dependence on excitation seems to be a common feature of photoreactive molecules. Such a variation does not, however, have a regular pattern that fits all molecules.

The pattern of the λ-effect on quantum yield can be considered as another property of the molecule, (registering various shapes, such as sigmoid, linear, and curved, that cannot be predicted without performing the experiment).

Finally, the strategy based on *SEM*-equations/monochromaticity is a good way to characterise and evidence the occurrence of fast- to ultra-fast excited-state processes by assessing the quantum yield variation with excitation, which might stand as an alternative method to time-resolved techniques.

References

1. Zimmerman G, Chow L-Y, Paik U-J (1958) The photochemical isomerization of azobenzene. J Am Chem Soc 80:3528–3531. https://doi.org/10.1021/ja01547a010

2. Schulte-Frohlinde D, Blume H, Gusten H (1962) Photochemical *cis-trans*-isomerization of substituted stilbene. J Phys Chem 66:2486–2491. https://doi.org/10.1021/j100818a039
3. Görner H, Kuhn HJ (1995) Cis-trans isomerization of stilbenes and stilbene-like molecules. In: Neckers DC, Volman DH, von Bunau G (eds) Advances in photochemistry, vol 19. Wiley, New York. ISBN 0-471-04912-3
4. Tønnesen HH (2004) Photostability of drugs and drug formulations.2nd edn. CRC Press, London. ISBN. 978-0415303231
5. Mielcarek J, Augustyniak W, Grobelny P, Nowacka G (2005a) Photoprotection of 1,4-dihydropyridine derivatives by dyes. Int J Pharm 304:145–151. https://doi.org/10.1016/j.ijpharm.2005.08.01
6. Langford CH, Achari G, Izadifard M (2011) Wavelength dependence of luminescence and quantum yield in dechlorination of PCBs. J Photochem Photobiol A Chem 222:40–46. https://doi.org/10.1016/j.jphotochem.2011.04.028
7. Cipolloni M, Gentili PL, Ortica F, Becker RS, Favaro G (2011) Effects of solvent, excitation wavelength, and concentration on the photobehavior of some diazonaphthoquinones. Arkivoc ix:205–220. https://doi.org/10.3998/ark.5550190.0012.915
8. Groeneveld I, Bagdonaite I, Beekwilder E, Ariese F, Somsen GW, van Bommel MR (2022) Liquid core waveguide cell with in situ absorbance spectroscopy and coupled to liquid chromatography for studying light-induced degradation. Anal Chem 94:7647–7654. https://doi.org/10.1021/acs.analchem.2c00886
9. Groeneveld I, Kanelli M, Ariese F, van Bommel MR (2023) Parameters that affect the photodegradation of dyes and pigments in solution and on substrate—an overview. Dyes Pigments 210:110999. https://doi.org/10.1016/j.dyepig.2022.110999
10. Maafi W, Maafi M (2013) Modelling nifedipine photodegradation, photostability and actinometric properties. Int J Pharm 456:153–164. https://doi.org/10.1016/j.ijpharm.2013.07.075
11. Maafi M, Maafi W (2015) Quantification of unimolecular photoreaction kinetics: determination of quantum yields and development of actinometers—the photodegradation case of cardiovascular drug nisoldipine. Int J Photoenergy:454895. https://doi.org/10.1155/2015/454895
12. Piechocki JT, Thoma K (2010) Pharmaceutical photostability and stabilisation technology. Informa Healthcare, London. ISBN. 978-0824759247
13. Huang C, Liu H, Ren J, Du X (1996) Studies of photodegradation stability of nisoldipine. Chin J Pharm 27:26–27
14. Thoma K, Klimek R (1991) Photostabilisation of drugs in dosage forms without protection from packaging materials. Int J Pharm 67:169–175. https://doi.org/10.1016/0378-5173(91)90429-R
15. Albini A, Fasani E (1998) Drugs photochemistry and photostability. The Royal Society of Chemistry, Cambridge, UK. ISBN. 9780854047437
16. Rowe RC, Sheskey PJ, Weller PJ (2003) Handbook of pharmaceutical excipients.3rd edn. Pharmaceutical Press and American Pharmaceutical Association. ISBN. 978-0917330964
17. Desai DS, Abdelnasser MA, Rubitski BA, Varia SA (1994) Photostabilization of uncoated tablets of sorivudine and nifedipine by incorporation of synthetic iron oxides. Int J Pharm 103: 69–76. https://doi.org/10.1016/0378-5173(94)90204-6
18. Mielcarek J, Grobelny P, Szamburska O (2005b) The effect of β-carotene on the photostability of nisoldipine. Methods Find Exp Clin Pharmacol 27:167–171. https://doi.org/10.1358/mf.2005.27.3.890873
19. Maafi M, Maafi W (2015) Quantitative assessment of photostability and photostabilisation of Fluvoxamine and its design for actinometry. Photochem Photobiol Sci 14:982–994. https://doi.org/10.1039/c5pp00022j
20. Teraoka R, Matsuda Y, Sugimoto I (1989) Quantitative design for photostabilization of nifedipine by using titanium dioxide and/or tartrazine as colourants in model film coating systems. J Pharm Pharmacol 41:293–297. https://doi.org/10.1111/j.2042-7158.1989.tb06459.x
21. Maafi M, Al-Qarni MA (2022) Mono- and polychromatic light diarylethene-actinometer for the visible range. Dyes Pigments 198.109942. https://doi.org/10.1016/j.dyepig.2021.109942

22. Sumi T, Takagi Y, Yagi A, Morimoto M, Irie M (2014) Photoirradiation wavelength dependence of cycloreversion quantum yields of diarylethenes. Chem Commun 50:3928–3930. https://doi.org/10.1039/c4cc00396a
23. Santos AR, Ballardini R, Belser P, Gandolfi MT, Iyer VM, Moggi L (2009) Photochemical investigation of a photochromic diarylethene compound that can be used as a wide range actinometer. Photochem Photobiol Sci 8:1734–1742. https://doi.org/10.1039/B9PP00037B
24. Tashiro M, Naito T, Yamamoto C, Katoh S-Y, Kawakami J (2019) Impact of light shielding on photo-degradation of dacarbazine during the preparation process. Biol Pharm Bull 42:2062–2068. https://doi.org/10.1248/bpb.b19-00537
25. Maafi M, Lee LY (2015) Determination of dacarbazine Φ-order photokinetics, quantum yields, and potential for actinometry. J Pharm Sci 104:3501–3509. https://doi.org/10.1002/jps.24568
26. Maafi M (2024) Excitation wavelength-dependent photochemistry. Photochem 4:233–270. https://doi.org/10.3390/photochem4020015
27. Fasani E, Albini A, Mella M (2008) Photochemistry of Hantzsch 1,4- dihydropyridines and pyridines. Tetrahedron 64:3190–3196. https://doi.org/10.1016/j.tet.2008.01.104
28. Görner H (2010a) Nitro group photoreduction of 4-(2-nitrophenyl)- and 4-(3- nitrophenyl)-1,4-dihydropyridines. Chem Phys 373:153–158. https://doi.org/10.1016/j.chemphys.2010.06.007
29. Krufurst A, Kuhan J (1983) Quantum chemical interpretation of electronic absorption spectra of the hantzsch dihydropyridines. Collect Czechoslov Chem Commun 43:1422–1428
30. Fasani E, Fagnoni M, Dondi D, Albini A (2006) Intramolecular electron transfer in the photochemistry of some nitrophenyldihydropyridines. J Org Chem 71:2037–2045. https://doi.org/10.1021/jo052463z
31. Maafi M, Al-Qarni MA (2018) Φ-order spectrophotokinetic characterisation and quantification of trans-cis oxyresveratrol reactivity, photodegradation and actinometry. Spectrochim Acta A Mol Biomol Spectrosc 188:64–71. https://doi.org/10.1016/j.saa.2017.06.06
32. Maafi M, Al-Qarni MA (2019) Dependence of pinosylvin photochemical quantum yield on excitation wavelength. Trends Photochem Photobiol 18:45–58
33. Maafi M, Lee LY (2015) Actinometric and Φ-order photodegradation of anti-cancer Sunitinib. J Pharm Biomed Anal 110:34–41. https://doi.org/10.1016/j.jpba.2015.02.026
34. Maafi M, Maafi W (2014) Montelukast photodegradation: elucidation of Φ-order kinetics, determination of quantum yields and application to actinometry. Int J Pharm 471:544–552. https://doi.org/10.1016/j.ijpharm.2014.05.016
35. Laanyi V, Dvorak P, Al-Anshori J, Vetrakova L, Wirz J, Heger D (2017) Azobenzene photoisomerization quantum yields in methanol redetermined. Photochem Photobiol Sci 16: 1757–1761. https://doi.org/10.1039/C7PP00315C
36. Becker RS, Dolan E, Balke DE (1969) Vibronic effects in photochemistry. Competition between internal conversion and photochemistry. J Chem Phys 50:239–245. https://doi.org/10.1063/1.1670783
37. Bishenden E, Donaldson DJ (1993) Mode-specific chemical branching ratios in the photodissociation of OClO. J Chem Phys 99:3129–3132. https://doi.org/10.1063/1.465167
38. Fidder H, Tschirschwitz F, Dühr O, Nibbering ETJ (2001) Vibrational mode-specific photochemical reaction dynamics of chlorine dioxide in solution. J Chem Phys 114:6781–6794. https://doi.org/10.1063/1.1357202
39. Turro NJ, Ramamurthy V, Cherry W, Farneth W (1978) The effect of wavelength on organic photoreactions in solution. Reactions from upper excited states. Chem Rev 78:125–145. https://doi.org/10.1021/cr60312a003
40. Greenberg JW, Malhotra V, Ennever JF (1987) Wavelength dependence of the quantum yield for the structural isomerization of Bilirubin. Photochem Photobiol 46:453–456. https://doi.org/10.1111/j.1751-1097.1987.tb04794.x
41. Dauben WG, Share PE, Ollmann RR (1988) Triene photophysics and photochemistry: previtamin D3. J Am Chem Soc 110:2548–2554. https://doi.org/10.1021/ja00216a030

42. Agati G, Fusi F, Pratesi R, McDonagh AF (1992) Wavelength-dependent quantum yield for Z → *E* isomerization of bbilirubin complexed with human serum albumin. Photochem Photobiol 55:185–190. https://doi.org/10.1111/j.1751-1097.1992.tb04226.x
43. Ern J, Bens AT, Martin H-D, Kuldova K, Trommsdorff HP, Kryschi C (2002) Ring-opening and -closure reaction dynamics of a photochromic dithienylethene derivative. J Phys Chem A 106:1654–1660. https://doi.org/10.1021/jp012614
44. Nakamura S, Kobayashi T, Takata A, Uchida K, Asano Y, Murakami A, Goldberg A, Guillaumont D, Yokojima S, Kobatake S, Irie M (2007) Quantum yields and potential energy surfaces: a theoretical study. J Phys Org Chem 20:821–829. https://doi.org/10.1002/poc.1245
45. Protti S, Ravelli D, Fagnoni M (2019) Wavelength-dependence and wavelength-selectivity in photochemical reactions. Photochem Photobiol Sci 18:2094–2101. https://doi.org/10.1039/C8PP00512E
46. Lewis FD, Weigel W, Zuo X (2001) Relaxation pathways of photoexcited diaminostilbenes. The *meta*-amino effect. J Phys Chem 105:4691–4696. https://doi.org/10.1021/jp0045282
47. Ho T-I, Su TM, Hwang T-C (1988) A convenient method of measuring quantum yields of photoisomerization of trans-stilbene. J Photochem Photobiol A Chem 41:293–298. https://doi.org/10.1016/1010-6030(88)87003-5
48. Lewis FD, Crompton EM (2003) Hydroxystilbene isomer-specific photoisomerization versus proton transfer. J Am Chem Soc 125:4044–4045. https://doi.org/10.1021/ja029873e
49. Lewis FD, Kalgutkar RS, Yang J-S (1999) The photochemistry of trans-ortho-, -meta-, and -Para-Aminostilbenes. J Am Chem Soc 121:12045–12053. https://doi.org/10.1021/ja992335q
50. Lewis FD, Kalgutkar RS (2000) The photochemistry of cis-ortho-, meta-, and Para-Aminostilbenes. J Am Chem Soc 105:285–291. https://doi.org/10.1021/jp002126o
51. Kaganowich M, Fischer G, Fischer E, Goedicke C, Stegemeyer H (1971) Temperature dependence of photoisomerisation VIII1. Phenyl-2-(2-naphthyl)-ethylene. Z Phys Chem 76:79–84. https://doi.org/10.1524/zpch.1971.76.1_2.079
52. Fischer G, Fischer E (1974) Oxygen enhancement of trans-cis photoisomerization in diarylethylenes. Mol Photochem 6:463–472
53. Lewis FD, Weigel W (2000) Excited state properties of donor–acceptor substituted trans-stilbenes: the meta-amino effect. J Am Chem Soc 104:8146–8153. https://doi.org/10.1021/jp001665q
54. Corredor CC, Belfield KD, Bondar MV, Przhonska OV, Hernandez FE, Kachkovsky OD (2006) One- and two-photon photochromism of 3,4-bis-(2,4,5-trimethyl-thiophen-3-yl)furan-2,5-dione. J Photochem Photobiol A Chem 184:177–183. https://doi.org/10.1016/j.jphotochem.2006.04.012
55. Favaro G, Matzzucato U, Ortica F, Smimmo P (2007) Thermal reversibility and bistability in photochromic diarylethenes. Inorg Chim Acta 360:995–999. https://doi.org/10.1016/j.ica.2006.07.007
56. Maafi M, Maafi W (2016) Modelling and elucidation of the kinetics of multiple consecutive photoreactions AB4(4Φ) with Φ-order kinetics. Application to the photodegradation of ribofavin. J Pharm Sci 105:3537–3548. https://doi.org/10.1016/j.xphs.2016.06.030
57. Moore WM, Ireton RC (1977) The photochemistry of riboflavin-V. The photodegradation of isoalloxazines in alcoholic solvents. Photochem Photobiol 25:347–356. https://doi.org/10.1111/j.1751-1097.1977.tb07354.x
58. Ahmad I, Fasihullah Q, Vaid FHM (2006a) Effect of light intensity and wavelengths on photodegradation reactions of riboflavin in aqueous solution. J Photochem Photobiol B 82:21–27. https://doi.org/10.1016/j.jphotobiol.2005.08.00459
59. Ahmad I, Fasihullah Q, Vaid FHM (2006b) Photolysis of formylmethylflavin in aqueous and organic solvents. Photochem Photobiol Sci 5:680–685. https://doi.org/10.1039/b602917e
60. Ahmad I, Ahmed S, Sheraz MA, Vaid FHM, Ansari IA (2010) Effect of divalent anions on photodegradation kinetics and pathways of riboflavin in aqueous solution. Int J Pharm 390:174–182. https://doi.org/10.1016/j.ijpharm.2010.01.042

61. Ahmad I, Sheraz MA, Ahmed S, Kazi SH, Mirza T, Aminuddin M (2011) Stabilizing effect of citrate buffer on the photolysis of riboflavin in aqueous solution. Results Pharma Sci 1:11–15. https://doi.org/10.1016/j.rinphs.2011.06.002
62. Koziol J (1966) Studies on flavins in organic solvents-I. Spectral characteristics of riboflavin, riboflavin tetrabutyrate and lumichrome. Photochem Photobiol 5:41–54. https://doi.org/10.1111/j.1751-1097.1966.tb05759.x
63. Heelis PF (1982) The photophysical and photochemical properties of flavins (isoalloxazines). Chem Soc Rev 11:15–39. https://doi.org/10.1039/CS9821100015
64. Chacon JN, McLearie J, Sinclair RS (1988) Singlet oxygen yields and radical contributions in the dye-sensitised photo-oxidation in methanol of esters of polyunsaturated fatty acids (oleic, linoleic, linolenic and arachidonic). Photochem Photobiol 47:647–656. https://doi.org/10.1111/j.1751-1097.1988.tb02760.x
65. Grodowski MS, Veyret B, Weiss K (1977) Photochemistry of flavins. II. Photophysical properties of alloxazines andisoalloxazines. Photochem Photobiol 26:341–352. https://doi.org/10.1111/j.1751-1097.1977.tb07495.x
66. Sikorska E, Khmelinskii E, Komasa A, Koput J, Ferreira LFV, Herance JR, Bourdelande JL, Williams SL, Worrall DR, Insińska-Rak M, Sikorski M (2005) Spectroscopy and photophysics of flavin related compounds: riboflavin and iso-(6,7)-riboflavin. Chem Phys 314:239–247. https://doi.org/10.1016/j.chemphys.2005.03.005
67. Görner H (2010b) Photocyclization of 2,6-dichlorodiphenylamines in solution. J Photochem Photobiol A 211:1–6. https://doi.org/10.1016/j.jphotochem.2010.01.010
68. Insinska-Rak M, Golczak A, Sikorski M (2012) Photochemistry of riboflavin derivatives in methanolic solutions. J Phys Chem A 116:1199–1207. https://doi.org/10.1021/jp2094593

Chapter 18
Development of Actinometers

18.1 Introduction

Actinometry is essential in photochemical studies because photoreactions depend on the amount of photons received by the considered reactive species. Around 70 chemical actinometers are listed in the IUPAC report [1] as standard actinometers. Most of these have known variable effective usage in experimental investigations. The most popular among these is the ferrioxalate actinometer. Limitations as well as the dynamic ranges of application of these actinometers were discussed. In general, the suitability of a photoactive chemical compound for actinometry has been linked to possessing a wavelength-invariant quantum yield. Almost none of the actinometric methodologies of the known actinometers used photokinetic data obtained through a kinetic analysis. In the present chapter, the foundations for kinactinometry are laid out for isosbestic and non-isosbestic monochromatic irradiations. The actinometric method is based on the photokinetic analyses developed in the previous chapters. The photosystems studied in Chap. 17 are experimentally tested here for kinactinometry. These represent the first examples of a series of common chemical compounds to be turned into efficient actinometers by using their respective kinetic data.

18.2 The Photokinetic Data

The solution of the considered reaction system is subjected to irradiation with monochromatic light at a given wavelength, $\lambda_{irr,\ q}$ (among a series, $1 \leq q \leq \mu$). The experiment will be conducted successively at "i" different radiation intensities, $P_{o,i}^{\lambda_{irr,q}}$, as permitted by the instrumentation employed. The values of each $P_{o,i}^{\lambda_{irr,q}}$ are determined by a physical actinometer (e.g. a photometer (Sect. 2.18)). A series of i total absorbance traces are collected for the $P_{o,i}^{\lambda_{irr,q}}$ values. The traces must include

M. Maafi, *Photokinetics*, https://doi.org/10.1007/978-3-031-98985-8_18

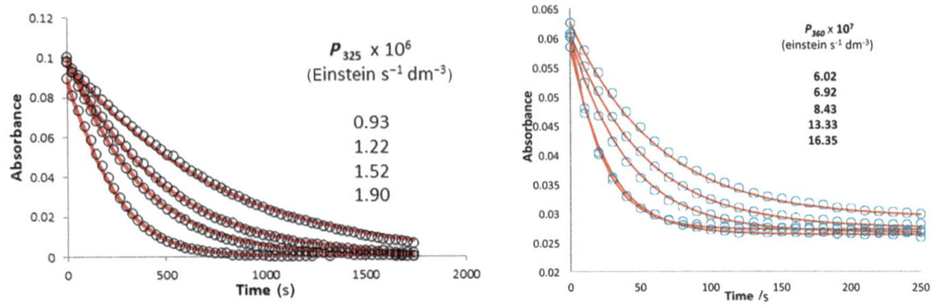

Fig. 18.1 Experimental total absorbance traces change with radiation intensity variation. The experimental data are represented by circles, and the best-fitting model equations are given by solid lines. Irradiation is performed in ethanol (22 ° C) by a monochromatic light, at the indicated values of $P_{o,i}^{\lambda_{irr}}$. *Left*: irradiation/observation (325/325) wavelengths for the traces of *DBZ* at 5.47×10^{-6} M. *Right*: 360/360 traces of *MONTE* at 1.85×10^{-6} M. (Reproduced with permissions from [2], *Journal of Pharmaceutical Sciences* (*left*), and from [3], *International Journal of Pharmaceutics* (*right*))

the whole variation with time, i.e. from $t = 0$ and until no variation is observed (at $t = \infty$). Each trace is individually fitted to the appropriate equation (Eqs. 15.1 or 15.23). A similar process will be undertaken for each of the μ selected irradiation-wavelength series of the considered reactive system. The initial concentrations must fall within the linearity range of the calibration graph of the studied system. The initial concentrations for a series of i irradiations are preferably selected to have the same or at least very close values (to avoid any underlying initial concentration effects).

The correspondence between experimental data and *SEM*-model equations was verified in all studied cases, where excellent fittings were observed (Fig. 18.1).

18.3 The Key Role of the Reaction Rate-Constant

Higher incident radiation intensities ($P_0^{\lambda_{irr}}$) are known to speed up photoreactions. The reaction's acceleration corresponds to steeper traces where the half-life time is reached sooner than for smaller values of $P_0^{\lambda_{irr}}$. In Fig. 18.1, an increased radiation intensity causes faster kinetics, indicated by shorter times to reach either complete deletion (*DBZ*) or the *pss* (*MONTE*).

The quantification of this phenomenon is best performed by the rate-constant of the reaction. In this perspective, $k_{Y_j \to Y_j'}^{\lambda_{irr}}$ value is expected to increase with increasing $P_0^{\lambda_{irr}}$ whether or not the monochromatic light has the wavelength of an isosbestic point ($\lambda_{irr} = \lambda_{isos}$ or λ_{n-isos}).

Such a concept is conveyed by the mathematical formulations of $k_{Y_j \to Y_j'}^{\lambda_{irr}}$, which have, so far, been seen for the different reactions studied in the previous chapters (Eqs. 8.39, 10.4, 11.21, 15.5 and 15.30). These expressions are made up of constant reaction parameters. In the context of actinometry, the rate-constant can be broken down into a general function of the incident radiation intensity (multiplying a constant, $\beta_r^{\lambda_{irr},T}$).

$$k_r^{\lambda_{irr}} = \beta_r^{\lambda_{irr},T} P_o^{\lambda_{irr}} \quad (18.1)$$

The $\beta_r^{\lambda_{irr},T}$ factor takes a specific expression, appropriate to the reaction considered and the type (isosbestic or non-isosbestic) of the light used for irradiation. It has the units of einstein^{-1} dm^3. The index "r" corresponds either to a reaction, e.g. *XY*(2*Φ*), or to a particular reaction-step $Y_j \rightarrow Y_{j'}$.

18.4 Trace Fitting

The fitting of either or both concentration and total absorbance traces was achieved by using the adequate model equation (Eqs. 15.1 or 15.23) for the reactive system seen in the previous chapter. It is preferable to perform the fitting process after finalising the elucidation. This way the rate-constant becomes the only fitting parameter of the model equation (i.e. no identifiability issues for the fitting). In fact, for all the reactive systems considered here, the fitting of the total absorbance traces is sufficient to determine the correct value of $k_{Y_j \rightarrow Y_j{'}}^{\lambda_{irr}}$, since Eq. 15.1 has the rate-constant as the only fitting parameter (for *XY*(2*Φ*, 1*k*) and its sub-mechanisms), and we have seen that the complete elucidation of XY_2(2*Φ*) can be achieved on the basis of the total absorbance traces only (meaning that, for this case, the equation of the total absorbance (Eq. 15.23) has two rate-constants as fitting parameters when it includes the values of the constant reaction parameters that are generated by the elucidation method (Sect. 16.8); these parameters are also required for the equations of the individual species concentrations that make part of Eq. 15.23).

18.5 The $\beta_r^{\lambda_{irr},T}$ Factor

In the equation of the reaction rate-constant (Eq. 18.1), the $\beta_r^{\lambda_{irr},T}$ factor represents the slope of the line established between the variation of $k_{r,i}^{\lambda_{irr}}$ with that of $P_{o,i}^{\lambda_{irr}}$. The experimental application of the equation shows, in general, a small intercept ($\ll 1$). Thus, for practical purposes, the equation for the correlation $k_r^{\lambda_{irr}} = f\left(P_o^{\lambda_{irr}}\right)$ should be written in the following form:

$$k_r^{\lambda_{irr}} = \beta_r^{\lambda_{irr},T} P_o^{\lambda_{irr}} + \text{intercept} \quad (18.2)$$

The experimental data for all the investigated molecules have shown good linearity between the fitting values of $k_r^{\lambda_{irr}}$ and those of the corresponding radiation intensity (Table 18.1). The $P_o^{\lambda_{irr}}$-independent $\beta_r^{\lambda_{irr},T}$ factor seems to be sensitive to reactivity and hence to molecular features.

Good linear correlations characterise plots of the experimental values of $\beta_r^{\lambda_{irr},T}$ factors determined by varying the radiation intensity (Table 18.1) and those calculated using both the expressions of the $\beta_r^{\lambda_{irr},T}$ from Eqs. 15.5 and 15.30 and the corresponding parameter values found by the elucidation method (Fig. 18.2). Such a correspondence supports the validity of the *SEM*-equations and the methodology applied for the kinetic treatment of the data.

It is interesting to notice that the determination of the $\beta_r^{\lambda_{irr},T}$ factor for a given *XY*(2Φ) reaction does not require a prior knowledge of the reaction attributes, i.e. it can be obtained without proceeding with elucidation since the fitting equation of the total absorbance trace only requires experimentally accessible coefficients.

18.6 The Initial Reaction-Rate

It is also interesting to envisage the quantification of the impact of the radiation intensity on reactivity by changes occurring on values of the initial reaction-rate. The *SEM*-equations have been established for such a reaction parameter both as a theoretical formulation (Eqs. 15.10, 15.11, 15.39 and 15.41), and as an expression using the trace fitting data (Eqs. 15.12–15.14, 15.40 and 15.42). According to these formulae, the initial reaction-rate should have a linear correlation with the radiation intensity, e.g. $r_{0,A}^{\lambda_{irr},T/\lambda_{obs}} = f\left(P_o^{\lambda_{irr}}\right)$. The experimental application of such a relationship indicates that the appropriate equation includes an intercept that should have a very small value (Eq. 18.3), which is given here for the total absorbance trace (with $Fit : r_{0,A}^{\lambda_{irr},T/\lambda_{obs}}$ are used). The initial rate and the intercept, and $\delta_A^{\lambda_{irr},T/\lambda_{obs}}$ factor are expressed in s^{-1} and einstein^{-1} dm^3, respectively.

$$r_{0,A}^{\lambda_{irr},T/\lambda_{obs}} = \delta_A^{\lambda_{irr},T/\lambda_{obs}} P_o^{\lambda_{irr}} + \text{intercept} \tag{18.3}$$

Obviously, the equation for the concentration traces has a similar form to that of Eq. 18.3, but with the initial reaction-rate and the constant being $r_{0,X}^{\lambda_{irr},T}$ and $\delta_X^{\lambda_{irr},T}$. To facilitate application to experimental data, the above formula using total absorbance data is recommended.

The reliability of the fitting values of the initial reaction-rate is assessed by comparing their experimental values to the corresponding ones calculated by the theoretical equation (the latter uses the relevant values of the parameters delivered by the elucidation method applied to the considered reaction). An illustration of the excellent correspondence between the two parameters is shown in Fig. 18.3 for *FLU*. The errors between the two sets of data were $<10\%$. The best-fit line is characterised

Table 18.1 Parameters of the linear correlations according to Eq. 18.2 established for several molecules

Parameters[a] → Molecule ↓ $C_X^{\lambda_{irr},T}(0)$	Mechanism	Irradiation wavelength, λ_{irr} /nm	$\beta_r^{\lambda_{irr},T}$ /einstein^{-1} dm^3	Intercept /s^{-1}	r^2	$P_o^{\lambda_{irr}} \times 10^7$ /einstein dm^{-3} s^{-1}
DBZ	*XY*(1Φ)	275	326.26	1.0 10^{-4}	0.9991	0.78–1.64
		300	2499.3	1.7 10^{-3}	0.9987	1.05–1.31
		325	4013.2	2.4 10^{-3}	0.9499	0.93–1.90
		345	2589.8	3.0 10^{-4}	0.9830	0.95–2.20
		350	2120.6	2.0 10^{-4}	0.9969	1.01–2.51
NIS	*XY*(1Φ)	320	7558.4	2.0 10^{-5}	0.996	1.34–4.35
1.56 10^{-5} M		345	8391.1	−7 10^{-6}	0.998	2.42–7.20
		370	8850.8	5.0 10^{-7}	0.999	5.55–11.9
		390	7902.7	−5 10^{-6}	0.996	3.81–10.9
NIF	*XY*(1Φ)	321	9131.8	−3 10^{-4}	0.997	1.20–3.46
		345	8116.9	4.0 10^{-4}	0.987	2.39–5.19
		370	9775.9	2.0 10^{-4}	0.988	2.30–7.37
		390	8489.2	−3 10^{-4}	0.993	2.87–9.11
MONTE	*XY*(2Φ)	258	2250.7	1.1 10^{-3}	0.9839	4.53–10.0
		328	17,064	7.9 10^{-3}	0.9994	5.67–13.1
		345	24,932	4.0 10^{-3}	0.9803	4.99–14.1
		360	28,069	8.0 10^{-4}	0.9957	6.02–16.3
SUT	*XY*(2Φ)	320	1096.2	1.0 10^{-5}	0.9954	0.61–1.33
		340	1560.1	−7 10^{-5}	0.9881	0.85–1.98
		345	1938.1	−4 10^{-4}	0.9941	1.42–2.04
		360	5866.2	−6 10^{-3}	0.9915	1.58–2.67
		380	3740.9	1.9 10^{-3}	0.9959	1.19–2.97
		400	5432.6	2.3 10^{-3}	0.9718	1.45–3.48
		420	5846.6	6.0 10^{-3}	0.9851	1.66–3.93

(continued)

Table 18.1 (continued)

Parameters[a] → Molecule ↓ $C_X^{\lambda_{irr},T}(0)$	Mechanism	Irradiation wavelength, λ_{irr} /nm	$\beta_r^{\lambda_{irr},T}$ /einstein^{-1} dm^3	Intercept /s^{-1}	r^2	$P_o^{\lambda_{irr}} \times 10^7$ /einstein dm^{-3} s^{-1}
		440	5521.9	6.0 10^{-4}	0.9882	1.77–4.07
		460	4301.4	1.0 10^{-3}	0.9887	1.80–4.30
		480	3437.7	−1 10^{-3}	0.9980	2.04–4.85
FLU	$XY(2\Phi)$	260	818.4	3.0 10^{-7}	0.99	2.93–4.60
2.95 10^{-5} M		270	1249	3.0 10^{-5}	0.96	2.51–5.21
		280	1584	−1 10^{-6}	0.99	2.70–5.51
		285	1998	2.0 10^{-5}	0.98	2.63–4.81
		290	2077	5.0 10^{-5}	0.99	2.56–4.42
RBF	$XY_2(2\Phi)$ $(X \rightarrow Y_1)$	420	302.2	−3 10^{-5}	0.96	6.27–10.3
1.8 10^{-5} M		445	421.9	3.0 10^{-5}	0.95	6.05–12.5
		460	305.1	−1 10^{-6}	0.94	7.46–11.4
		480	225.8	2.0 10^{-5}	0.98	8.17–19.2
FMF	$XY_2(2\Phi)$ $(Y_1 \rightarrow Y_2)$	420	3.520	−5 10^{-7}	0.93	6.27–10.3
		445	43.71	−3 10^{-5}	0.91	6.05–12.5
		460	41.92	−2 10^{-5}	0.98	7.46–11.4
		480	39.74	−2.10^{-5}	0.95	8.17–19.2

[a] $l_{obs} = 1$ cm, $l_{irr} = 2$ cm for RBF, FLU

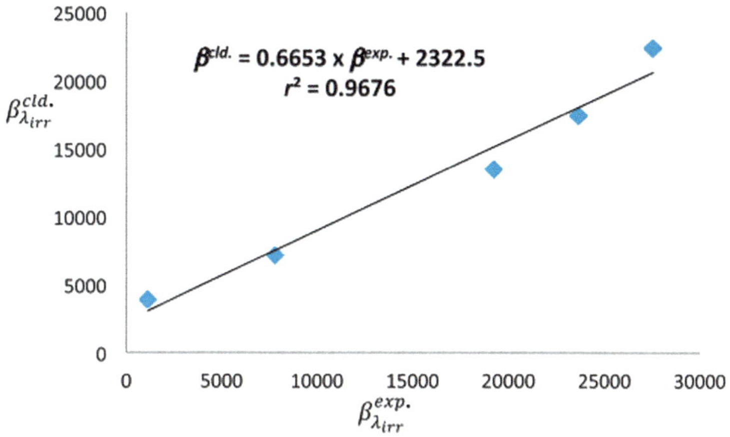

Fig. 18.2 Linear plot between experimentally determined and calculated $\beta_r^{\lambda_{irr},T}$-factor values for *MONTE*. (Reproduced with permission from [3], *International Journal of Pharmaceutics*)

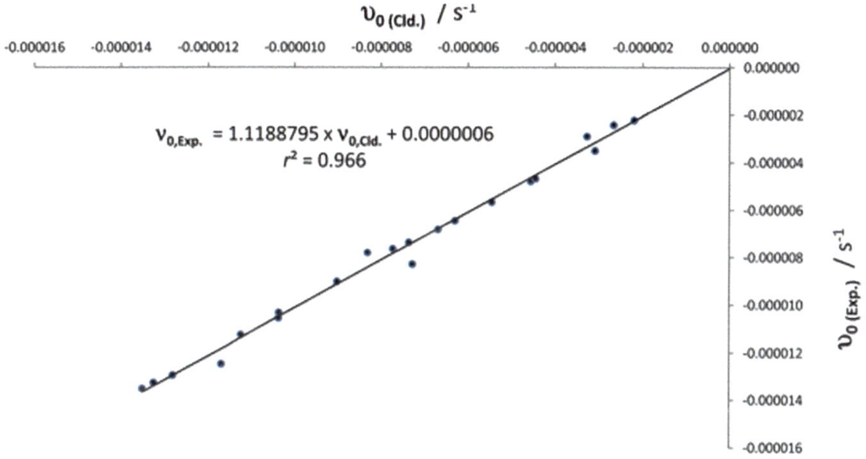

Fig. 18.3 Fitting (*Fit* : $r_{0,A}^{\lambda_{irr}/\lambda_{obs},T}$) against calculated (*Theo* : $r_{0,A}^{\lambda_{irr}/\lambda_{obs},T}$) values of the initial reaction-rate of the photoreversible *FLU* (both have negative values). The initial rates are labelled here as $\upsilon_{0\,(Exp.)}$ and $\upsilon_{0(Cld.)}$, respectively. (Reproduced with permission from [2], *Photochemical and Photobiological Sciences*)

by a high correlation coefficient ($r = 0.983$) and a negligible value of the intercept (6×10^{-7}).

For the investigated molecules, the experimental absolute values of the initial rate increased with increasing radiation intensity, proving a faster depletion of the reactant. The correlation meant that the end of the reaction or the time necessary to reach *pss* is shorter with increasing $P_o^{\lambda_{irr}}$.

Table 18.2 Linear correlation equations and parameters for the relationship, $Fit : r_{0,A}^{\lambda_{irr}/\lambda_{obs},T} = f\left(P_o^{\lambda_{irr}}\right)$, for a set of molecules of different reaction mechanisms

Parameters → Molecule ↓	mechanism	Irradiation Wavelength, λ_{irr} /nm	$\delta_A^{\lambda_{irr},T/\lambda_{obs}}$ /einstein^{-1} dm^3	Intercept /s^{-1}	r^2	$P_o^{\lambda_{irr}} \times 10^7$ / einstein dm^{-3} s^{-1}
NIS	$XY(1\Phi)$	320	−130.26	$-4\ 10^{-6}$	0.996	1.34–4.35
		345	−124.00	$-1\ 10^{-5}$	0.984	2.42–7.20
		370	−226.05	$7.0\ 10^{-5}$	0.956	5.55–11.9
		390	−97.980	$-5\ 10^{-6}$	0.944	3.81–10.9
FLU	$XY(2\Phi)$	260	−10.58	$4.8\ 10^{-7}$	0.93	2.93–4.60
		270	−14.57	$-9\ 10^{-7}$	0.98	2.51–5.21
		280	−25.16	$1.3\ 10^{-6}$	0.99	2.70–5.51
		285	−28.11	$1.8\ 10^{-7}$	0.99	2.63–4.81
		290	−7.21	$1.0\ 10^{-6}$	0.99	2.56–4.42
RBF	$XY_2(2\Phi)$	420	−2.88	$6.0\ 10^{-7}$	0.95	6.27–10.3
		445	−4.15	$-3\ 10^{-7}$	0.99	6.05–12.5
		460	−2.49	$-1\ 10^{-6}$	0.94	7.46–11.4
		480	−7.21	$5.0\ 10^{-5}$	0.97	8.17–19.2

The linearity of the plots, as shown in Table 18.2, represents compelling evidence of the usefulness of the *SEM*-equations, the effectiveness of the quantification of $P_o^{\lambda_{irr}}$ -effects on reactivity, and the dependence of the $\delta_{A\ or\ X}^{\lambda_{irr},T/\lambda_{obs}}$ on both reaction-specific features and conditions. The negative values of $r_{0,A}^{\lambda_{irr},T/\lambda_{obs}}$ (or those of $\delta_A^{\lambda_{irr},T/\lambda_{obs}}$) reported in Table 18.2 are due to the magnitude of the absorptivity of the immediate product to X (Y or Y_1), being smaller than that of X for the molecules displayed therein.

Similarly to what has been stipulated for the $\beta_r^{\lambda_{irr},T}$ factors, the construction of an $Fit : r_{0,A}^{\lambda_{irr},T/\lambda_{obs}} = f\left(P_o^{\lambda_{irr}}\right)$ plot does not require the prior knowledge of the reaction intrinsic parameters of the reaction investigated. All that is needed is a good fit of the total absorbance trace, whose fitting data will serve the determination of the $Fit : r_{0,A}^{\lambda_{irr},T/\lambda_{obs}}$. This is valid for both $XY(2\Phi)$ and $XY_2(2\Phi)$. Hence, the quantification of the $P_o^{\lambda_{irr}}$-effect is affordable even when the quantum yields and the absorptivities of the species are unknown. It is a considerable advantage for actinometry.

18.7 Kinactinometry

The purpose is to measure the unknown radiation intensity of a light source, delivering a monochromatic irradiation of the same wavelength used above to construct $k_r^{\lambda_{irr}} = f\left(P_o^{\lambda_{irr}}\right)$ plot. The light-source, *Lp*, of unknown radiation intensity tested here might or might not be the same as the one used for the actinometric experiment leading to the latter plot.

Given that, in the calibration experiment, the effect of the radiation intensity was performed for a given molecule (*Act*), in a solvent *Svt*, at an initial concentration $C_{Act}^{\lambda_{irr,Cal.}}(0)$, at an irradiation wavelength ($\lambda_{irr,\ Cal.}$), and at a temperature *T*, and the corresponding equation (Eq. 18.4) has been established for S_{irr}, V_{irr}, l_{obs} and l_{irr}.

$$k_{Act}^{\lambda_{irr,Cal.},T} = \beta_r^{\lambda_{irr},T}\ P_o^{\lambda_{irr}} + \text{intercept} \tag{18.4}$$

This equation, Eq. 18.4, is a calibration equation for the selected chemical actinometer.

The kinactinometric method can then be set out. The implementation of this method will be performed in stages.

The first stage of the method is to prepare a solution of *Act* in *Svt* at $C_{Act}^{\lambda_{irr,Cal.}}(0)$ (or a concentration very close to that value). The solution is exposed to a monochromatic irradiation at $\lambda_{irr,\ Cal.}$, from a light-source *Lp*, of an unknown intensity, $P_{o,unk}^{\lambda_{irr}}$ (with S_{irr} and V_{irr} are the same as those used for the experiments using Eq. 18.4).

The photoreaction of *Act* in these conditions is monitored, in a second stage, and its experimental total absorbance trace recorded up to the end of the reaction (i.e. either to a complete depletion of *Act*, or up to reaching *pss*). The trace is, in a third stage, fitted to the appropriate model equation (either Eq. 15.1 or Eq. 15.23), and the reaction rate-constant ($k_{Act,unk}^{\lambda_{irr,Cal.},T}$) value or values determined. The fourth and final stage consists of measuring the actual unknown radiation intensity $P_{o,unk}^{\lambda_{irr}}$ of the source *Lp*. This is readily achieved from the calibration formula $k_{Act}^{\lambda_{irr,Cal.}T}$ (Eq. 18.4), as follows:

$$P_{o,unk}^{\lambda_{irr}} = \frac{k_{Act,unk}^{\lambda_{irr,Cal.},T} - \text{intercept}}{\beta_r^{\lambda_{irr},T}} \tag{18.5}$$

Equation 18.5 uses the values of $\beta_r^{\lambda_{irr},T}$ and intercept characterising the original calibration equation (Eq. 18.4), and the value of $k_{Act,unk}^{\lambda_{irr,Cal.},T}$ obtained under the unknown $P_{o,unk}^{\lambda_{irr}}$.

For the procedure to be widely applied, it is important to consider the situations where the experimental conditions are not exactly the same between calibration and kinactinometry.

When the irradiation optical path lengths employed for calibration (l_{irr}) and kinatinometry (l_{irr}^{unk}) are not equal, the rate-constant obtained from the kinactinometric experiment needs to be corrected in order to apply Eq. 18.5 (based on the calibration equation, Eq. 18.4), i.e. $k_{Act,unk}^{\lambda_{irr,Cal.},T}$(corrected) should replace $k_{Act,unk}^{\lambda_{irr,Cal.},T}$ in that equation. This is achieved for *XY*(1 or 2Φ) systems (based on Eq. 15.5), by the following expression.

$$k_{Act,unk}^{\lambda_{irr},Cal.,T}(\text{corrected}) = k_{Act,unk}^{\lambda_{irr},Cal.,T} \frac{l_{irr}}{l_{irr}^{unk}} \frac{PKF^{\lambda_{irr},Cal.}}{PKF_{unk}^{\lambda_{irr},Cal.}} \tag{18.6}$$

and for $XY_2(2\Phi)$ reactions (based on Eq. 15.30), by

$$k_{Act,unk}^{\lambda_{irr},Cal.,T}(\text{corrected}) = k_{Act,unk}^{\lambda_{irr},Cal.,T} \frac{l_{irr}}{l_{irr}^{unk}} \frac{PKF_{Y_j}^{\lambda_{irr},Cal.}}{PKF_{Y_j,unk}^{\lambda_{irr},Cal.}} \tag{18.7}$$

In this case, the values of S_{irr}^{unk} and V_{irr}^{unk} employed for the actinometric experiment are different from those of the calibration (S_{irr} and V_{irr}), the $P_{o,unk}^{\lambda_{irr}}$ value obtained from Eq. 18.5 corresponds to calibration conditions and not to the real amount of photons entering the reactor in the kinactinometric experiment. Therefore, it is necessary to adjust the calculated $P_{o,unk}^{\lambda_{irr}}$ as stipulated in Sect. 2.17 (Eq. 2.25), to correspond to the number of photons entering the reactor per second and per area and volume irradiated during the actinometric experiment, as follows:

$$P_{o,unk}^{\lambda_{irr}}(\text{correced}) = P_{o,unk}^{\lambda_{irr}} \frac{S_{irr}^{unk}}{S_{irr}} \frac{V_{irr}}{V_{irr}^{unk}} \tag{18.8}$$

Hence, $P_{o,unk}^{\lambda_{irr}}(\text{correced})$ (Eq. 18.8) is the actual amount of photons entering the reactor in the actinometric experiment.

The above actinometric procedure will likewise work when the initial reaction-rate is used for calibration of the actinometer.

The reliability of the kinactinometric method is indicated by good correlations between the experimental and the calculated values of the radiation intensities (Fig. 18.4).

Experimentally, the actinometric experiment can be achieved by using a physical or a chemical actinometer. Reference and precise values of $P_0^{\lambda_{irr}}$ for the calibration, at given wavelengths, can be determined by a set of reliable chemical actinometers that were calibrated by the kinactinometric methodology developed here [2–12]. The large spectral region covered by these actinometers (ranging between 250 and 580 nm, Fig. 18.5) is practically useful for virtually any organic photoactive molecule.

The kinactinometric methodology developed here is certainly not limited to the aforementioned molecules. The list of these actinometers (Fig. 18.5) can be extended, in the future, by new (cheaper, water soluble, nanocarriers, etc.) chemical actinometer candidates employing the kinactinometric method proposed in the present chapter. In fact, such an extended list of chemical actinometers may concern photo-systems obeying similar types of mechanisms that are covered by the *SEM*-equations (Chap. 15).

The straightforward actinometric method proposed here is a good alternative to the existing approaches, particularly for the one proposed by the international

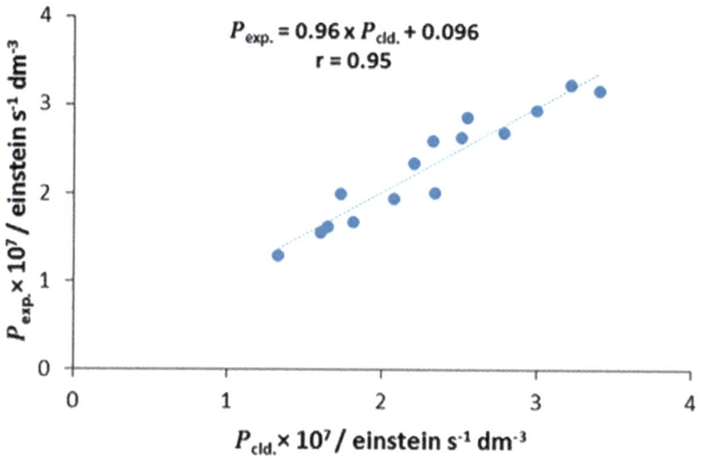

Fig. 18.4 A linear relationship characterising experimental ($P^{\lambda_{irr}}_{0,exp.}$) versus calculated ($P^{\lambda_{irr}}_{0,cld.}$) values of the radiant intensity for *OXY* over the 310 – 350 nm wavelength range. (Reproduced with permission from [11], *Spectrochimica Acta A Molecular and Biomedical Spectroscopy*)

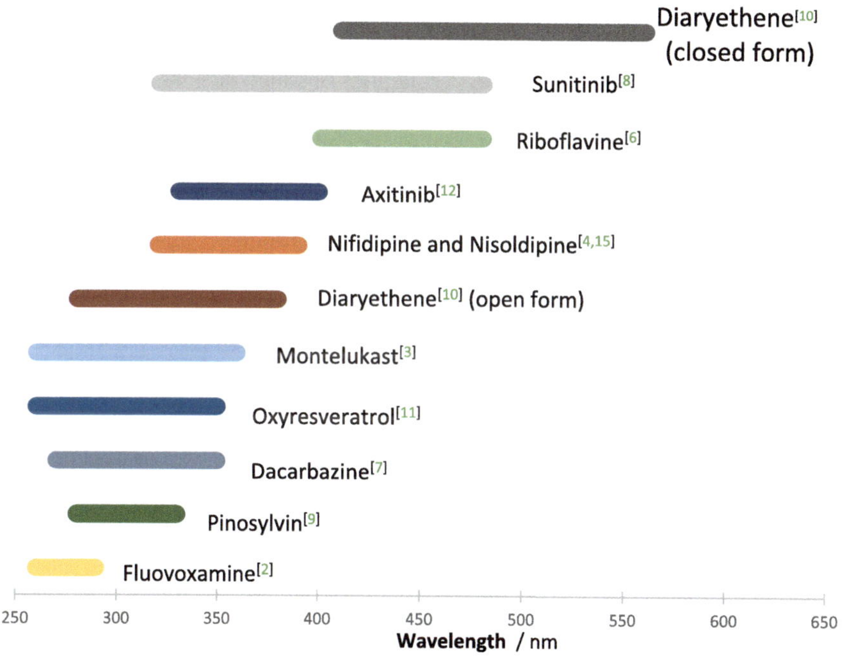

Fig. 18.5 Specific spectral ranges for kinactinometry by a series of efficient and reliable actinometers. The corresponding references are shown in the figure. (Reproduced with permission from [13], *Frontiers in Chemistry*)

council for harmonisation of technical requirements for pharmaceuticals for human use (*ICH*), recommending a procedure using quinine hydrochloride as a chemical actinometer [14], whose limitations and drawbacks have been signalled in a number of reports [4, 15–21].

The present general kinactinometric procedure is also an alternative to classical chemical actinometry procedures, as we shall discuss in Chap. 22.

18.8 Conclusion

The *SEM*-equations turned out to apply very well to experimental data. The prediction and expectations built on the *RK*-simulated data, and these equations were all verified experimentally. Overall, the strategy involving *RK*-calculated traces to develop useful equations (as the ones delivered by *SEM*) is rational and leads to objective simulated results that are confirmed by experiment. Owing to their potential, *RK*-simulations can be exploited further to rationalise photokinetics.

The procedures and methods proved efficient in quantifying several effects in photokinetics, and establishing a kinactinometric methodology offering a reasonable possibility for expanding, practically at wish, the set of reliable chemical actinometers.

Even though very efficient, the *SEM* equations obtained thus far have two major limitations: their development is time-consuming, and, so far, they apply to only a few reactions. Further developments could be envisaged for more (complicated) reactive systems if some computing element is introduced in the development stages with the aim of saving processing time.

Also, the *SEM* equations are limited to monochromatic light. Methods for polychromatic-light actinometry shall be discussed in Part VII.

References

1. Braslavsky SE (2007) Glossary of terms used in photochemistry, 3rd Ed. Pure Appl Chem 79: 293–465. https://doi.org/10.1351/pac200779030293
2. Maafi M, Maafi W (2015a) Quantitative assessment of photostability and photostabilisation of fluvoxamine and its design for actinometry. Photochem Photobiol Sci 14:982–994. https://doi.org/10.1039/c5pp00022j
3. Maafi M, Maafi W (2014) Montelukast photodegradation: elucidation of F-order kinetics, determination of quantum yields and application to actinometry. Int J Pharm 471:544–552. https://doi.org/10.1016/j.ijpharm.2014.05.016
4. Maafi W, Maafi M (2013) Modelling nifedipine photodegradation, photostability and actinometry properties. Int J Pharm 456:153–164. https://doi.org/10.1016/j.ijpharm.2013.07.075
5. Maafi M, Maafi W (2014) Φ-order kinetics of photoreversible-drug reactions. Int J Pharm 471: 536–543. https://doi.org/10.1016/j.ijpharm.2014.05.017

6. Maafi M, Maafi W (2016) Modelling and elucidation of the kinetics of multiple consecutive photoreactions AB4(4Φ) with Φ-order kinetics. Application to the photodegradation of ribofavin. J Pharm Sci 105:3537–3548. https://doi.org/10.1016/j.xphs.2016.06.030
7. Maafi M, Lee LY (2015) Determination of dacarbazine Φ-order photokinetics, quantum yields, and potential for actinometry. J Pharm Sci 104:3501–3509. https://doi.org/10.1002/jps.24568
8. Maafi M, Lee LY (2015) Actinometric and Φ-order photodegradation of anti-cancer Sunitinib. J Pharm Biomed Anal 110:34–41. https://doi.org/10.1016/j.jpba.2015.02.026
9. Maafi M, Al-Qarni MA (2019) Dependence of pinosylvin photochemical quantum yield on excitation wavelength. Trends Photochem Photobiol 18:45–58. http://www.researchtrends.net/tia/abstract.asp?in=0&vn=18&tid=15&aid=6484&pub=2019&type=
10. Maafi M, Al-Qarni MA (2022) Mono-and polychromatic light diarylethene-actinometer for the visible range. Dyes Pigments 198:109942. https://doi.org/10.1016/j.dyepig.2021.109942
11. Maafi M, Al-Qarni MA (2018) Φ-order spectrophotokinetic characterisation and quantification of trans-cis oxyresveratrol reactivity, photodegradation and actinometry. Spectrochimica Acta Mol Biomed Spectroscopy 188:64–71. https://doi.org/10.1016/j.saa.2017.06.067
12. Lee LY (2016) PhD Thesis. De Montfort University
13. Maafi M (2023) On photokinetics under monochromatic light. Front Chem 11:1233151. https://doi.org/10.3389/fchem.2023.1233151
14. ICH, International Conference on Harmonisation (1996) Stability testing: photostability testing of new drug substances and products, Q1b. ICH, 1996. Guidelines for industry: Q1B photostability testing of new substances and products. Fed Regist 62:27115–27122
15. Maafi M, Maafi W (2015) Quantification of unimolecular photoreaction kinetics. Determination of quantum yields and development of actinometers—The photodegradation case of cardiovascual Nisoldipine. Int J Photoenergy 2015:1–12. https://doi.org/10.1155/2015/454895
16. Baertschi SW (1997) Commentary on the quinine actinometry system described in the ICH draft guidelines on photostability testing of new drug substances and products. Drug Saf 1:193–195
17. Baertschi SW, Alsante KM, Tonnesen HH (2010) A critical assessment of the ICH guidelines on photostability testing of new drug substances and products (Q1B): recommendation for revision. J Pharm Sci 99:2934–2940. https://doi.org/10.1002/jps.22076
18. Azevedo Filho CA, De Filgueiras Gomes D, De Melo Guedes JP, Batista RMF, Santos BS (2011) Considerations on the quinine actinometry calibration method used in photostability testing of pharmaceuticals. J Pharm Biomed Anal 54:886–888. https://doi.org/10.1016/j.jpba.2010.11.012
19. Kester TC, Zhan Z, Bergstrom DH (1996) Quinine actinometry studies under two light sources specified by the ICH guideline on photostability testing. In: Presented at the National Meeting AAPS, Seattle, Washington
20. Tønnesen HH (2004) Photostability of drugs and drug formulations.2nd edn. CRC Press, London. ISBN. 978–0415303231
21. Piechocki JT, Thoma K (2010) Pharmaceutical photostability and stabilisation technology. Informa Healthcare, London. ISBN. 978–0824759247

Part VII
A General Photokinetic Model Equation for Photo-thermal Reactions Under Monochromatic and Polychromatic Lights

Chapter 19
A General Model Equation for Photokinetics

19.1 Introduction

Our journey, so far, has led us to develop a better understanding of photokinetics through the development of various strategies. Nonetheless, the efficiency of solving photokinetics remains limited, mainly useful for relatively simple reaction mechanistic options. And among other matters, two important ideas have become self-evident. The first relates to a lack, in photokinetics, of integrated rate-laws, and the second, inferred from the first, stipulates that when the integrated rate-equations are available, they considerably simplify both data treatment and kinetic elucidation.

Unfortunately, the photochemistry literature does not offer a way out of the dilemma posed by, on one hand, the impossibility of achieving analytical solutions for quasi all photochemical rate-laws and, on the other hand, the endeavour of reliability and consistency of the results sought for the reaction parameters and/or features.

Hence, the matter relative to the integrated rate-law remains central to the advancement of photokinetics and the standardisation of procedures. A corollary to this problem resides in the partial or specific character of the proposals accomplished so far in this field, whereas a generalisation of the integrated rate-law formulation seems to be the best way forward. Part VII is dedicated to tackling such issues and attempts to bring generality and consistency to photokinetics.

The findings discussed in this Part VII represent a synopsis of previously published works [1–3].

M. Maafi, *Photokinetics*, https://doi.org/10.1007/978-3-031-98985-8_19

19.2 Basic Terminology and Context

From a kinetic viewpoint, thermal and photochemical processes are quantified differently. The latter obey Φ-order kinetics, whereas the former reactions are usually described, in thermal kinetics, by first-order models. This feature might be considered important enough to distinguish the photoreactions that involve exclusively photochemical reaction-steps, from those sequences, photothermal reactions, that encompass both photochemical as well as thermal reaction-steps. It seems, hence, more reasonable to adopt the term “photothermal reactions” instead of “photoreactions” as a generic name for light-induced reactions (since, within this framework, it is understood that photothermal reactions include the sub-group consisting of photoreactions).

The photo- and thermal reaction-steps considered here are those which are experimentally observed in common laboratory conditions. The processes considered here evolve within a time frame compatible with measurements performed by spectrophotometry (a required technique for photokinetic studies). The species hence detected are relatively long-lived since light excitation and most of the subsequent molecular transformations are much faster events to be detected by the routine analytical/spectroscopic techniques that are usually available in the majority of research labs.

The reactions are branded by $XY_q(uk, v\Phi)$, as previously adopted (Sect. 2.3). The $n_{sp} = q + 1$ species of a given photothermal reaction mechanism are labelled as Y_j ($n_{sp} \geq j \geq 0$), with the reactant being $X = Y_0$ (i.e. $j = 0$), and the products of the reaction indexed by $n_{sp} \geq j > 0$. The overall mechanism of a given photothermal reaction involves "v" photochemical and "u" thermal reaction-steps (a total of $n_{rs} = u + v$). Each photochemical reaction-step is characterised by an individual quantum yield (Φ), and each thermal reaction-step is associated with an individual rate-constant (k). Each species Y_j, in the overall mechanism, is the start and/or the end of one or several reaction-steps, including v_{Φ}^{j} photochemical and u_{k}^{j} thermal reaction-steps. Accordingly, there are n_{sp} sets of u_{Φ}^{j} and v_{k}^{j} characterising the species of the mechanism at hand. For instance, for the sets of the photothermal $XY_3(1k, 4\Phi)$ reaction ($n_{sp} = 4$), $X \rightleftharpoons Y_1 \rightarrow Y_2 \rightleftharpoons Y_3$ (with the thermal step being the central link, $Y_1 \rightarrow Y_2$), we have: $v_{\Phi}^{0} = 1$, $v_{\Phi}^{1} = 0$ and $u_{\mathrm{k}}^{1} = 2$, $v_{\Phi}^{2} = 3$, $v_{\Phi}^{3} = 2$, (the numbers indicated here refer to the species labels, e.g. 0 for $Y_0 = X$). The rest of u_{k}^{j} sets are empty. Each element of either v_{k}^{j} or u_{Φ}^{j} sets must correspond, for each Y_j species, to the end of the reaction-steps (one where the species considered (e.g. $j = 0$) sits at the start, *e. g.* , $Y_0 \rightarrow Y_1$ and $v_{\Phi}^{0} = 1$).

In order to prove the concepts presented hereafter, these are tested against calculated traces using the fourth-order Runge–Kutta numerical integration method. The reliability of $RK4$-NIM for describing photokinetics [4] makes it the best testing means for any new concepts and/or equations, allows prediction of reaction behaviour, gives access to reaction parameters that are difficult to reach experimentally, and offers to generate a large set of traces for reactions considered in a wide range of conditions. The $RK4$-NIM is used here in a similar way to that employed for the semi-method (Chap. 14), even though there is no determination of the explicit formulae of the parameters occurring in the general explicit equations.

The assessment of the model equations proposed further in this Part VII is achieved by fitting them to the *RK*-generated traces and plots. The fitting of the *RK*-data is here performed by a Levenberg–Marquardt Algorithm (*LMA*), run on the curve fitting tool option offered by the *R*2020*b* MatLab software (Sect. 2.22). The quality of fit is assessed by three parameters, namely, the correlation coefficient (r^2) corresponding to the linear plot of *LMA*-calculated against *RK*-simulated data (of each species), the sum of squares error (*SSE*), and the root-mean-square deviation (*RSMD*) of the two data sets.

The *RK*-simulations are developed for a vigorously stirred liquid reaction medium, contained in a slab-shaped reactor of volume V_{irr}, and irradiated over an area S_{irr} by a collimated light beam.

The light can either be mono- or polychromatic. The optical lengths l_{irr} and l_{obs}, respectively, across the sample directions of irradiation and observation (or monitoring) lights, might or might not be equal.

Furthermore, the rate-equation implemented for *RK*-calculations does not include terms corresponding to the reflection, scattering and/or emission of light [5], in the sense that only the simplest reaction case was considered here (where those contributions are assumed negligible).

19.3 The OSIA Wavelength Domain

The irradiation wavelength (λ_{irr}) selected for the monochromatic experiment is independent of the light source supplying the beam; only its radiation intensity ($P_0^{\lambda_{irr}}$) is relevant. The situation is rather different when the irradiation light is polychromatic. The lamp (*Lp*) emitted light spans the spectral domain ($\Delta\lambda_{Lp}$) between two wavelengths λ_a^{LP} and λ_b^{LP}, $\Delta\lambda_{Lp} = \lambda_b^{LP} - \lambda_a^{LP}$. Here, the particular properties of the lamp are as relevant as the radiation intensity due to the specificity of the lamp's emitted light profile (i.e. the plot of intensity vs. wavelength). In addition, there is another important feature to take into account; that is, the only useful wavelength section of the lamp profile is the one that is absorbed by the reactive system. In other words, the spectral domain $\Delta\lambda = \lambda_b - \lambda_a$ corresponds to the overlap between the absorption spectrum of the reactive medium and the profile of the lamp (there might also be more than one such spectral domain; see the example of a spiky lamp profile in Fig. 22.2). This overlap section(s) of irradiation (lamp) and absorption (reactive medium) wavelengths, $\Delta\lambda$, is dubbed the *OSIA*.

Examples of the disposition of the lamp profile relative to the absorption spectra (given here in an absorptivity scale) of reactive species are shown in Fig. 19.1. Depending on the situation, the two absorption spectra in Fig. 19.1 may well be assigned, for instance, to two different reaction media, to an actinometer and an investigated species, or to reactants and products. The illustration shows that the

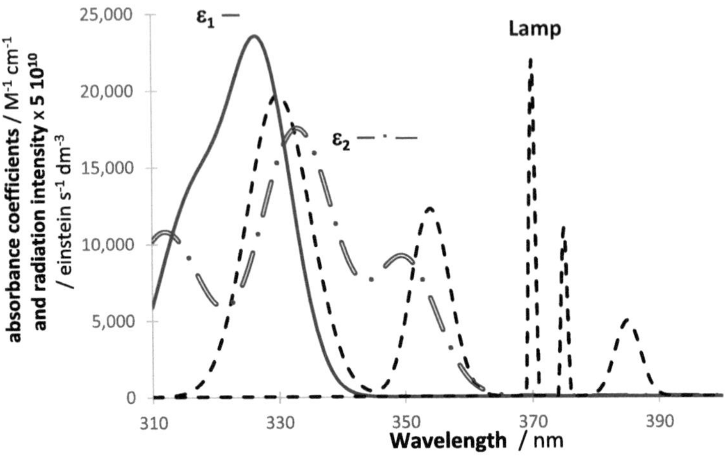

Fig. 19.1 Examples of two species absorption spectra, ε_1 (plain line) and ε_2 (discontinued line), and a lamp profile (dashed line). (Reproduced with permission [2], *Frontiers in Chemistry*)

OSIA of species (ε_1) is different from that of the second species (ε_2), and $\Delta\lambda_{Lp}$ is larger than the *OSIA*s for both species. The consequences of such a difference and other aspects of *OSIA*, notably in actinometry, will be reviewed in Sect. 23.4. Incidentally, *OSIA* might have a similar meaning to the overlap integral adopted in some literature [6]. However, this terminology, invoking an integral, might perhaps lead to confusion owing to the rate (Eq. 19.1) being itself an integral and normally solved by integration. Such a lack of specificity discourages the usage of the "the overlap integral" terminology.

19.4 The General Integro-Differential Rate-Equation

The kinetic behaviour of a $XY_q(uk, v\Phi)$ photothermal reaction subjected to polychromatic irradiation (provided by a lamp, *Lp*, over an *OSIA*, $\Delta\lambda$) and maintained at a constant temperature (T) is fully captured by the general integro-differential rate-equation ($r_{Y_j}^{Lp,\Delta\lambda,T}(t)$) expressing the variation of species Y_j concentration, $C_{Y_j}^{Lp,\Delta\lambda,T}(t)$ (expressed in units, M or mol dm^{-3}) with reaction time (t in s). $r_{Y_j}^{Lp,\Delta\lambda,T}(t)$ (in $M\,s^{-1}$) is expressed as a linear combination of the individual photochemical ($r_{Y_j}^{\Phi}(t)$) and thermal ($r_{Y_j}^{T}(t)$) rates. Each of the latter individual rates is, itself, a linear combination of the individual rates corresponding to the reaction-steps starting or ending at species Y_j. The rate $r_{Y_j}^{Lp,\Delta\lambda,T}(t)$ must also convey the light effects at each wavelength of the full *OSIA* for the polychromatic irradiation range ($\Delta\lambda = \lambda_b - \lambda_a$). A feature that is implemented by the occurrence of an extra integral (and an extra integration variable, $d\lambda$) that amends the rate-equation of a photochemical reaction ($r_{Y_j}^{\Phi}(t)$)

subjected to each monochromatic light within $\Delta\lambda$. As a result, the left-hand side of the general rate-law is the classical differential, $dC_{Y_j}^{Lp,\Delta\lambda}(t)/dt$, and the right-hand side is an integral over λ (Eq. 19.1). Also, owing to the mixed kinetic nature of the photothermal reaction, it is conjectured that its kinetic behaviour must be mapped out by a combination of both Φ- and first-order (exponential) types of mathematical functions, respectively, relaying pure photochemical and pure thermal kinetics.

$$r_{Y_j}^{Lp,\Delta\lambda,T}(t) = \frac{dC_{Y_j}^{Lp,\Delta\lambda,T}(t)}{dt} = r_{Y_j}^{\Phi}(t) + r_{Y_j}^{T}(t)$$

$$= \int_{\lambda_a}^{\lambda_b} \left(\sum_{j';j' \neq j}^{v_\Phi^j} \left(-\Phi_{Y_j \to Y_{j'}}^{\lambda_{irr}} P_{a_{Y_j}}^{\lambda_{irr}}(t) + \Phi_{Y_{j'} \to Y_j}^{\lambda_{irr}} P_{a_{Y_{j'}}}^{\lambda_{irr}}(t) \right) \right) d\lambda$$

$$+ \sum_{j';j' \neq j}^{u_k^j} \left(-k_{Y_j \to Y_{j'}}^{T} C_{Y_j}^{Lp,\Delta\lambda,T}(t) + k_{Y_{j'} \to Y_j}^{T} C_{Y_{j'}}^{Lp,\Delta\lambda,T}(t) \right) \tag{19.1}$$

The photochemical rates of the individual v_Φ^j photochemical reaction-steps are given by the product of each dimensionless absolute quantum yield ($\Phi_{Y_j \to Y_{j'}}^{\lambda_{irr}}$ or $\Phi_{Y_{j'} \to Y_j}^{\lambda_{irr}}$), and the time-dependent ($P_{a_{Y_{j\ or\ J'}}}^{\lambda_{irr}}(t)$, in eisntein dm^{-3} s^{-1}) amount of light that is specifically absorbed by the photoreactive species starting the targeted photochemical reaction-step. These two quantities must be defined at each of the individual irradiation wavelengths (λ_{irr}) of the *OSIA* $\Delta\lambda$-domain.

The light absorbed by a given species is worked out from its Beer–Lambert law at each λ_{irr}, according to the derivation given in Sect. 2.18, as

$$P_{a_{Y_{j\ or\ j'}}}^{\lambda_{irr}}(t) = \frac{A_{Y_{j\ or\ j'}}^{\lambda_{irr},\ T/\lambda_{irr}}(t)}{A_{tot}^{\lambda_{irr},\ T/\lambda_{irr}}(t)} P_0^{\lambda_{irr}} \left(1 - 10^{-\frac{l_{irr}}{l_{obs}} A_{tot}^{\lambda_{irr},\ T/\lambda_{irr}}(t)}\right)$$

$$= A_{Y_{j\ or\ j'}}^{\lambda_{irr},\ T/\lambda_{irr}}(t)\, P_0^{\lambda_{irr}}\, PKF^{\lambda_{irr}}(t) \tag{19.2}$$

$P_0^{\lambda_{irr}}$ (in einstein s^{-1} dm^{-3}) is the incident number of photons, provided by lamp *Lp* at wavelength λ_{irr}, that enters the reactor and reaches the reactive medium per second and per irradiated area and volume of the investigated sample. The photokinetic factor, defined for each λ_{irr} of $\Delta\lambda$ as, $PKF^{\lambda_{irr}}(t) = \left(1 - 10^{-A_{tot}^{\lambda_{irr},T/\lambda_{irr}}(t)}\right)/A_{tot}^{\lambda_{irr},T/\lambda_{irr}}(t)$, is dimensionless and time-dependent.

Both total ($A_{tot}^{\lambda_{irr},T/\lambda_{irr}}(t)$) and individual ($A_{Y_{j\ or\ j'}}^{\lambda_{irr},T/\lambda_{irr}}(t)$) dimensionless absorbances are measured at the specific irradiation wavelength λ_{irr} that must belong to the absorption *OSIA* wavelength-domain (Eq. 19.3). As stated by the Beer–Lambert law, $A_{tot}^{\lambda_{irr},T/\lambda_{irr}}(t)$ is the sum of the individual absorbances of those species absorbing at λ_{irr} that are present in the reactive medium at time *t* (among the overall set of the

n_{sp} reaction-species). The absorptivity of each species Y_j, $\varepsilon_{Y_j}^{\lambda_{irr}}$, is expressed in mol^{-1} dm^3 cm^{-1}, and the optical path length of the irradiation light, l_{irr}, in *cm*. The experimentally measured total absorbance (across l_{obs}) must be corrected by the ratio l_{irr}/l_{obs} when used in the rate-equation (Eq. 19.2). This satisfies the general case where l_{irr} and l_{obs} are not equal.

$$A_{tot}^{\lambda_{irr},\,T/\lambda_{irr}}(t)\,\frac{l_{irr}}{l_{obs}} = \sum_{j=0}^{n_{sp}} A_{Y_j}^{\lambda_{irr},\,T/\lambda_{irr}}(t)\,\frac{l_{irr}}{l_{obs}} = \sum_{j=0}^{n_{sp}} \varepsilon_{Y_j}^{\lambda_{irr}}\, l_{irr}\, C_{Y_j}^{Lp,\Delta\lambda,T}(t) \tag{19.3}$$

Incidentally, the validity of Eq. 19.1 to describe photothermal reactions requires that Eq. 19.3 is, itself, valid. It is then mandatory that the absorbance data that are allowed to be used for the above formulae must obey the Beer–Lambert law. This means that only the absorbance data belonging to the linearity range of the absorbance calibration graphs of the species is useful (Fig. 2.9) (i.e. beyond the limits of the linearity ranges, Eq. 19.3, and by ricochet, also Eq. 19.1 are corrupt and invalidated). Because this holds for the whole n_{sp} set of species in the reactive medium, whose calibration graphs and/or absorption spectra might not necessarily all be accessible, it has been recommended that the measured total absorbance of the reactive medium (with $l_{obs} = 1$ cm), at any λ_{irr}, should not exceed a value of 0.5, as a safety measure, in order to satisfy the linearity range condition. According to the analytical features of organic materials, an absorbance value equal to or less than 0.5 might well translate for the species of many reactive systems into each individual species' absorbance to fall within its linearity range [1, 2].

The mass balance, Eq. 19.4, is verified at all times for the n_{sp} species involved in the overall reaction.

$$C_{Y_0}^{Lp,\Delta\lambda,T}(0) = C_{X}^{Lp,\Delta\lambda,T}(0) = \sum_{j=0}^{n_{sp}} C_{Y_j}^{Lp,\Delta\lambda,T}(t) = \sum_{j=0}^{n_{sp}} C_{Y_j}^{Lp,\Delta\lambda,T}(\infty) \tag{19.4}$$

A dimension analysis of the products $\left(\Phi_{Y_j\to Y_{j'}}^{\lambda_{irr}}\, P_{aY_j}^{\lambda_{irr}}(t)\right)$ and $\left(k_{Y_j\to Y_{j'}}^{\Delta}\, C_{Y_j}^{Lp,\Delta\lambda,T}(t)\right)$, making up the right-hand side of Eq. 19.1, reveals the same unit (mol dm^{-3} s^{-1}) as that of the rate on the left-hand side of the equation. That dimension coherence is implied by the equivalence of units of thermal and photochemical terms, mol dm^{-3} s^{-1} = einstein s^{-1} dm^{-3}, as an einstein is a mole of photons.

For generalisation purposes, the quantum yields as well as all the kinetic and spectroscopic parameters in Eqs. 19.1 to 19.3 are considered, de facto, to be wavelength-dependent.

Eq. 19.1 applies to both situations where either mono- or polychromatic light is used. For the latter experimental condition, the equation applies as stated above. However, when monochromatic light is employed, the integral and the integration variable ($d\lambda$) are omitted from the right-hand side of Eq. 19.1 corresponding to the photochemical contribution [1], and hence, the concentration and rate are labelled as $C_{Y_j}^{\lambda_{irr},T}(t)$ and $r_{Y_j}^{\lambda_{irr},T}(t)$, respectively. Equation 19.1 reduces to Eq. 11.12 for the basic photoreaction.

A purely thermal rate-equation is deduced from Eq. 19.1 whenever any of its parameter, $\Phi_{Y_j \to Y_{j'}}^{\lambda_{irr}}$ and $\Phi_{Y_{j'} \to Y_j}^{\lambda_{irr}}$, $P_0^{\lambda_{irr}}$ and/or $A_{tot}^{\lambda_{irr},T/\lambda_{irr}}(t)$, has a zero value, resulting in $r_{Y_j}^{\Phi}(t) = 0$, as in the case of a photothermal reaction observed in the dark after having been irradiated in a previous period.

A purely photochemical rate-law is obtained from Eq. 19.1 when the reaction mechanism does not include any thermal reaction-steps, i.e. $r_{Y_j}^{T}(t) = 0$ and all $k_{Y_j \to Y_{j'}}^{T} = 0$ [2].

19.5 A Necessary Approximation

Despite that photokinetically, the general rate-law (Eq. 19.1) describes fully the physical system represented by the photothermal reaction, its mathematical complexity poses a difficulty to solving it analytically. In fact, this is a common situation for integro-differential equations, for which standardised analytical methods only exist for specific types of such equations [7, 8]. The number and, when known, the complicated mathematical functions describing the wavelength-dependent parameters (Φ, ε, P_0) of the photochemical reaction-rate ($r_{Y_j}^{\Phi}(t)$) are an important and extenuating factor in the way of reaching a closed-form solution for Eq. 19.1. Such a particular formulation meant that, thus far, Eq. 19.1 did not even benefit from any published methods offering its evaluation by numerical methods, except for a recent example employing numerical methods and high-order positivity-preserving approximation-schemes for the integro-differential equations of photoreactions, where the time evolutions of the concentrations were assumed to obey first-order kinetics (i.e. the rate does not involve a *PKF*, but dC/dt = $\sum$ k C) [9]. The lack of numerically solving techniques of Eq. 19.1 is a critical gap in photokinetics that certainly deserves to be explored in future research.

To circumvent such a hurdle, it is here assumed that the integral of Eq. 19.1 is equivalent to a summation over the wavelength range at a 1-nm iteration step [1–3, 10]. In practice, this assumption transforms Eq. 19.1 into a formulation (Eq. 19.5) that accepts to be readily evaluated by numerical integration as by using *RK*-NIM. Such an approach, using the Euler numerical method, has been proven satisfactory to replicate experimental data [10, 11]. The *RK*-NIM option is implemented in the following sections to develop the photokinetic description and procedures for photothermal reactions [1–3].

$$r_{Y_j}^{Lp,\Delta\lambda,T}(t) = \sum_{\lambda_{irr}=\lambda_a}^{\lambda_b} \left(\sum_{j';j'\neq j}^{v_\Phi^j} \left(-\Phi_{Y_j\to Y_{j'}}^{\lambda_{irr}} P_{a_{Y_j}}^{\lambda_{irr}}(t) + \Phi_{Y_{j'}\to Y_j}^{\lambda_{irr}} P_{a_{Y_{j'}}}^{\lambda_{irr}}(t) \right) \right)$$

$$+ \sum_{j';j'\neq j}^{u_k^j} \left(-k_{Y_j\to Y_{j'}}^{T} C_{Y_j}^{Lp,\Delta\lambda,T}(t) + k_{Y_{j'}\to Y_j}^{T} C_{Y_{j'}}^{Lp,\Delta\lambda,T}(t) \right) \quad (19.5)$$

The expression of the theoretical initial rate of the reactant, $X = Y_0$, is worked out from Eq. 19.5 as

$$Theo : r_{0,X}^{Lp,\Delta\lambda,T} = -\sum_{\lambda_{irr}=\lambda_a}^{\lambda_b} \left(\sum_{j=1}^{v_k^0} \Phi_{X\to Y_j}^{\lambda_{irr}} P_0^{\lambda_{irr}} \left(1 - 10^{-\frac{l_{irr}}{l_{obs}} A_{tot}^{\lambda_{irr}/\lambda_{irr},T}(0)} \right) \right)$$

$$- C_X^{Lp,\Delta\lambda,T}(0) \sum_{j=1}^{u_k^0} k_{X\to Y_j}^{T} \quad (19.6)$$

For practical reasons, it is also useful to define the equation of the total absorbance that is measured at a given observation wavelength, λ_{obs}, and the observation optical path length, l_{obs} (Eq. 19.7), where the actual values of λ_{obs} and l_{obs} are not necessarily equal to those of their counterparts, λ_{irr} and l_{irr}.

$$A_{tot}^{Lp,\Delta\lambda,T/\lambda_{obs}}(t) = \sum_{j=0}^{n_{sp}} A_{Y_j}^{\lambda_{obs}}(t) = \sum_{j=0}^{n_{sp}} \varepsilon_{Y_j}^{\lambda_{obs}} l_{obs} C_{Y_j}^{Lp,\Delta\lambda,T}(t) \quad (19.7)$$

The corresponding reaction-rate at time t is

$$r_A^{Lp,\Delta\lambda,T}(t) = \left(A_{tot}^{Lp,\Delta\lambda,T/\lambda_{obs}}(t) \right)' = \sum_{i=0}^{n_{sp}} \varepsilon_{Y_j}^{\lambda_{obs}} l_{obs} r_{Y_j}^{Lp,\Delta\lambda,T}(t) \quad (19.8)$$

And the theoretical initial reaction-rate is derived from the latter equation as

$$Theor : r_{0,A}^{Lp,\Delta\lambda,T} = r_A^{Lp,\Delta\lambda,T}(0) = \sum_{i=0}^{v_k^0} \varepsilon_{Y_j}^{\lambda_{obs}} l_{obs} r_{Y_j}^{Lp,\Delta\lambda,T}(0) \quad (19.9)$$

For reactions conducted under monochromatic light excitation, the sum over λ_{irr} is discarded in Eq. 19.6, and the "*Lp*, $\Delta\lambda$" labelling of the parameters of the above equations is replaced by λ_{irr}.

It is perhaps important to mention that the trace of the total absorbance must vary with the conditions of measurement (i.e. $\varepsilon_{Y_j}^{\lambda_{obs}}$ are wavelength-dependent as λ_{obs} varies).

19.6 A Proposed General Integrated Rate-Law

It was established (Chap. 11) that the kinetic traces of the basic primary photoprocess (i.e. whose $\varepsilon_{Y}^{\lambda_{irr}}=0$) subjected to monochromatic light obeyed Φ-order kinetics with its characteristic, analytically derived equation [12]. Furthermore, we have shown that expressions bearing the mathematical character of Φ-order kinetics (obtained by the semi-empirical method (Chap. 15) fit with high accuracy the experimental as well as the *RK*-traces of several photoreactions investigated under monochromatic light [13–15]. The semi-empirical method was employed for the latter photoreactions because their rate-laws cannot be solved by closed-form integrations.

Qualitatively, the literature data strongly point to a substantial similarity of the overall shape of experimental and *RK*-calculated traces obtained for both photo- and some photothermal reactions subjected to either mono- or polychromatic irradiation [10, 11, 16–21].

Since neither the integro-differential rate-equation Eq. 19.1 (or Eq. 19.5) has an analytical solution nor does literature provide an explicit equation for the kinetics traces of photothermal reactions, the above observations can be used to put forward a conjecture. That is, the photochemical kinetics of photoreactions may well have a Φ-order character. If such a suggestion is valid, then the integrated rate-law of photothermal reactions would encompass both the Φ-order kinetic type relative to the photochemical processes (expressed by *log-exp* functions, similar to Eq. 11.41), and the first-order kinetics of the thermal reaction-steps (as exponential functions). This will supposedly hold for the kinetics of any species Y_j of any photothermal reaction, irrespective of the reaction mechanism in play and/or the nature of the irradiation light used (either monochromatic at λ_{irr}, or polychromatic over *OSIA*).

The explicit unified global integrated rate-equation (coined here as *GIREC*(*j*)) describing the timely evolution of a species concentration takes the following general form.

$$C_{Y_j}^{Lp,\Delta\lambda,T}(t)=\omega_j^{\infty}+\sum_{i=1}^{i_{\Phi,j}}\omega_{ij}^{\Phi}\,Log\left(1+cc_j^{\Phi}e^{-k_{ij}^{\Phi}\,t}\right)+\sum_{i=1}^{i_{\Delta,j}}\omega_{ij}^{T}\,e^{-k_{ij}^{T}\,t} \tag{19.10}$$

The coefficients of Eq. 19.10 (ω_j^{∞}, ω_{ij}^{Φ}, ω_{ij}^{T}, cc_j^{Φ}, k_{ij}^{Φ} and k_{ij}^{T}), which are constant for a given reaction, are extracted from the fitting of that equation to the trace corresponding to the considered species Y_j. *Log* and *e* are the base 10 logarithm and the exponential functions, respectively. The parameters $i_{\Phi,\ j}$ and $i_{\Delta,\ j}$ are, respectively, the numbers of mono-Φ-order ($\omega\ Log(1\ +\ cc\ e^{-k\ t})$) and first-order ($\omega\ e^{-k\ t}$) terms that will be included in Eq. 19.10 of the considered species Y_j. These numbers will, at most, be respectively equal to the numbers of photochemical, $v_{\Phi}^{j}\geq i_{\Phi,j}$, and thermal, $u_{k}^{j}\geq i_{\Delta,j}$, reaction-steps starting or ending at the considered species Y_j.

The differentiation of Eq. 19.10 gives a formula for the rate of species Y_j from the fitting data (Eq. 19.11) at reaction time *t*. The meaning of this expression is equivalent to that of the theoretical Eq. 19.5 (with, in Eq. 19.11, *ln* being the natural logarithm).

$$Fit: r_{Y_j}^{Lp,\Delta\lambda,T}(t) = -\frac{1}{ln(10)}\sum_{i=1}^{i_{\Phi,j}} \frac{\omega_{ij}^{\Phi}\, cc_j^{\Phi}\, k_{ij}^{\Phi}\, e^{-k_{ij}^{\Phi}\, t}}{1 + cc_j^{\Phi} e^{-k_{ij}^{\Phi}\, t}} - \sum_{i=1}^{i_{\Delta,j}} \omega_{ij}^{T}\, k_{ij}^{T}\, e^{-k_{ij}^{\Delta}\, t} \tag{19.11}$$

For quantification purposes, the expression for the reactant initial rate, $Fit: r_{0,X}^{Lp,\Delta\lambda,T}$, is derived from Eq. 19.11, as

$$Fit: r_{0,X}^{Lp,\Delta\lambda,T} = -\frac{1}{ln(10)}\sum_{i=1}^{i_{\Phi,0}} \frac{\omega_{i0}^{\Phi}\, cc_0^{\Phi}\, k_{i0}^{\Phi}}{1 + cc_0^{\Phi}} - \sum_{i=1}^{i_{\Delta,0}} \omega_{i0}^{T}\, k_{i0}^{T} \tag{19.12}$$

A further important equation is the one describing the total absorbance traces of the reactive medium (dubbed *GIREA*).

$$A_{tot}^{Lp,\Delta\lambda,T/\lambda_{obs}}(t) = \omega_A^{\infty} + \sum_{i=1}^{i_{\Phi,A}} \omega_{i,A}^{\Phi}\, Log(1 + cc_A^{\Phi}\, e^{-k_{iA}^{\Phi}\, t}) + \sum_{i=1}^{i_{\Delta,A}} \omega_{i,A}^{T}\, e^{-k_{iA}^{T}\, t} \tag{19.13}$$

In analogy with Eq. 19.10, the coefficients of the latter equation are constants, with $i_{\Phi,\ A}$ and $i_{\Delta,\ A}$, respectively, at most, equal to the total number of the photochemical (v) and thermal (u) reaction-steps occurring in the overall mechanism of the investigated reaction. However, $i_{\Phi,\ A}$ and $i_{\Delta,\ A}$ might well be less than v and u, respectively, depending on the shape of the total absorbance trace (provided that an excellent fit of that trace is achieved). Also, as $A_{tot}^{Lp,\Delta\lambda,T/\lambda_{obs}}$ is depending on which of these combinations (i) $\lambda_{irr} = \lambda_{obs}$ and $l_{irr} = l_{obs}$, (ii) $\lambda_{irr} = \lambda_{obs}$ and $l_{irr} \neq l_{obs}$, (iii) $\lambda_{irr} \neq \lambda_{obs}$ and $l_{irr} = l_{obs}$, (iv) $\lambda_{irr} \neq \lambda_{obs}$ and $l_{irr} \neq l_{obs}$, is considered, the properties of the total absorbance traces will change, and consequently, the values of the fitting parameters of Eq. 19.13 will also change for the investigated system.

The expression of the rate of the $A_{tot}^{Lp,\Delta\lambda,T/\lambda_{obs}}(t)$ trace is obtained from the differentiation of Eq. 19.13, as

$$Fit: r_A^{Lp,\Delta\lambda,T/\lambda_{obs}}(t) = \left(A_{tot}^{Lp,\Delta\lambda,T/\lambda_{obs}}(t)\right)'_{t=0}$$

$$= -\frac{1}{ln(10)}\sum_{i=1}^{i_{\Phi,A}} \frac{\omega_{i,A}^{\Phi}\, cc_A^{\Phi}\, k_{iA}^{\Phi}\, e^{-k_{iA}^{\Phi}\, t}}{1 + cc_A^{\Phi}\, e^{-k_{iA}^{\Phi}\, t}} - \sum_{i=1}^{i_{\Delta,A}} \omega_{i,A}^{T}\, k_{iA}^{T}\, e^{-k_{iA}^{T}\, t} \tag{19.14}$$

From which the initial reaction-rate is derived, as

$$Fit : r_{0,A}^{Lp,\Delta\lambda,T/\lambda_{obs}} = -\frac{1}{\ln(10)}\sum_{i=1}^{i_{\Phi,A}} \frac{\omega_{i,A}^{\Phi}\, cc_{A}^{\Phi}\, k_{iA}^{\Phi}}{(1+cc_{A}^{\Phi})\ln(10)} - \sum_{i=1}^{i_{\Delta,A}} \omega_{i,A}^{T}\, k_{iA}^{T} \tag{19.15}$$

The same equations (Eqs. 19.10 to 19.15) apply when the reaction is exposed to monochromatic light, with λ_{irr} replaces "$Lp,\Delta\lambda$" in the exponents of the various parameters. The formulae relative to pure thermal reactions are derived from the above equations wherever the pre-logarithmic coefficients are set to zero ($\omega_{ij}^{\Phi} = \omega_{i,A}^{\Phi} = 0$).

The expressions for the individual initial rates of the reactant and its immediate photo- and thermal products cannot be obtained from the fitting value of $Fit : r_{0,A}^{Lp,\Delta\lambda,T/\lambda_{obs}}$ (Eq. 19.15) because the theoretical equation of that initial-rate (Eq. 19.9) is a combination of those species initial rates. Nonetheless, r_{0A} has a valuable advantage; that is, its determination and subsequent use for quantification of the reaction kinetics does not require the experimentalist to obtain the individual species concentration, $C_{Y_j}^{\lambda_{irr}}(t)$, traces ($A_{tot}^{Lp,\Delta\lambda,T/\lambda_{obs}}(t)$, and $A_{tot}^{\lambda_{irr},T/\lambda_{obs}}(t)$ can be acquired on a routine spectrophotometer, whether or not performed using an in-situ irradiation).

The handy explicit model equations (Eqs. 19.10 and 19.13), which stand for the integrated rate-laws of photothermal reactions, will be exploited in subsequent chapters for photokinetic analysis and quantification. But first, these equations have to be validated, as discussed in the following section.

19.7 Validation of the Model Equations (GIREC(j) and GIREA)

The validation of both Eqs. 19.10 and 19.13 (*GIREC(j)* and *GIREA*) is performed by testing their performance to fit *RK*-calculated traces. For generality purposes, both concentrations of individual species and total absorbance *RK*-traces were generated for a variety of reaction mechanisms, reaction conditions and intrinsic parameter values, including variation over wavelength of both species' quantum yield patterns and/or values and absorption coefficient for each species, the lamp profile/intensity, and the irradiation and observation optical path lengths. Also, the application of the model equations concerned photo- and photothermal reactions subjected to both mono- and polychromatic light.

The trace-fitting procedure was implemented, in each case, according to simple protocols. One way is to first fit the concentration trace that shows most inflection points indicating a major number of kinetic regimes. The obtained fitting parameters' values of the exponential factors (k_{ij}^{Φ} and k_{ij}^{T}, or k_{iA}^{Φ} and k_{iA}^{T}) are used as such and incorporated in the different equations fitting the remaining traces, which might have fewer terms (showing fewer kinetic regimes). Another way starts by fitting the thermal reaction traces and then the photothermal reaction traces. Alternatively, a

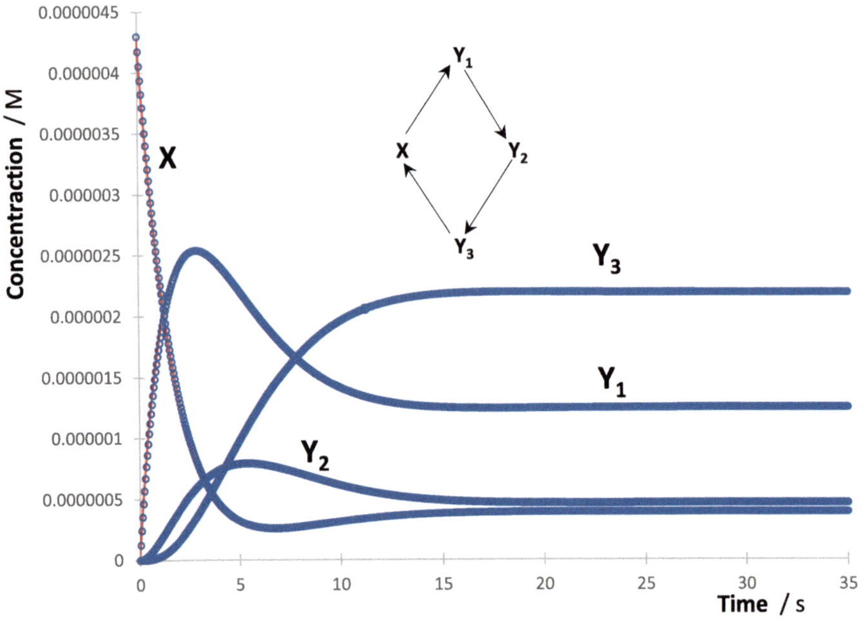

Fig. 19.2 An excellent fit of the concentration *RK*-generated traces of an $XY_3(4\Phi)$ cyclic reaction (circles) by their corresponding Eq. 19.10 (lines). The mechanism of the reaction is indicated. (Reproduced with permission from [1], *Frontiers in Chemistry*)

fitting option is to proceed by fitting the reactant's trace, then the trace of the adjacent product in the reaction mechanism, followed by the trace of the next nearest Y_j species, and so forth. The exponential fitting parameters of an earlier species (e.g. $k_{10}^{\lambda_{irr}}$ and $k_{20}^{\lambda_{irr}}$ for X and $j = 0$) are used in the subsequent species equations, which might involve new terms (e.g. $k_{11}^{\lambda_{irr}} = k_{10}^{\lambda_{irr}}$, $k_{21}^{\lambda_{irr}} = k_{20}^{\lambda_{irr}}$, and $k_{31}^{\lambda_{irr}}$ for species Y_1). The process is repeated until the last species trace is fitted. In both ways, the features of a given species' trace are dependent on the rate-constants of those of previous reaction steps generating that species. Accordingly, the numbers $i_{\Phi,j}$ and $i_{\Delta,j}$ of mono-Φ-order and thermal first-order terms in Eq. 19.10 is minimal for the reactant and would increase (or at least remain the same) for the subsequent species of the mechanism. In principle, the maximum number of terms in Eq. 19.10 should be equal to $n_{rs}^j = i_{\Delta,j} + i_{\Phi,j}$, the number of reaction-steps starting or ending at the considered Y_j species. However, in some cases and depending on the trace-shape, Eq. 19.10 may encompass fewer terms than n_{rs}^j but still provide a good fit of the trace. Interestingly, the equation (Eq. 19.10) of each species belonging to a cyclic reaction mechanism involving three or more molecules is most likely to have the same number of terms (e.g. Fig. 19.2). In practice, it has also been observed that the value of the factor, cc_j^{Φ}, remains invariant ($cc_j^{\Phi} = cc^{\Phi}$) for all Eq. 19.10 describing each of the species occurring in a given reaction mechanism (though, this should not be considered as a fitting condition). The equation parameter ω_j^{∞} (or ω_A^{∞}) takes the value of the final concentration of the considered species Y_j measured at $t = \infty$, $C_{Y_j}^{Lp,\Delta\lambda,T}(\infty)$ (or the final total absorbance, $\omega_A^{\infty} = A_{tot}^{Lp,\Delta\lambda,T/\lambda_{obs}}(\infty)$), and the $\omega_{ij}^{\lambda_{irr}}$ coefficients have, generally, different values for different traces of the n_{sp} species of the investigated mechanism.

Over 500 kinetic traces were *RK*-simulated to constitute a large enough sample to be representative of photothermal reactions. Excellent fittings of both concentration and total absorbance traces were achieved by the model equations, as indicated by high correlation coefficient values (> 0.995) for *RK*-simulated vs. Eqs. 19.10 and 19.13 data, relatively low sums of squared errors and root-mean-square errors values ($SSE < 10^{-10}$, and $RMSE < 10^{-9}$ for Eq. 19.10 and a little higher, $RMSE < 10^{-5}$, for Eq. 19.13).

Two illustration examples are shown hereafter. The first one relates the data of a cyclic photoreaction involving four species interlinked by four photochemical steps, $XY_3(4\Phi)$ (Fig. 19.2). The reaction is investigated under monochromatic light irradiation, with the latter being absorbed by the four reaction species. The quantum yields of the four transitions have different values at λ_{irr}. The reactive medium was maintained at constant temperature. The timely evolutions of the species concentrations are shown in Fig. 19.2.

The individual *RK*-generated traces of the species have been well fitted by Eq. 19.10 encompassing each four mono-Φ-order terms ($i_{\Phi,\,j} = n_{\mathrm{rs}} = 4$), possessed no thermal kinetic terms ($\omega_{ij}^{T} = 0$), shared a unique set of four k_{ij}^{Φ} values, and had a single common value for $cc^{\lambda_{irr}}$ (Table 19.1).

The second example corresponds to a tetramolecular photothermal reaction involving six photochemical and five thermal reaction-steps, $XY_3(6\Phi, 5k)$ (see reaction mechanism in the inset of Fig. 19.2a). The reaction medium is continuously exposed to a polychromatic light that variously overlaps the absorption spectra of the four species (see lamp profile and ε-spectra on Fig. 19.3a). The four ε-spectra overlap over the *UV* (310 – 370 *nm*) region, whereas only the photoproducts absorb beyond 370 *nm*. The *OSIA* of the reactant equals the full width of the wavelength span of the lamp emission (i.e. ε_X-spectrum is completely overlapped by the lamp profile). Hence, under irradiation, this reactive system reaches a photostationary state (because, in the equilibrated mechanism, the four species concomitantly absorb the incident light). In the dark, the thermal reaction ultimately regenerates the thermally stable reactant.

Four out of the six photochemical quantum yields are wavelength-dependent (Fig. 19.3b). These have increased ($\Phi_{X\to Y_1}^{\lambda_{irr}}$) and decreased ($\Phi_{Y_1\to Y_3}^{\lambda_{irr}}$, $\Phi_{Y_2\to Y_1}^{\lambda_{irr}}$ and $\Phi_{Y_3\to Y_1}^{\lambda_{irr}}$) trends over the *UV*-region. The absolute photochemical quantum yield values for the six phototransformations are situated within the 0.027–0.005 range.

Excellent fittings of each individual concentration trace were achieved with high precision (Fig. 19.3c and Table 19.1). The four model equations emerging from the fitting process (with Eq. 19.10) possessed the same template of two mono-Φ-order ($i_{\Phi,\,j} = 2$) and one first-order thermal ($i_{\Delta,\,j} = 1$) term. For this kinetics, the number of terms in the equations is much less than the number of reaction steps $i_{\Phi,\,j} + i_{\Delta,\,j} = 3 < n_{\mathrm{rs}} = 11$. It is also an illustration example for a quasi-total dissimilarity

Table 19.1 Fitting parameter sets of Eqs. 19.10 characterising the concentration kinetic trace parameter values for the photochemical $XY_3(4\Phi)$ and photothermal $XY(5k, 6\Phi)$ reactions as depicted in, respectively, Figs. 19.2 and 19.3c

	Reaction $XY_3(4\Phi)$				Reaction $XY_3(5k, 6\Phi)$			
	$C_{Y_0}(t)$[a]	$C_{Y_1}(t)$	$C_{Y_2}(t)$	$C_{Y_3}(t)$	$C_{Y_0}(t)$	$C_{Y_1}(t)$	$C_{Y_2}(t)$	$C_{Y_3}(t)$
ω_j^{∞}	$3.91\ 10^{-7}$	$1.24\ 10^{-6}$	$4.67\ 10^{-7}$	$2.19\ 10^{-6}$	$4.26\ 10^{-6}$	$7.39\ 10^{-6}$	$1.12\ 10^{-5}$	$1.30\ 10^{-5}$
ω_{1j}^{Φ}	$1.98\ 10^{-5}$	$-4.14\ 10^{-5}$	$3.69\ 10^{-5}$	$1.51\ 10^{-5}$	−0.293	$1.49\ 10^{-4}$	0.8017	0.987
ω_{2j}^{Φ}	$1.01\ 10^{-3}$	$-1.91\ 10^{-3}$	$-1.39\ 10^{-3}$	$2.29\ 10^{-3}$	0.4244	$-5.01\ 10^{-4}$	1.036	0.3964
ω_{3j}^{Φ}	$-1.25\ 10^{-3}$	$2.61\ 10^{-3}$	$1.72\ 10^{-3}$	$-3.09\ 10^{-3}$	–	–	–	–
ω_{4j}^{Φ}	$2.27\ 10^{-4}$	$-6.81\ 10^{-4}$	$-3.69\ 10^{-4}$	$7.79\ 10^{-4}$	–	–	–	–
cc_j^{Φ}	0.1679	0.1679	0.1679	0.1679	$4.26\ 10^{-6}$	0.1632	$5.37\ 10^{-6}$	$-8.25\ 10^{-4}$
k_{1j}^{Φ}	0.8265	0.8265	0.8265	0.8265	0.0668	0.04866	0.04983	0.02614
k_{2j}^{Φ}	0.3405	0.3405	0.3405	0.3405	0.0162	0.04461	0.04983	0.03177
k_{3j}^{Φ}	0.3182	0.3182	0.3182	0.3182	–	–	–	–
k_{4j}^{Φ}	0.2771	0.2771	0.2771	0.2771	–	–	–	–
ω_{1j}^{T}	–	–	–	–	$2.35\ 10^{-5}$	$1.58\ 10^{-5}$	$-1.54\ 10^{-5}$	$4.55\ 10^{-4}$
k_{1j}^{T}	–	–	–	–	0.06217	0.01594	0.01275	0.02862
r^2	0.9998	0.9999	0.9999	0.9998	0.9999	0.9999	0.9999	0.9999
SSE	$3.21\ 10^{-14}$	$2.83\ 10^{-13}$	$6.35\ 10^{-15}$	$4.97\ 10^{-14}$	$3.31\ 10^{-15}$	$1.58\ 10^{-15}$	$1.14\ 10^{-13}$	$2.33\ 10^{-14}$
RMSE	$2.84\ 10^{-9}$	$1.63\ 10^{-9}$	$5.14\ 10^{-8}$	$8.44\ 10^{-9}$	$1.63\ 10^{-9}$	$2.60\ 10^{-9}$	$2.21\ 10^{-8}$	$3.16\ 10^{-8}$

[a] The concentrations are expressed in mol dm^3

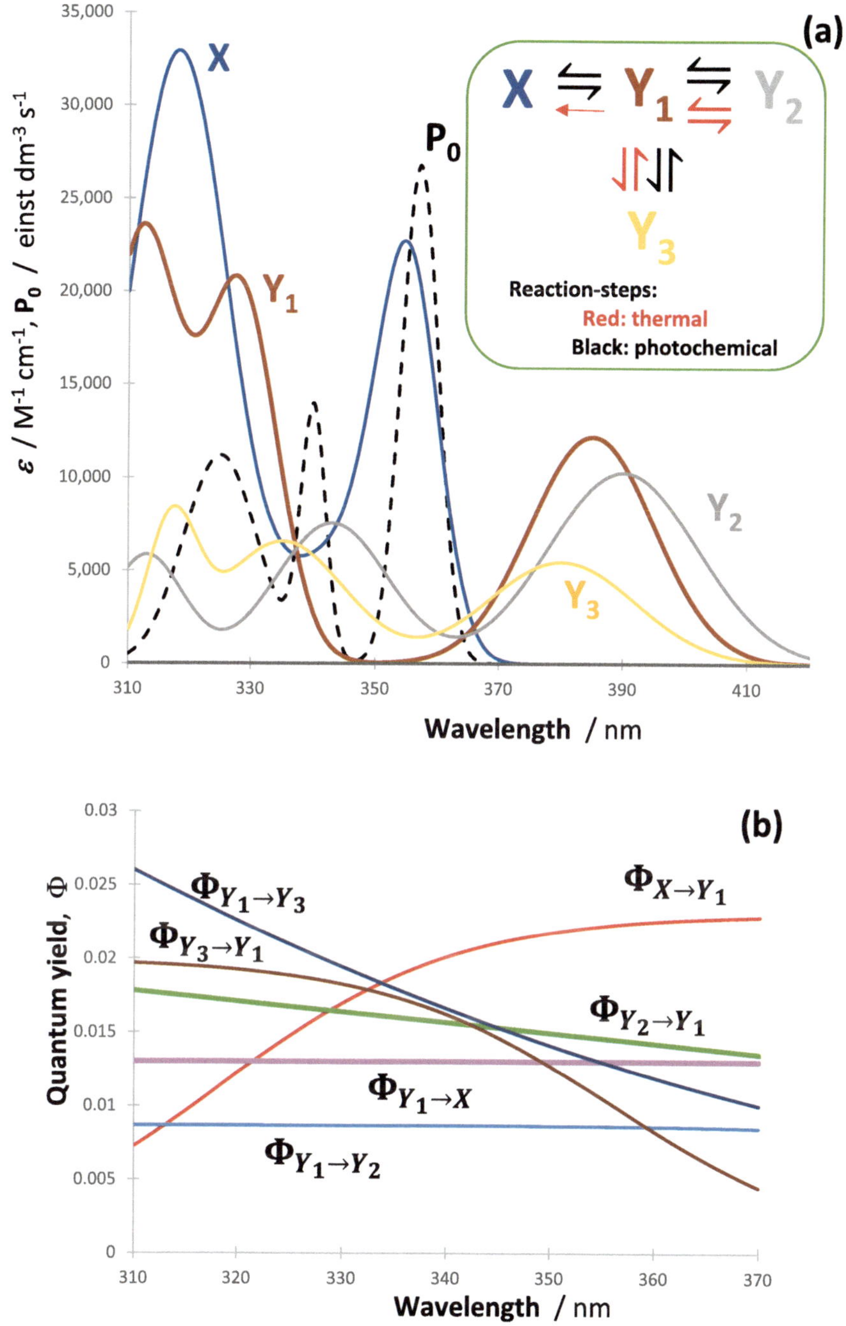

Fig. 19.3 (**a**) Lamp profile (P_0) and absorptivity spectra (ε_{Y_j}) of the four reactive species of an $XY_3(6\Phi, 5k)$ photothermal reaction (whose mechanism is shown in the inset). (**b**) Wavelength-dependent patterns of the quantum yields ($\Phi_{Y_j \to Y_j'}$) of the six photochemical reaction-steps. (**c**) *RK*-generated concentration traces of the four species (circles) and their fittings (red lines) by the corresponding Eq. 19.10. (Reproduced with permission from [3], *Molecules*)

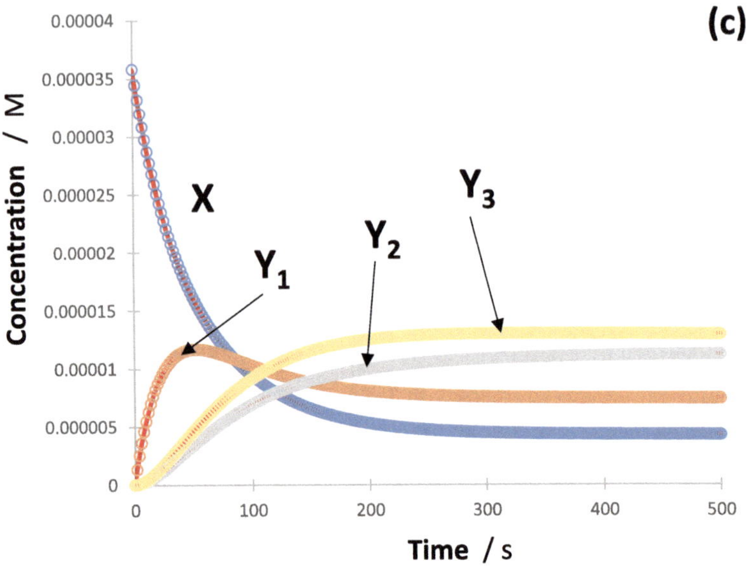

Fig. 19.3 (continued)

between the sets of parameters' values obtained for each trace (Table 19.1). Nonetheless, the precision criteria were good in all cases, as shown by the fitting curves (Fig. 19.3c). Also, the sum of the species concentrations reached at the photostationary state (parameters ω_j^{∞} in Table 19.1) is equal to the initial reactant concentration (3.58 10^{-5} M).

The overall findings point to several important conclusions. It is clear that the explicit model equations (Eqs. 19.10 and 19.13) satisfy the photokinetic behaviour of photo- and photothermal reactions. In a sense, this stands for a confirmation of our conjecture, indicating that the Φ-order equation is the seminal model for photoreactions in spite of the fact that it has only been analytically derived for the primary photoprocess whose photoproduct is transparent to the monochromatic radiation driving the reaction [12]. Such a situation might be seen as analogous to the one observed for thermal reactions, for which the mono-exponential function of the first-order kinetics, characterising the kinetics of the simplest thermal unimolecular reaction, is the basis for a wider description of thermal kinetic data generated for more complex reactions. Incidentally, this is also supported by the typical Φ-order equations that were set out by the semi-empirical method (Chap. 15), for the concentration traces of several photoreactions [13–15, 22].

The two model equations (Eqs. 19.10 and 19.13) stand, therefore, for the first example of general, explicit formulae for photokinetic data that have ever existed in the wider photochemistry literature. Their inclusiveness comes from their applicability, irrespective of the photo- or photothermal reaction mechanism at hand, the nature (mono- or polychromatic) of the irradiation light, the profile of the employed lamp (or light source), and the dependency of the quantum yields on the wavelength. The equations also strongly suggest that the Φ-order character is invariably preserved regardless of the nature of the light shone on the sample and whether or not concomitant thermal processes are active in the reaction (confirming a distinctability issue).

The generality of the explicit equation Eq. 19.10 for all mechanisms means that distinguishability problems are unavoidable.

Retrospectively, and from both practical and quantification viewpoints, as it will become clear in subsequent chapters, and for simpler kinetic analyses, it is relevant to observe that the application of the model equations is preferably used with prior knowledge of the mechanism governing the reaction investigated (see Chap. 20).

The pertinence of the model equations, Eqs. 19.10 and 19.13, might be challenged either because they have not been derived analytically or because there might be a possibility of fitting the traces of a given reaction by using other equations (for instance, by using a series of exponential functions). A change of Φ-order terms by an exponential function is unlikely to reach similar fitting performance and/or to keep the kinetic meaning of the fitting results. The inefficiency of such an interchange of equations might not always work, as has been proven for the unimolecular reaction (Chap. 11, Fig. 11.5 [12]). Also, and in principle, a justification for using Φ-order terms instead of exponential functions stems from the evident difference between the mathematical structures of the standard rate-laws describing photoreactions and thermal reactions (Sects. 3.5 and 3.6). Replacing the model equations by other formulae might work for a set of reactions, but overall, this approach remains limited and cannot be generalised to all photothermal reactions. In any case, unless Eq. 19.1 is solved analytically, introducing any other expression to replace the model equations will likewise lack the precision of a closed-form equation. For the time being, the model equations proposed here remain the only example in the literature that benefit from a proven generality of application in the field of photokinetics.

Looking forward, the conjecture is to stipulate an extension of applicability of the model equations and their validity for the description of photokinetic data irrespective of the spatial distribution of the incident light on the sample and/or the reactor geometry, as usually found in engineering and real-life setups.

Furthermore, an attempt to express the factors of Eqs. 19.10 and 19.13 by explicit equations remains a challenging prospect to consolidate these models.

19.8 Identifiability and Distinguishability Issues

Despite the success of the model equation Eq. 19.10 (and Eq. 19.13) in fitting photothermal reaction traces, it is important to consider an aspect relevant to any practical photokinetic investigation. That is, a good fit of the traces is a necessary condition of the kinetic treatment, but that does not shed sufficient light on many features of the reaction (Chaps. 6 and 9). We will, here, highlight two essential limitations.

The generality of the model equations meant that they equally apply to reactions of very different mechanisms. The handiness and ability of the model equations rather pose a problem when it comes to gaining information about the reaction mechanism. On one hand, the evidence collected from the fitting of reaction traces only offers minimal knowledge on the mechanistic option governing the reaction. For instance, the number of mono-Φ-order and mono-exponential terms in the fitting Eq. 19.10 is considered, a priori, as the minimum number for, respectively, photochemical and thermal reaction-steps. In the case of $XY_3(5k, 6\Phi)$ studied in the previous section, the fitting of the traces led to two mono-Φ-order and one mono-exponential terms for Eqs. 19.10 (Table 19.1), whereas the true mechanism effectively includes 11 reaction-steps. So, in general and as a part of this minimal indication, the fitting information remains mostly blind to the minute mechanistic details of the reaction. On the other hand, the features of the fitting equations obtained for a given reaction (i.e. number of its individual terms) can easily be applied to fit the data of another, completely different reaction. For example, the three terms of the equation found for the $XY_3(5k, 6\Phi)$ can well describe the kinetics of an $XY_2(1k, 2\Phi)$ reaction. Therefore, there is an unsolvable distinguishability problem when using the model equations.

The second limitation relates to the values of individual fitting parameters of the model equations (ω_{ij}^{Φ}, k_{ij}^{Φ}, cc_j^{Φ}, ω_{ij}^{T}, and k_{ij}^{T} for Eq. 19.10). These are hugely dependent on the initial values adopted for these parameters, from where the fitting iteration starts. As a matter of fact, different sets of parameters' values for the trace ($\omega_{ij}^{\Phi}, k_{ij}^{\Phi}, cc_j^{\Phi}, \omega_{ij}^{T}$, and k_{ij}^{T}) parameters can lead to an equal goodness of fit of the trace at hand.

The total number of such sets might not be infinite, but it is certainly large enough not to be discarded as a simple variation generated by the fitting algorithm. The diversity of sets of parameters for a given trace means that the extraction of the "*true*" set of the kinetic parameters for that trace is impossible to single out. Hence, while one is certain that the formulation taken by the fitting Eq. 19.10 (or Eq. 19.13) is correct, an identification of the set of parameters that exactly corresponds to the physical system studied is, however, out of reach. This is a typical case of an identifiability issue (for the fitting of a trace).

The occurrence of unidentifiability impedes the usage of the rate-constant (k_{ij}-factors in the model equations) to be used for the quantification of either a species' photoreactivity or thermal evolution, i.e. the true value of a rate-constant assigned to a given reaction-step, is certainly not identifiable from the actual fitting by either or both Eqs. 19.10 and 19.13 (except for very simple mechanisms). The unavoidable occurrence of such identifiability issues recommends both an extra care in the treatment of the kinetic data and the development of new ways to solve them.

Incidentally, this identifiability issue cannot be resolved for the case where monochromatic irradiation is used by simply considering a combination of a set of traces that would be obtained at a number of different irradiation wavelengths for the same reaction, as might have been previously claimed [23–25]. The reason for the inefficiency of that approach is an increase in the number of unknown parameters, proportionally with an increasing number of traces (Sect. 10.6.5). The seeming success of that approach came at a cost: the procedure solving for the reaction parameters must imperatively consider the quantum yields of each of the individual photoreaction-step as invariants with λ_{irr} (i.e. for any λ_{irr}, $\Phi_{Y_j}^{\lambda_{irr}} = \Phi_{Y_j}$). A condition that applies to some molecules, but which, in any case and in accordance with a large body of evidence (Chaps. 4 and 18), must crucially be proven experimentally and not just assumed (the description of the latter approach has, unfortunately, neither proven experimentally such a λ_{irr}-independence of the photoreaction quantum yields nor has it recommended it before envisaging the application of the approach).

A previous study [26] introduced detailed spectrokinetic methods aiming at solving an identifiability problem in relation to photoreversible systems. The limits of such an approach were critically shown by the treatment of the multi-trace data corresponding to the exposure of the photoreactive system to irradiation beams of different monochromatic wavelengths.

19.9 A Kinetic Metric Tool

We have seen in Sect. 2.4 the kinetic parameters that fully describe a photothermal reaction are of two types: extrinsic and intrinsic. The former coefficients, representing the experimental conditions, such as $C_X(0)$, $P_0^{\lambda_{irr}}$, l_{irr}, and l_{obs}, are selected by the experimentalist and are assumed known. The latter type of coefficients are specific features of each of the different species involved in the photothermal reaction at hand, i.e. $\varepsilon_{Y_j}^{\lambda_{irr}}$, $\Phi_{Y_j \to Y_{j'}}^{\lambda_{irr}}$ and $k_{Y_j \to Y_{j'}}^{T}$ (with $j \neq j'$), which should be delivered by the kinetic investigation.

Unfortunately, a general approach is compromised by the non-solvability of the identifiability problem of the trace-fitting (see previous section) itself due to the non-specificity of Eqs. 19.10 and 19.13 coefficients.

Therefore, an alternative strategy needs to be built up. One of the elements of this strategy is to define a metric, such as the reaction rate-constants (k), that is generally used in thermal kinetics.

In this context, the initial reactant-rate (measured on the reactant trace) and/or that of the reaction (measured on the total absorbance trace) represents an appropriate means to quantify the variability of the reaction speed. In fact, not only is this technique widely employed and tested in thermal kinetics, but also the initial-rate value is the only objective reaction-parameter value that is directly accessible from either the fitting of the traces ($Fit : r_{0,X}^{\lambda_{irr}}$ or $Fit : r_{0,A}^{\lambda_{irr}}$) or from the *NIM* calculation ($RK : r_{0,X}^{\lambda_{irr}}$ or $RK : r_{0,A}^{\lambda_{irr}}$). Conversely, the rate-constants of the different reaction regimes are not possibly calculated by *NIM*s and not accessible to the experimentalist from either the fitting of the traces or from the *NIM*s-fed parameters (because of identifiability issues), as the explicit formulae of the coefficients of both Eqs. 19.10 and 19.13, linking the reaction parameters in the fashion of Eq. 15.1, are unknown.

It is also important to underline here that the choice of r_0 is justified by the fact that the values of $Fit : r_0$ (Eqs. 19.12 and 19.15, respectively) will not vary due to the occurrence of the identifiability issue relative to the trace fitting, i.e. a single value of Fit : $r_{0,X}^{\lambda_{irr}}$ (or $Fit : r_{0,A}^{\lambda_{irr}}$) is obtained whatever the set of fitting parameters used (among those possibly generated for the fitting of the corresponding *RK*-trace), provided that each of the parameter sets, when inputted in Eq. 19.10 (or Eq. 19.13) produces an excellent fit of the *RK*-trace at hand. Incidentally, the initial reactant-rate is the highest value of the reactant initial rates recorded at any other times (e.g. $r_{0,X} \geq r_X(t)$). The same considerations apply for $r_{0,X}^{Lp,\Delta\lambda,T}$ and $r_{0,A}^{Lp,\Delta\lambda,T}$, obtained on reactant concentration or total absorbance traces when the reactive medium is exposed to polychromatic light.

The initial rate is an essential kinetic metric tool because it is proportional to the set of reaction parameters that contribute to the depletion of the reactant (Eq. 19.6), and hence, it is indicative of the reaction performance. It, also, informs on the reactivity of the rest of the species. For photoreactions, a speed-up (or a slowdown) recorded for the reactant photoconversion (for instance, by a change of an extrinsic parameter value) is expected to be followed by a similar reactivity trend for the other species in relative proportions, i.e. the variation of $r_{0,X}^{Lp,\Delta\lambda}$ in one direction is indicative of the variation of every $r_{0,Y_j}^{Lp,\Delta\lambda}$ in the same direction. Such an overview holds in general but must be considered, for given reaction conditions, in the context of Φ-order kinetics specificity, as described in Sect. 11.5.

The reliability of the metric must be confirmed, in all circumstances, by the following equalities.

$$Theo : r_{0,Y_j}^{Lp,\Delta\lambda,T} = RK : r_{0,Y_j}^{Lp,\Delta\lambda,T} = Fit : r_{0,Y_j}^{Lp,\Delta\lambda,T} \tag{19.16}$$

and

$$Theo : r_{0,A}^{Lp,\Delta\lambda,T} = RK : r_{0,A}^{Lp,\Delta\lambda,T} = Fit : r_{0,A}^{Lp,\Delta\lambda,T} \tag{19.17}$$

These conditions also hold for monochromatic irradiation, where λ_{irr} is used instead of "$Lp,\Delta\lambda$" in the labelling of the initial rates in Eqs. 19.16 and 19.17.

A confirmation of the validity of the above methodology is given by the linear plot (Fig. 19.4) obtained for the $Theo : r_0$ against both $RK : r_0$ and $Fit : r_0$ that are determined for photo- and photothermal reactions of different mechanistic sequences and various intrinsic and extrinsic parameter values, considered under mono- and polychromatic irradiations [1–3].

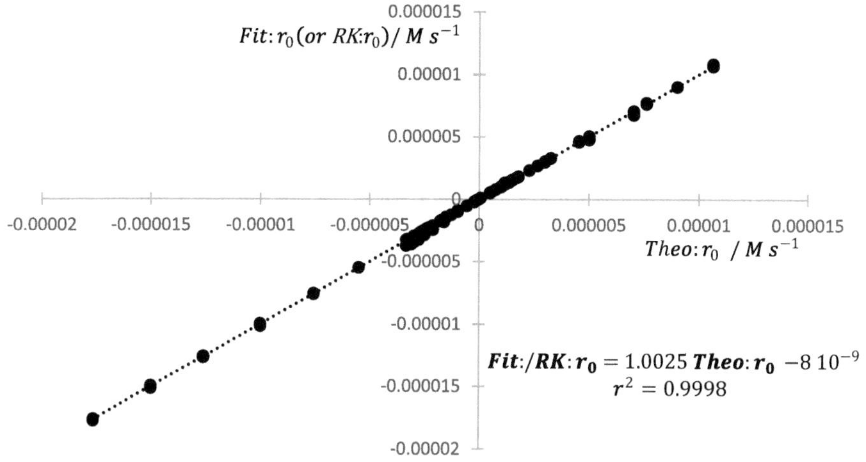

Fig. 19.4 An example of a linear correlation between the values of *Theo* : r_{0,Y_j} against both *RK* : r_{0,Y_j} and *Fit* : r_{0,Y_j}. (Reproduced with permission from [1], *Frontiers in Chemistry*)

19.10 Conclusion

The explicit general equation (Eq. 19.10) is shown to describe the concentration traces of any photothermal reaction, irrespective of whether the light used is monochromatic or polychromatic. It represents an important breakthrough that opens a wide field of characterisation and quantification of photokinetics for both photo- and photothermal reactions. The following chapters convey examples of the rationalisation of several kinetic aspects.

The method combining *RK*-data and explicit equations has proven to be of interest in the present chapter, as this basic idea was also successful for the construction of the semi-empirical method (Chaps. 14 and 15). The exploitation of data generated by numerical integration methods may contribute to further enlarging the scope of standardisation of photokinetics on a mathematically coherent basis.

References

1. Maafi M (2023) On photokinetics under monochromatic light. Front Chem 11:1233151. https://doi.org/10.3389/fchem.2023.1233151
2. Maafi M (2024) On photokinetics under polychromatic light. Front Chem 12:1233151. https://doi.org/10.3389/fchem.2024.1367276
3. Maafi M (2025) Photokinetics of Photothermal reactions. Molecules 30:330. https://doi.org/10.3390/molecules30020330
4. Mauser H, Gauglitz G, Compton RG, Hancock G (1998) Comprehensive chemical kinetics, photokinetics: theoretical fundamentals and applications, vol 36. Elsevier, Amsterdam. ISBN-13. 978–0444825360

5. Zalazar CS, Labas MD, Martin CA, Brandi RJ, Alfano OM, Cassano AE (2005) The extended use of actinometry in the interpretation of photochemical reaction engineering data. Chem Eng J 109:67–81. https://doi.org/10.1016/j.cej.2005.03.011
6. Tonnesen HH (ed) (2004) Photostability of drugs and drug formulations. CRC Press, Boca Raton. ISBN. 0–415–30323-0
7. Brunner H (2017) Volterra integral equations. Cambridge University Press. https://doi.org/10.1017/9781316162491
8. Lewis BJ, Onder EN, Prudil AA (2022) Advanced mathematics for engineering students. Butterworth-Heinemann, Elsevier. https://doi.org/10.1016/B978-0-12-823681-9.00002-2
9. Pezzella M (2025) High order positivity-preserving numerical methods for a non-local photochemical model. https://doi.org/10.48550/arXiv.2501.04573
10. Fabian I, Lente G (2010) Light-induced multistep redox reactions: the diode-array spectrophotometer as a photoreactor. Pure Appl Chem 82(1957–1973):2010. https://doi.org/10.1351/PAC-CON-09-11-16
11. Lente G, Espenson JH (2004) Photoreduction of 2,6-dichloroquinone in aqueous solution use of a diode array spectrophotometer concurrently to drive and detect a photochemical reaction. J Photochem Photobiol A Chem 163:249–258. https://doi.org/10.1016/j.jphotochem.2003.12.005
12. Maafi M, Brown RG (2007) The kinetic model for AB(1Φ) systems: a closed-form integration of the differential equation with a variable photokinetic factor. J Photochem Photobiol Chem 187:319–324. https://doi.org/10.1016/j.jphotochem.2006.10.030
13. Maafi W, Maafi M (2013) Modelling nifedipine photodegradation, photostability and actinometry properties. Int J Pharm 456:153–164. https://doi.org/10.1016/j.ijpharm.2013.07.075
14. Maafi M, Maafi W (2014) Φ-order kinetics of photoreversible-drug reactions. Int J Pharm 471: 536–543. https://doi.org/10.1016/j.ijpharm.2014.05.017
15. Maafi M, Maafi W (2016) Modelling and elucidation of the kinetics of multiple consecutive photoreactions AB4(4Φ) with Φ-order kinetics. Application to the photodegradation of ribofavin. J Pharm Sci 105:3537–3548. https://doi.org/10.1016/j.xphs.2016.06.030
16. Roibu A, Fransen S, Leblebici ME, Meir G, Van Gerven T, Kuhn T (2018) An accessible visible-light actinometer for the determination of photon flux and optical pathlength in flow photo microreactors. Sci Rep 8:5421. https://doi.org/10.1038/s41598-018-23735-2
17. Volfova H, Hu Q, Riedle E (2019) Determination of reaction quantum yields: LED based setup with better 5% precision. Eur Photochem Ass Newslett 96:51–69
18. Yutani R, Haku R, Teraoka R, Tode C, Koide T, Kitagawa S et al (2019) Comparative evaluation of the photostability of carbamazepine polymorphs and cocrystals. Crystals 9:553. https://doi.org/10.3390/cryst9110553
19. Maafi M, Al-Qarni MA (2022) Mono-and polychromatic light diaryletheneactinometer for the visible range. Dyes Pigments 198:109942. https://doi.org/10.1016/j.dyepig.2021.109942
20. Maafi M, Al-Qarni MA (2022) Photokinetics of Dacarbazine and Nifedipine under polychromatic light irradiation and their application as new reliable actinometers for the ultraviolet range. Sci Rep 12:7622. https://doi.org/10.1038/s41598-022-11570-5
21. Weingartz T, Nagorny S, Adams J, Eitzeroth A, Schewe M, Rembe C et al (2023) Bis(thienyl)-ethenes with α-methoxymethyl groups. Syntheses, spectroscopic Hammett plots, and stabilities in PMMA films. RSC Adv 13:25704–25716. https://doi.org/10.1039/D3RA04444K
22. Maafi M, Maafi W (2014) Montelukast photodegradation: elucidation of F-order kinetics, determination of quantum yields and application to actinometry. Int J Pharm 471:544–552. https://doi.org/10.1016/j.ijpharm.2014.05.016
23. Crano JC, Guglielmetti RJ (eds) (2003) Organic photochromic and Thermochromic compounds. Plenum, New York. https://doi.org/10.1007/b114211
24. Delbaere S, Vermeersh G, Micheau JC (2011) Quantitative analysis of the dynamic behaviour of photochromic systems. J Photochem Photobiol C Chem Photochem Rev 12:74–105. https://doi.org/10.1016/j.jphotochemrev.2011.05.004

25. Micheau JC, Coudret C, Kobeleva OI, Barachevsky VA, Yarovenko VN, Ivanov SN, Lichitsky BV, Krayushkin MM (2014) Quantitative study of photochromic transformations of diarylethene derivatives with either perhydrocyclopentene or oxazolone or lactone units. Dyes Pigments 106:32–38. https://doi.org/10.1016/j.dyepig.2014.02.013
26. Maafi M, Brown RG (2008) Kinetic analysis and elucidation options for AB($1k$,2Φ) systems. New spectrokinetic methods for photochromes. Photochem Photobiol Sci 7:1360–1372. https://doi.org/10.1039/B807556E

Chapter 20
Standardisation of Photokinetic Elucidation for Photothermal Reactions

20.1 Introduction

One of the main objectives of a photokinetic investigation is to determine the absolute values of the reaction parameters. A general standard procedure for such a goal is still lacking in the literature, even though a few methods have been proposed for solving the kinetics of particular reactions, as those reviewed in Chaps. 10 and 16. In this chapter, a new method is introduced for the elucidation of the photokinetics of any photothermal reaction. It is based on the general model equations proposed in Chap. 19.

20.2 A Standard Elucidation Method

The standard method discussed here aims at resolving the identifiability problem for photothermal reactions that have no distinguishability issues. Hence, the mechanism of the $XY_q(uk, v\Phi)$ reaction must be known with certainty, with a clear assignment of the reaction-steps. Also, as previously discussed, the determination of the absolute values of the quantum yields is only possible if the data used were obtained by monochromatic irradiation of the reactive medium. The development of the actual standard elucidation method further necessitates the availability of the concentration traces of the individual species involved in the reaction, as well as that of the total absorption at one or more selected observation wavelengths, λ_{obs}. Similarly, the species absorbing at λ_{irr} must also be known. The values of the extrinsic parameters are all known too.

Assuming the above conditions are satisfied, a step-by-step procedure for the elucidation of photothermal reaction kinetics is presented hereafter. (NB: the identifiability considered here concerns the determination of the kinetic parameters of the reaction, i.e. quantum yields, thermal rate-constants and absorptivitics, and

M. Maafi, *Photokinetics*, https://doi.org/10.1007/978-3-031-98985-8_20

hence is different from the identifiability of the trace fitting discussed in the previous chapter).

The Trace Fitting

1. Concentration and total absorbance traces for both photothermal (under irradiation) and thermal (in the dark) are individually fitted by Eqs. 19.10 and 19.13, and the fitting parameters of each equation are determined.

The Determination of the Absorptivities

2. n_{sp} theoretical equations of the total absorbance (of thermal or photothermal reaction) at the t_θ (*with* $\theta = 1, 2, \ldots, n_{sp}$) selected time intervals are written down in the form of Eq. 19.7 ($n_{sp} = q + 1$ is the number of species).
3. The values of the concentrations, calculated from the fitting equations (Eq. 19.10) of each species at the t_θ selected time intervals, are introduced in the appropriate total absorbance equations (Eqs. 19.7). Each time interval of the t_θ set corresponds to n_{sp} concentration equations (e.g. when $\theta = 1$ and $t_\theta = t_1$, we write n_{sp} concentration, $C_{Y_j}^{\lambda_{irr}}(t_1)$, equations, one for each of the n_{sp} species).
4. The values of the total absorbances in Eqs. 19.7 are calculated from the fitting equation of the total absorbance equation (Eq. 19.13) at each t_θ and introduced in the appropriate Eq. 19.7.
5. The system of these (n_{sp}) linear equations, Eqs. 19.7, is solved for the n_{sp} unknown absorption coefficients, $\varepsilon_{Y_j}^{\lambda_{obs}}$, and their respective values are determined.

The Determination of the Thermal Rate-Constants

6. The individual thermal rate-equations of each species in the reaction progressing in the dark ($P_{a_{Y_j}}^{\lambda_{irr}}(t) = P_{a_{Y_{j'}}}^{\lambda_{irr}}(t) = 0$), are written in the form of Eq. 19.5 at the t_σ selected times (with $\sigma = 1, 2, \ldots, \frac{u}{n_{sp}}$, with u being the total number of thermal reaction-steps in $XY_q(uk, v\Phi)$). Each time interval of the t_σ set corresponds to n_{sp} concentration equations (e.g. when $\sigma = 1$ and $t_\sigma = t_1$, we write n_{sp} concentration, $C_{Y_j}^{T}(t_1)$, equations, one for each of the n_{sp} species). In some cases, t_θ (*Point (2)*) is large enough and can be employed here. Otherwise, if $t_\theta < t_\sigma$, select $t_\sigma - t_\theta$ new additional time intervals to gather as many $C_{Y_j}^{T}(t)$ equations as the total number of thermal reaction-steps (u) in the mechanism.
7. The concentration of each species is calculated at each of the selected t_σ time interval by the corresponding Eq. 19.10 (whose $\omega_{ij}^{\Phi} = 0$).
8. The rate value of each species is calculated using Eq. 19.11 (where $\omega_{ij}^{\Phi} = 0$) at each t_σ time interval.
9. The values of concentrations and rates are introduced in the adequate Eq. 19.5 (Point (**6**)).

10. The system of the u linear equations, Eqs. 19.5, is solved for the u unknown rate-constants of the thermal reaction-steps, $k^{\Delta}_{Y_j \to Y_{j'}}$ and $k^{\Delta}_{Y_{j'} \to Y_j}$, and their respective values are determined.

The Determination of the Quantum Yields

11. The theoretical photothermal rate-equation of each species (exposed to monochromatic light) is written in the form of Eq. 19.5 at t_v selected times (with $v = 1, 2, \ldots, \frac{v}{n_{sp}}$ and v is the total number of photochemical reaction-steps in $XY_q(uk, v\Phi)$). Each time interval of the t_v set corresponds to n_{sp} concentration equations (i.e. one for each of the n_{sp} species). If $t_v > t_\theta$, select $t_v - t_\theta$ new additional time intervals to gather as many $C^{\lambda_{irr},T}_{Y_j}(t)$ equations as the total number of photochemical reaction-steps (v) in the mechanism at hand, $XY_q(uk, v\Phi)$.
12. The concentration of each species is calculated at each of the selected t_v time intervals by the corresponding Eq. 19.10.
13. The values of the absorbed light by each species are calculated at the selected v time interval by the appropriate Eq. 19.2, using the relevant values of the absorptivities (Point (**5**)), concentrations (Point (**12**)), and absorbances (Point (**4**)).
14. The rate value of each species is calculated using Eq. 19.11 at each t_v time intervals.
15. The values of concentrations and rates are introduced in the adequate Eq. 19.5 (Point (**11**)).
16. The system of the v linear equations (Eqs. 19.5) is solved for the v unknown quantum yields of the photochemical reaction-steps, $\Phi^{\lambda_{irr}}_{Y_j \to Y_{j'}}$ and $\Phi^{\lambda_{irr}}_{Y_{j'} \to Y_j}$, and their respective values are determined.

An application example of the above elucidation method is provided in the following section.

20.3 The Case of the $XY_2(2k, 3\Phi)$ Naphthopyran Photothermal Reaction

The family of photochromic materials has benefited from a large research effort. Naphthopyrans (*NPY*), a subgroup of that family, have been extensively studied, notably to advance understanding of their photokinetics. They obey photothermal reactions, where *UV*-irradiation of the thermally stable reactant leads most probably to the generation of two products (*CT* and *TT*, see Scheme 9.1 and Fig. 9.2). The latter *NPY* products absorb both in the *UV* and the visible spectral regions (whereas the reactant only absorbs in the *UV*), and they relax thermally to fully regenerate the initial reactant. Nonetheless, the true mechanism of naphthopyrans has not been

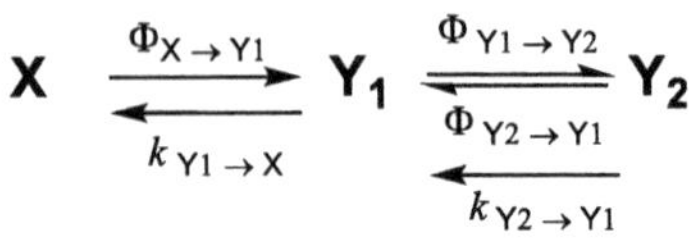

Scheme 20.1 An example $XY_2(2k, 3\Phi)$ mechanism proposed for naphthopyrans

definitively identified in detail. In place, the literature proposes many preconceived mechanisms for these materials. One of these proposals assigns an $XY_2(2k, 3\Phi)$ photothermal reaction (Scheme 20.1) to some *NPY* derivatives [1, 2], where under *UV* irradiation, the reactant only photo-produces Y_1 (the *cis-trans* isomer), which in turn establishes a photoreversible reaction with its product Y_2 (the *trans-trans* isomer). In the dark, and after the photothermal reaction had reached the photostationary state, a consecutive thermal relaxation reaction leads Y_2 to relax into Y_1, and the latter into *X* (the initial *NPY* reactant). This reaction is illustrated here as an example of the elucidation of photochromes' photokinetics.

An *RK*-simulation of the above reaction was performed under the condition of monochromatic light irradiation, at a $\lambda_{irr} = \lambda_{n\ -\ isos}$ where the three species of the reaction absorb. The parameters of the reaction that were fed to the *RK*-iteration are given in Table 20.1. The temperature of the vigorously stirred reaction medium is considered constant, the reactor has a slab-shaped form, and the monochromatic irradiation-light beam is considered collimated.

The elucidation of the $XY_2(3\Phi, 2k)$ photokinetics follows the procedure laid out in the previous section.

20.3.1 *Fitting of the Traces*

The *RK*-generated traces for the concentrations of the three species involved in the photothermal reaction $XY_2(3\Phi, 2k)$, based on the data of Table 20.1, are presented in Fig. 20.1. Thermal and photothermal reactions' traces were fitted separately.

The concentration traces of the photothermal reaction were fitted to Eqs. 19.10 that encompassed two mono-Φ-order ($i_{\Phi,j} = 2$) and two mono-exponential ($i_{\Delta,j} = 2$) terms, as given by Eq. 20.1. Table 20.2 shows the values of the different equations' parameters.

$$C_{Y_j}^{\lambda_{irr},T}(t) = \omega_j^{\infty} + \omega_{1j}^{\Phi}\, Log\left(1 + cc^{\Phi} e^{-k_{1j}^{\Phi}\, t}\right) + \omega_{2j}^{\Phi}\, Log\left(1 + cc^{\Phi} e^{-k_{2j}^{\Phi}\, t}\right) + \omega_{1j}^{T}\, e^{-k_{1j}^{T}\, t} + \omega_{2j}^{T}\, e^{-k_{2j}^{T}\, t} \quad (20.1)$$

The simulated total absorbance trace of the photothermal reaction is presented in Fig. 20.2. The fitting of the trace was performed by Eq. 19.13, which also included four terms ($i_{\Phi,\,A} = i_{\Delta,\,A} = 2$), as given by Eq. 20.2 (which has a similar template to that of Eq. 20.1). The values of the equation parameters are shown in Table 20.2.

Table 20.1 Parameters used to *RK*-simulate the photothermal $XY_2(3\Phi, 2k)$ reaction (Scheme 20.1)

Parameter	$C_X^{\lambda_{irr}}(0)/\mathrm{M}$	$\varepsilon_X^{\lambda_{irr}}/\mathrm{M}^{-1}\ \mathrm{cm}^{-1}$	$\varepsilon_{Y_1}^{\lambda_{irr}}/\mathrm{M}^{-1}\ \mathrm{cm}^{-1}$	$\varepsilon_{Y_2}^{\lambda_{irr}}/\mathrm{M}^{-1}\ \mathrm{cm}^{-1}$	$P_0^{\lambda_{irr}}/\mathrm{M\ s}^{-1}$	l_{irr}/cm	l_{obs}/cm
Value	$1.33\ 10^{-5}$	8879.20	8651.31	13770.67	$3.13\ 10^{-6}$	2	2
Parameter	$\Phi_{X\to Y_1}^{\lambda_{irr}}$	$\Phi_{Y_1\to Y_2}^{\lambda_{irr}}$	$\Phi_{Y_2\to Y_1}^{\lambda_{irr}}$	λ_{obs}/nm	$k_{Y_2\to Y_1}/\mathrm{s}^{-1}$	$k_{Y_1\cdot\to X}/\mathrm{s}^{-1}$	h[a]
Value	0.02895	0.03039	0.02885	λ_{irr}	0.007	0.015	3

[a] h is the iteration step of the *RK*-calculation

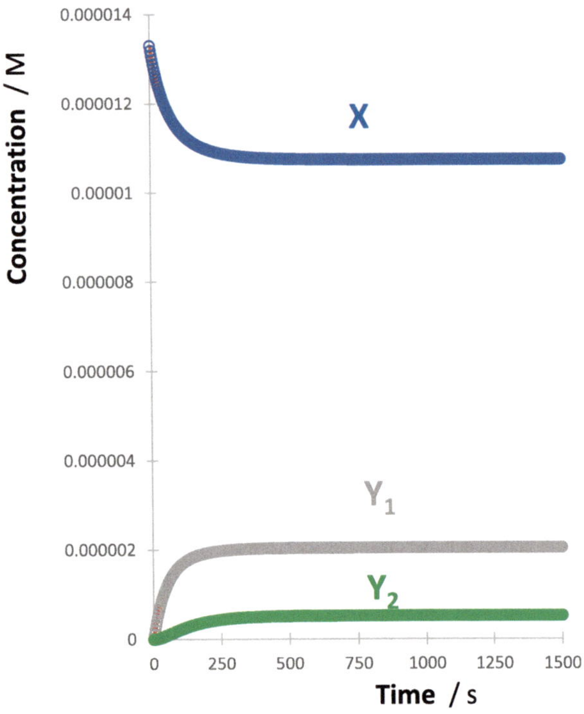

Fig. 20.1 Simulated species concentration traces for the trimolecular $XY_2(2k, 3\Phi)$ reaction (circles) and their fittings to Eqs. 20.1 (red lines)

$$A_{tot}^{\lambda_{irr},T/\lambda_{obs}}(t) = \omega_A^{\infty} + \omega_{1,A}^{\Phi}\, Log\left(1 + cc^{\Phi}\, e^{-k_{iA}^{\Phi}\, t}\right)$$
$$+ \omega_{2,A}^{\Phi}\, Log\left(1 + cc^{\Phi}\, e^{-k_{iA}^{\Phi}\, t}\right) + \omega_{1,A}^{T}\, e^{-k_{iA}^{T}\, t} + \omega_{2,A}^{T}\, e^{-k_{iA}^{T}\, t} \tag{20.2}$$

The equations relative to the fitting of the thermal traces (Eqs. 20.3) are derived from Eq. 19.10, as Eqs. 20.1 and 20.2 for which $\omega_{1j}^{\Phi} = \omega_{2j}^{\Phi} = \omega_{1,A}^{\Phi} = \omega_{2,A}^{\Phi} = 0$. The traces are shown in Fig. 20.3 and the equations fitting parameters are given in Table 20.2.

$$C_{Y_j}^{T}(t) = \omega_j^{\infty} + \omega_{1j}^{T}\, e^{-k_{1j}^{T}\, t} + \omega_{2j}^{T}\, e^{-k_{2j}^{T}\, t} \tag{20.3}$$

It is also possible to fit the total absorbance trace, $A_{tot}^{T/\lambda_{obs}}(t)$, of the thermal reaction by Eq. 20.2 whose photochemical coefficients are zero ($\omega_{1,A}^{\Phi} = \omega_{2,A}^{\Phi} = 0$). However, this is not necessary for the present photokinetics, as it shall become evident from the following elucidation steps.

The fitting data show that the four traces of the photothermal reaction (under irradiation) share the same values of five parameters, cc^{Φ}, k_{1j}^{Φ}, k_{2j}^{Φ}, k_{1j}^{T}, and k_{2j}^{T} (Table 20.2), and differ by the values of the other five fitting parameters ω_j^{∞}, ω_{1j}^{Φ},

Table 20.2 The fitting parameters of Eqs. 20.1 and 20.2, respectively, corresponding to Figs. 20.1 and 20.2

	Photothermal reaction $XY_2(2k, 3\Phi)$				Thermal reaction $XY_2(2k)$			
	$C_{Y_0}(t)$[a]	$C_{Y_1}(t)$	$C_{Y_2}(t)$	$A_{tot}(t)$	$C_{Y_0}(t)$	$C_{Y_1}(t)$	$C_{Y_2}(t)$	$A_{tot}(t)$
ω_j^{∞}	$1.07\ 10^{-5}$	$2.04\ 10^{-6}$	$5.20\ 10^{-7}$	0.240	$1.33\ 10^{-5}$	$1.02\ 10^{-24}$	$1.41\ 10^{-23}$	0.2362
ω_{1j}^{Φ}	$5.38\ 10^{-5}$	$-6.23\ 10^{-5}$	$8.48\ 10^{-6}$	0.129	–	–	–	–
ω_{2j}^{Φ}	$1.62\ 10^{-4}$	$-1.95\ 10^{-4}$	$3.28\ 10^{-5}$	0.426	–	–	–	–
cc^{Φ}	1.1	1.1	1.1	1.1	–	–	–	–
k_{1j}^{Φ}	0.01173	0.01173	0.01173	0.01173	–	–	–	–
k_{2j}^{Φ}	0.006935	0.006935	0.006935	0.006935	–	–	–	–
ω_{1j}^{T}	$1.36\ 10^{-5}$	$-1.64\ 10^{-5}$	$2.79\ 10^{-6}$	0.0386	$-1.59\ 10^{-6}$	$1.59\ 10^{-6}$	$-8.98\ 10^{-21}$	0.006233
ω_{2j}^{T}	$-8.07\ 10^{-5}$	$9.72\ 10^{-5}$	$-1.66\ 10^{-5}$	−0.233	$-9.79\ 10^{-7}$	$4.57\ 10^{-7}$	$5.224\ 10^{-7}$	−0.002074
k_{1j}^{T}	0.015	0.015	0.015	0.015	0.015	0.02287	0.01129	0.007572
k_{2j}^{T}	0.007	0.007	0.007	0.007	0.007	0.009325	0.007	0.01142
r^2	0.9999	0.9999	0.9999	0.9999	1	1	1	1
SSE	$6.52\ 10^{-15}$	$6.43\ 10^{-15}$	$1.34\ 10^{-13}$	$2.45\ 10^{-9}$	$8.50\ 10^{-18}$	$6.60\ 10^{-18}$	$1.35\ 10^{-19}$	$4.47\ 10^{-8}$
RMSE	$3.63\ 10^{-9}$	$3.61\ 10^{-9}$	$1.65\ 10^{-8}$	$2.22\ 10^{-6}$	$1.31\ 10^{-10}$	$1.15\ 10^{-10}$	$1.65\ 10^{-11}$	$9.48\ 10^{-6}$

[a] The concentrations are expressed in mol dm^3, and the absorbance is dimensionless

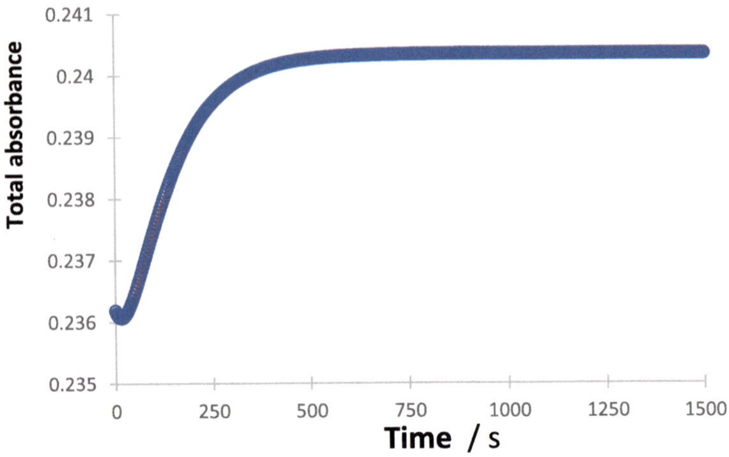

Fig. 20.2 Total absorbance *RK*-generated traces of the $XY_2(2k, 3\Phi)$ reaction (circles) and it fitting with Eq. 20.2 (line)

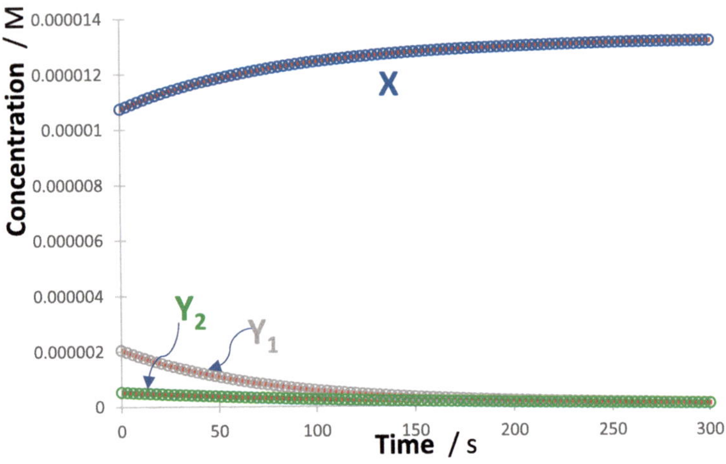

Fig. 20.3 Concentration traces of the species thermal relaxation from *pss* (circles) and their *RK*-fitting with Eq. 20.2 (lines)

ω_{2j}^{Φ}, ω_{1j}^{T}, and ω_{2j}^{T}. Even if this is not required by the formulation of the original Eqs. 19.10 and 19.13, it is here used for illustration.

Conversely, the values of the four fitting parameters of the thermal reaction traces do not seem to have any consistent communality, where it is important to notice that the rate-constants values obtained for the four thermal reaction traces are all different, despite the trend being only bi-exponential (a feature which is relevant and which should be taken into account when fitting experimental or simulated data).

This can be considered as evidence for the identifiability issues relative to the fitting as was discussed in Sect. 19.8.

The selected equations (Eqs. 20.1–20.3), all corroborated the *RK*-generated data for the studied reaction. The goodness of the fittings of all traces was indicated by high values of the correlation coefficients (r^2), and relatively low values of both *SSE* and *RMSE* coefficients (Table 20.2).

This represents the completion of the first step (*Point (1)*) of the elucidation procedure laid out in Sect. 20.2.

20.3.2 The Absorption Coefficients

The total absorbance equation (Eq. 20.4) is worked out from Eq. 19.7 for an observation wavelength where the three species absorb. It is here written for the photothermal reaction trace (as an additional option to that given in the elucidation procedure example in Sect. 20.2, Point (3)).

$$A_{tot}^{\lambda_{irr},\,T/\lambda_{obs}}(t) = \varepsilon_X^{\lambda_{obs}}\, l_{obs}\, C_X^{\lambda_{irr},T}(t) + \varepsilon_{Y_1}^{\lambda_{obs}}\, l_{obs}\, C_{Y_1}^{\lambda_{irr},T}(t) + \varepsilon_{Y_2}^{\lambda_{obs}}\, l_{obs}\, C_{Y_2}^{\lambda_{irr},T}(t) \quad (20.4)$$

The values of total absorbance and concentration at different time intervals (*Points (3)* and *(4)* in the elucidation procedure) can be calculated from their respective fitting equations and the values in Table 20.2 of their fitting parameters.

In our case, there are three species, i.e. $n_{sp} = 3$, and so, total absorbance readings at three different time intervals ($\theta = n_{sp}$, $t_\theta = 3$, i.e. t_1, t_2, and t_3) will be sufficient to attain the quorum required for three equations (*Point (2)* in the elucidation procedure).

However, only two time intervals (t_1 and t_2) are necessary, as the reactant in Scheme 20.1 is thermally stable and $C_{Y_1}^{\lambda_{irr},T}(0) = C_{Y_2}^{\lambda_{irr},T}(0) = 0$, so that the reactant absorption coefficient can be deduced from its calibration graph (Eq. 20.5).

$$\varepsilon_X^{\lambda_{obs}} = \frac{A_{tot}^{\lambda_{irr},T/\lambda_{obs}}(0)}{l_{obs}\, C_X^{\lambda_{irr},T}(0)} \quad (20.5)$$

Therefore, two total absorbance equations are needed to solve for the two remaining absorption coefficients (*Point (5)*). These will correspond to the two time intervals, $t_1 = 129$ and $t_2 = 348$ *s*, on the photothermal reaction traces (Table 20.3). The extracted values of the species absorptivities are very close to those given in Table 20.1, as indicated by the small percentage errors (% Err.).

Table 20.3 The fitting parameters of Eq. 20.4, extracted from Figs. 20.1 and 20.2 and their fitting equations

	Data from trace fitting			Calculated absorptivities	
Time/s	$C_{Y_1}^{\lambda_{irr},T}(t)/\text{M}$	$C_{Y_2}^{\lambda_{irr},T}(t)/\text{M}$	$A_{tot}^{\lambda_{irr},T/\lambda_{obs}}(t)$	$\varepsilon_{Y_1}^{\lambda_{obs}}$ (% Err.)/M^{-1}cm^{-1}	$\varepsilon_{Y_2}^{\lambda_{obs}}$ (% Err.)/M^{-1}s^{-1}
129	1.780 10^{-6}	2.797 10^{-7}	2.380	8491.64	14298.52
348	2.011 10^{-6}	4.874 10^{-7}	2.399	(1.85)	(3.83)

20.3.3 The Thermal Rate-Constants

The determination of the rate-constants associated with the thermal reaction-steps, requires the theoretical equations of the thermal rates for each species, which can be worked out from Eq. 19.10 and the reaction sequence (Scheme 20.1). The obtained system of equations is solved for the rate-constants (the number of equations must be the same as that of the number of unknown rate-constants of the reaction ($u = 3$), and hence, because we have three species, we require an observation at a single time interval, t_1 (i.e. $n_{sp} = u = 3$, so that $u/n_{sp} = 1 = \sigma$ and $t_\sigma = t_1$).

For our reaction, the rates are given by the following expressions (*Point (6)* in the elucidation procedure of Sect. 20.2).

$$r_X^T(t) = k_{Y_1\to X}^T\, C_{Y_1}^T(t) \tag{20.6}$$

$$r_{Y_1}^T(t) = -k_{Y_1\to X}^T\, C_{Y_1}^T(t) + k_{Y_2\to Y_1}^T\, C_{Y_2}^T(t) \tag{20.7}$$

$$r_{Y_2}^T(t) = -k_{Y_2\to Y_1}^T\, C_{Y_2}^T(t) \tag{20.8}$$

The values of the concentrations $C_{Y_1}^T(t)$ and $C_{Y_2}^T(t)$ are calculated from the fitting equations (Eq. 20.3) for species Y_1 and Y_2, respectively (*Point (7)*). The rates' values at the same time intervals are obtained from Eq. 20.9 (derived from Eq. 19.11) and the fitting data of Table 20.2 (*Point (8)*).

$$Fit : r_{Y_j}^T(t) = -\omega_{1j}^T\, k_{1j}^T\, e^{-k_{1j}^T\, t} - \omega_{2j}^T\, k_{2j}^T\, e^{-k_{2j}^T\, t} \tag{20.9}$$

In the present kinetic case, only one time interval is necessary since the rate-constants $k_{Y_1\to X}^T$ and $k_{Y_2\to Y_1}^T$, can directly be obtained from Eq. 20.6 and 20.7, respectively (*Point (9) and (10)*). Eq. 20.7 can be used for confirmation of the results found. However, two different time intervals on the thermal reaction traces were selected to test the approach. The findings are given in Table 20.4. Each of the calculated rate-constants have very similar values at the two selected time intervals, and they are very close to the original values used to feed the *RK*-calculation of the traces (Table 20.1). This means that the true values of the rate-constant are readily

Table 20.4 The fitting parameters of Eqs. 20.1 and 20.2, respectively, corresponding to Fig. 20.3

	Data from trace fitting				Calculated rate-constants	
Time /s	$C^T_{Y_1}(t)/\mathrm{M}$	$C^T_{Y_2}(t)/\mathrm{M}$	$r^T_{Y_1}(t)/\mathrm{M\,s^{-1}}$	$r^T_{Y_2}(t)/\mathrm{M\,s^{-1}}$	$k^T_{Y_1\to X}/\mathrm{s^{-1}}$	$k^T_{Y_2\to Y_1}/\mathrm{s^{-1}}$
15	$1.684\ 10^{-6}$	$4.703\ 10^{-7}$	$-2.197\ 10^{-8}$	$-3.292\ 10^{-9}$	0.014999	0.006999
33	$1.334\ 10^{-6}$	$4.146\ 10^{-7}$	$-1.711\ 10^{-8}$	$-2.902\ 10^{-9}$	0.014998	0.006999

determined by this method. It is interesting to notice that these excellent results were found despite the differences in the fitting parameters of the thermal reaction traces, especially the differences recorded for k^{T}_{1j} and k^{T}_{2j} (Table 20.1), i.e. the identifiability problem occurring in the fitting results does not affect the determination of precise values of concentrations and rates from the fitting equations (as long as these have an excellent fit to the traces). The values of these rate-constants ($k^T_{Y_1\to X}$ and $k^T_{Y_2\to Y_1}$) were included in the equations (Eqs. 20.1 and 20.2) when subsequently fitting the traces of the photothermal reaction (where $k^T_{Y_1\to X} = 0.15\ s^{-1} = k^{\mathrm{T}}_{1j}$ and $k^T_{Y_2\to Y_1} = 0.007\ s^{-1} = k^{\mathrm{T}}_{2j}$, in those equations).

20.3.4 The Quantum Yields

The values of the three quantum yields of the photothermal reaction-steps (i.e. $v = 3 = n_{sp}$) are determined by solving three theoretical rate-equations of the photothermal reaction-rates of the three species (Eqs. 20.10–20.12) that are worked out from Eq. 19.5 (as recommended by the elucidation procedure of Sect. 20.2 in *Point (11)*).

$$r_X^{\lambda_{irr},T}(t) = -\Phi^{\lambda_{irr}}_{X\to Y_1} P^{\lambda_{irr}}_{a_X}(t) \tag{20.10}$$

$$\begin{aligned} r_{Y_1}^{\lambda_{irr},T}(t) &= \Phi^{\lambda_{irr}}_{X\to Y_1} P^{\lambda_{irr}}_{a_X}(t) - \Phi^{\lambda_{irr}}_{Y_1\to Y_2} P^{\lambda_{irr}}_{a_{Y_1}}(t) + \Phi^{\lambda_{irr}}_{Y_2\to Y_1} P^{\lambda_{irr}}_{a_{Y_2}}(t) \\ &\quad - k^T_{Y_1\to X} C^{\lambda_{irr},T}_{Y_1}(t) + k^T_{Y_2\to Y_1} C^{\lambda_{irr},T}_{Y_2}(t) \end{aligned} \tag{20.11}$$

$$r_{Y_2}^{\lambda_{irr},T}(t) = -\Phi^{\lambda_{irr}}_{Y_2\to Y_1} P^{\lambda_{irr}}_{a_{Y_2}}(t) + \Phi^{\lambda_{irr}}_{Y_1\to Y_2} P^{\lambda_{irr}}_{a_{Y_1}}(t) - k^T_{Y_2\to Y_1} C^{\lambda_{irr},T}_{Y_2}(t) \tag{20.12}$$

The rates' values are calculated (*Point (14)*) from the fitting rate-equations (Eq. 19.11) and their parameters laid out in Table 20.2. They are given by the following general equation, Eq. 20.13.

$$Fit : r_{Y_j}^{\lambda_{irr},T}(t) = -\frac{1}{ln(10)}\frac{\omega_{1j}^{\Phi}\, cc_j^{\Phi}\, k_{1j}^{\Phi}\, e^{-k_{1j}^{\Phi}\, t}}{1 + cc^{\Phi}e^{-k_{1j}^{\Phi}\, t}} - \frac{1}{ln(10)}\frac{\omega_{2j}^{\Phi}\, cc_j^{\Phi}\, k_{2j}^{\Phi}\, e^{-k_{2j}^{\Phi}\, t}}{1 + cc_j^{\Phi}e^{-k_{2j}^{\Phi}\, t}}$$
$$- \omega_{1j}^{T}\, k_{1j}^{T}\, e^{-k_{1j}^{T}\, t} - \omega_{2j}^{T}\, k_{2j}^{T}\, e^{-k_{2j}^{T}\, t} \quad (20.13)$$

The values of the amount of absorbed light of each species are obtained from Eq. 19.2 (*Point (13)*).

$$P_{a_X}^{\lambda_{irr}}(t) = \frac{\varepsilon_X^{\lambda_{irr}}\, l_{irr}\, C_X^{\lambda_{irr},T}(t)}{A_{tot}^{\lambda_{irr},T/\lambda_{irr}}(t)}\, P_0^{\lambda_{irr}}\left(1 - 10^{-A_{tot}^{\lambda_{irr},T/\lambda_{irr}}(t)}\right) \quad (20.14)$$

$$P_{a_{Y_1}}^{\lambda_{irr}}(t) = \frac{\varepsilon_{Y_1}^{\lambda_{irr}}\, l_{irr}\, C_{Y_1}^{\lambda_{irr},T}(t)}{A_{tot}^{\lambda_{irr},T/\lambda_{irr}}(t)}\, P_0^{\lambda_{irr}}\left(1 - 10^{-A_{tot}^{\lambda_{irr},T/\lambda_{irr}}(t)}\right) \quad (20.15)$$

$$P_{a_{Y_2}}^{\lambda_{irr}}(t) = \frac{\varepsilon_{Y_2}^{\lambda_{irr}}\, l_{irr}\, C_{Y_2}^{\lambda_{irr},T}(t)}{A_{tot}^{\lambda_{irr},T/\lambda_{irr}}(t)}\, P_0^{\lambda_{irr}}\left(1 - 10^{-A_{tot}^{\lambda_{irr},T/\lambda_{irr}}(t)}\right) \quad (20.16)$$

The total absorbance in Eqs. 20.14–20.16 is calculated by Eq. 20.4 where λ_{obs} is replaced by λ_{irr} (for our case, $\lambda_{irr} = \lambda_{obs}$, Table 20.1).

The absorption coefficients ($\varepsilon_{Y_j}^{\lambda_{irr}}$) values have been obtained in the previous sub-section (Table 20.4). The concentration ($C_{Y_j}^{\lambda_{irr},T}(t)$) values in all the above equations are determined from the fitting equations of the photothermal reactions (*Point (12)*).

For the particular case of the reaction investigated here, it is clear that $\Phi_{X\to Y_1}^{\lambda_{irr}}$ can directly be obtained from Eq. 20.10, using the data in Table 20.5 at $t = 0$, as

$$\Phi_{X\to Y_1}^{\lambda_{irr}} = -\frac{r_X^{\lambda_{irr},T}(0)}{P_{a_X}^{\lambda_{irr}}(0)} = -\frac{Fit : r_{0,X}^{\lambda_{irr},T}}{P_0^{\lambda_{irr}}\left(1 - 10^{-A_X^{\lambda_{irr},T/\lambda_{irr}}(0)}\right)} = 0.02945 \quad (20.17)$$

Several ways could be implemented to solve for the two other quantum yields (*Point (16)*). One of these delivers the required pair of linear equations by using the rate-law $r_{Y_2}^{\lambda_{irr},T}(t)$ (Eq. 20.12) considered at two time intervals (e.g. 129 and 348 *s*). The equations' parameters are listed in Table 20.5. The calculated quantum yields have less than a 5% difference with those originally used to feed the *RK*-simulation of the traces (Table 20.1).

Table 20.5 The fitting parameters of Eq. 20.11 and the calculated quantum yield values

	Data from trace fitting				Calculated quantum yields		
Time/s	$P_{aY_1}^{\lambda_{irr}}(t)/\mathrm{M\ s}^{-1}$	$P_{aY_2}^{\lambda_{irr}}(t)/\mathrm{M\ s}^{-1}$	$r_{Y_1}^{\lambda_{irr},T}(t)/\mathrm{M\ s}^{-1}$	$r_{Y_2}^{\lambda_{irr},T}(t)/\mathrm{M\ s}^{-1}$	$\Phi_{Y_1 \to X}^{\lambda_{irr}}$ (%Err.)	$\Phi_{Y_2 \to Y_1}^{\lambda_{irr}}$ (%Err.)	$\Phi_{X \to Y_1}^{\lambda_{irr}}$ (%Err.)
129	1.678 10^{-7}	4.439 10^{-8}	3.583 10^{-9}	2.005 10^{-9}	0.03094 (1.78)	0.02765 (4.15)	0.02945 (1.74)
348	1.891 10^{-7}	7.719 10^{-8}	1.874 10^{-10}	3.054 10^{-10}			

20.4 Conclusion

The standardisation of the procedures is one of the important objectives sought for photokinetics. The procedure (*Points* (**1**) to (**16**)) proposed in this chapter is a general elucidation methodology that fits that purpose. The illustration example treats the case of an *NPY*-suggested $XY_2(3\Phi, 2k)$ reaction, whereby the step-by-step implementation of the procedure is detailed. The performance of the procedure is indicated by the low differences recorded between the true intrinsic parameters of the reaction (Table 20.1) and those obtained by the treatment (Tables 20.3, 20.4, 20.5, and Eq. 20.16). As such, the reliability of the methodology is demonstrated. It represents a new and invaluable tool for photokinetics investigation.

References

1. Frigoli M, Marrot J, Gentili PL, Jacquemin D, Vagnini M, Pannacci D, Ortica F (2015) P-type Photochromism of new helical Naphthopyrans: synthesis and photochemical, Photophysical and theoretical study. Chem Photo Chem 16:2447–2458. https://doi.org/10.1002/cphc.201500251
2. Gierczyk B, Rode MF, Burdzinski G (2022) Mechanistic insights into photochromic 3H-naphthopyran showing strong photocoloration. Sci Rep 12:10781. https://doi.org/10.1038/s41598-022-14679-9

Chapter 21
Revisiting Some Concepts

21.1 Introduction

A fresher look at previously discussed concepts might be useful in light of the new general equations of rates (Eq. 19.5) and concentrations (Eq. 19.10) established in Chap. 19. The target is to evaluate if these equations can help in characterising further or allow to establish general equations for the quantum yield and the photonic yield at low and high initial concentrations, irrespective of the nature (monochromatic or polychromatic) of the light used.

21.2 The Quantum Yield Equation

It is perhaps useful to revisit the definition of the quantum yield discussed in Chap. 5, but now within the framework set out in Chap. 19.

Since the general rate-law, given by Eq. 19.5, also corresponds to photothermal reactions exposed to polychromatic light, there is a need to review the meaning of the quantum yield relative to this particular irradiation condition. In reference to the *IUPAC* classification [1], as stipulated earlier, such an evaluation of the quantum yield should not, at all, be envisaged when the incident light is polychromatic. Indeed, the term *quantum yield* is therein stated to only be valid for reactions subjected to *strictly* monochromatic irradiation. Whereas, for polychromatic light irradiation, the *photonic yield* is a more convenient concept.

In photokinetics, when the light is polychromatic, there is little physical sense in seeking a single value for the quantum yield of the reaction, whether or not based on the analysis of the reaction traces. The quantum yield is a quantity that should, a priori, be considered wavelength-dependent, in reference to the data and discussion offered in Chaps. 4 and 18. This aspect is conveyed by Eq. 19.5, which also gives an indication of the extended complexity of the quantum yield issue since the rate-law

M. Maafi, *Photokinetics*, https://doi.org/10.1007/978-3-031-98985-8_21

of an individual species Y_j may well also be expressed as a function of the wavelength-dependent quantum yields of several other reaction species.

In the case where the quantum yield of the reactant in the reaction-step, $X \rightarrow Y_1$, is proven (or supposed) to be wavelength-invariant ($\Phi_{X\rightarrow Y_1}^{\lambda_{irr}} = \Phi_{X\rightarrow Y_1}$), a formula can be worked out from Eq. 19.5 for $\Phi_{X\rightarrow Y_1}$ as given by Eq. 21.1. In this expression, the right-hand side is time-dependent (due to the variation of $C_{Y_j}^{Lp,\Delta\lambda,T}(t)$ and $P_{a_X}^{\lambda_{irr}}(t)$), whereas the left-hand side is constant, which makes Eq. 21.1 valid for any reaction time regardless of the actual shape of the reactant concentration trace (which is usually expected to be non-linear). The involvement of other species parameters (including those relative to the thermal reaction-steps in the right-hand side of the equation) stems from the fact that the species-rate ($dC_{Y_j}^{Lp,\Delta\lambda,T}(t)/dt$) value varies in relation to these parameters' values (which, in turn, is a combination ensuring that $\Phi_{X\rightarrow Y_1}$ has a constant value whatever t).

$$\Phi_{X\rightarrow Y_1} = \frac{\frac{dC_X^{Lp,\Delta\lambda,T}(t)}{dt}}{\sum\limits_{\lambda_{irr}=\lambda_a}^{\lambda_b}\left(P_{a_X}^{\lambda_{irr}}(t)\right)} =$$

$$\frac{1}{\sum\limits_{\lambda_{irr}=\lambda_a}^{\lambda_b}\left(P_{a_X}^{\lambda_{irr}}(t)\right)} \times \left(-\sum_{\lambda_{irr}=\lambda_a}^{\lambda_b}\sum_{j'}^{v_\Phi^0-1}\left(\Phi_{X\rightarrow Y_{j'}}^{\lambda_{irr}} P_{a_X}^{\lambda_{irr}}(t)\right) + \sum_{\lambda_{irr}=\lambda_a}^{\lambda_b}\sum_{j'}^{v_\Phi^0-1}\left(\Phi_{Y_{j'}\rightarrow X}^{\lambda_{irr}} P_{a_{Y_{j'}}}^{\lambda_{irr}}(t)\right)\right)$$

$$+\frac{1}{\sum\limits_{\lambda_{irr}=\lambda_a}^{\lambda_b}\left(P_{a_X}^{\lambda_{irr}}(t)\right)}\sum_{j'}^{u_k^0}\left(-k_{X\rightarrow Y_{j'}}^{T}\, C_X^{Lp,\Delta\lambda,T}(t) + k_{Y_{j'}\rightarrow X}^{\Delta}\, C_{Y_{j'}}^{Lp,\Delta\lambda,T}(t)\right) \quad (21.1)$$

The middle (ratio) term of Eq. 21.1 somewhat resembles the equation set out for a differential quantum yield (Eqs. 1.5 and 5.13, as a rate over $P_{a_X}^{\lambda_{irr}}$), but overall, this equation (Eq. 21.1) is significantly different from those assigned in the literature for the definition of the differential quantum yield of reactants under either monochromatic [1] or polychromatic [2] irradiation.

Equation 21.1, however, reduces down to those general equations proposed for the photoreaction ($X \rightarrow Y_1$) exposed to either monochromatic [3] or polychromatic [4] light. In this context, it might be useful to amend the literature concerned with such aspects with Eq. 21.1 and the considerations and assumptions leading to its derivation when either mono- or polychromatic light is used to drive the reaction.

Practically, it is not realistically possible to experimentally evaluate Eq. 21.1 unless more assumptions are introduced. Eventually, the rate and the amount of light absorbed by X are evaluated using similar equations and procedures employed for

Eqs. 20.13 and 20.14–20.16. However, such a task is only possible if both the reactant concentration traces are available at each individual wavelength of the *OSIA* domain and the other photochemical reaction-steps' quantum yields ($\Phi_{X \to Y_{j'}}^{\lambda_{irr}}$ and $\Phi_{Y_{j'} \to X}^{\lambda_{irr}}$) are available. Also, Eq. 21.1 could be calculated if all the values of all its parameters are known for all the *OSIA* wavelengths. This supposes that the quantum yields ($\Phi_{X \to Y_j}^{\lambda_{irr}}$ and $\Phi_{Y_j \to X}^{\lambda_{irr}}$) are known, which incidentally does not justify why only $\Phi_{X \to Y_1}$ would not be known. In any case, the knowledge of all these parameters at the required level of detail is usually not available in real-life situations. This type of data can be obtained by applying, to each wavelength of the *OSIA*-domain, the elucidation method proposed in Chap. 20.

Finally, under polychromatic light irradiation, it is possible to propose Eq. 21.2 to accurately determine the quantum yield if the seven following conditions are satisfied: the reactant undergoes a single photoreaction (but no thermal reaction) and is not regenerated by the immediate product to the reactant (i.e. a simple $X \to Y_1$); the quantum yield is wavelength invariant (i.e. $\Phi_{X \to Y_1}^{\lambda_{irr}} = \Phi_{X \to Y_1}$); the profile of the light source is available; the *OSIA* of the reactant is known; the reactant concentration trace is available; the latter is fitted to Eq. 19.11 whose fitting parameters were determined; and the initial rate of the concentration trace is calculated by Eq. 19.12.

$$\Phi_{X \to Y_1} = -\frac{Fit : r_{0,X}^{Lp,\Delta\lambda,T}}{\sum\limits_{\lambda_{irr}=\lambda_a}^{\lambda_b} \left(P_0^{\lambda_{irr}} \left(1 - 10^{-A_X^{\lambda_{irr},T/\lambda_{irr}}(0)} \right) \right)} \tag{21.2}$$

Unfortunately, for the general case where $\Phi_{X \to Y_1}^{\lambda_{irr}} \neq \Phi_{X \to Y_1}$, there is no known single expression capable of evaluating the quantum yield of a photoreaction under polychromatic light.

The situation is much simpler when the reaction $X \to Y_1$ is driven by a monochromatic light. Here, the absolute value of the quantum yield is easily obtained from Eq. 21.2, whose denominator does not include the sum over wavelength. An application of such a formula is given by Eq. 20.17. Hence, for photothermal reactions, under monochromatic light, that are characterised by a sequence where the thermally stable reactant engages in a single reaction-step, $X \to Y_1$, Eq. 20.17 represents a handy tool to unravel the trend of the quantum yield values with wavelength. For that purpose, the equation is applied to the available reactant concentration traces obtained at individual wavelengths of the employed monochromatic lights, which would successively be used in independent experiments for the irradiation of the sample. It is an easy way to check the variability of the reactant quantum yield with irradiation.

The concentration and absorbance equations derived from the semi-empirical method being all of Φ-order character, they can be written in the format of Eq. 19.10 (or Eq. 19.13), with a procedure based on Eq. 20.17 leading to the same results for $\Phi_{X \to Y_1}^{\lambda_{irr}}$ as those described in Chap. 16.

21.3 The Photonic Yield

When polychromatic light, covering a range of wavelength ($\Delta\lambda$), is employed for photolysis, the *IUPAC* recommended quantity for reactivity assessment is the *photonic yield* or *quantum efficiency*. The photonic yield refers to the amount or number of monochromatic photons, counted for each wavelength of the emission spectrum of the lamp, $\Delta\lambda_{Lp}$, that reach the internal surface of the reactor before absorption by the reactive medium. This definition of the photonic yield clearly violates the "first law of photochemistry" in the sense that the photons counted are those incident from the lamp and not those effectively absorbed by the reactive medium, and hence, cannot be responsible for photoreaction; yet, they are used to determine a feature of photoreactivity. For that reason, it was suggested to interpret the photonic yield as the amount of light-energy impinging on the surface of the reactor and available for the reaction [2]. In addition to this specificity, it is important to emphasise that the formula assigned to the photonic yield (vide infra Eqs. 21.3 and 21.4) has neither been established on the basis of empirical data nor has it been derived analytically from the rate-equation. Therefore, the precise kinetic meaning of the photonic yield is unclear. It has, rather, to be understood as a quantity estimating the effect of a given number of incident photons (or energy) that reaches the reactor from the light source on the photoreactivity of the species of the photothermal system present inside that reactor.

The commonly used photonic yield (*PHYD*) equation (Eq. 21.3) is a ratio of the reacted reactant concentration measured at time *t* over the total number of photons of the light source entering the reactor in the same period of time. This number is measured as the product $\left(P_0^{Lp,\Delta\lambda_{LP}} \times t\right)$. $P_0^{Lp,\Delta\lambda_{Lp}}$ is the incident total photon flux per unit time and per area and volume of the irradiated sample over the emission spectrum of the lamp. It is expressed as a sum, $P_0^{Lp,\Delta\lambda_{em}} = \sum_{\lambda_{irr}=\lambda_a}^{\lambda_b} P_0^{\lambda_{irr}}$. $P_0^{Lp,\Delta\lambda_{Lp}}$ is measured by chemical (or physical) actinometry, and the relevant species' concentrations are determined by routine analytical techniques.

$$PHYD_X^{Lp,\Delta\lambda_{Lp},T}(t) = \frac{C_X^{Lp,\Delta\lambda,T}(0) - C_X^{Lp,\Delta\lambda,T}(t)}{P_0^{Lp,\Delta\lambda_{Lp}}\ t} \tag{21.3}$$

The *PHYD* can also be measured for a selected end product, Y_{end}, of the photothermal reaction (using its formed concentration at time *t*) by the classical formula,

$$PHYD_{Y_{end}}^{Lp,\Delta\lambda_{Lp},T}(t) = \frac{C_{Y_{end}}^{Lp,\Delta\lambda,T}(t)}{P_0^{Lp,\Delta\lambda_{Lp}}\ t} \tag{21.4}$$

The *PHYD* concept can, in fact, be generalised to include any species of the photothermal reaction.

$$PHYD_{Y_j}^{Lp,\Delta\lambda_{Lp},T}(t) = \pm \frac{C_{Y_j}^{Lp,\Delta\lambda,T}(0) - C_{Y_j}^{Lp,\Delta\lambda,T}(t)}{P_0^{Lp,\Delta\lambda_{Lp}}\ t} \quad (21.5)$$

The *PHYD* is a dimensionless quantity, as given by the dimension analysis $[PHYD] = (\text{mol} \times \text{L}^{-1})/(\text{einstein} \times \text{dm}^{-3} \times \text{s}^{-1} \times \text{s}) = 1$.

A closer examination of these equations reveals that the value of either $PHYD^{Lp,\Delta\lambda_{Lp},T}(t)$ will differ depending on whether the lamp profile covers entirely *OSIA* ($\Delta\lambda_{Lp} \leq \Delta\lambda$) or when it exceeds it ($\Delta\lambda_{Lp} > \Delta\lambda$). For the same photothermal reaction, the value of $PHYD^{Lp,\Delta\lambda_{Lp},T}(t)$ will be smaller if $\Delta\lambda_{Lp} > \Delta\lambda$ than when $\Delta\lambda_{Lp} \leq \Delta\lambda$, because of the $P_0^{Lp,\Delta\lambda_{Lp}}$ value being larger than $P_0^{Lp,\Delta\lambda}$ (i.e. $PHYD^{Lp,\ \Delta\lambda,\ T}(t)$, wherein Eqs. 21.3–21.5, $\Delta\lambda_{Lp}$ is replaced with $\Delta\lambda$), as in each case the only useful photons to induce photoreactions are those absorbed within *OSIA* (i.e. $\Delta\lambda$).

In Fig. 19.1, if the spectra ε_1 or ε_2 are, respectively, assigned to the reactant and the end product, the photons emitted by the lamp over the 365 *nm* wavelength are accounted for in $P_0^{Lp,\Delta\lambda_{Lp}}$ but they do not, effectively, contribute to the photoreactions undergone by either species (the species in Fig. 19.1 absorb the light emitted in the range $310 > \lambda > 365$ nm). This subtlety makes it mandatory, when reporting *PHYD*(*t*) data and irrespective of the investigation, to indicate which of $P_0^{Lp,\Delta\lambda}$ or $P_0^{Lp,\Delta\lambda_{Lp}}$ was used to calculate the photonic yield (by either Eqs. 21.3 or 21.4). Unless this feature is specified, any comparison between a species *PHYD*(*t*) between two or more reaction conditions will, most likely, be inaccurate.

Another interesting property of the photonic yield, as given by Eqs. 21.3 or 21.4, can be observed. Indeed, unlike the photochemical quantum yield, these equations indicate that $PHYD^{Lp,\Delta\lambda_{Lp},T}(t)$ (or $PHYD^{Lp,\ \Delta\lambda,\ T}(t)$) is undeniably time-dependent. The concentration of both the reactant and the end product are both time-dependent and non-linear functions, but the total number of photons entering the reactor ($P_0^{Lp,\Delta\lambda_{Lp}} \times t$) is necessarily increasing linearly with time. Hence, mathematically, the value of their ratio (e.g. $PHYD^{Lp,\Delta\lambda_{Lp},T}(t)$) must be variable through reaction-time. This adds a new and mandatory specification to characterise the photonic yield, namely, the sampling time (*t*) at which *PHYD*(*t*) is calculated must be indicated (otherwise, the value reported will not be informative).

Furthermore, the value of *PHYD*(*t*) must, in principle, be sensitive to any variation of the reaction experimental conditions or any changes occurring in the reactive medium composition, owing to the close dependency of the species concentrations on these factors. The *PHYD*(*t*) is, therefore, a relative quantity which should be used with caution.

Incidentally, it will be helpful to amend the current reference documents that are dedicated to the evaluation of photochemical reactions (such as the *IUPAC* reports, e.g [2].), with an emphasis on the non-linear character of the photonic yield when given by Eqs. 21.3 or 21.4.

Scheme 21.1 Photochromic $XY_2(4\Phi, 2k)$ reaction exposed to simultaneous UV and Vis irradiation

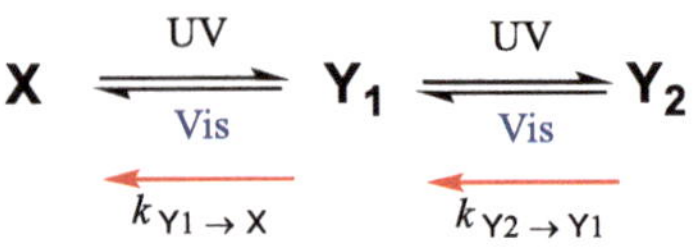

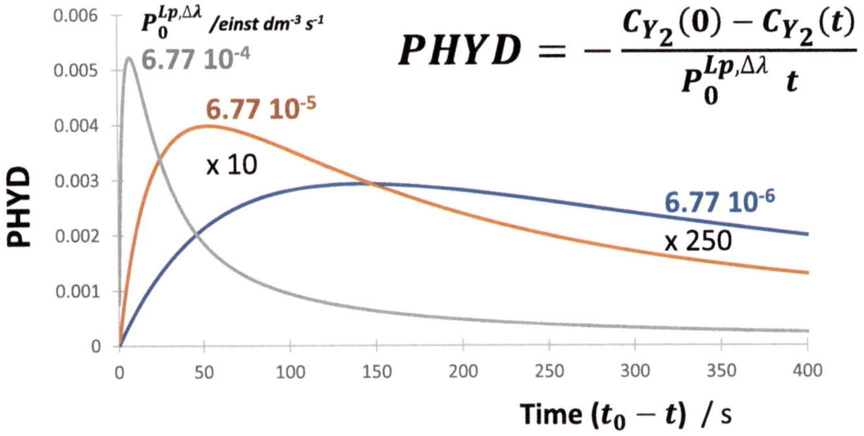

Fig. 21.1 Simulated plot of the $XY_2(4\Phi, 2k)$ reaction photonic yield (Eq. 21.5) variations with both reaction time (t-t$_0$) and radiation intensity. The curves have been rescaled with the multiplicative factors indicated. $C_{Y_2}^{Lp,\Delta\lambda,T}(0) = 0$. Reproduced with permission from [5], *Molecules*

An example of $PHYD_{Y_{end}}^{Lp,\Delta\lambda_{Lp},T}(t)$ (Eqs. 21.5) for the end product (Y_2) of a photochromic $XY_2(4\Phi, 2k)$ reaction (Scheme 21.1) exposed to a polychromatic beam that provides both ultraviolet (UV) and visible (Vis) lights over the *OSIA* range.

A change in the radiation intensity ($6.77\ 10^{-6} \le P_0^{Lp,\Delta\lambda} \le 6.77\ 10^{-4}$ einstein dm^{-3} s^{-1}) caused a dramatic variation of the photonic yield measured at a given time (Fig. 21.1). The values of $PHYD^{Lp,\ \Delta\lambda,\ T}(t)$ undergo up to a 250-fold decrease between the higher and the lower values of the radiation intensity. Furthermore, a non-linear behaviour of $PHYD_{Y_2}^{Lp,\Delta\lambda,T}(t)$ with reaction time is observed irrespective of the radiation intensity.

Many literature examples have reported the variation of $PHYD^{Lp,\Delta\lambda_{Lp},T}(t)$, such as with initial reactant concentration [6, 7] or with time or radiation intensity, Fig. 21.2 [8].

In the context of using the initial rate as a metric for the quantification of photokinetic quantities, the photonic yield equations can be rewritten for the evaluation of the efficiency of photothermal reactions subjected to polychromatic light. The dimensionless photonic yield of the reactant, now labelled as $PY_X^{Lp,\Delta\lambda_{Lp},T}$ (Eq. 21.6), is positive and time-independent. (for all cases, $r_{0,X}^{Lp,\Delta\lambda,T} = r_{0,X}^{Lp,\Delta\lambda_{Lp},T}$).

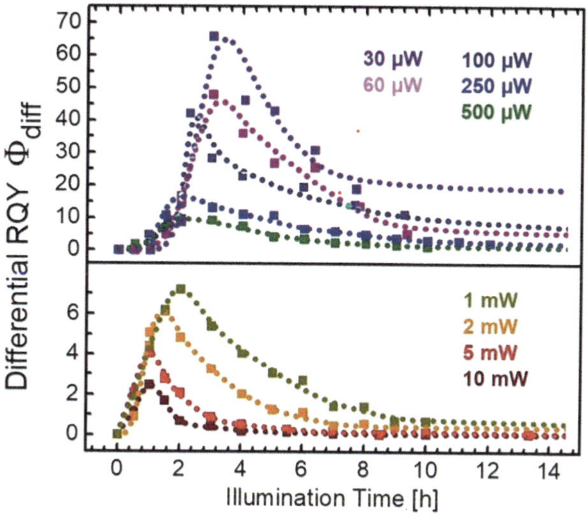

Fig. 21.2 Variation of the photonic yield (dubbed here the differential reaction quantum yield, $\Phi_{diff} = \Delta N_{product}/\Delta N_{photons}$) with irradiation time and excitation power (ranging between 30 μW and 10 mW) of an LED emitting between 410 and 460 nm, for the photocatalytic conversions of two different species. Reproduced with permission from [8], *ChemPhotoChem*

$$PY_X^{Lp,\Delta\lambda_{Lp},T} = -\frac{r_{0,X}^{Lp,\Delta\lambda,T}}{P_0^{Lp,\Delta\lambda_{Lp}}} \quad \left(or \quad PY_X^{Lp,\Delta\lambda,T} = -\frac{r_{0,X}^{Lp,\Delta\lambda,T}}{P_0^{Lp,\Delta\lambda}} \right) \tag{21.6}$$

Similar to *PHYD(t)*, PY_X is sensitive to any changes in the reaction conditions (e.g. $C_X^{Lp,\Delta\lambda_{Lp}}(0)$, $P_0^{Lp,\Delta\lambda_{Lp}}$, *SPM*, l_{irr}, …etc), and it is suited for comparison between investigations.

An analogous formula, $PY_A^{Lp,\Delta\lambda_{Lp},T}$, is also devised for the total absorbance trace, where $r_{0,\,A}$ replaces $r_{0,\,X}$ in Eq. 21.6 (the negative sign will be removed whenever $r_{0,\,A}$ is positive). Using $PY_A^{Lp,\Delta\lambda_{Lp},T}$ has the advantage of relieving the treatment from the necessity of acquiring the individual concentration trace of the reactant. Overall, $r_{0,\,X}$ and $r_{0,\,A}$ are easily obtained from either Eqs. 19.12 or 19.15 and the corresponding trace fitting parameters. The shape of PY_X (Eq. 21.6) adopts the same pattern of $r_{0,\,X}$ (or $r_{0,\,A}$ for PY_A) when any individual reaction parameter is varied. However, no expressions of *PY* can be assigned to the end products (see an example in Fig. 22.1).

21.4 The High Reactant Initial Concentration Issue for Quantum Yield

In principle, the species concentrations that lie outside the linearity range are not suitable for photokinetic investigation (Sects. 2.10–2.13). Accordingly, a high concentration in photokinetics is that falling beyond the higher limit of the linearity range of the considered species (belonging to the curved part of the calibration graph of that species, Fig. 2.11). The calibration graph deviation from linearity is proof of the non-applicability of the Beer–Lambert law. It is obvious that this deviation from linearity points to the likely occurrence of some phenomena different from just a quantitative light absorption by the individual molecules of the species present in the medium. Several interpretations have been proposed for the experimentally observed curvature of the calibration graph, including the absorptivity not being an inherent and/or invariant property of the species (Sect. 2.12).

Also, it is important to remember that the application of the Beer–Lambert law requires monochromatic light and does not apply at all when the light is polychromatic.

It is further interesting to notice that the literature does not seem to raise a concern about whether the quantum yield value at a given wavelength is a molecular property, i.e. whether it varies when the concentration is high. Despite the lack of evidence, it might be suggested that such a variation might happen in a way analogous to the variation postulated for the absorption coefficient considered in the same (high concentration) condition. This would mean that the quantum yield value measured at high species concentration might well be different from that measured at low concentration. Conceptually, only the latter should be considered as a photokinetic reference (since the former would be dependent on the concentration). If the quantum yield is proven not to be a molecular property, then working within the linear range of the calibration graph becomes a mandatory condition (as was set out from the onset in Chap. 2). It would be interesting if a future research project was dedicated to bringing evidence on this aspect.

In any case, a current and relatively popular practice advocates an evaluation of the reactant quantum yield value ($X \rightarrow Y_1$) by using a high initial concentration of that species. Such a condition is justified by the conversion of the rate of the reactant into a constant. That is, when $A_X(t) \gg 1$, the power number $10^{-A_X(t)} \cong 0$ and the factor $\left(1 - 10^{-A_X(t)}\right)$ are equal to unity, leading to a reactant-rate that is invariant with time (Eq. 21.7, which is a ubiquitous formula in the literature). The mathematical formulation of Eq. 21.7 sets the reaction to, presumably, obey zeroth-order kinetics [9–11].

$$\frac{dC_X(t)}{dt} = -\Phi_{X \rightarrow Y_1}\, P_0 \left(1 - 10^{-A_X(t)}\right) = -\Phi_{X \rightarrow Y_1}\, P_0 \tag{21.7}$$

The procedure goes on to determine the quantum yield value by the ratio,

$$\Phi_{X \rightarrow Y_1} = \frac{C_X(0) - C_X(t_1)}{P_0\, t_1} \tag{21.8}$$

The use of the incident radiation intensity (P_0) in Eq. 21.8 was usually justified by a total absorption by the reactant of all the photons reaching the reactor (due to its high concentration). In practice, the procedure does not set any particular conditions on t_1, so that, according to this approach, the time at which the measurement is performed could virtually take any value. Also, it does not impose any particular specifications on the light and/or on the lamps used (could be either mono-or polychromatic light, irrespective of the profile of the light source relative to the *OSIA* of the molecule).

Even though the above procedure is widely employed in the literature, it is relevant to notice that Eq. 21.7 can only be established if, at least, six conditions are satisfied:

(i) The rate-law (Eq. 19.1) perfectly applies at high concentrations as it does at low concentrations.
(ii) The incident light is totally absorbed by the reactant only, i.e. it exclusively serves the photochemical transformation of the reactant up to time t_1 (at which the concentration measurement is performed).
(iii) The quantum yield must be invariant with the wavelength.
(iv) The light must be considered as monochromatic.
(v) The profile of the light delivered matches the reactant *OSIA* $((\Delta\lambda_{Lp} \leq \Delta\lambda)$.
(vi) The observed concentration trace of the reactant should be a line.

However, a careful review of the above conditions shows that none of them could be fully justified.

Our previous discussion about the non-applicability of the Beer–Lambert in its classical linear form, when the concentration becomes high, should, in principle, impose an alteration of the rate-equation of the reaction to include the non-linearity of the calibration graph, which consequently invalidates condition (i). The precise amount of light absorbed by the reactant from the incident light over the time-interval t_1 cannot be known with the necessary accuracy due to competitive light absorptions by possible emerging new species (static or excited-state complexes) in the highly concentrated medium and/or by the direct reactant's photoproducts. Therefore, the second condition is no longer verified. As a postulate, the quantum yield might or might not be invariant with wavelength, so that, until proven by experimental evidence (either way), it is not consistent to claim the validity of condition (iii). The next condition (iv) must be discarded whenever the experimental irradiation setup does not include a monochromator (Sect. 2.5). When polychromatic light is used, the mathematical description of the system is provided by an integro-differential rate-equation, which does not correspond to the formulation set out in Eq. 21.7 and hence weakens further a rational adoption of condition (i). Finally, the lamp profile must be known beforehand to test condition (v), which is rarely the case in the published literature using this approach.

It is hence clear that, despite its popularity, the above practice (high reactant initial concentration and usage of Eqs. 21.7 and 21.8) is an approximation procedure that is not able to provide a unique value for the quantum yield of the reactant. Also, objectively, Eq. 21.7 is nothing more than Eq. 21.3, which was assigned to the photonic yield of the reactant $(PHYD_X^{Lp,\Delta\lambda_{Lp},T}(t))$. As such, Eq. 21.7 does not calculate $\Phi_{X\to Y_1}$ as an absolute value, but rather a relative quantity that, moreover, depends on time (e.g. depends on the extent of the time-interval t_1).

Alternatively, when the reactant concentration trace is available and the reaction is conducted under strictly monochromatic irradiation, there is no practical need to work at high concentration since the reactant quantum yield value can readily be measured from its initial-rate, as given by Eq. 20.17. It can be formulated for the general case (the reactant reacts through several photo-and thermal reaction-steps), from the initial reactant-rates (Eqs. 21.9 and 21.10, obtained from Eqs. 19.5 and 19.12, respectively), and the fitting of the traces by Eq. 19.10, as

$$Theor: r_{0,X}^{\lambda_{irr},T} = -P_{a_X}^{\lambda_{irr}}(t) \sum_{j'}^{v_\Phi^0} \left(\Phi_{X \to Y_{j'}}^{\lambda_{irr}} \right) - \sum_{j'}^{u_k^0} \left(k_{X \to Y_{j'}}^{T} \, C_X^{Lp,\Delta\lambda,T}(t) \right) \tag{21.9}$$

$$Fit: r_{0,X}^{\lambda_{irr},T} = -\frac{1}{ln(10)} \sum_{i=1}^{i_{\Phi,0}} \frac{\omega_{i0}^{\Phi} \, cc_0^{\Phi} \, k_{i0}^{\Phi}}{1 + cc_0^{\Phi}} - \sum_{i=1}^{i_{\Delta,0}} \omega_{i0}^{T} \, k_{i0}^{T} \tag{21.10}$$

And considering the equality, $Theor: r_{0,X}^{\lambda_{irr},T} = Fit: r_{0,X}^{\lambda_{irr},T}$, the sum of the quantum yields (Eq. 21.11) of the photochemical reaction-steps depleting X is given by

$$\sum_{j'}^{v_\Phi^0} \left(\Phi_{X \to Y_{j'}}^{\lambda_{irr}} \right) = -\frac{Fit: r_{0,X}^{\lambda_{irr},T} + C_X^{Lp,\Delta\lambda,T}(0) \sum_{j'}^{u_k^0} \left(k_{X \to Y_{j'}}^{T} \right)}{P_0^{\lambda_{irr}} \left(1 - 10^{-A_X^{\lambda_{irr},T/\lambda_{irr}}(0) \frac{l_{irr}}{l_{obs}}} \right)} \tag{21.11}$$

21.5 Concluding Remarks

The aspects reviewed in this chapter might bring relevant elements to a discussion about the procedures and formulae adopted in the literature for the determination of the quantum and photonic yields. The alternatives proposed are sustainably effective tools for photokinetic investigation. The latter options have been derived from the general equations established for photothermal reactions. The model equations (Eqs. 19.5, 19.10 and 19.13) turn out to be good means of analysis, contributing to the rationalisation/standardisation of photokinetics. Further application of these equations to other photokinetic aspects shall be discussed in the following chapters.

References

1. Braslavsky SE (2007) Glossary of terms used in photochemistry. IUPA Pure Appl Chem 79: 293–465. https://doi.org/10.1351/pac200779030293
2. Braslavsky SE, Braun AM, Cassano AE, Emeline AV, Litter MI, Palmisano L, Parmon VN, Serpone N (2011) Glossary of terms used in photocatalysis and radiation catalysis (IUPAC recommendations 2011). Pure Appl Chem 83:931–1014. https://doi.org/10.1351/PAC-REC-09-09-36

3. Maafi M (2023) On photokinetics under monochromatic light. Front Chem 11:1233151. https://doi.org/10.3389/fchem.2023.1233151
4. Maafi M (2024) On photokinetics under polychromatic light. Front Chem 12:1233151. https://doi.org/10.3389/fchem.2024.1367276
5. Maafi M (2025) Photokinetics of Photothermal reactions. Molecules 30:330. https://doi.org/10.3390/molecules30020330
6. Jakusová K, Cigáň M, Donovalová J, Gáplovský M, Sokolík R, Gáplovský A (2014) Light initiated E–Z and Z–E isomerization of isatinphenylsemicarbazones: Tautomeric equilibrium effect. J Photochem Photobiol A Chem 288:60–69. https://doi.org/10.1016/j.jphotochem.2014.05.004
7. Menzel JP, Noble BB, Blinco JP, Barner-Kowollik C (2021) Predicting wavelength-dependent photochemical reactivity and selectivity. Nat Commun 12:1691. https://doi.org/10.1038/s41467-021-21797-x
8. Reiß B, Hu Q, Riedle E, Wagenknecht H-A (2021) The dependence of chemical quantum yields of visible light Photoredox catalysis on the irradiation power. ChemPhotoChem 5:1009–1019. https://doi.org/10.1002/cptc.202100090
9. Stranius K, Borjesson K (2017) Determining the photoisomerization quantum yield of photoswitchable molecules in solution and in the solid state. Sci Rep 7:41145. https://doi.org/10.1038/srep41145
10. Stadler E, Eibel A, Fast D, Freißmuth H, Holly C, Wiech M, Moszner N, Gescheidt G (2018) A versatile method for the determination of photochemical quantum yields *via* online UV-vis spectroscopy. Photochem Photobiol Sci 17:660–669. https://doi.org/10.1039/C7PP00401J
11. Amouroux B, Roux C, Micheau J-C, Gauffre F, Coudret C (2019) A photochemical determination of luminescence efficiency of upconverting nanoparticles. Beilstein J Org Chem 15:2671–2677. https://doi.org/10.3762/bjoc.15.260

Chapter 22
Quantification of Photothermal Reactions

22.1 Introduction

We will now explore a few ways to use the explicit general model equations (Eqs. 19.5, 19.10, and 19.13) for the quantification of some kinetic properties of photo- and photothermal reactions when exposed to either mono- or polychromatic light. In each case, an interpretation of the phenomenon is provided on the basis of the observed behaviour of the simulated data in relation to the fundamental kinetic equations.

22.2 Quantum Yield Variation with Wavelength

One of the interesting tools offered by the trace fitting by Eq. 19.10 is the use of the initial reactant-rate for the calculation of the quantum yield of the reactant photoconversion into its photoproduct or photoproducts. This is readily achieved by Eq. 21.11 for a photothermal reaction successively exposed, in separate experiments (maintained at temperature T), to different monochromatic irradiations (λ_{irr}). For each photothermal reaction experiment, the reactant concentration trace is acquired and then fitted by an appropriate Eq. 19.10 and its initial rate ($Fit : r_{0,X}^{\lambda_{irr},T}$) calculated by Eq. 19.12. If the reactant is thermally active, then its thermal concentration trace is fitted by the adequate Eq. 19.10 and the reaction rate-constant determined.

Here we need to review some details relative to a few possible situations.

When the reactant is thermally stable ($k_{X\rightarrow Y_{j'}}^{T} = 0$), it undergoes a single photochemical reaction and produces a unique photoproduct, $X \rightarrow Y_1$; the value of the quantum yield of this reaction-step, $\Phi_{X\rightarrow Y_1}^{\lambda_{irr}}$, is determined by an equation analogous to Eq. 20.17 for each individual wavelength of the irradiation beam used.

M. Maafi, *Photokinetics*, https://doi.org/10.1007/978-3-031-98985-8_22

$$\Phi^{\lambda_{irr}}_{X \to Y_1} = -\frac{r^{\lambda_{irr},T}_{X}(0)}{P^{\lambda_{irr}}_{a_X}(0)} = -\frac{Fit : r^{\lambda_{irr},T}_{0,X}}{P^{\lambda_{irr}}_{0}\left(1 - 10^{-A^{\lambda_{irr},T/\lambda_{irr}}_{X}(0)\,\frac{l_{irr}}{l_{obs}}}\right)} \tag{22.1}$$

If the reactant produces more than one product ($X \to Y_j$) photochemically, Eq. 21.11 can only deliver a value for the sum of the quantum yields of the photoreaction-steps leading to these products. It is not possible to know from Fit : $r^{\lambda_{irr},T}_{0,X}$ the individual quantum yields of the different photoreactions of the reactant. In this situation (i.e. occurrence of multiple photoproducts from X), further analysis is required to unravel the quantum yield values of the individual reaction-steps. This can be achieved by the elucidation methodology discussed in Sect. 20.2.

For the case where the reactant ($X = Y_0$) depletes thermally ($k^T_{X \to Y_{j'}} \neq 0$), the sum of the rate-constants of the concurrent thermal reaction-steps ($X \to Y_{j'}$), measured in the absence of irradiation, is obtained as a single constant (k^T_{10}). The corresponding fitting equation is a mono-exponential function, derived from Eq. 19.10, where $\omega_0 = \omega^{\Phi}_{i0} = 0$.

$$C^T_X(t) = \omega^T_{10}\, e^{-k^T_{10}\, t} \tag{22.2}$$

From a practical point of view, Eq. 22.2 can be applied to the reactant thermal trace, even if its actual $C^T_X(t)$ value is not exactly equal to $C^T_X(0)$, e.g. $C^T_X(t) = 0.9\, C^T_X(0)$. This consideration comes from the fact that such thermal reactions are bound to start as soon as the solution is made so that, by the time the kinetic analysis is performed, it is certain that $C^T_X(t) \neq C^T_X(0)$. The recommendation is to proceed with the kinetic measurement at the earliest after preparing the solution at a medium temperature, preferably equal to room temperature. However, in order to determine the quantum yield (assuming a unique $X \to Y_1$ photoreaction-step), the exact value of $C^{Lp,\Delta\lambda,T}_X(0)$ must be known, as stated by Eq. 21.11, an information that might not be available for all reactions, and therefore, the value of the quantum yield of $X \to Y_1$ or the value of the sum of quantum yields (for multi-photoconversions of X) will not be accessible (when X is thermally active).

When obtained, future data of the quantum yield variation with wavelength will certainly contribute to expand the group of molecules and systems reviewed in Chap. 18.

22.3 Auto-Photostabilisation by $C^{Lp,\Delta\lambda,T}_X(0)$

The effect of the initial reactant concentration $C^{Lp,\Delta\lambda,T}_X(0)$ on the reactivity of a photothermal reaction is discussed in the following.

Analytically, a change of the initial reactant concentration is expected to induce a variation of reactivity when the photochemical term of the concentration rate-equation of the reactant reaction includes the total absorbance (A_{tot}) factor. In other words, $C_X^{Lp,\Delta\lambda,T}(0)$ will have an effect on reactivity if several species concomitantly absorb at the irradiation wavelength (λ_{irr}) or over the *OSIA* domain (i.e. the overlap section of irradiation and absorbance).

From our study of the primary photoreaction in Chap. 11, when X is the only absorbing species of the incident excitation light in the reactive medium, we have seen that an increase in the initial concentration of the reactant causes an increase in both the initial rate and the half-life time of the reaction (Sect. 11.5). However, for this particular photoreaction considered under a monochromatic irradiation, the rate-constant is independent of $C_X^{\lambda_{irr},T}(0)$, as it is evident from its formulation (Eq. 11.21) in the analytically derived integrated rate-law (Eq. 11.20).

In all other reaction cases, there is a quantitative and measurable impact of a variation of $C_X^{\lambda_{irr},T}(0)$ values on the photoreactivity of Φ-order reactions when all remaining reaction attributes are kept the same. Theoretically, the Φ-order kinetic character of concentration and total absorbance traces should be preserved when $C_X^{\lambda_{irr},T}(0)$ changes (however, the reaction-rate should be affected due to the presence of multiple absorbing species of the excitation light).

Under monochromatic irradiation, it was experimentally proven, in Chap. 17, that photoreversible reactions gradually slow down for higher values of $C_X^{\lambda_{irr},T}(0)$.

Equally, *RK*-simulated data demonstrate that the initial reactant-rate increases with increasing initial reactant concentration for photoreactions obeying a variety of mechanisms [1]. The fundamental explanation of this pattern, observed for photoreactions under monochromatic irradiation, stems from an increase in the term $\left(1-10^{-A_{tot}^{\lambda_{irr},T/\lambda_{irr}}(0)\,(l_{irr}/l_{obs})}\right)$ when $C_X^{\lambda_{irr},T}(0)$ increases (owing to a decrease in $10^{-A_{tot}^{\lambda_{irr},T/\lambda_{irr}}(0)\,(l_{irr}/l_{obs})}$ value), which increases the absolute value of the theoretical reactant initial-rate ($-r_{0,X}^{\lambda_{irr},T}$), as given by Eq. 22.3 (the latter is derived from Eq. 19.6). The concomitant retardation of the reactant-rate is also proven for photoreactions [1], as shown in the example proposed in Fig. 22.1. The slowdown of the reactant reaction with an increase in $C_X^{\lambda_{irr},T}(0)$ impacts the reactivity of immediate and far away photoproducts, irrespective of the photoreaction mechanism in play. Indeed, the reduction observed for the reactant rate with an increase in $C_X^{\lambda_{irr},T}(0)$ is also proportionally recorded for all the products of the photoreaction [1].

$$Theo : r_{0,X}^{\lambda_{irr},T} = -P_0^{\lambda_{irr}}\left(1-10^{-A_{tot}^{\lambda_{irr},T/\lambda_{irr}}(0)\,\frac{l_{irr}}{l_{obs}}}\right)\sum_{j=1}^{v_\Phi^0}\left(\Phi_{X\to Y_j}^{\lambda_{irr}}\right) \tag{22.3}$$

The initial-rate of the total absorbance trace, whose equation is worked out from Eq. 19.9 (as expressed in Eq. 22.4) stipulates that its variation with an increasing $C_X^{\lambda_{irr},T}(0)$ is dependent, on the one hand, on the increasing value of the factor $\left(1-10^{-A_{tot}^{\lambda_{irr},T/\lambda_{irr}}(0)\,(l_{irr}/l_{obs})}\right)$ and, on the other hand, on both the value and sign of a second factor involving the absorptivities of the reactant and its immediate photoproducts (as given by the last bracket in Eq. 22.4)

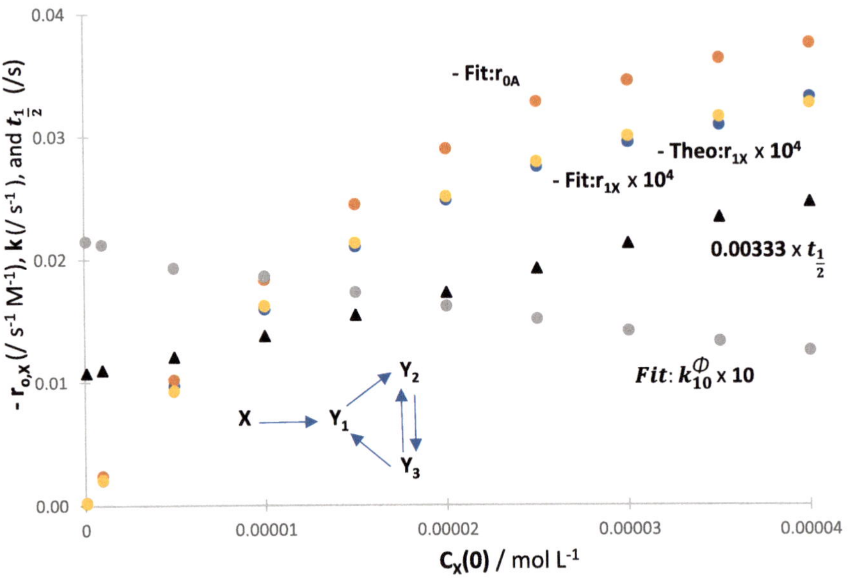

Fig. 22.1 Variation of the initial reactant rates (in M s^{-1}) with an increasing reactant initial concentration. Reported are the negative values of $RK: r_{0X}^{\lambda_{irr}}$, $Fit: r_{0X}^{\lambda_{irr}}$, and $Fit: r_{0A}^{\lambda_{irr}}$ (the latter being the initial rate of the total absorbance), as well as the fitting rate-constant (k_{10}^{Φ}) of the adequate Eq. 19.10 and the half-life time of the reactant (plain triangles) for the RK-simulated $XY_3(5\Phi)$ reaction indicated in the inset. (Reproduced with permission from [1], *Frontiers in Chemistry*)

$$Theor: r_{0,A}^{\lambda_{irr},T/\lambda_{obs}} = \sum_{i=1}^{v_{\Phi}^{0}} \varepsilon_{Y_j}^{\lambda_{obs}}\, l_{obs}\, r_{Y_j}^{\lambda_{irr},T}(0)$$

$$= P_0^{\lambda_{irr}}\left(1 - 10^{-A_{tot}^{\lambda_{irr},T/\lambda_{irr}}(0)\,\frac{l_{irr}}{l_{obs}}}\right) l_{obs}\left(\sum_{i=1}^{v_{\Phi}^{0}} \varepsilon_{Y_j}^{\lambda_{obs}}\, \Phi_{X\to Y_j}^{\lambda_{irr}} - \varepsilon_X^{\lambda_{obs}} \sum_{i=1}^{v_{\Phi}^{0}} \Phi_{X\to Y_j}^{\lambda_{irr}}\right) \tag{22.4}$$

Figure 22.1 shows an example of the initial rates for an $XY_3(5\Phi)$ reaction. This reaction includes a depleting reactant and a cyclic sequence between the three products (with the four species absorbing at λ_{irr}). The negative values of the initial rates are shown to increase with an increase in $C_X^{\lambda_{irr},T}(0)$. These correspond to the concentration trace of the reactant ($-Theor: r_{0,X}^{\lambda_{irr},T}$ and $-Fit: r_{0,X}^{\lambda_{irr},T}$) and that of the total absorbance trace ($-Fit: r_{0,A}^{\lambda_{irr},T}$).

Figure 22.1 also includes the variation of the fitting parameter, $k_{1,0}^{\Phi}$, that belongs to the single mono-Φ-order term of the equation fitting the reactant concentration trace (Eq. 19.10, where $i_{\Phi,j} = 1$ and $\omega_j = \omega_{ij}^{T} = 0$). The diminishing value of $k_{1,0}^{\Phi}$ with increasing $C_X^{\lambda_{irr},T}(0)$ clearly indicates a slowdown of the reaction. This is confirmed by longer half-life times when $C_X^{\lambda_{irr},T}(0)$ increases.

In another instance, let us look at the initial rate of the total absorbance traces of an $XY_2(2\Phi)$ sequence, where the reactant undergoes concurrent photoreactions that generate two products, $Y_2 \leftarrow X \rightarrow Y_1$. The initial rates of the species, in terms of concentration, are worked out from Eq. 19.5. We consider that the reaction is driven by a monochromatic light of wavelength λ_{irr}, which is absorbed by all the species in the medium.

$$r_{0,X}^{\lambda_{irr},T} = -\left(\Phi_{X\rightarrow Y_1}^{\lambda_{irr}} + \Phi_{X\rightarrow Y_2}^{\lambda_{irr}}\right) P_0^{\lambda_{irr}} \left(1 - 10^{-A_{tot}^{\lambda_{irr},T/\lambda_{irr}}(0)\,\frac{l_{irr}}{l_{obs}}}\right) \tag{22.5}$$

$$r_{Y_1}^{\lambda_{irr},T}(t) = \Phi_{X\rightarrow Y_1}^{\lambda_{irr}} P_0^{\lambda_{irr}} \left(1 - 10^{-A_{tot}^{\lambda_{irr},T/\lambda_{irr}}(0)\,\frac{l_{irr}}{l_{obs}}}\right) \tag{22.6}$$

$$r_{Y_2}^{\lambda_{irr},T}(t) = \Phi_{X\rightarrow Y_2}^{\lambda_{irr}} P_0^{\lambda_{irr}} \left(1 - 10^{-A_{tot}^{\lambda_{irr},T/\lambda_{irr}}(0)\,\frac{l_{irr}}{l_{obs}}}\right) \tag{22.7}$$

The initial rate of the total absorbance trace takes into account the individual absorption coefficients $\left(\varepsilon_{Y_j}^{\lambda_{obs}}\right)$ of the species, together with the products of the observation optical path length (l_{obs}) and the corresponding initial rates, which are then added together into a sum (with $v_\Phi^0 = 1, 2$), as

$$\begin{aligned} Theor: r_{0,A}^{\lambda_{irr},T/\lambda_{obs}} &= P_0^{\lambda_{irr}} \left(1 - 10^{-A_{tot}^{\lambda_{irr},T/\lambda_{irr}}(0)}\right) l_{obs} \\ &\times \left(\varepsilon_{Y_1}^{\lambda_{obs}} \Phi_{X\rightarrow Y_1}^{\lambda_{irr}} + \varepsilon_{Y_2}^{\lambda_{obs}} \Phi_{X\rightarrow Y_2}^{\lambda_{irr}} - \varepsilon_X^{\lambda_{obs}} \left(\Phi_{X\rightarrow Y_1}^{\lambda_{irr}} + \Phi_{X\rightarrow Y_2}^{\lambda_{irr}}\right)\right) \end{aligned} \tag{22.8}$$

The sign of the initial rate, $r_{0,A}^{\lambda_{irr},T}$, for this reaction under monochromatic light will depend on the sign of the second bracket in the right-hand-side term of Eq. 22.8, but its magnitude always evolves towards smaller values when $C_X^{\lambda_{irr},T}(0)$ increases (because of a diminishing value of the coefficient, $1 - 10^{-A_{tot}^{\lambda_{irr},T/\lambda_{irr}}(0)\,\frac{l_{irr}}{l_{obs}}}$).

Conceptually, the pattern induced by the variation of $C_X^{Lp,\Delta\lambda,T}(0)$ on the initial reactant-rate should be preserved when the excitation radiation changes from a monochromatic to a polychromatic light. The reason resides in the fact that the rate-equation of reactions under polychromatic light is no more than just a sum of individual rates for a series of monochromatic irradiations.

Irrespective of the mechanism of the photo- or the photothermal reaction investigated, the analyses of *RK*-simulated traces show that the reactant initial-rate absolute value ($-\,r_{0,X}^{Lp,\Delta\lambda,T}$) increases, whereas the reaction slows down when the initial reactant concentration value increases [2, 3].

Illustrations of this behaviour are provided hereafter for a photoreaction and a photothermal sequence. Both of them are subjected to polychromatic irradiations, and for both the concentrations of the light-absorbing reactive species fall within the

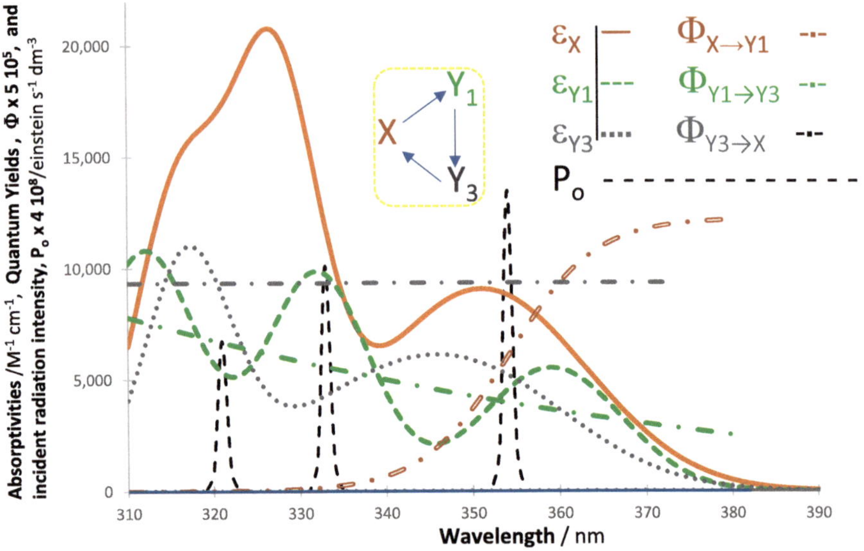

Fig. 22.2 Species (ε and Φ) and lamp features for the studied trimolecular, cyclic $XY_2(3\Phi)$ photoreaction (whose mechanism is shown in the inset). (Reproduced with permission from [2], *Frontiers in Chemistry*)

species' respective spectrophotometric linearity ranges of the calibration graphs (i.e. where Eq. 19.5 is consistent and the basis for *RK*-numerical calculations).

The spectroscopic features of the species of a trimolecular cyclic $XY_2(3\Phi)$ photoreaction, as well as the profile of the lamp used for its irradiation and the patterns of each species' quantum yield with wavelength, are shown in Fig. 22.2. The *OSIA*s of the species are very narrow due to the properties of the lamp that has spiked emissions at 321, 333, and 354 nm, where the three species absorb (their electronic spectra overlap over a 310 – 390 nm region). The reactant is virtually inactive for irradiation below 335 nm, as indicated by its quantum yield pattern (whereas the products are photoactive for the three emission regions of the lamp). The observation wavelength was set to 326 nm (corresponding to the short-wavelength maximum on the reactant spectrum).

Evidence of the occurrence of an auto-photostabilisation of the reaction, resulting from an increase in the initial reactant concentration, was supplied by monitoring the variations of the initial rates of both the reactant and its immediate photoproduct (Y_1) and the initial-rate of the reaction that is worked out from the total absorbance trace. The value of each initial rate obtained for each $C_X^{Lp,\Delta\lambda,T}(0)$ has been defined theoretically by *RK*-calculation and from the fitting data. In Fig. 22.3, the correlation between these three quantities is well-indicated by linear relationships. It is a confirmation of the required rule stated by Eqs. 22.5–22.7. The data points making up the lines correspond to initial rates obtained for the reaction starting with different $C_X^{Lp,\Delta\lambda,T}(0)$. It is clear that the higher the initial concentration (ranging between $1.3\ 10^{-6}$ and 10^{-4} M), the higher the absolute value of the initial reaction-rate, which

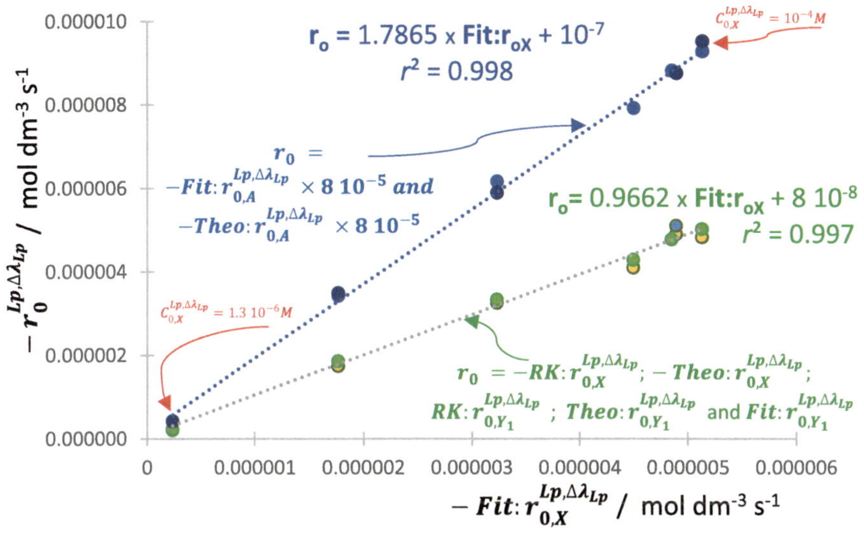

Fig. 22.3 Variation of the initial rates of the reactant, $RK : r_{0X}^{Lp,\Delta\lambda}$ and $Theo : r_{0X}^{Lp,\Delta\lambda}$ and the photoproduct Y_1, $Fit : r_{0Y_1}^{Lp,\Delta\lambda}$, $RK : r_{0Y_1}^{Lp,\Delta\lambda}$, and $Theo : r_{0Y_1}^{Lp,\Delta\lambda}$, and that of the total absorbance trace, $Fit : r_{0A}^{Lp,\Delta\lambda}$ and $Theo : r_{0A}^{Lp,\Delta\lambda}$, with the initial reactant rate, $Fit : r_{0X}^{Lp,\Delta\lambda}$, obtained from the fitting of traces corresponding to initial reactant concentrations of 1.3 10^{-6}, 1.3 10^{-5}, 3 10^{-5}, 5 10^{-5}, 7 10^{-5}, 9 10^{-5}, and 10^{-4} *M* (all belong to the linearity range of the reactant calibration graph). (Reproduced with permission from [2], *Frontiers in Chemistry*)

testifies of an auto-photostabilisation of the reaction (this is corroborated by longer half-life time values with higher $C_X^{Lp,\Delta\lambda,T}(0)$ values).

The correlation established between $r_{0,X}^{Lp,\Delta\lambda,T}$ and $r_{0,A}^{Lp,\Delta\lambda,T}$ in Fig. 22.3 can readily be explained by the inherent proportionality that exists between the initial velocities of the reactant and the photoproduct $\left(r_{0,X}^{Lp,\Delta\lambda,T}/r_{0,Y_1}^{Lp,\Delta\lambda,T} = \text{constant}\right)$ and that of the reaction ($r_{0,A}^{Lp,\Delta\lambda,T}$) because the latter is a linear combination of $r_{0,X}^{Lp,\Delta\lambda,T}$ and $r_{0,Y_1}^{Lp,\Delta\lambda,T}$. This is a general concept that should hold for any overall photomechanism of the reaction, irrespective of the experimental conditions selected for the reactive system. In this context, it is interesting that the subsequent photoproducts (appearing in the reactive medium after *X*) should see proportionality between their reactivities and that of the reactant, i.e. they evolve in the same direction so that an increase in $r_{0,X}^{Lp,\Delta\lambda,T}$ and its slower reactivity (as the half-life time is longer) imply a reduction in the reactivity of the photoproducts (and vice versa). Finally, the measured initial rates ($r_{0,X}^{Lp,\Delta\lambda,T}$, $r_{0,A}^{Lp,\Delta\lambda,T}$, etc.) are independent of the total profile of the lamp ($\Delta\lambda_{Lp}$) but are only a function of the *OSIA* ($\Delta\lambda$), i.e. whatever the difference between the profile of the lamp (e.g. $\Delta\lambda_{Lp} \neq \Delta\lambda$), we always have $r_{0,X}^{Lp,\Delta\lambda,T} = r_{0,X}^{Lp,\Delta\lambda_{Lp},T}$.

A corollary advantage of studying the initial concentration effect on the initial rate relates to the pattern, $-r_{0,X}^{Lp,\Delta\lambda,T} = f\left(C_X^{Lp,\Delta\lambda,T}(0)\right)$, which happens to be that of the photonic yield, $PY_X^{Lp,\Delta\lambda_{em},T}$ (Eq. 21.6), since division of $r_{0,X}^{Lp,\Delta\lambda,T}$ by the constant incident radiation intensity ($P_0^{Lp,\Delta\lambda_{em}}$) affects the magnitude but not the shape of the pattern.

Similar behaviour and conclusions are expected for photothermal reactions. The addition of pure thermal reaction-steps does not, in principle, alter the mathematical framework adopted for photoreactions (except that the theoretical expressions of $r_{0,X}^{Lp,\Delta\lambda,T}$ or $r_{0,A}^{Lp,\Delta\lambda,T}$ might now involve thermal terms, and their fitting values are calculated by equations (Eqs. 19.12 and 19.15) that include the fitting parameters of the thermal processes).

However, due to the presence of thermal processes in the mechanism of the reaction, the auto- photostabilisation might not be clearly noticeable on the overall photokinetics of the reaction. Therefore, the slowing down of the photoreaction is more evident for some photothermal reactions than for others.

In all cases, an equivalent slowdown/speeding up of the reaction-rate, comparable to the one shown for higher/lower values of $C_X^{Lp,\Delta\lambda,T}(0)$, should be expected when the irradiation optical path length (l_{irr}) takes higher/lower values. This follows from the fact that a variation of the total absorbance is due to the product of these parameters ($C_X^{Lp,\Delta\lambda,T}(0) \times l_{irr}$). Furthermore, a similar effect would be observed when proportionally changing the value of the intrinsic parameter, $\varepsilon^{\lambda_{irr}}$, also because of its presence in the formula of $A_{tot}^{\lambda_{irr},T/\lambda_{irr}}$. The latter variation would correspond to investigating a series of derivatives whose absorption coefficients differ, a feature that can be useful for the optimisation of reactive systems to tune their properties for a suited application.

Incidentally, the ratios of final concentrations of pairs of species that either reach photostationary states or are end products remain constant and invariant with $C_X^{Lp,\Delta\lambda}(0)$ values, both under mono- and polychromatic light conditions. Also, the shape of the traces will vary to a more linear pattern when the value of $A_{tot}^{\lambda_{irr},T/\lambda_{irr}}$ absorption increases (i.e. through an increase in either $C_X(0)$, l_{irr}, or $\varepsilon^{\lambda_{irr}}$, as shown in Fig. 11.3). An example of the variation of the initial velocity resulting from an increase in the initial reactant concentration is illustrated in Fig. 22.4 for a $XY_2(4\Phi, 2k)$ reaction.

The patterns obtained on simulated data are corroborated by experimental results relative to the degradation of itraconazole and ibuprofen in solutions exposed to polychromatic light, where the degradation rates decreased as the initial concentrations increased [4, 5]. The quantification of reactivity in these studies was performed by fitting the traces to first-order equations.

For instance, itraconazole rate-constants recorded under *UV-A* radiation for initial concentrations of 0.1 and 1 mg mL^{-1} were 0.0822 and 0.0685 hour^{-1}, respectively, i.e. its degradation rate decreased as its initial concentration increased [5]. The first-order rate-constant decrease was attributed to a reduction in the obtained photon quantum per unit molecule when the initial concentration is increased, hence, diminishing photodegradation levels [4]. In this interpretation, the ratio of photons supplied to the reaction medium to the number of molecules present in solution is put forward as the main reason for the observed reactivity decrease. Such an explanation

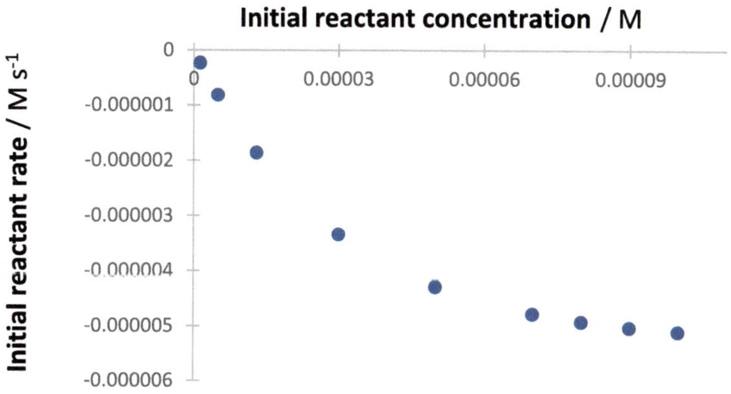

Fig. 22.4 Trend of the initial rate ($r_{0,X}^{Lp,\Delta\lambda,T}$) towards more negative values when increasing the initial concentration ($C_X^{Lp,\Delta\lambda,T}(0)$) for the $XY_2(4\Phi, 2k)$ photothermal reaction. (Reproduced with permission from [3], *Molecules*)

does not account for all cases because the phenomenon persists when the light intensity is very high, i.e. the number of photons >> number of molecules.

22.4 Effect of Spectator Molecules on Photoreactivity

In many instances, the reactive medium contains the photothermal reaction as well as one or several spectator molecules (*SPM*s), which are both thermally and photochemically inert and do not interfere with the reaction components (Chaps. 2 and 16). However, the *SPM*s will affect photoreactivity if they happen to absorb in the *OSIA* region. We have seen that the light absorption by *SPM*s results in depriving the reactive system of a fraction of the incident light (Chap. 16). The reduction in the amount of light available to the reaction induces a reduction in reactivity.

The total absorbance cumulates the individual time-variable absorbances of the photothermal reaction species and the constant absorbance of the "n_{SPM}" *SPM*s ($1 \leq r \leq n_{SPM}$).

$$A_{tot}^{\lambda_{irr},T/\lambda_{irr}}(t) = \sum_{r=1}^{n_{SPM}} A_{SPM_r}^{\lambda_{irr},T/\lambda_{irr}} + \sum_{j=0}^{n_{sp}} A_{Y_j}^{\lambda_{irr},T/\lambda_{irr}}(t) \tag{22.9}$$

The total absorbance is considered at each wavelength to conform with the formulation of the rate-equation (which is a sum of individual rates at each λ_{irr}).

Formally, for otherwise constant experimental conditions, the higher the value of $A_{SPM_r}^{\lambda_{irr},T/\lambda_{irr}}$ over *OSIA*, the lower the value of $PKF^{\lambda_{irr}}(t)$ and, hence, the lower the individual rates at the concerned λ_{irr}, which means an overall reduction in the rates of the reaction species (e.g. a lower global rate of the photothermal reaction).

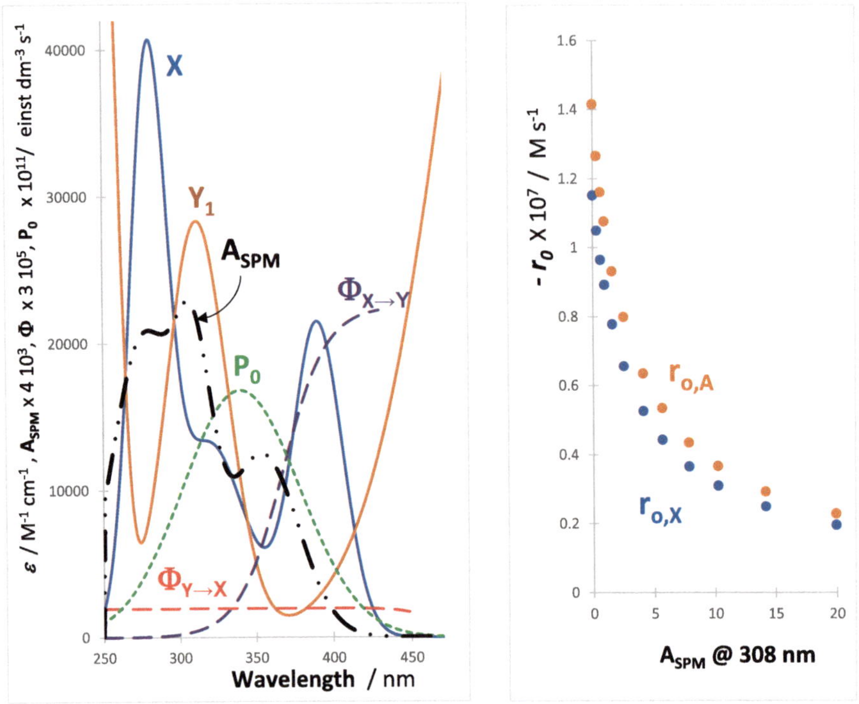

Fig. 22.5 Left: properties of the species of the $XY_1(2\Phi, 1k)$ photothermal reaction, the *SPM* absorption spectrum, and the lamp profile used for irradiation. Right: plots of the initial rates decrease with increasing *SPM* absorbance. (Reproduced with permission from [3], *Molecules*)

The phenomenon is illustrated by the behaviour of the species reactivities of an $XY_1(2\Phi, 1k)$ photothermal reaction (with $k^T_{Y_1 \rightarrow X} = 0.0028\ s^{-1}$) and an $XY_2(2\Phi)$ divergent photoreaction, $Y_1 \leftarrow X \rightarrow Y_3$. For both reactions, the electronic spectra of the species overlap over a large section.

The lamp profile covers the whole of the absorption of the species for the $XY_2(2\Phi)$ reaction (Fig. 22.5) but variably for the other reaction (Fig. 22.4). The quantum yield has various patterns, including sigmoid, constant, and curved outlines with the irradiation wavelength.

The higher the value of the *SPM* absorbance (or concentration), the smaller the reactant $(-r_{0,X}^{Lp,\Delta\lambda,T})$ and the total absorbance $(-r_{0,A}^{Lp,\Delta\lambda,T})$ initial rates and the concomitant increase in the thermal reaction-step contribution to the kinetics of the overall reaction kinetics (i.e. the value of the reactant concentration at *pss*, $C_X^{Lp,\Delta\lambda,T}(pss)$, increases with increasing $A_{SPM_r}^{\lambda_{irr},T/\lambda_{irr}}$).

The photochemical reaction-steps of the divergent $XY_2(2\Phi)$ reaction are assumed to activate two separate functional groups within the reactant molecule (Fig. 22.6), a situation analogous to the reaction depicted in Scheme 4.4. Each of the functional groups of X is characterised by one of the absorption bands of the reactant electronic

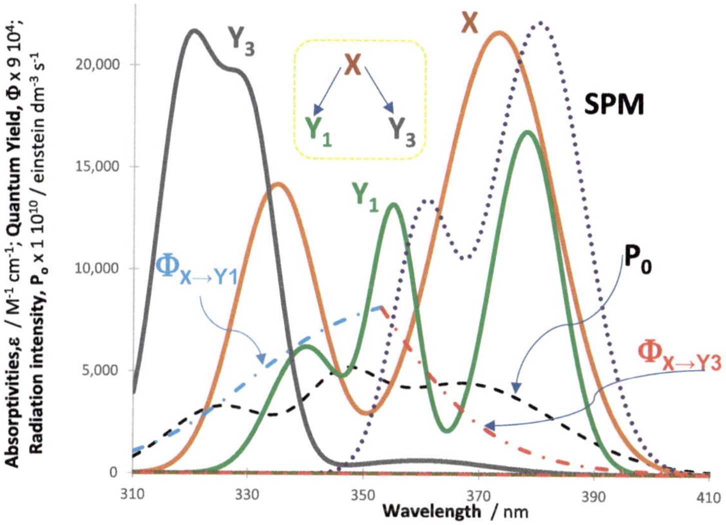

Fig. 22.6 Absorptivities (ε, solid lines) and quantum yields (Φ, dashdotted lines) of the divergent reaction's species, the incident light profile (P_0, dashed line), and an example of the absorptivity envelope of the spectator molecules (*SPM*, dotted purple line) present in the reactive medium (the *SPM* concentration takes various values). (Reproduced with permission from [2], *Frontiers Chemistry*)

absorption spectrum, whose maxima are situated at 333 and 374 nm. The lamp profile (Fig. 22.6), hence, indicates that the irradiation light impinges simultaneously on both functional groups. In the case where the *SPM*s' absorption is set to only overlap one absorption band of X (covering the 350 – 410 nm range, which presumably corresponds to the $X \rightarrow Y_3$ branch of the reaction) but not the other absorption band (Fig. 22.6). This means that the effect of *SPM*s is selectively targeting the formation of Y_3. Indeed, an increased absorption value of the spectator molecules impedes, virtually fully, the production of Y_3 but does not seem to significantly affect that of Y_1 (whose absorption lies between 310 and 350 nm). Quantitatively, the percentage reduction in the initial rate of Y_3, recorded between its values in the presence and absence of *SPM*, exceeds 75%, whereas that of Y_1, for the same concentration of *SPM*, does not reach 10% (Fig. 22.7). The combined effects are measured by the percentage reduction observed on the reaction total absorbance initial rate, monitored at an observation wavelength ($\lambda_{obs} = 374$ nm, where the three reaction species absorb).

Such a selective irradiation is typically exploited in chromatic orthogonality [6–9], mostly when the reaction is performed under polychromatic light. The interpretation offered for the quantitative effect of *SPM*s, together with the use of both numerical integration of Eq. 19.5 and the explicit model equations (Eqs. 19.10 and 19.13), provides a consistent explanation and powerful predictive and quantitative tools for such and similar photokinetic situations and applications when the reactive

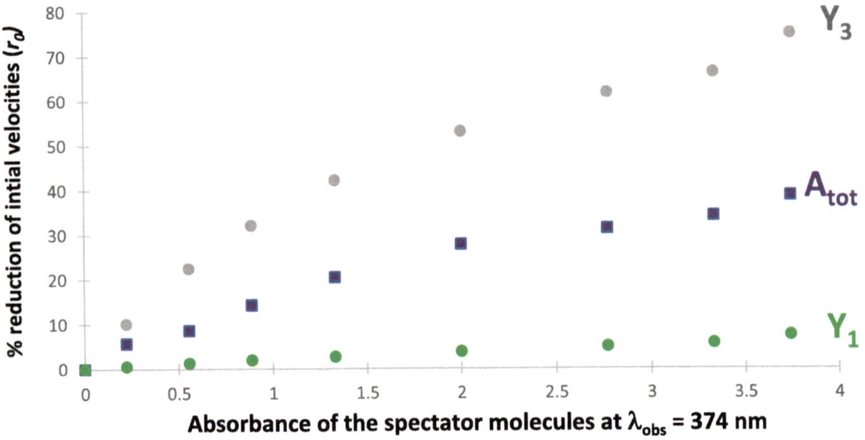

Fig. 22.7 Percentage reduction in the initial speeds of both reaction products Y_1 and Y_3, as well as that of the total absorbance of the medium, A_{tot}, observed at 380 nm, with increasing absorbance of the spectator molecules measured at 374 nm. The values corresponding to the purple squares represent the initial rate of the total absorbance trace multiplied by a scaling factor ($-10^{-5}\ r_{o,\ A}$). (Reproduced with permission from [2], *Frontiers Chemistry*)

media and materials are exposed to either monochromatic or polychromatic irradiations.

22.5 Temperature Impact on Photothermal Reactions

The influence of temperature is meant to affect only the thermal processes of the photothermal reaction but not the photochemical ones (i.e. excited-state thermal barriers are supposed to be negligible). The balance between thermal and photochemical contributions to the kinetics of the reaction is then expected to shift towards the former at high temperatures. The rate-constants of the thermal reaction-steps ($k^T_{Y_j \to Y_{j'}}$) uniformly increase with increments of temperature, making the rates of the corresponding species linearly correlate with the values of the second sum on the right-hand side of Eq. 19.5.

The concentration traces of the reactant and the second product (Y_2) of the photothermal $XY_2(4\Phi, 2k)$ reaction (Scheme 21.1) were *RK*-simulated for the reaction performed at a series of medium temperatures (Fig. 22.8). For this case, an increase in the temperature favours the reverse thermal reaction-steps towards the reactant. Accordingly, the reactant's *pss* concentration is highest for the highest temperature, whereas that of Y_2 is the lowest for the same temperature (red lines in Fig. 22.8) and vice versa for the lowest temperature (blue lines in Fig. 22.8). The alteration of the *pss* composition means that the observed values of the *pss* total absorbance is dependent on temperature. It is interesting to notice that the initial

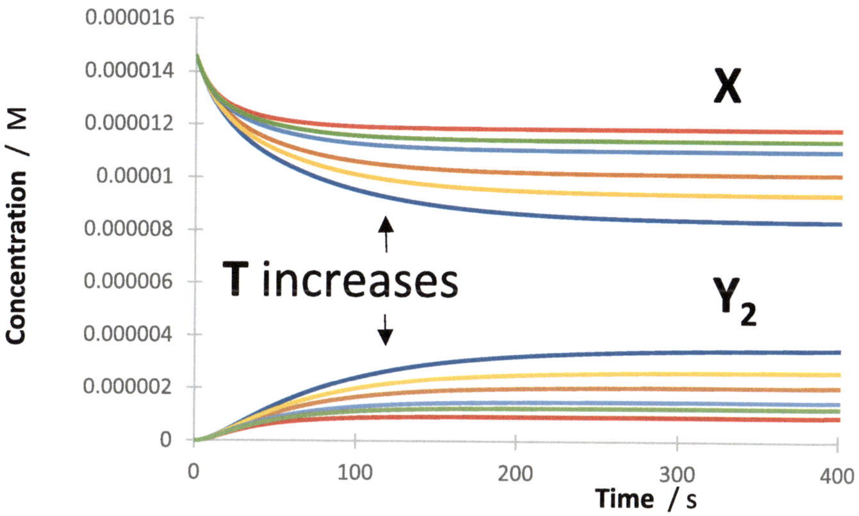

Fig. 22.8 Concentration traces of X and Y_2 of the $XY_2(4\Phi, 2k)$ reaction (Scheme 21.1) performed at different medium temperatures. Temperature increases in the order of the colour code: blue, yellow, orange, light blue, green, and red. (Reproduced with permission from [3], *Molecules*)

reactant-rate would be invariant with temperature if this species is thermally stable. Therefore, the effect of temperature, for such cases, cannot be quantified by monitoring the initial rates, either $r_{0,X}^{Lp,\Delta\lambda,T}$ or $r_{0,A}^{Lp,\Delta\lambda,T}$. This represents the only case where the metric (r_0) selected for photokinetics is not pertinent. However, the impact of temperature is rendered well by the photonic yield in terms of concentration ($PHYD_{Y_j}^{Lp,\Delta\lambda,T}$, given by Eq. 21.5), as depicted in Fig. 21.1. The variation of $PHYD_{Y_2}^{Lp,\Delta\lambda,T}$ with both time and temperature is shown in Fig. 22.9. The maximum of the $PHYD_{Y_2}^{Lp,\Delta\lambda,T}$ plot varies both in magnitude and position, a feature that might help set the optimal reaction conditions for the production of a given species.

22.6 Wavelength Region of Highest Photothermal Reactivity

Polychromaticity of the excitation light supplied by a lamp ($\Delta\lambda_{Lp}$) over the *OSIA* ($\Delta\lambda$) draws an interrogation on whether there is a particular wavelength section of the incident light within $\Delta\lambda$ ($\Delta\lambda_{WROR} \leq \Delta\lambda$) that induces a relatively higher reactivity of the photothermal system (i.e. where the values for the rate are the highest) than other wavelength sections within $\Delta\lambda$. This wavelength section is coined here the wavelength range of optimal reactivity (*WROR*). .

Despite this concept having merit in defining an important kinetic criterion, which might be found useful for some applications and/or energy savings, the literature,

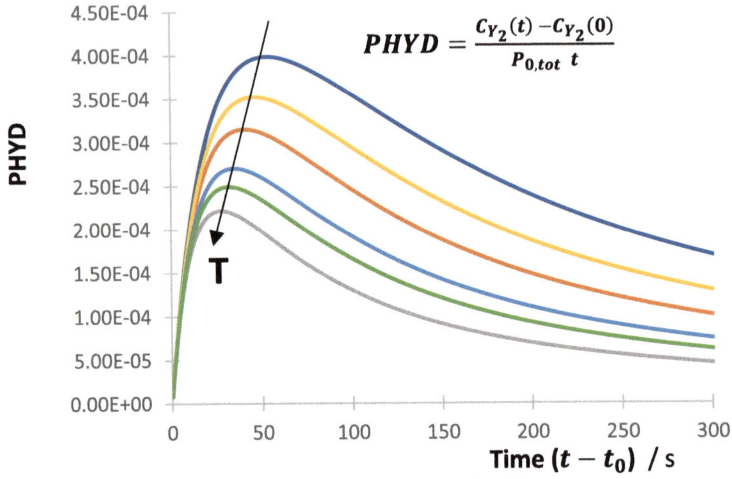

Fig. 22.9 Temperature and temporal evolution of the photonic yield, *PHYD*, of the product Y_2 of the $XY_2(4\Phi, 2k)$ reaction with irradiation time. (Reproduced with permission from [3], *Molecules*)

unfortunately, does not provide the means to address such a matter. We will explore the potential of establishing a rate-law-based tool for that purpose.

For any given reaction conditions, the value of the initial reactant rate, $r_{0,X}^{Lp,\Delta\lambda,T}$, encompasses some of the features of the reaction that extend to the whole $\Delta\lambda$ domain (Eq. 22.10). $r_{0,X}^{Lp,\Delta\lambda,T}$ can, hence, be seen as a global equation that is not capable of providing information on distinct segments within $\Delta\lambda$, e.g. $r_{0,X}^{Lp,\Delta\lambda,T} = r_{0,X}^{Lp,\Delta\lambda_{WROR},T} + r_{0,X}^{Lp,\Delta\lambda_1,T} + r_{0,X}^{Lp,\Delta\lambda_2,T} + \ldots,$ with $\Delta\lambda = \Delta\lambda_{WROR} + \Delta\lambda_1 + \Delta\lambda_2 + \ldots$).

$$Theo : r_{0,X}^{Lp,\Delta\lambda,T} = \sum_{\lambda_{irr}=\lambda_a}^{\lambda_b} r_{0,X}^{\lambda_{irr},T} = -C_X^{Lp,\Delta\lambda,T}(0)\, l_{irr} \sum_{\lambda_{irr}=\lambda_a}^{\lambda_b} \times \left(\sum_{j=1}^{v_k^0} \left(\Phi_{X\to Y_j}^{\lambda_{irr}} P_0^{\lambda_{irr}} \varepsilon_X^{\lambda_{irr}} PKF^{\lambda_{irr}}(0) \right) \right) - C_X^{Lp,\Delta\lambda,T}(0) \sum_{j=1}^{u_k^0} k_{X\to Y_j}^T \quad (22.10)$$

As the photochemical part of Eq. 22.10 is a sum of $r_{0,X}^{\lambda_{irr},T}$, corresponding to individual wavelengths, $r_{0,X}^{\lambda_{irr},T}$ must be variable over $\Delta\lambda$, in proportion to the values of its component parameters at each wavelength. And since most of the latter parameters are expressed by non-linear patterns with wavelength, the profile of $r_{0,X}^{\lambda_{irr},T}$ over $\Delta\lambda$ is most likely to be non-linear or curved and eventually present maxima and/or minima. The highest value on $r_{0,X}^{\lambda_{irr},T}$ profile (the maximum) would correspond to the centre of the *WROR*, thus allowing us to identify or estimate $\Delta\lambda_{WROR}$.

An examination of Eq. 22.10 reveals that the profile of $r_{0,X}^{\lambda_{irr},T}$ is a function of, on the one hand, intrinsic parameters ($\varepsilon_X^{\lambda_{irr}}$ and $\Phi_{X\to Y_j}^{\lambda_{irr}}$), which are inherent properties of the reactant that cannot be modified by the experimentalist and, on the other hand, the extrinsic parameter, $P_0^{\lambda_{irr}}$, whose pattern depends on the lamp employed for irradiation. The initial concentration, $C_X^{Lp,\Delta\lambda}(0)$; optical path length of irradiation, l_{irr}; and the rate-constants at *T*, $k_{X\to Y_j}^T$, play no role in defining the profile of $r_{0,X}^{\lambda_{irr},T}$ with wavelength (assuming a thermally stable reactant).

The necessary dependence of $r_{0,X}^{\lambda_{irr},T}$ pattern on the light source emission profile predicts a variability of the photothermal system reactivity with the distribution of the light source emission over the *OSIA* wavelength domain. Thus, lamps possessing different emission profiles might well lead to defining, for the same reactant *X*, different *WROR*s. Nonetheless, a plot of $r_{0,X}^{\lambda_{irr},T}$ versus wavelength remains, thus far, the best tool to delimitating the wavelength domain, *WROR*, of the reactant. Alternatively, the $r_{0,A}^{\lambda_{irr},T}$ can be used for an inclusive view of the reactivity of *X*, as this initial rate involves a contribution of the immediate photoproducts generated from the reactant.

Practically, the *WROR* definition requires the traces at individual wavelengths to be known, an information that is unavailable from the kinetic data of polychromatic light kinetics but is exclusively provided by monochromatic irradiation experiments. Hence, performing experiments with monochromatic light (of the same or normalised radiation intensity) would provide a pattern for $r_{0,X}^{\lambda_{irr},T}$ and, therefore, will allow us to determine the wavelength section of the *WROR*.

A precise definition of *WROR* finds importance in enabling the control of efficiency, safety, and economic issues in relation to various materials' applications [10].

22.7 Conclusion

The quantification of phenomena related to the photokinetics of photothermal reactions is shown in this chapter to be readily achieved by the data facilitated by the explicit general model equation (Eq. 19.10). The initial rate is proven to be a very useful metric. The *RK*-calculations with the aim to investigate, predict, and quantify the photokinetics of photoreactive systems and their behaviours are, once more, revealing an acute efficiency, which sets them to becoming a valuable tool in the field.

The photochemical reaction of photothermal sequences slows down with an increase in the reactant initial concentration value. Photostabilisation is also observed for photothermal reactions with increments of the absorbance values of appropriate spectator molecules. Both phenomena may have utility in tuning photoreactivity to the requested needs of applications.

The straightforward method for screening the evolution of the reactant quantum yield with wavelength can turn out to be pivotal in expanding this particular theme.

Finally, the definition of the wavelength range of optimal reactivity (*WROR*) can be useful to enable selective control of reactions and reaction conditions.

References

1. Maafi M (2023) On photokinetics under monochromatic light. Front Chem 11:1233151. https://doi.org/10.3389/fchem.2023.1233151
2. Maafi M (2024) On photokinetics under polychromatic light. Front Chem 12:1233151. https://doi.org/10.3389/fchem.2024.1367276
3. Maafi M (2025) Photokinetics of Photothermal reactions. Molecules 30:330. https://doi.org/10.3390/molecules30020330
4. Li FH, Yao K, Lv WY, Liu GG, Chen P, Huang HP, Kang YP (2015) Photodegradation of ibuprofen under UV–vis irradiation: mechanism and toxicity of photolysis products. Bull Environ Contam Toxicol 94:479–483. https://doi.org/10.1007/s00128-015-1494-8
5. Kryczyk A, Zmudzki P, Hubicka U (2016) Determination of itraconazole and its photodegradation products with kinetic evaluation by ultra-performance liquid chromatography/tandem mass spectrometry. Biomed Chromatogr 11:1733–1743. https://doi.org/10.1002/bmc.3747
6. Bochet CG (2006) Photochemical release of various functional groups. Pure Appl Chem 78: 241–224. https://doi.org/10.1351/pac200678020241
7. Corrigan N, Boyer C (2019) Photochemical reaction orthogonality in modern macromolecular science. ACS Macro Lett 8:812–818. https://doi.org/10.1021/acsmacrolett.9b00292
8. Kumar GS, Sha MS, Yempally S, Cabibihan JJ, Sadasivuni KK (2023) A practical perspective for chromatic orthogonality for implementing in photolithography. Sci Rep 13:694. https://doi.org/10.1038/s41598-023-27869-w
9. Sansalone L, Zhao J, Nguyen LTB, Gupta S, Benson DL, Abe M, Ellis-Davies GCR (2024) Bidirectional neuronal actuation by uncaging with violet and green light. Angew Chem Int Ed 63:e202315726. https://doi.org/10.1002/anie.202315726
10. Walden SL, Carroll JA, Unterreiner A-N, Barner-Kowollik C (2023) Photochemical action plots reveal the fundamental mismatch between absorptivity and photochemical reactivity. Adv Mater 11:2306014. https://doi.org/10.1002/advs.202306014

Chapter 23
Kinactinometry

23.1 Introduction

Actinometry is a measure of $P_0^{\lambda_{irr}}$ (the incident light flux) at each λ_{irr} that enters the reactor (a slab-shaped reactor). A chemical actinometer is a standardised photothermal reactive system that is able to deliver $P_0^{\lambda_{irr}}$. In principle, any photoactive molecular system can be turned into an actinometer, so much so that a large supply of potential actinometers should be expected in photochemistry. However, as a matter of convenience and to avoid the difficulty that might be posed by a thermal instability of the reactant in precise quantification, the actinometer systems that will be envisaged in this chapter undergo photothermal processes that are initiated by a thermally stable reactant. The general conditions relative to the reactor shape and collimation of the light are supposed to apply (Sect. 2.6). Concentration and total absorbance traces are fitted to the adequate equations, Eqs. 19.10 and 19.13, and the fitting parameters are used to calculate the initial rates. When photokinetic data and methods form the basis for the development of an actinometric method, the latter is termed kinactinometry (Sects. 2.4, 2.6 and 2.21, 18.7).

In the subject of actinometry, despite the number of actinometers proposed in the literature, the standardisation of the actinometric methodology remains a challenge in the field, as evidenced by the variety of procedures used for the different actinometers (Chap. 17). The potential of the model equations (Eqs. 19.5 and 19.10), together with the metric (initial reaction-rate), is exploited here for that standardisation purpose.

M. Maafi, *Photokinetics*, https://doi.org/10.1007/978-3-031-98985-8_23

23.2 Some General Aspects

Qualitatively, an increase in the radiation intensity (higher values of $P_0^{\lambda_{irr}}$ at each wavelength of the irradiation domain) has the effect of increasing the overall reaction photoreactivity. The quantitative assessment of such an effect was previously provided by a proportional change in the rate-constants' values of $k_{ij}^{\lambda_{irr}}$, as discussed in Chap. 18 and related studies [1–8]. Also, the increase in $P_0^{\lambda_{irr}}$ shortens the overall time for the reaction to reach completion and, therefore, a reduction in the half-life time, $t_{1/2}$, of the reactant. In the same perspective, a variation of $P_0^{\lambda_{irr}}$ induces a proportional change in the reactant's initial rate value, $r_{0,X}^{\lambda_{irr},T}$ (the latter increases with $P_0^{\lambda_{irr}}$).

Alteration of the $P_0^{\lambda_{irr}}$ value, as defined in Sect. 2.18, can be achieved by a variation of either $flx_{h\nu}^{\lambda_{irr}}$ (the incident flux of the lamp), S_{irr}, or V_{irr}(the irradiated area or volume of the sample exposed to irradiation, respectively). Practically, the experimentalist controls the variations of $flx_{h\nu}^{\lambda_{irr}}$, S_{irr}, and V_{irr} (which are extrinsic parameters) by changing the electric supply of the lamp, the irradiated (not covered) surface, or irradiated volume of the sample, respectively. For consistency with the value of V_{irr}, it is important to perform a vigorous stirring of the reactor during irradiation. Also, the ratio (V_{irr}/S_{irr}) is not equal to l_{irr}, unless the whole volume of the reactive medium is exposed to the irradiation light through one of its full sides (in general, however, $V_{irr}/S_{irr} \geq l_{irr}$). It is then recommended to report these three properties in the investigation to allow an adequate photokinetic analysis. Unfortunately, some or all of these quantities are most often missing in published studies, which might, at least in part, explain the usually encountered difficulties in comparing photokinetic results obtained in different laboratories.

23.3 Kinactinometry When Irradiation Is Monochromatic

The standardisation of actinometers, activated by monochromatic light, is kinetically achieved here by the quantification of the linearly correlated variation of the initial rate ($r_{0X}^{\lambda_{irr},T}$ or $r_{0A}^{\lambda_{irr},T/\lambda_{obs}}$) with $P_0^{\lambda_{irr}}$. Such a correlation stems from the proportionality of the initial rate to the number of photons absorbed by the reactant (Eq. 23.1). The initial rates (e.g. $r_{0,X}^{\lambda_{irr},T}$ or $r_{0,Y_1}^{\lambda_{irr},T}$) worked out from the concentration traces of the reactant and its immediate products are exclusively expressed in quantities corresponding to the irradiation wavelength, whereas the initial rate of the total absorbance ($r_{0A}^{\lambda_{irr},T/\lambda_{obs}}$) includes, in addition to the initial rates of the reactant and its products, their absorptivities measured at λ_{obs} and the observation optical path length, l_{obs}.

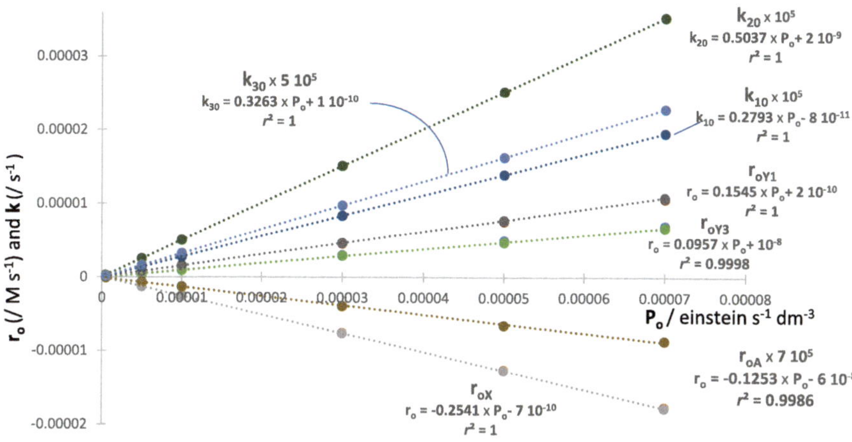

Fig. 23.1 Linear correlations of $k_{i0}^{\lambda_{irr}}$ and $r_{0,X}^{\lambda_{irr},T}$, $r_{0,Y_1}^{\lambda_{irr},T}$, $r_{0,Y_3}^{\lambda_{irr},T}$, and $r_{0,A}^{\lambda_{irr},T/\lambda_{obs}}$ with increasing $P_0^{\lambda_{irr}}$ values (5 10^{-7} – 7 10^{-5} einstein dm^{-3} s^{-1}) for a $XY_7(12\Phi)$ reaction, under monochromatic light irradiation, governed by the Φ-shaped mechanism (Fig. 2.1). Every point of $r_{0,Y_j}^{\lambda_{irr},T}$ in the plot combines *RK*, *Theor*, and *Fit* values (fitting $C_X^{\lambda_{irr},T}(t)$, $C_{Y_1}^{\lambda_{irr},T}(t)$, and $C_{Y_3}^{\lambda_{irr},T}(t)$ traces). The values of the rate-constants k_{i0}^{Φ} were obtained from appropriate Eq. 19.10 (or Eq. 19.13) fittings of the individual traces. The rates $r_{0,Y_j}^{\lambda_{irr},T}$ were calculated from the differentiated fitting equations at $t = 0$. (Reproduced with permission from [9], *Frontiers in Chemistry*)

$$Theo : r_{0,A}^{\lambda_{irr},T/\lambda_{obs}} = \varepsilon_X^{\lambda_{obs}}\, l_{obs}\, Theo : r_{0,X}^{\lambda_{irr},T} + \sum_{j=1}^{v_k^0} \left(\varepsilon_{Y_j}^{\lambda_{obs}}\, l_{obs}\, Theo : r_{0,Y_j}^{\lambda_{irr},T} \right)$$
$$P_0^{\lambda_{irr}} \left(1 - 10^{-A_{tot}^{\lambda_{irr},T/\lambda_{irr}}(0)\,\frac{l_{irr}}{l_{obs}}} \right) l_{obs}\, \Phi_{X}^{\lambda_{irr}}{}_{Y_j} \left(\sum_{j=1}^{v_k^0} \left(\varepsilon_{Y_j}^{\lambda_{obs}}\right) - \varepsilon_X^{\lambda_{obs}} \right) \tag{23.1}$$

RK-simulations performed for a large number (>400) of photo- and photothermal reactions show excellent linear correlations of $P_0^{\lambda_{irr}}$ with $RK : r_{0,X}^{\lambda_{irr},T}$ and $Theo : r_{0,X}^{\lambda_{irr},T}$. An example of these is the $XY_7(12\Phi)$ photoreaction obeying the Φ-shaped mechanism depicted in Scheme 2.1 (the reactant being thermally stable, generating a concurrent sequence, $Y_3 \leftarrow X \rightarrow Y_1$).

Figure 23.1 shows a number of correlations with $P_0^{\lambda_{irr}}$, including the initial rates of the reactant and those of its immediate photoproducts ($r_{0,X}^{\lambda_{irr},T}$, $r_{0,Y_1}^{\lambda_{irr},T}$, and $r_{0,Y_3}^{\lambda_{irr},T}$), the reaction's initial rate ($r_{0,A}^{\lambda_{irr},T/\lambda_{obs}}$), and the fitting rate-constants (k_{10}^{Φ}, k_{20}^{Φ}, and k_{30}^{Φ}) obtained for the reactant concentration traces by fitting it to an Eq. 19.10 that involves three mono-Φ-order terms. The latter feature is not necessarily true for all cases (due to the identifiability issues relative to the fitting that might affect the rate-constants' values, as discussed in Chap. 19); it is presented here for illustration purposes only, underlining the possibility of its occurrence in relatively simple reactions' traces.

Technically, the values of the rate-constant obtained when varying $P_0^{\lambda_{irr}}$ might be favoured by fitting the traces (either Eqs. 19.10 and 19.13) of the different considered species, whose model equations all share the same $P_0^{\lambda_{irr}}$-independent factors, ω_j^{∞}, ω_{ij}^{Φ}, and $cc^{\lambda_{irr}}$ for the concentration traces and ω_A^{∞}, $\omega_{i,A}^{\Phi}$, and $cc^{\lambda_{irr}}$ for the total absorbance traces. In these conditions, only the k_{ij}^{Φ} factors vary with $P_0^{\lambda_{irr}}$. The same understanding can be reached by considering that the final species concentrations ($\omega_j^{\infty}=C_j^{\lambda_{irr},T}(pss)$), reached at $t=\infty$, of the $XY_7(12\Phi)$ reactive system that reaches a photostationary state are invariant with increasing $P_0^{\lambda_{irr}}$, but they are reached faster (shorter $t_{1/2}$). Hence, changing the value of $P_0^{\lambda_{irr}}$ will automatically induce a proportional change in the values of the k_{ij}^{Φ} factors, which are the only changing fitting parameters.

Beyond the detail about whether or not it is possible to achieve $k_{ij}^{\Phi}=f\left(P_0^{\lambda_{irr}}\right)$, in practice, for a given experiment, the value of an unknown value of $P_{0,unk}^{\lambda_{irr}}$ supplied by a light source is worked out by using the equation of the actinometric equation built on "i" known values of $P_0^{\lambda_{irr}}$ (Eq. 23.2) and the corresponding initial rates (e.g. one of the lines given in Fig. 23.1),

$$Fit : r_{0,Y_j\ or\ A}^{\lambda_{irr},T,i} = \pm \alpha_{0,Y_j\ or\ A}^{\lambda_{irr}} P_0^{\lambda_{irr},i} \tag{23.2}$$

The slope ($\alpha_{0,Y_j\ or\ A}^{\lambda_{irr}}$) of the specific equation (e.g. $\alpha_{0,X}^{\lambda_{irr}}=0.2541$, Fig. 23.1) is used together with the initial rate measured for the trace obtained under the unknown $P_{0,unk}^{\lambda_{irr}}$ to determine the value of the latter. For instance, for the reactant concentration trace, we have

$$P_{0,unk}^{\lambda_{irr}} = -\frac{Fit : r_{0,X,unk}^{\lambda_{irr},T}}{\alpha_{0,X}^{\lambda_{irr}}} = -\frac{Fit : r_{0,X,unk}^{\lambda_{irr},T}}{0.2541} \tag{23.3}$$

The strategy of using the initial rates correlation with $P_0^{\lambda_{irr}}$ (i.e. the linear relationships, $r_{0,X,\ Y_1\ or\ Y_3}^{\lambda_{irr},T,i}=f\left(P_0^{\lambda_{irr},i}\right)$) is a simple means to perform reliable actinometry and recruit a wide range of new actinometers. This methodology lays down a strong basis for the standardisation of actinometry.

Experimentally, one needs to determine reliable values of $P_0^{\lambda_{irr}}$ to develop new actinometers. For that purpose, precise values of $P_0^{\lambda_{irr}}$ at selected wavelengths within the spectral range $260-560\ nm$ can readily be determined by using the adequate actinometer from the set proposed in Sect. 18.7 (Fig. 18.5), developed on the basis of semi-empirical equations.

23.4 Lamp Emission and Measured Amount of Photons

The monochromatic irradiation is very specific to the wavelength of irradiation, which is obviously absorbed by the reactant and other species of the reaction. This makes the light flux ($P_0^{\lambda_{irr}}$) specific, even if its value can vary.

The situation for an irradiation with polychromatic light is quite different and needs to be discussed further to specify what number of photons is measured by actinometry. The issue here relates to the fact that the span of the emission spectrum of a lamp, *Lp*, might be different from that of the electronic absorption spectra of the reactive species. We can distinguish four domains, namely, the emission span of the lamp, $\Delta\lambda_{Lp}$; the absorption of the species, $\Delta\lambda_{sp}$; the absorption of the chemical actinometer, $\Delta\lambda_{Act}$; and the *OSIA*s (overlap section of irradiation and absorption), $\Delta\lambda$, of the lamp with either the actinometer ($OSIA_{Act}$) or the species investigated ($OSIA_{sp}$).

In any case, the light flux for polychromatic light is expressed as a sum of the photons delivered by *Lp* at each individual wavelength of the wavelength domain considered. For instance, the flux for an *OSIA* covering the domain $\Delta\lambda$ situated between λ_a and λ_b is

$$P_0^{Lp,\Delta\lambda} = \sum_{\lambda_a}^{\lambda_b} P_0^{\lambda_{irr}} \tag{23.4}$$

It is important to underline that the numerical value of $P_0^{Lp,\Delta\lambda}$, worked from Eq. 23.4 or obtained by actinometry, informs neither on the profile of the lamp over $\Delta\lambda$ nor on the range $\Delta\lambda$ itself (if not clearly indicated on the labelling $P_0^{Lp,\Delta\lambda}$, which is, most often, not accessible from the actinometric measurement).

Also, according to Eq. 19.5, the value of the initial rate must necessarily correspond to the *OSIA* of the compound investigated (either the actinometer, the reactant, or the reactant and its photoproducts for $r_{0,A}^{\lambda_{irr},T/\lambda_{obs}}$). The reason for that being the terms of the summation in Eq. 19.5 will be having non-zero values only and only if all the coefficients $P_0^{\lambda_{irr}}$, $\varepsilon^{\lambda_{irr}}$, and $\Phi_{Y_j \to Y_{j'}}^{\lambda_{irr}}$ all have non-zero values. Therefore, the determination of the initial rate from Eq. 19.10 (using Eq. 19.12) will exclusively translate the kinetics of the relevant lamp/reaction *OSIA*.

We ought, then, to consider the situations that dictate what number of photons will be counted for the reaction at hand.

Let us first consider a lamp and the reactive species, where actinometry serves the determination of the number of photons incident from the lamp. This number of photons is counted over the $\Delta\lambda_{Lp}$ domain. We can separate, in this case, between two situations, with respect to the range of absorption relative to the reaction studied.

First is a situation (sit_1), where $\Delta\lambda_{Lp}$ encompasses but is larger than $OSIA_{sp}$. An example for sit_1 is represented in Fig. 19.1, where the species $OSIA_{sp} = 315 - 360$ nm, corresponding to ε_2-spectrum, is narrower than the lamp emission region, $\Delta\lambda_{Lp} = 310 - 390$ nm. It is obvious that the reactor receives an excess number of photons compared to that potentially absorbed by the reaction. This means that the photons occurring at wavelengths outside $OSIA_{sp}$ are counted by actinometry but are not used by the reaction.

However, in a second situation sit_2, if $\Delta\lambda_{Lp}$ is shorter than or equal to $\Delta\lambda_{sp}$, where, for instance, $\Delta\lambda_{sp} = 315 - 360$ nm for ε_2-species in Fig. 19.1, and assuming the lamp profile spans only the $\Delta\lambda_{Lp} = 310 - 340$ nm, all the counted photons serve the reaction.

Such observations indicate that when chemical actinometry is used, it is mandatory to take into account the *OSIAs* of the actinometer ($OSIA_{Act}$) and/or the reaction ($OSIA_{sp}$), together with the light source profile ($\Delta\lambda_{Lp}$). The situation sit_2 is realised when the species and actinometer *OSIAs* are equal to each other and larger than the emission domain of the lamp ($\Delta\lambda_{Lp} < OSIA_{act} = OSIA_{sp}$), whereas, sit_1 will be observed when the species has a narrower *OSIA* compared to that of the actinometer, but the latter is narrower than the emission wavelength span of the lamp, i.e. $\Delta\lambda_{Lp} \geq OSIA_{act} > OSIA_{sp}$ (e.g. in Fig. 19.1, $OSIA_{sp} = 310 - 340\ nm$ for ε_1-spectrum, $OSIA_{Act} = 310 - 365\ nm$ for ε_2-spectrum, and the lamp profile covers the $\Delta\lambda_{Lp} = 310 - 390\ nm$ region).

From these settings, one can postulate a third situation (sit_3), even if it is rather unrealistic. In this case, the $OSIA_{Act}$ of the chemical actinometer is narrower than that of the investigated species, and both are shorter than the wavelength range covered by the lamp emission, i.e. $\Delta\lambda_{Lp} \geq OSIA_{sp} > OSIA_{act}$. Here, the species absorbs more photons than those counted by the actinometer, e.g. in Fig. 19.1, $OSIA_{Act} = 310 - 340\ nm$ with the actinometer now has ε_1-spectrum, whereas the species with $OSIA_{sp} = 310 - 365$ nm has ε_2-spectrum, and finally, the lamp is characterised by a $\Delta\lambda_{Lp} = 310 - 390$ nm.

This discussion leads to the conclusion that the number of photons accounted for by chemical actinometry (e.g. using the ferrioxalate actinometer; see Sect. 23.7 below) for situations sit_1 and sit_3 is not consistent with Eq. 19.5. Such actinometry procedures will undeniably introduce errors and discrepancies, both on the subsequent determination of any photokinetic features (e.g. the evaluation of the quantum yield) and/or on any built actinometric methods (which, effectively, cannot be deemed standardised in such a case).

In this context, employing Eq. 19.10 that conforms totally with Eq. 19.5 guarantees that only the true reaction/lamp *OSIA* is accounted for, i.e. the photons outside the *OSIA* region do not contribute to the experimentally observed or *RK*-simulated traces. Therefore, a greater reliability of kinactinometry based on Eq. 19.10 (or Eq. 19.13) should be expected over existing actinometric methods. A useful recommendation for a precise actinometry is to define the full lamp profile by physical actinometry before developing the new actinometers in order to avoid the problems discussed above in relation to the *OSIAs*.

23.5 Kinactinometry When Irradiation Is Polychromatic

The kinactinometry goal, when the irradiation light is polychromatic, is the determination of the total amount of photons (as set in Eq. 23.4) delivered by a lamp, *Lp*, to the reactor, over the whole or part of the emission wavelength span of *Lp* (either the *OSIA*, $\Delta\lambda$, or $\Delta\lambda_{Lp_1}$ ranges for a given lamp, Lp_1). Such a consideration might be important in developing new actinometers, as the accessible level of detail on the total number of photons obtained by a given method depends on the method itself. The method used to count the number of photons can either measure all the light emerging from the lamp with no more details (and gives $\Delta\lambda_{Lp_1}$ that might be different from the *OSIA*, $\Delta\lambda$), or it can only measure the light useful for the reaction ($\Delta\lambda$). In summary, the number of incident photons entering the reactor (before absorption) is not necessarily the same as the number of photons that are absorbed by the reactant. We must also keep in mind that the absorption of light is a necessary but not a sufficient condition of reactivity.

The standardisation of polychromatic-light kinactinometry can be set out through a straightforward methodology. Let us consider a given species intended to become an actinometer, ACT_1, to be standardised when exposed to a polychromatic light of lamp, Lp_1. Let us also consider here the most common situations, sit_1 and sit_2 (see previous section). These situations are distinguished by the comparison between the wavelength domains of the lamp and actinometer absorption. $\Delta\lambda$ will be used for sit_2, where $\Delta\lambda_{Lp_1} = \Delta\lambda = OSIA_{act_1}$, and $\Delta\lambda_{Lp_1} > \Delta\lambda$ for sit_1. The step-by-step kinactinometric method is as follows:

(i) Choose a lamp Lp_1, the actinometer ACT_1, the initial concentration of ACT_1 within its calibration linearity range, and the irradiated area and volume of the reactor.

(ii) Irradiate the sample with Lp_1 at n_{irr} different intensities, $P_0^{Lp_1,\Delta\lambda,i}$ (or $P_0^{Lp_1,\Delta\lambda_{Lp_1},i}$), with $1 \leq i \leq n_{irr}$.

(iii) Determine the values of $P_0^{Lp_1,\Delta\lambda,i}$ (or $P_0^{Lp_1,\Delta\lambda_{Lp_1},i}$) by a physical actinometry.

(iv) Record the n_{irr} traces of the species concentrations or the total absorbance at a selected observation wavelength, λ_{obs}.

(v) Fit each of the available traces by Eq. 19.10 (or Eq. 19.13).

(vi) Calculate the n_{irr} initial rates ($r_{o,\ X}$ or $r_{o,\ A}$) from the respective n_{irr} sets of fitting parameters using Eq. 19.12 or 19.15.

(vii) Draw the linear plot, e.g. $r_{o,A}^{Lp_1,\Delta\lambda_{Lp_1},i,T/\lambda_{obs}} = \alpha_{Lp_1}^{\Delta\lambda_{Lp_1}} P_0^{Lp_1,\Delta\lambda_{Lp_1},i}$, and record the equation of the line (the slope of the line, e.g. $\alpha_{Lp_1}^{\Delta\lambda_{Lp_1}}$, is dimensionless).

(viii) The unknown lamp intensity $P_0^{Lp_1,\Delta\lambda_{Lp_1},unk}$ (or $P_0^{Lp_1,\Delta\lambda,unk}$) of Lp_1 is measured in a separate experiment (e.g. by chemical actinometry). Fit the trace generated by $P_0^{Lp_1,\Delta\lambda_{Lp_1},unk}$ (or $P_0^{Lp_1,\Delta\lambda,unk}$) irradiation to the adequate equation, Eqs. 19.10 or 19.13, and determine $r_o^{Lp_1,\Delta\lambda_{Lp_1},unk,T/\lambda_{obs}}$ (or $r_o^{Lp_1,\Delta\lambda,unk,T/\lambda_{obs}}$). Calculate the values of lamp intensity from the equation of the line constructed in point (*vii*), together with the corresponding $\alpha_{Lp_1}^{\Delta\lambda_{Lp_1}}$ (or $\alpha_{Lp_1}^{\Delta\lambda}$) as

$$P_0^{Lp_1,\Delta\lambda_{Lp_1},unk} = \frac{r_{o,A}^{Lp_1,\Delta\lambda_{Lp_1},unk,T/\lambda_{obs}}}{\alpha_{Lp_1}^{\Delta\lambda_{Lp_1}}} \quad \left(or \;\; P_0^{Lp_1,\Delta\lambda,unk} = \frac{r_{o,A}^{Lp_1,\Delta\lambda,i,T/\lambda_{obs}}}{\alpha_{Lp_1}^{\Delta\lambda}} \right) \tag{23.5}$$

(ix) If both *OSIA* and the lamp profile are known, calculate the n_{irr} $P_0^{Lp_1,\Delta\lambda_{Lp_1},i}$ from $P_0^{Lp_1,\Delta\lambda,i}$ using Eq. 23.12 (vide infra) and Eq. 23.5 (and vice versa, $P_0^{Lp_1,\Delta\lambda,i}$ from $P_0^{Lp_1,\Delta\lambda_{Lp_1},i}$ and Eqs. 23.5 and 23.12).

The explanation of the linearity obtained for both plots referring to either $\Delta\lambda$ and $\Delta\lambda_{Lp_1}$ can be delivered if the total number of photons entering the reactor is viewed as a sum,

$$P_0^{Lp_1,\Delta\lambda_{Lp_1},i} = P_0^{Lp_1,\Delta\lambda,i} + P_0^{Lp_1,\left(\Delta\lambda_{Lp_1} - \Delta\lambda\right),i} \tag{23.6}$$

with $P_0^{Lp_1,\left(\Delta\lambda_{Lp_1} - \Delta\lambda\right),i}$ being the extra number of photons (since $\Delta\lambda_{Lp_1} > \Delta\lambda$) not absorbed by the actinometer. However, the initial rate values obtained from the fitting of the traces in the two cases are invariant, i.e. $r_{o,A}^{Lp_1,\Delta\lambda_{Lp_1},i} = r_{o,A}^{Lp_1,\Delta\lambda,i}$. The extra photons do not affect the value of the initial rates because Eq. 19.10 corresponds, in each case, to a reaction induced by the absorbed number of photons (collected on the Lp_1/ACT_1 *OSIA*, i.e. $\Delta\lambda$).

Furthermore, it is obvious that any increase recorded on $P_0^{Lp_1,\left(\Delta\lambda_{Lp_1} - \Delta\lambda\right),i}$ should be proportional to the increase sustained by $P_0^{Lp_1,\Delta\lambda,i}$, as both belong to the same lamp. The following equation stands for this case,

$$\frac{P_0^{Lp_1,\Delta\lambda,i+1}}{P_0^{Lp_1,\Delta\lambda,i}} = \frac{P_0^{Lp_1,\left(\Delta\lambda_{Lp_1} - \Delta\lambda\right),i+1}}{P_0^{Lp_1,\left(\Delta\lambda_{Lp_1} - \Delta\lambda\right),i}} \tag{23.7}$$

Which leads to the dimensionless and necessarily constant ratio (Eq. 23.8), irrespective of the actual value of the intensity of the radiation ($1 \leq i \leq n_{irr}$).

$$\frac{P_0^{Lp_1,\left(\Delta\lambda_{Lp_1} - \Delta\lambda\right),i}}{P_0^{Lp_1,\Delta\lambda,i}} = \beta_{Lp_1}^{\Delta\lambda_{Lp_1}/\Delta\lambda} \tag{23.8}$$

The $\beta_{Lp_1}^{\Delta\lambda_{Lp_1}/\Delta\lambda}$ factor is a property of the lamp (Lp_1) and the actinometer (Act_1) used in the experiment. Rearranging Eq. 23.8 gives

$$P_0^{Lp_1,(\Delta\lambda_{Lp_1}-\Delta\lambda),i}=\beta_{Lp_1}^{\Delta\lambda_{Lp_1}/\Delta\lambda}\ P_0^{Lp_1,\Delta\lambda,i} \tag{23.9}$$

which is then used, in combination with Eq. 23.7, to derive the relationship,

$$P_0^{Lp_1,\Delta\lambda_{Lp_1},i}=P_0^{Lp_1,\Delta\lambda,i}+\beta_{Lp_1}^{\Delta\lambda_{Lp_1}/\Delta\lambda}\ P_0^{Lp_1,\Delta\lambda,i}=\left(1+\beta_{Lp_1}^{\Delta\lambda_{Lp_1}/\Delta\lambda}\right)P_0^{Lp_1,\Delta\lambda,i} \tag{23.10}$$

Hence, for a given lamp, Lp_1, we have

$$\begin{aligned}r_{o,A}^{Lp_1,\Delta\lambda_{Lp_1},i,T/\lambda_{obs}}&=r_{o,A}^{Lp_1,\Delta\lambda,i,T/\lambda_{obs}}=\alpha_{Lp_1}^{\Delta\lambda_{Lp_1}}\ P_0^{Lp_1,\Delta\lambda_{Lp_1},i}\\&=\alpha_{Lp_1}^{\Delta\lambda_{Lp_1}}\left(1+\beta_{Lp_1}^{\Delta\lambda_{Lp_1}/\Delta\lambda}\right)P_0^{Lp_1,\Delta\lambda,i}=\alpha_{Lp_1}^{\Delta\lambda}\ P_0^{Lp_1,\Delta\lambda,i}\end{aligned} \tag{23.11}$$

and, therefore, to a proportionality relationship between $\alpha_{Lp_1}^{\Delta\lambda_{Lp_1}}$ and $\alpha_{Lp_1}^{\Delta\lambda}$,

$$\alpha_{Lp_1}^{\Delta\lambda_{Lp_1}}=\left(\frac{1}{1+\beta_{Lp_1}^{\Delta\lambda_{Lp_1}/\Delta\lambda}}\right)\alpha_{Lp_1}^{\Delta\lambda} \tag{23.12}$$

This analysis clearly shows that Act_1 can be employed to perform actinometric measurements, irrespective of whether the profile of the lamp is lager or partially covers the electronic absorption spectrum of Act_1. When the lamp profile is equal to or narrower than the span of the absorption spectrum of the system studied (sit_2), i.e. $\Delta\lambda \cong \Delta\lambda_{Lp_1}$, there is no extra number of photons, i.e. $P_0^{Lp_1,(\Delta\lambda_{Lp_1}-\Delta\lambda),i}\cong 0$, and the factor $\beta_{Lp_1}^{\Delta\lambda_{Lp_1}/\Delta\lambda} \ll 1$, which means that $\alpha_{Lp_1}^{\Delta\lambda}=\alpha_{Lp_1}^{\Delta\lambda_{Lp_1}}$. Otherwise, when $\Delta\lambda < \Delta\lambda_{Lp_1}$, we have $\alpha_{Lp_1}^{\Delta\lambda_{Lp_1}} < \alpha_{Lp_1}^{\Delta\lambda}$. The same reasoning holds for sit_1, where $\Delta\lambda_{Lp_1}$ is supposed to be equal to the actinometer's $OSIA_{Act_1}$ and $\Delta\lambda$ is equal to the investigated species' $OSIA_{sp}$.

Anyway, this proves that the initial rate, e.g. $r_{o,A}^{Lp_1,\Delta\lambda_{Lp_1},i,T/\lambda_{obs}}$, is linearly proportional to both $P_0^{Lp_1,\Delta\lambda_{Lp_1},i}$ and $P_0^{Lp_1,\Delta\lambda,i}$. Therefore, whenever the value of $\beta_{Lp_1}^{\Delta\lambda_{Lp_1}/\Delta\lambda}$ is known, knowing one proportionality factor (e.g. $\alpha_{Lp_1}^{\Delta\lambda_{Lp_1}}$ obtained graphically) will allow calculating the value of the other ($\alpha_{Lp_1}^{\Delta\lambda}$). Henceforward, knowing the value of $\alpha_{Lp_1}^{\Delta\lambda}$ will allow the determination of $P_0^{Lp_1,\Delta\lambda,i}$ (and vice versa). The value of $\beta_{Lp_1}^{\Delta\lambda_{Lp_1}/\Delta\lambda}$ can easily be determined if the lamp profile, $P_0^{\lambda_{irr}}$, and absorption spectrum of ACT_1 are known (the latter, i.e. the absorption spectrum of the actinometer, ACT_1, must be known since the latter should be a thermally stable reactant).

The relevant implication of $\beta_{Lp_1}^{\Delta\lambda_{Lp_1}/\Delta\lambda}$ in polychromatic-light kinactinometry strongly recommends that the full set of numeric values for each lamp profile (values of P_0^{λ} against λ) is published. The construction of a commonly shared database of the profiles of lamps is acutely needed in this context. Such a database would offer the

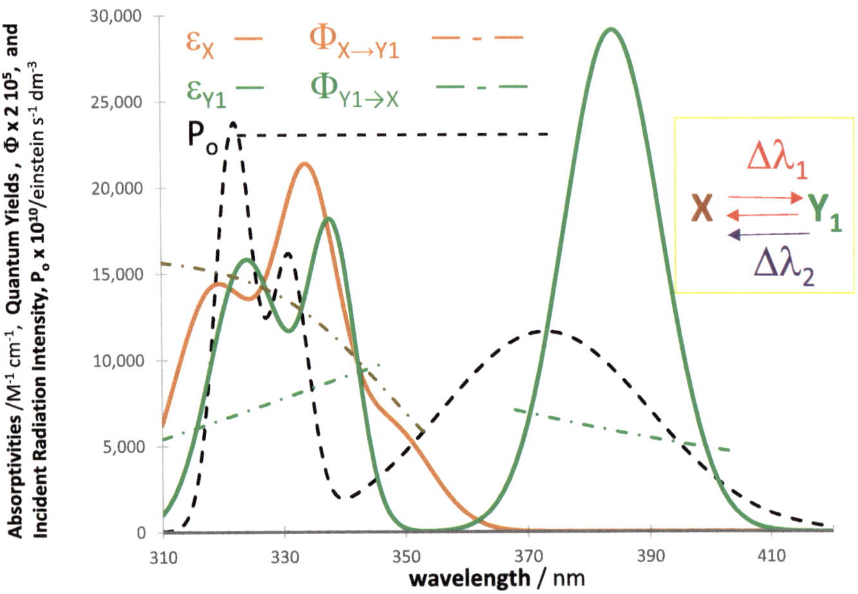

Fig. 23.2 Profiles of absorption coefficients, quantum yields, and radiation intensities, respectively, of the species of a typical reaction of diarylethenes and lamp used for irradiation. The photomechanism (in the inset) depicts the reaction-steps and the light section they are sensitive to. The reactive system is continuously under $\Delta\lambda = \Delta\lambda_1 + \Delta\lambda_2$ irradiation. (Reproduced with permission from [10], *Frontiers in Chemistry*)

community a useful tool to expand the library of accurate polychromatic-light actinometers.

The methodology for polychromatic kinactinometry presented above readily applies to photo- and photothermal systems, proving their effective conversion into efficient actinometers. In a recent publication [10], the methodology was successfully applied to the reactions of the most commonly used organic actinometers employed in the field.

For illustration, let us consider the particular reaction of a photoreversible and photochromic system involving three reaction-steps, $XY(3\Phi)$ (Fig. 23.2). The reaction is continuously exposed to a large-span lamp, $\Delta\lambda_{Lp_1} = 310 - 420$ nm, whose polychromatic light covers the whole absorption region of the species of the reactive system. The properties of the species and the lamp are depicted in Fig. 23.2.

The three reaction-steps are variably sensitive to the different sections of the excitation light. The photoreversible reaction occurring between the reactant and its photoproduct is driven by a polychromatic light emitted in the wavelength range $\Delta\lambda_1$ (e.g. 310 – 360 nm, in Fig. 23.2), where both species absorb. Simultaneously, the lamp induces a back reaction of the photoproduct alone through the light spanning another $\Delta\lambda_2$ region (e.g. $\Delta\lambda_2 = 360 - 420$ nm, in Fig. 23.2), where X does not absorb. Hence, the reactant is only sensitive to light within $\Delta\lambda_1$, whereas the photoproduct reacts to the lights covering both $\Delta\lambda_1$ and $\Delta\lambda_2$ ($\Delta\lambda_{Lp_1} = \Delta\lambda_1 + \Delta\lambda_2$).

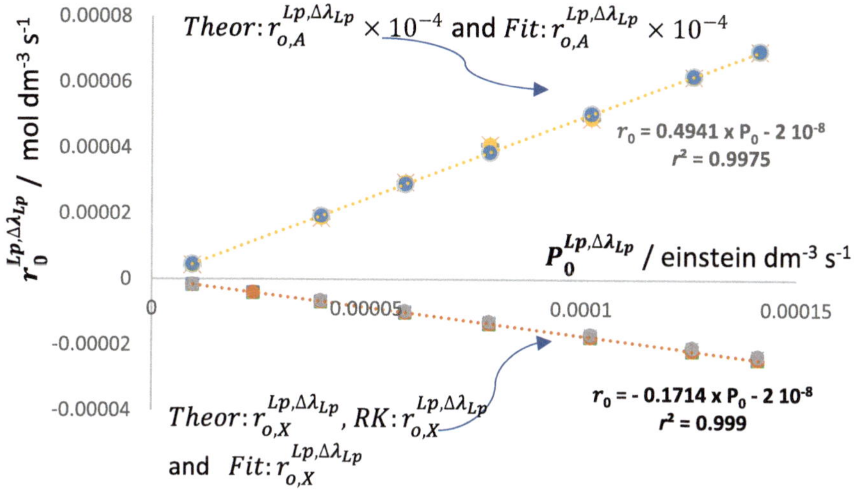

Fig. 23.3 Linear relationships between initial velocity ($r_{o,X}^{Lp,\Delta\lambda_{Lp1}}$ or $r_{o,A}^{Lp,\Delta\lambda_{Lp1}}$) and incident radiation intensity ($P_o^{Lp,\Delta\lambda_{Lp1}}$) for an *XY*(3*Φ*) photoreaction exposed to both *UV* and *Vis*. (Reproduced with permission from [10], *Frontiers in Chemistry*)

This reactive system mimics the reaction of diarylethene photochromes. In real-life situations, these compounds would simultaneously be exposed to ultraviolet (UV) (e.g. $\Delta\lambda_1$) and visible (Vis) (e.g. $\Delta\lambda_2$) radiations from sunlight. It is interesting to describe this reaction since the photokinetics of diarylethenes has never been studied under concomitant irradiations by *UV* and visible lights. Also, there are no known photokinetic or actinometric investigations of diarylethenes or equivalent systems while under simultaneous *UV* and visible radiations. The available descriptions mainly concerned their deployment in visible light actinometry, as seen in Chap. 18 and some publications [11, 12].

RK-traces, generated for the *XY*(3*Φ*) photoreaction, were fitted by an Eq. 19.10 (and Eq. 19.13) holding a single mono-Φ-order term. The initial rates for the reactant and the total absorbance were calculated with Eqs. 19.12 and 19.15 from the trace-fitting data obtained at a given radiation intensity. The total incident radiation intensity impinging onto the reaction varied in the range $9.43\ 10^{-6} - 1.41\ 10^{-4}$ *einstein* $dm^{-3}\ s^{-1}$.

Excellent linear relationships characterised the $r_{0,X\ or\ A}^{Lp,\Delta\lambda_{Lp1},T} = f\left(P_0^{Lp,\Delta\lambda_{Lp1}}\right)$ plots (Fig. 23.3), where $Theo: r_{0,X}^{Lp,\Delta\lambda_{Lp1},T.}$, $RK: r_{0,X}^{Lp,\Delta\lambda_{Lp1},T}$, $Fit: r_{0,X}^{Lp,\Delta\lambda_{Lp1},T}$, $Theo: r_{0,A}^{Lp,\Delta\lambda_{Lp1},T/\lambda_{obs}}$, and $Fit: r_{0,A}^{Lp,\Delta\lambda_{Lp1},T/\lambda_{obs}}$ have been used for the correlations. Altogether, such findings irrevocably indicate that the *XY*(3*Φ*) reaction is a good actinometer when irradiated by both *UV* and *Vis*.

Practically, the actinometric experiment only delivers the values $P_0^{Lp,\Delta\lambda_{Lp1},i}$ at n_{irr} radiation intensities ($1 \le i \le n_{irr}$) but not the corresponding numbers of photons,

Scheme 23.1 UV-activated photochromic reaction in naphthopyran

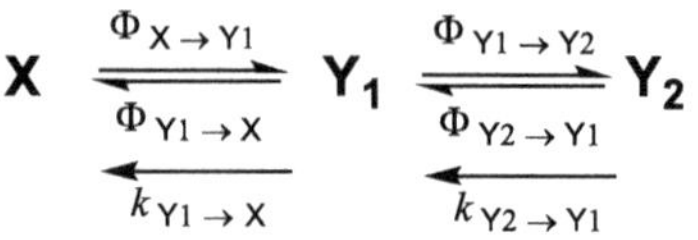

$P_0^{Lp,\Delta\lambda,i}$, responsible for the reactant photoconversion (because, technically, there is usually no possibility of selectively filtering the polychromatic light of the lamp to have an output precisely equal to the section of wavelengths useful for the experiment, e.g. $\Delta\lambda = \Delta\lambda_1$ in Fig. 23.2). In our case, irradiation with both *UV* and *Vis* intervenes in the reactions of both species, though in different proportion.

Anyway, using the framework developed above, $P_0^{Lp,\Delta\lambda,i}$ can be determined if the lamp profile is known. The factor $\beta_{Lp}^{\Delta\lambda_{Lp_1}/\Delta\lambda} = P_0^{Lp,\left(\Delta\lambda_{Lp_1} - \Delta\lambda\right),i} / P_0^{Lp,\Delta\lambda,i}$ for the considered reactant of the *XY*(3*Φ*) system is calculated as equal to 1.49 from Fig. 23.2. Henceforward, the series of $P_0^{Lp,\Delta\lambda,i}$ values are obtained from Eq. 23.10 as $P_0^{Lp,\Delta\lambda,i} = P_0^{Lp,\Delta\lambda_{Lp_1},i} / \left(1 + \beta_{Lp}^{\Delta\lambda_{Lp_1}/\Delta\lambda}\right)$. In addition, $\beta_{Lp}^{\Delta\lambda_{Lp_1}/\Delta\lambda}$ can serve the calculation of $\alpha_{Lp}^{\Delta\lambda}$ value (for $r_{o,X}^{\Delta\lambda}$) by using both Eq. 23.12 and the slope ($\alpha_{Lp}^{\Delta\lambda_{Lp_1}} = -0.1714$) of the initial reactant-rate vs. intensity plot (Fig. 23.3), which is obtained as $\alpha_{Lp}^{\Delta\lambda} = -0.427$. As predicted above, for the present reaction with $\Delta\lambda < \Delta\lambda_{Lp_1}$ and $\beta_{Lp}^{\Delta\lambda_{Lp_1}/\Delta\lambda} > 1$, the absolute value of $\alpha_{Lp}^{\Delta\lambda_{Lp_1}}$ is smaller than that of $\alpha_{Lp}^{\Delta\lambda}$ (or $-\alpha_{Lp}^{\Delta\lambda_{Lp_1}} < -\alpha_{Lp}^{\Delta\lambda}$). This procedure can also apply when a plot $r_{o,A}^{Lp,\Delta\lambda_{Lp_1},T/\lambda_{obs}}$ vs. $P_0^{Lp,\Delta\lambda_{Lp_1}}$ is considered. Accordingly, these results and the kinactinometric methodology promote, in particular, diarylethenes and, in general, photochromic materials, to become excellent and reliable actinometers that possess large dynamic ranges covering both *UV* and *Vis* spectral regions.

It is noticeable that no examples exist in the literature of organic photothermal systems used as standardised actinometers. The notorious case of the photothermal reversible *E*/*Z* azobenzenes is generally considered for actinometry, provided their reverse thermal reaction is negligible (experimentally, the photoreversible reaction is performed at a relatively low temperature) [13]. The occurrence of thermal reactions in the overall reaction of the ferrioxalate actinometer is considered a limitation [14] despite this actinometer's popularity (also see Sect. 23.7).

For an illustration of actinometry using a photothermal reaction, let us explore the case of the $XY_2(4\Phi, 2k)$ sequence (Scheme 23.1). The mechanism has been proposed for naphthopyran molecular switches endowed with emergent mechanochemical reactivity [15].

The authors indicated that the forward and reverse photoreactions are differently activated by *UV* and *Vis*, respectively.

Here, the reaction involves two consecutive photoreversible reactions between the three species (a reactant and two photoproducts). Forward and reverse photoreaction-steps are all simultaneously activated by the lamp's light (Fig. 23.4). The absorption spectra of the three species overlap completely (310 – 420 nm) and with the lamp emission, $\Delta\lambda_{Lp_1} = \Delta\lambda = OSIA_{act_1}$, ($Sit_2$). The reaction also includes a

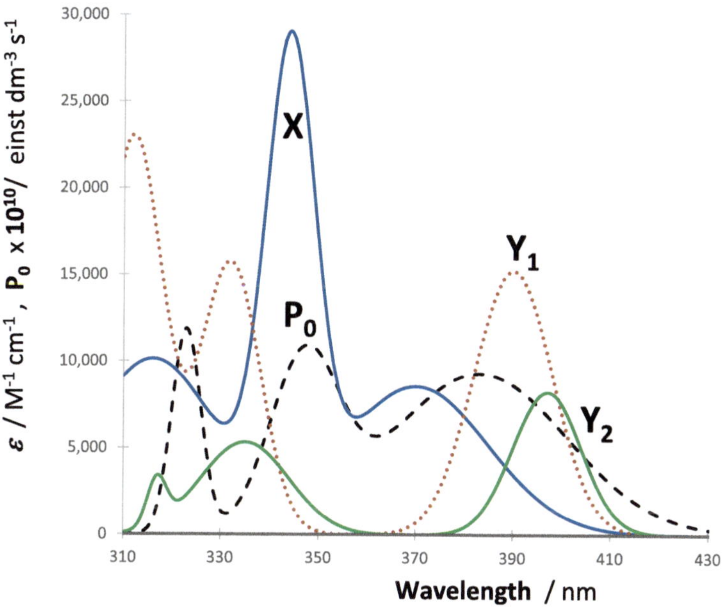

Fig. 23.4 Characteristics of a photothermal reaction, $XY_2(4\Phi, 2k)$ (Scheme 23.1), and irradiation lamp profile. The polychromatic light is absorbed by the three species of the photothermal system. (Reproduced with permission from [16], *Molecules*)

two-step, consecutive, purely thermal reaction, starting at the last product, Y_2. In the present simulation, while under irradiation, the six reaction-steps are active, and the quantum yields of the photoreactions are wavelength-dependent. The initial concentration of X belongs to its calibration linearity range.

Kinetic (concentration and total absorbance) traces are obtained under irradiation with several incident lamp intensities, $P_0^{Lp,\Delta\lambda,i}$ $(1 \le i \le n_{irr} = 6)$. They are fitted to either Eq. 19.10 or 19.13, composed of three mono- Φ-order ($i_{\Phi,\,j} = 3$) and two mono-exponential ($i_{\Delta,\,j} = 2$) terms, as given in the example of Eq. 23.13, for the concentration traces.

$$C_{Y_j}^{Lp,\Delta\lambda,T}(t) = \omega_j^{\infty} + \omega_{1j}^{\Phi}\, Log\left(1 + cc_j^{\Phi} e^{-k_{1j}^{\Phi}\, t}\right) + \omega_{2j}^{\Phi}\, Log\left(1 + cc_j^{\Phi} e^{-k_{2j}^{\Phi}\, t}\right)$$
$$+ \omega_{3j}^{\Phi}\, Log\left(1 + cc_j^{\Phi} e^{-k_{3j}^{\Phi}\, t}\right) + \omega_{1j}^{T}\, e^{-k_{1j}^{T}\, t} + \omega_{2j}^{T}\, e^{-k_{2j}^{T}\, t} \qquad (23.13)$$

The above kinactinometric method has been applied to the simulated data of the photothermal $XY_2(4\Phi, 2k)$ sequence (Scheme 23.1). The fittings of the traces are characterised by excellent statistical features (r^2, *SSE*, and *RMSE* values; see Fig. 23.5a). The values of the fitting parameters facilitated the calculation of the initial rates (Eq. 19.12 or 19.15). The values $(-Fit : r_{0,\,i})$ linearly correlated with the corresponding $P_0^{Lp,\Delta\lambda,i}$ (Fig. 23.5b). The $XY_2(4\Phi, 2k)$ sequence is used here as an

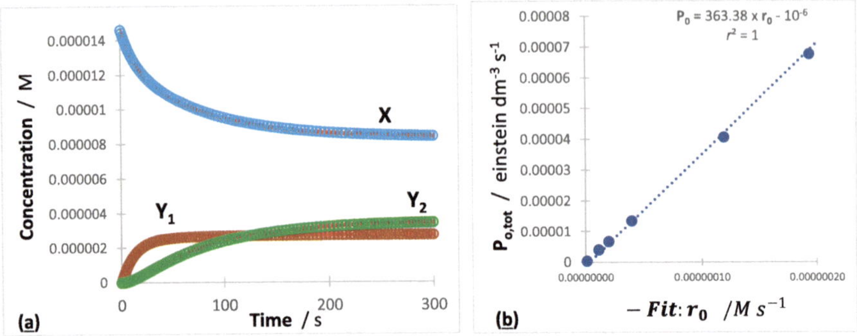

Fig. 23.5 (**a**) Example of concentration traces of the $XY_2(4\Phi, 2k)$ system (coloured circles) and their excellent fit with Eq. 23.13 (red lines). (**b**) A linear $P_0 = f(r_{0,\ x})$ relationship of the photothermal $XY_2(4\Phi, 2k)$ reaction exposed to a (*UV* and *Vis*) polychromatic light. (Reproduced with permission from [16], *Molecules*)

example of photothermal reactions' kinactinometry under mixed (*UV* and *Vis*) irradiation.

Photothermal chromic materials incite special interest towards their potential application in actinometry, owing to the advantages they offer, including reversibility, a quite large spectral window, and/or a variety of already available molecular structures.

23.6 Actinometer, Lamp, Reaction Conditions, and Kinactinometry

The general assumption in polychromatic chemical actinometry is that a standardised actinometer should be useful to determine the number of photons, P_0, of any lamp or light source. Unfortunately, kinactinometry states otherwise. Our discussions in *Part VII* clearly indicate that the rate-law (Eq. 19.5) becomes specific when the reaction is defined. This stems from the uniqueness of the attributes of the reaction and the conditions of the experiment (e.g. reaction mechanism, the experimental conditions, the lamp used, etc.). Therefore, we can set up the triple criteria "*actinometer/lamp/reaction conditions*" as a unique property belonging to the actinometric measurement.

Here, the general concept conveys the meaning that when an actinometer, ACT_1, is standardised by using a lamp, Lp_1, as laid out in Sect. 23.5, it cannot, subsequently, be employed for the determination of the unknown intensity of a different lamp, Lp_2 (to which ACT_1 would be subjected). This is true, as long as the light profiles of the two lamps are different (even when assuming that all other experimental conditions are the same). As a result, the value of the factor $\alpha_{Lp_1}^{\Delta\lambda}$ characterising ACT_1 with Lp_1 should be different from that of its counterpart, $\alpha_{Lp_2}^{\Delta\lambda}$, obtained when using Lp_2 and ACT_1. This remains valid even if ACT_1 has the same *OSIA* ($\Delta\lambda$) with both lamps. But in real-life situations, the profiles of the lamps, the *OSIA*s of the actinometer with each lamp, and often, some or all of the experimental conditions will most likely differ between the experiments performed for the standardisation of ACT_1 and the actinometric measurement of the unknown radiation intensity of Lp_2. Therefore, *no chemical actinometer is universal* (but all actinometers are efficient for their specific triple criteria, actinometer/lamp/reaction conditions).

Now, we can draw a similar conclusion if two actinometers (ACT_1 and ACT_2) are standardised by a single lamp, Lp_1, that delivers the same total number of photons for the successive $P_0^{Lp,\Delta\lambda_{Lp_1},i}$. Owing to the actinometers' attributes being different (i.e. their intrinsic parameters $\Phi^{\lambda_{irr}}$ and $\varepsilon^{\lambda_{irr}}$), the actinometric calibration lines of the two actinometers will be different ($\alpha_{Lp_1}^{\Delta\lambda}(ACT_1) \neq \alpha_{Lp_1}^{\Delta\lambda}(ACT_2)$, even when assuming that the *OSIA* for the actinometers is the same with Lp_1).

Another observation might have an additional indirect but still an important impact on experimental photochemistry. That is, it is commonly considered that the incident radiation intensity entering the reactor, obtained by chemical actinometry, is to be used for direct calculations of photokinetic parameters (such as the quantum yield) of the investigated system. By doing so, two assumptions are implicitly supposed: on the one hand, that the incident number of photons determined by actinometry (using a given ACT_1) is automatically valid for the lamp used for the investigated reaction (i.e. assuming the same lamp *OSIA* for both ACT_1 and reaction) and, on the other hand, that the number of photons absorbed is the same for both the actinometer and the studied reaction. These assumptions are not necessarily consistent for all situations [17, 18] due to differences in *OSIA*s and intrinsic parameters between the actinometer and reaction species.

The above argument forces us to admit that the classical way of polychromatic-actinometric measurements, performed by a given actinometer that was standardised by and for a specific lamp, is not necessarily always useful for the study of photothermal reactions. Consequently, one comes to the conclusion that, when polychromatic light is used, the best actinometer for the reaction at hand would be that reaction itself. However, such a goal remains a dream target for polychromatic-light actinometry. As a matter of fact, an approach can be proposed using the framework developed in the previous section to turn a thermally stable reactant into an actinometer if the profile of the lamp ($P_0^{\lambda_{irr}} = f(\lambda_{irr})$) is known. Incidentally, this again emphasises the usefulness and the need for a database of lamp profiles. This reactant, labelled here as $X - ACT$, stands for ACT_1 in the kinactinometric procedure of Sect. 23.5, which is supposed to be standardised by physical actinometry. In case a chemical actinometer (ACT_{ref}) is required for standardisation, it is mandatory that the *OSIA* of the standardised ACT_{ref} for the lamp used is larger than that of $X - ACT$ and that the $\beta_{Lp}^{\Delta\lambda_{Lp_1}/\Delta\lambda}$ of $X - ACT$ is worked out from $P_0^{\lambda_{irr}} = f(\lambda_{irr})$. Then, by applying the above kinactinometric method, it will be possible to construct the actinometric plots $P_0^{Lp,\Delta\lambda_{Lp_1},i}$ vs. $r_{0,X}^{Lp,\Delta\lambda_{Lp_1},T}$ and from there the $P_0^{Lp,\Delta\lambda,i}$ vs. $r_{0,X}^{Lp,\Delta\lambda_{Lp_1},T}$ plot for $X - ACT$, which practically means that $X - ACT$ becomes an actinometer.

23.7 Observations on the Ferrioxalate Actinometer

In principle, the non-universality of chemical actinometers (Sect. 23.6) should also apply to ferrioxalate actinometry. The number of photons recorded by this actinometer is equal to the number of ferrous ions produced in solution (as a result of the photoreduction of the initial solution containing ferric ions). The produced number of ferrous ions depends on both the actinometer's wavelength-dependent quantum yield, Fig. 4.4b [14], and on the lamp emission profile initiating the photoreduction reaction. It is, hence, predictable that different lamps will, most likely, engender in the actinometric solution, different numbers of ferrous ions. Incidentally, the lamps used for the irradiation of the solution are certainly different from the light sources adopted in the original study [19] for the calibration of this actinometer.

Anyway, in ferrioxalate actinometry, the number of photons absorbed by this actinometer is experimentally determined by a global number of ferrous ions produced in the solution. This number corresponds to the photons emitted by the lamp used for irradiation, over its whole spectral range. If the lamp is unfiltered, the ferrioxalate absorbs up to 500 nm ($200 < \lambda < 500$ nm; this actinometer does not absorb at $\lambda > 500$ nm). Practically, this would be the case of almost all classical unfiltered lamps traditionally employed in experimental photochemistry and, therefore, the measured $\Delta\lambda_{Lp_1} > \Delta\lambda$. LED lights and filtered lamps might be exceptions whenever $\Delta\lambda_{Lp_1}$ does not exceed $\Delta\lambda$.

Each time the emission domain of the (unfiltered) lamp is larger than the absorption spectrum of the species studied or the actinometer in development, the measured number of photons by the ferrioxalate actinometer, $P_0^{Lp,\Delta\lambda_{Lp_1}}$, is higher than the number of photons ($P_0^{Lp,\Delta\lambda}$) absorbed by that species. This observation invalidates the usage of the number of photons measured by ferrioxalate actinometry ($P_0^{Lp,\Delta\lambda_{Lp_1}}$), when irradiation is polychromatic, to evaluate the reactant quantum yield by using the differential quantum yield equation, as usually practised in the literature (Sects. 21.1 and 21.2). Such a quantum yield (or, more precisely, photonic yield) value must be underestimated compared to the true quantum yield (photonic yield) value (because $P_0^{Lp,\Delta\lambda_{Lp_1}} > P_0^{Lp,\Delta\lambda}$).

In this context, the latter observation also discards an implicit assumption of the ferrioxalate actinometry procedure. It is held that the photons absorbed by the ferrioxalate actinometer are absorbed by the species studied. The inequality $P_0^{Lp,\Delta\lambda_{Lp_1}} > P_0^{Lp,\Delta\lambda}$ invalidates such an assumption [18].

Furthermore, there is another observation about the ferrioxalate procedure, which imposes that the quantum yield of investigated species is necessarily invariant with wavelength for the differential quantum yield equation to be applicable. This remains an approximation until otherwise proven experimentally (by implementing a screening of the reactive species by strictly monochromatic light irradiations at different wavelengths [9]). It is very rare that published works report on performing such individual wavelength screenings before proceeding with photolysis by polychromatic light. A supposed Φ-invariance might lead to inaccuracy of the values determined for the quantum yield.

As a matter of fact, most often, these observations are overlooked in the literature. Taking such observations into consideration, as well as specifying the details of the experiment, is an essential requirement for consistent research.

23.8 Conclusion

The kinactinometric methodology presented in the present chapter is a general method for the standardisation of actinometers obeying any photo- or photothermal reaction mechanism. It is an unprecedented useful tool to be added to photokinetic investigation.

The method has many advantages. It clearly eludes the limitations encountered for ferrioxalate and classical actinometries, facilitates the determination of both the number of photons emitted by the lamp and the one absorbed by the organic actinometer within its *OSIA*, and opens up an avenue to turn an investigated system into its own actinometer.

References

1. Maafi M, Lee LY (2015) Actinometric and Φ-order photodegradation of anti-cancer Sunitinib. J Pharm Biomed Anal 110:34–41. https://doi.org/10.1016/j.jpba.2015.02.026
2. Maafi M, Lee LY (2015) Determination of dacarbazine Φ-order photokinetics, quantum yields, and potential for actinometry. J Pharm Sci 104:3501–3509. https://doi.org/10.1002/jps.24568
3. Maafi W, Maafi M (2013) Modelling nifedipine photodegradation, photostability and actinometry properties. Int J Pharm 456:153–164. https://doi.org/10.1016/j.ijpharm.2013.07.075
4. Maafi M, Maafi W (2014a) Φ-Order kinetics of photoreversible-drug reactions. Int J Pharm 471: 536–543. https://doi.org/10.1016/j.ijpharm.2014.05.017
5. Maafi M, Maafi W (2014) Montelukast photodegradation: elucidation of F-order kinetics, determination of quantum yields and application to actinometry. Int J Pharm 471:544–552. https://doi.org/10.1016/j.ijpharm.2014.05.016
6. Maafi M, Maafi W (2015) Quantitative assessment of photostability and photostabilisation of fluvoxamine and its design for actinometry. Photochem Photobiol Sci 14:982–994. https://doi.org/10.1039/c5pp00022j
7. Maafi M, Maafi W (2016) Modelling and elucidation of the kinetics of multiple consecutive photoreactions AB4(4Φ) with Φ-order kinetics. Application to the photodegradation of ribofavin. J Pharm Sci 105:3537–3548. https://doi.org/10.1016/j.xphs.2016.06.030
8. Maafi M, Al-Qarni MA (2018) Φ-order spectrophotokinetic characterisation and quantification of trans-cis oxyresveratrol reactivity, photodegradation and actinometry. Spectrochim Acta A Mol Biomol Spectrosc 188:64–71. https://doi.org/10.1016/j.saa.2017.06.067
9. Maafi M (2023) On photokinetics under monochromatic light. Front Chem 11:1233151. https://doi.org/10.3389/fchem.2023.1233151
10. Maafi M (2024) On photokinetics under polychromatic light. Front Chem 12:1367276. https://doi.org/10.3389/fchem.2024.1367276

11. Roibu A, Fransen S, Leblebici ME, Meir G, Van Gerven T, Kuhn T (2018) An accessible visible-light actinometer for the determination of photon flux and optical pathlength in flow photo microreactors. Sci Rep 8:5421. https://doi.org/10.1038/s41598-018-23735-2
12. Maafi M, Al-Qarni MA (2022) Mono-and polychromatic light diarylethene-actinometer for the visible range. Dyes Pigments 198:109942. https://doi.org/10.1016/j.dyepig.2021.109942
13. Zimmerman G, Chow L-Y, Paik U-J (1958) The photochemical isomerization of azobenzene. J Am Chem Soc 80:3528–3531. https://doi.org/10.1021/ja01547a010
14. Montalti M, Credi A, Prodi L, Gandolfi MT (2020) Handbook of photochemistry.3rd edn. CRC Press, p 1251. ISBN-10. 0367577909. ISBN-13. 978-0367577902
15. McFadden ME, Barber RW, Overholts AC, Robb MJ (2023) Naphthopyran molecular switches and their emergent mechanochemical reactivity. Chem Sci 14:10041–10067. https://doi.org/10.1039/D3SC03729K
16. Maafi M (2025) Photokinetics of phothermal reactions. Molecules 30:330. https://doi.org/10.3390/molecules30020330
17. Zalazar CS, Labas MD, Martin CA, Brandi RJ, Alfano OM, Cassano AE (2005) The extended use of actinometry in the interpretation of photochemical reaction engineering data. Chem Eng J 109:67–81. https://doi.org/10.1016/j.cej.2005.03.011
18. Lente G (2015) Deterministic kinetics in chemistry and systems biology: the dynamics of complex reaction networks. Springer, Cham. https://doi.org/10.1007/978-3-319-15482-4
19. Hatchard CG, Parker CA (1956) A new sensitive chemical Actinometer. II. Potassium Ferrioxalate as a standard chemical Actinometer. Proceedings of the Royal Society of London. Series A, Math Phys Sci 235(1203):518–536. https://doi.org/10.1098/rspa.1956.0102

Index

M. Maafi, *Photokinetics*, https://doi.org/10.1007/978-3-031-98985-8

Q

R

S

GPSR Compliance

The European Union's (EU) General Product Safety Regulation (GPSR) is a set of rules that requires consumer products to be safe and our obligations to ensure this.

If you have any concerns about our products, you can contact us on ProductSafety@springernature.com

In case Publisher is established outside the EU, the EU authorized representative is:

Springer Nature Customer Service Center GmbH
Europaplatz 3
69115 Heidelberg, Germany

Batch number: 10755329

Printed by Printforce, the Netherlands